Applied Mineralogy

Applications in Industry and Environment

Applied Mineralogy

Applications in Industry and Environment

Swapna Mukherjee

Formerly at
Geological Survey of India
Kolkata, India

Springer

CP

A C.I.P. Catalogue record for this book is available from the Library of Congress.

ISBN 978-94-007-1161-7 (HB)
ISBN 978-94-007-1162-4 (eBook)

Copublished by Springer,
P.O. Box 17, 3300 AA Dordrecht, The Netherlands
with Capital Publishing Company, New Delhi, India.

Sold and distributed in North, Central and South America by Springer,
233 Spring Street, New York 10013, USA.

In all other countries, except SAARC countries–Afghanistan, Bangladesh, Bhutan, India, Maldives, Nepal, Pakistan and Sri Lanka–sold and distributed by Springer, Haberstrasse 7, D-69126 Heidelberg, Germany.

In SAARC countries–Afghanistan, Bangladesh, Bhutan, India, Maldives, Nepal, Pakistan and Sri Lanka–sold and distributed by Capital Publishing Company, 7/28, Mahaveer Street, Ansari Road, Daryaganj, New Delhi, 110 002, India.

www.springer.com

Cover photos credit: *Top:* http://www.allaboutgemstones.com/mineral_classification.html, *bottom (left to right):* http://uscdn.creamermedia.co.za/assets/articles/images/resized/52823_resized_asbestos_ground_pollution.jpg, http://prodimg.exportersindia.com/bc-small/746691/mica-sheet-793314.jpg, http://www.surfaceng.com/photos/large/testing_xrd1.jpg and www.wordpress.com

Printed on acid-free paper

FOREWORD

The book "Applied Mineralogy: Applications in Industry and Environment" by Dr. Swapna Mukherjee is a major stride forward due to several reasons. First, the content is rather unique and combines very elegantly theoretical aspects of mineralogy with applied aspects, such as industrial and environmental applications. In line with the requirements of the specific scientific discipline in this century and of the societal relevance, students need to be exposed to both the sides. The book can be used both for undergraduate and post-graduate studies in Indian universities/institutes, as well as in any similar course all over the world. Secondly, it will be readily available for Indian students, and to the best of my knowledge, there is no other similar book available in the Indian market. Finally, the book reflects the author's depth of understanding of the diverse issues. The author has vast experience, as holder of key positions in one of India's international standard laboratories, in research related to both theoretical mineralogy and its applications. The first part of the book deals essentially with theoretical basis and developments in the subject, and the same trend continues in the first section of the second part. The later sections in part two and the whole of the part three deal with applications, and these sections are exceptionally well dealt with. This is perhaps reflecting the exposure of the author in these fields and her expertise.

Lastly, I would like to congratulate the author for bringing out this much-needed book, which certainly fills up a void.

Somnath Dasgupta
J.C.Bose National Fellow & Director
National Centre of Experimental Mineralogy & Petrology
University of Allahabad
Allahabad-211 002
Uttar Pradesh, India

PREFACE

Mineralogy, a multidisciplinary subject, is essential for understanding Earth science. In microscopic scale, this subject involves both physics and chemistry in understanding the formation processes of minerals, ore deposits, and other similarly related aspects. This book introduces widely different aspects of mineralogy, explaining each of those features with examples to the post-graduate and graduate level students. Particular emphasis is given for the general understanding of mineralogy for average students, and for specially motivated students it creates a reference platform for higher learning of mineralogy and its importance in geology, materials science and environmental science. This book approaches the subject of mineralogy in a simple way so that no special prior training (except the basic knowledge of physics and chemistry) is needed for its understanding. Subsequently it takes the reader to the higher levels explaining the utility of the subject in various fields, including the industrial, gemological and medicinal uses.

Although a number of well referred introductory books on mineralogy (Perkins, 1998; Putnis, 1992; Blackburn and Dennen, 1994; Nesse, 2000 and Klien, 2002) are available, this book is intended to accommodate the curricula of our universities which is extended considerably than earlier years and is expanding gradually with incorporation of newly developed ideas at interdisciplinary level. Emphasis is given in the descriptive part rather than mathematically oriented things, with only few exceptions where crystal structure and thermodynamics of minerals are discussed. Most of the descriptions are accompanied with figures and diagrams so as to make things more appealing and easily understandable.

The book starts with few introductory remarks on minerals and the importance of their study. The main text is divided into three broad parts.

Part I deals with the basic properties of minerals, including their structural and chemical classifications. This includes a detailed description of crystallographic studies of minerals, their optical and other physical properties like hardness, cleavage, surface properties etc. related with the structure.

Part II deals with the origin and transformation of minerals, their stability under different pressure-temperature conditions, mineral deposit characteristics and special features of marine minerals. Application of the mineral formation and transformation processes in estimation of pressure-temperatures, i.e. geothermo-barometry is also discussed with examples.

Part III describes mainly the application aspects of mineralogy. It starts with the brief summary of common analytical techniques used for mineral studies. Some of these techniques are described in detail. Applications of magnetic, electrical and gravitational properties of minerals as common tools for exploration of mineral deposits have been described. Properties of precious and semi-precious stones, the distinction between the mineral and gem is well explained with the various cuttings and technical enhancement data and figures. The synthetic minerals preparation, necessary for both academic and industrial purposes, are described shortly as the details are too technical and beyond the scope of this book. The utilization of minerals in variety of industries are well documented with the different beneficiation processes and other relevant data for industrial mineralogy.

In discussing the environmental problems generated from pollution and other problems creating health hazards, relevant suggestions are mentioned for handling the environmental hazards to optimize it. A very special emphasis is given to understand the environmental mineralogy, distinguishing it from environmental geology. Lastly, the concept of medicinal mineralogy has been discussed providing useful information on the newly emerging field.

To summarize as a whole the book has been written with the objective to explain the basic concepts of some higher branches of mineralogy like precious and semiprecious stones, industrial mineralogy, medical mineralogy for the beginners, to elaborate the details of analytical methods which are of immense use in every branch of science, including mineralogy and to serve as an informative guide medium with respect to the generalized aspect of mineralogy. Hope the purpose of the book will be achieved and the readers will be benefited.

This book has resulted from the immense amount of help and encouragement from many of my friends, colleagues and scientists working in the related areas. In particular, I am very much indebted to the members of the Geology Departments of Jadavpur University, Kolkata University and Presidency College, Kolkata; Centre for Advanced Studies and Geology department of University of Lucknow; members of the Geological Survey of India, Northern Region, Lucknow; The M.S. University of Baroda, Vadodara and NCEMP, Allahabad. I am thankful to Geological Survey of India (C.H.Q. and E.R.), Kolkata and N.R. Lucknow and my colleagues there of different departments and laboratories for providing various kinds of help and suggestions. I also appreciate the availability of the excellent library facility of

G.S.I. My sincere thanks to senior teachers in the field for their academic discussions, suggestions and help.

Interactions from the students, while teaching in different universities, helped me to frame the structure of the course to fulfill their current need in this field. Lastly, the success of the book will depend on how the students for whom this is written are benefited and how this is received by the teachers who will guide them in this subject.

Last but not the least, this book is the outcome of continuous support, encouragement and immense help from my best friend and husband, Prof. P.K. Mukherjee.

This book is dedicated to our daughter, Swagata.

May 2011 **Swapna Mukherjee**

CONTENTS

INTRODUCTION

Minerals have been used in a number of various ways from time immemorial. Man has been using minerals even when he didn't know its definition. Hence with the progress of civilization the concern about minerals slowly increased and it is this concern which gave birth to the science of mineralogy. Mineralogy is narrowly defined as the science of minerals. Moreover the scope of this subject covers a much wider range which encompasses mineral definition, origin, structure, properties and their impact on environment and applications in modern industry. Now human civilization is just indispensable without minerals. So before investigating the various aspects of mineralogy in details, let's have a closer look into the history and development of the science of mineralogy.

BACKGROUND OF MINERALOGY

Minerals started to be a part of human life prior to the Bronze Age in Neolithic times, at the dawn of human civilization. It was referred to as the use of copper and was known as Chalcolithic Age. It appears that copper was not widely exploited at first and those efforts in alloying it with tin and other metals began quite soon, which led to the onset of Bronze Age. It dated back from 3500-1200 BC. The Bronze Age refers to a period of time in prehistoric societies where metallurgy had advanced to the point of making bronze—an alloy of tin and copper—from natural ores, but not yet to the point of the systematic production of iron (the Iron Age). Smelted copper was rarely pure. In fact, it is clear that by 2500 BC, the Sumerians had recognized that if different ores were blended together in the smelting process, a different type of copper—which flowed more easily, was stronger after forming and was easy to cast—could be made. An axe head from 2500 BC revealed that it contained 11% tin and 89% copper. This was of course the discovery of Bronze. The emergence of metallurgy occurred first in the Fertile Crescent, where it gave rise to the Bronze Age in the 4th millennium BC. There was an independent and limited invention of copper and bronze smelting by the Incas in South America and Mesoamerican civilization in West Mexico.

Bronze was primarily used because its melting point is lower than iron. Bronze is preferable to stone for a wide variety of applications—whether you're making a knife, an axe, armour, pottery, or artwork. Bronze is harder and longer-lived. A more durable capital base enhances not only the potential for sustained economic activity, but also warfare. With further progress, man understood that a harder material was needed for the utilization of their purpose; hence the use of iron came into being. By the Middle Bronze Age, increasing number of smelted iron objects appeared throughout Anatolia, Mesopotamia, the Indian subcontinent, the Levant, the Mediterranean, and Egypt. The Iron Age began with the development of higher temperature smelting techniques. It dated between 1200-1000 BC. During the Iron Age, the best tools and weapons were made from steel, an alloy consisting of iron with carbon content between 0.02% and 1.7% by weight. Steel weapons and tools were nearly of the same weight as those of bronze, but stronger. Due to less persistence of methods, steel was not properly made and wrought iron was more in use. But it should be kept in mind that, till that time, the recognition of minerals was not yet possible. Man was just using materials from different rocks according to his convenience. Gradually the idea of metals established and the domain of minerals conceptualized.

BEGINNING OF MINERALOGY

In the mid 1400-1500 AD, the term mineral was started to be used. The word 'mineral' originated from the Latin word meaning 'something mined'. Primarily the search for minerals was on a localized scale, as the need was limited. But day by day the process of mining and exploitation of minerals rapidly increases due to

- an increase in the world population, and
- an increase in the needs of the population.

This implies an increase in production or a relaunching of the investigations. However, the evolution of technology in addition creates new needs. For example, the appearance and start up of the motors, of explosion and fuel derived from petroleum, has made necessary the creation of more resistant alloys that are possible with the increase of new mineral prospecting, such as antimony, chrome, nickel and cadmium, besides petroleum. But it should be kept in mind that it is an exhaustible resource. So by now man should get cautious about his needs and demands. All of this implies the cycle of mineral economy.

The first mineral used by man is quartzite or flint. In the beginning, a prehistoric man, who was more observant than others, accidentally discovered the origin of the properties of quartzite and its advantages compared with the wood and the hardened clay in heat as it had more hardness, a minimum amount of wearing down, sharp cutting edges etc.

That was the start. From then onwards there is no looking back. Minerals and the science of mineralogy are progressing with each day. Let us start with soil and its mineral content. The quality of soil controls the vegetation it supports. This factor is predominantly controlled by the minerals it contains. The difference between rich, fertile soil and poor, infertile soil is essentially the mineral composition of the soil. The flavour and nutrition in fruit, grains and vegetables are based on the soil minerals available, not the amount or type of organic matter. If a needed mineral is missing, no amount of organic matter will make up for it. No amount or combination of the air elements—carbon, oxygen and hydrogen—will add zinc to soil, and all plants and animals need zinc. Plants also need copper and iron as also manganese, calcium, magnesium and phosphorus, plus at least seven other essential minerals (animals need at least fourteen more). Moreover, they need them in an available and balanced form. Rotten organic matter may release growth stimulants like nitrogen and potassium, but the other minerals are what create sweetness, flavour and nutrition in food. Without these other minerals one may achieve high production, but the highest quality cannot be achieved. Some essential minerals of soil are evaporite, saline minerals, and silica-alumina clay minerals, which are polymerized secondary minerals with silica tetrahedra and alumina octahedra form in soil environments.

Among them, the clay minerals form the most important group. Clays and clay minerals have been mined since the Stone Age; today they are among the most important minerals used by manufacturing and environmental industries. The studies of the properties of clays, the mechanisms of clay formation, and the behaviour of clays during weathering are greatly supported nowadays. These studies can tell us how and where these minerals form and provide industry and land-planning agencies with the information necessary to decide how and where clay and clay mineral deposits (to be discussed in details later) can be developed safely with minimal effects on the environment.

Mineralogy takes a crucial role in extraction of metals. Man initiated his progress in civilization with the use of metals. Starting from tin, copper and iron, man finds uncountable uses from metals. The matter has been elaborated in next section.

Steady and appreciating advancement of mineralogy led to global industrialization of minerals. Various objects in everyday life like aggregates, abrasives, ceramics, cement, fertilizers, fillers and absorbents, foundry sands, fluxes, drilling fluids, electronics, insulation, lubricants, glass, pigments, refractories, plasters etc. are constituted by minerals. This has been discussed in details in later chapters.

Next comes the concept of 'mineral fibres'. Mineral wool, mineral fibres or man-made mineral fibres are fibres made from natural or synthetic minerals or metal oxides. The latter term is generally used to refer solely to synthetic materials including fibreglass, ceramic fibres and rock or stone wool. Industrial applications of mineral wool include thermal insulation, filtration, soundproofing, and germination of seedlings. Slag wool was first made in 1840 in Wales by Edward Parry but the harmful effects on the workers caused production to be abandoned. It was first produced commercially in 1871 at the Georgsmarienhütte in Osnabrück Germany.

Nowadays, minerals also show its progress in geomedicinal concepts. Various minerals and elements have diverse health effects. The excess and deficiency of these substances affect the human health in some or the other unique way. This has been discussed in details in later chapters.

This over-emphasis on minerals and their rapid modernization have some global adverse impact on the environment. So the subject of environment is a matter of concern nowadays. The subject deals with all aspects of interactions between humans, ecosystems and the earth. Coverage includes water and soil contamination; environmental problems associated with transportation; geological processes affecting biosystems and people; remediation of man-made or geological hazards; environmental problems associated with mining and abstraction activities for industrial minerals, coal and ores as well as for oil and gas, water and energy; environmental impacts of exploration and recultivation; impacts of hazardous facilities and activities; land use management; and management of environmental data and information in data banks and information systems. Through dissemination of knowledge on techniques, methods, approaches and experiences, the study aims to improve and remediate the environment as habitat for life on earth.

FIRST USE OF METALS

Currently there are 86 known metals. Before the 19th century only 24 of these metals had been discovered and, of these 24 metals, 12 were discovered in the 18th century. Therefore, from the discovery of the first metals, gold and copper, until the end of the 17th century, only 12 metals were known. Four of these metals, arsenic, antimony, zinc and bismuth, were discovered in the thirteenth and fourteenth centuries, while platinum was discovered in the 16th century. The other seven metals, which were known as the Metals of Antiquity, were the metals upon which civilization was based. These seven metals were: Gold (ca) 6000 BC; Copper (ca) 4200 BC; Silver (ca) 4000 BC; Lead (ca) 3500 BC; Tin (ca) 1750 BC; Iron, smelted (ca) 1500 BC; Mercury (ca) 750 BC. These metals were known to the Mesopotamians, Egyptians, Greeks and the Romans. Of the seven metals, five can be found in their native states, e.g., gold, silver, copper, iron (from meteors) and mercury. However, the occurrence of these metals was not abundant and the first two metals to be used widely were gold and copper. And, of course, the history of metals is closely linked to that of coins and gemstones.

Metal	*Discovery*	*Uses*
Gold	Archaeological digs suggest that the use of gold began in the Middle East in the first known civilizations. The oldest pieces of gold Egyptian jewellery were found in the tomb of Queen Zer and that of Queen Pu-abi of Ur in Sumeria and are the oldest examples of any kind of jewellery in a find from 3000 BC.	Gold articles are found extensively in antiquity mainly as jewellery e.g. bracelets, rings etc. Early gold artifacts are rarely pure and most contain significant silver contents. This led to the ancients naming another metal i.e. electrum, which was an alloy of gold and silver, pale yellow and similar in colour to amber. Therefore, early gold varied from pure through electrum to white gold. Stone Age man learned to fashion gold into jewellery and ornaments, learning that it could be formed into sheets and wires easily.
Copper	From 4000 to 6000 BC was the Chalcolithic period when copper came into common use by peasants. Malachite, a green friable stone, was the source of copper in the early smelters. Originally it was thought that the smelting of copper was by chance dropping of malachite into campfires. However, campfire temperatures are normally in the region of 600-650 °C, whereas 700-800 °C is necessary for reduction. It is more probable that early copper smelting was discovered by ancient potters whose clay firing furnaces could reach temperatures of 1100-1200 °C. If malachite was added to these furnaces copper nodules would easily be found. Although the first smelted copper was found in the Nile valley, it is thought that this copper was brought to Egypt by the Gerzeans and copper smelting was produced first in Western Asia between 4000 and 4300 BC.	Initially copper was chipped into small pieces from the main mass. The small pieces were hammered and ground in a manner similar to the techniques used for bones and stones. However, when copper was hammered it became brittle and would easily break. The solution to this problem was to anneal the copper. This discovery was probably made when pieces were dropped in camp fires and then hammered. By 5000 BC, copper sheet was being made. By 3600 BC, the first copper smelted artifacts were found in the Nile valley and copper rings, bracelets, chisels were found. By 3000 BC, weapons, tools etc. were widely found. However it took another 500 years before they reached the peasants.
Silver	By 2500 BC, silver was being manufactured by cupellation process.	Pure silver was used for ornaments, jewellery and as a measure of wealth.

(*Contd.*)

(Contd.)

Metal	*Discovery*	*Uses*
Lead	Lead is not found free in nature but galena (lead sulfide) was used as an eye paint by the ancient Egyptians. The production of metallic lead from its ore is relatively easy and could have been produced by reduction of galena in a camp fire. The melting point of lead is 327 °C; therefore, it would easily flow to the lowest point in the fireplace and collect. At first lead was not used widely because it was too ductile and the first uses of lead were around 3500 BC.	Lead's use as a container and conduit was important and lead pipes bearing the insignia of Roman emperors can still be found. Lead is highly malleable, ductile and noncorrosive making it an excellent piping material.
Tin	Native tin is not found in nature. The first tin artifacts date back to 2000 BC. However, it was not until 1800 BC, that tin smelting became common in western Asia. Tin was reduced by charcoal and at first was thought to be a form of lead.	Tin was rarely used on its own and was most commonly alloyed with copper to form bronze. The most common form of tin ore is the oxide casserite. By 1400 BC, bronze was the predominant metal alloy.
Iron	Iron was available to the ancients in small amounts from meteors. This native iron is easily distinguishable because it contains 6-8% nickel. There is some indication that man-made iron was available as early as 2500 BC; however, iron making did not become an everyday process until 1200 BC. Hematite, an oxide of iron, was widely used by the ancients for beads and ornaments. It is also readily reduced by carbon. However, if reduced at temperatures below 700-800 °C it is not suitable for forging and must be produced at temperatures above 1100 °C. Wrought iron was the first form of iron known to man. The product of reaction was a spongy mass of iron intermixed with slag.	Iron weapons revolutionized warfare and iron implements did the same for farming. Iron and steel was the building block for civilization. Interestingly, an iron pillar dating 400 AD, remains standing today in Delhi, India. Corrosion to the pillar has been minimal—a skill lost to current iron-workers.
Mercury	Mercury was also known to the ancients and has been found in tombs dating back to 1500 and 1600 BC. Pliny, the Roman chronicler, outlined purification techniques by squeezing it through leather and also noted that it was poisonous. Mercury, also known as quicksilver, is the only metal which is liquid at room temperature. In 315 BC, Dioscorides mentions recovery of quicksilver (which he called hydrargyros, liquid silver) by distillation, stating "An iron bowl containing cinnabar is put into an earthenware container and sealed with clay. It is then set on a fire and the soot which sticks to the cover is quicksilver".	Mercury was widely used because of its ability to dissolve silver and gold (amalgamation) and was the basis of many plating technologies. There are also indications that it was prized and perhaps worshipped by the Egyptians.

ADVANCEMENT OF METALLURGY

The seven metals discussed in the earlier section marked the beginning of metallurgical development. Followed by them, the next metal to be discovered was arsenic in the 13th century by Albertus Magnus. Arsenicus (arsenious oxide) when heated with twice its weight of soap, became metallic. By 1641 arsenious oxide was being reduced by charcoal. The next metal to be isolated was antimony. Stibium or antimony sulphide was roasted in an iron pot to form antimony. Agricola reported this technique in 1560. By 1595, bismuth was produced by reduction of the oxide with carbon. However, it was not until 1753 when bismuth was classified as an element. Zinc was known to the Chinese in 1400; however, it was not until 1738, when William Champion patented the zinc distillation process that zinc came into common use. A Chinese text from 1637 stated the method of production was to heat a mixture of calamine (zinc oxide) and charcoal in an earthenware pot. The zinc was recovered as an incrustation on the inside of the pot. In 1781 zinc was added to liquid copper to make brass. This method of brass manufacture soon became dominant. One other metal was discovered in the 1500's in Mexico by the Spaniards. This metal was platinum. Early use of platinum was banned because it was used as a blank for coins which were subsequently gold coated, proving that the early metallurgists understood not only density but also economics. Although, platinum was known to the western world, it was not until the 1800's that platinum became widely used.

Several other metals were isolated during the 1700's. These were cobalt, nickel, manganese, molybdenum, tungsten, tellurium, beryllium, chromium, uranium, zirconium and yttrium. Only laboratory specimen were produced and all were reduced by carbon with the exception of tungsten which became the first metal to be reduced by hydrogen. Before 1805, all metals were reduced by either carbon or hydrogen. However, the majority of the metals once smelted were not pure. In 1817, cadmium was discovered. Stroymeyer noted that zinc carbonate had a yellowish tinge not attributable to iron. Upon reduction he thought that the alloy contained two metals. The metals were separated by fractional distillation. At 800 °C, as cadmium's boiling point is lower than zinc, the cadmium distilled first. In 1841, Charles Askin developed a method of separating cobalt and nickel when both metals are in solution. Chromium, although it had been produced by reduction with carbon, was the first metal to be extensively produced using another metal (zinc). Wohler in 1859 melted chromium chloride under a fused salt layer and attracted the chromium with zinc. The resulting zinc chloride dissolved in the fused salt and chromium produced. In 1828, Wohler produced beryllium by reducing beryllium chloride with potassium in a platinum crucible. Aluminium was first produced by Christian Oersted in 1825. However it was not until 20 years later that significant quantities were produced.

From this short review of metallurgical developments it can be seen that as the early metallurgists became more sophisticated their ability to discover and separate all the metals grew. However in all of their work it was necessary for all the basic steps to be carried out e.g. the ore had to be identified, separated from gangue, sized, concentrated and reduced in a manner which accomplished a phase separation.

FIRST USE OF COLOURED STONES

While searching for the usual metals man came across a different kind of rocks displaying coloured stones. This inaugurated the use of gemstones.

The lapislazuli mineral or "gem" of a sky blue colour was greatly appreciated and sought after ever since Neolithic times. In Iran, at Tell-i-Bakun and about 4500 years BC, the local gem industry was cutting lapislazuli into cubic polyhedral stones with rounded edges. In Mesopotamia, at Tepe Gawra, a centre of the Ubaid culture, lapislazuli made its appearance, together with turquoise, amethyst and beryl, round about 4500 BC in the form of stones that had been locally cut. From 3000 to 2000 BC the Sumerian civilization flourished here. The minerals found in the graves were in majority objects of lapislazuli, carnelian and agate. The Egyptians turned their attention to precious stones rather early and studied them from every possible point of view. But they remained famous in the history of mineralogy on account of the fact that they were the first to make artificial imitations of gems and hard stones and were more skillful than any other people in the difficult art of incising them: an art that came to be known as glyptics. There exist many legends about marvellous stones with which the Egyptians associated supernatural virtues, but on the whole they are far too vague and legendary.

Gemstones are not plentiful. Gemstones do not form "ore" deposits in the normal sense. Gems, when present at all, tend to be scattered sparsely throughout a large body of rock or to have crystallized as small aggregates or fill veins and small cavities. Even stream gravel concentrations tend to be small—a few stones in each of several bedrock cracks, potholes, or gravel lenses in a stream bed. The average grade of the richest diamond kimberlite pipes in Africa is about one part diamond in 40 million parts "ore." Kimberlite, a plutonic igneous rock, ascends from a depth of at least 100 kilometres (60 miles) to form a diatreme (narrow cone-shaped rock body or "pipe"). Moreover, because much diamond is not of gem quality, the average stone in an engagement ring is the product of the removal and processing of 200 to 400 million times its volume of rock.

Nowadays gem industry has been gaining wider importance. Previously gems after being discovered were started to be cut using various cutting styles. Slowly more complicated styles are invented. Various enhancement techniques are used to upgrade the look and quality of the gemstones. The modernization reaches its peak with the initiation of *synthetic* and *simulant* gemstone preparation. Laboratory grown synthetic gemstones have essentially the same appearance and optical, physical and chemical properties as the natural material that they represent. Laboratory grown simulants have an appearance similar to that of a natural gemstone but have different optical, physical and chemical properties. The year 1902 saw the first production of synthetic ruby using the Verneuil flame-fusion process. Later, sapphire, spinel, rutile and strontium titanate were grown with this technique. (This topic will be elaborated in later chapters.)

Concluding Remarks

So the aspects discussed above set the background for the science of mineralogy. Started in prehistoric times, the studies in mineralogy are still going on and each day it is broadening its realm of scope and approach. This book has been attempted to have a closer look in each of these aspects. Hope the readers will be benefitted.

PART I

Essentials of Mineralogy

CHAPTER

1

MINERALS AND THEIR CHEMICAL CLASSIFICATION

A mineral, by definition, is any naturally (not man-made) occurring inorganic (not a result of life plant or animal) substance. Its chemical structure can be exact, or can vary. All minerals belong to a chemical group, which represents their affiliation with certain elements or compounds. The science of mineralogy has spanned over several decades owing to its importance in various aspects. Knowledge of minerals of variable sources becomes essential for its application in metallurgy, gem-industry etc.

1.1 INTRODUCTION TO MINERALOGY

The first matter we should emphasise upon is why should we study mineralogy? While the general geology deals with the large scale processes like mountain-building and deformation in Earth crust, mineralogy focusses on the elemental parts that form the rocks and that determine their general appearance, the minerals. Mineralogy deals with the study of morphology, chemical composition, physical properties, impurities and internal structure of minerals. It thus plays a major role in geological exploration and prospecting. Mineralogy can be ascribed as a combination of geology and inorganic chemistry, which mainly highlights the origin, description and classification of minerals. The science of mineralogy can be sub-divided into four main aspects, which are:

- Paragenic mineralogy (study of mineral associations and occurrences)
- Crystal chemistry (composition, atomic arrangements)
- Descriptive mineralogy (study of physical properties and means of identification)
- Taxonomic mineralogy (classification, systematisation and nomenclature)

The origin of minerals is mainly dependant upon the elemental abundances in Earth crust, which are approximately O = 46.6%, Si = 27.7%, Al = 8.1%, Fe = 5.0%, Ca = 3.6%, Na = 2.8%, K = 2.6% and Mg = 2.1%. Since oxygen and silicon form the major abundant elements of the Earth, the most abundant minerals in the Earth crust are the silicates. This part will be covered in paragenic mineralogy. Because minerals are formed of chemical elements, an important aspect of mineralogy is the analysis of chemical elements. The analyses of atoms—the structural unit of elements—form the basis of this study. The electrons taking part in the chemical reactions lead to the formation of compounds, hence minerals. Such transferable atoms, which have an excess of positive or negative charges are called as ions. The ways by which electrons are shared between atoms result in the development of different types of bonds in minerals, which should be detailed in crystal chemistry. As a result of different types of bonding and different atomic arrangements, various types of crystals have different properties form, which will be discussed in descriptive and taxonomic mineralogy. Minerals are found to be extremely useful in various fields.

1.2 MINERALS

The word mineral comes from the Latin word *minare*, meaning mining. It denotes the strong historical relationship between mining and the history of studying ore. Minerals are naturally occurring inorganic, homogeneous, solid, having a definite but not necessarily fixed chemical composition and a fixed, ordered internal structure. According to the International Mineralogical Association, there are over 4000 known minerals on Earth and newer minerals are being discovered at the rate of 60 per year.

1.3 IMPORTANCE OF MINERALS

Importance of minerals has been proved in several ways nowadays and thus minerals have become an integral part not only in view of geological aspect, but also in respect to various fields and branches of technology and development.

1.4 CLASSIFICATION OF MINERALS

Minerals, like all the other natural entities, are required to be classified scientifically for the convenience of study and utilization. The physical and chemical properties of a mineral are primarily controlled by its chemical composition and internal structure. So these are the two most important parameters used in the scheme of classification. The complete hierarchical order in the classification of minerals is as follows.

Hierarchical Orders

Class: Based on the *dominant anions* and *anionic groups* present in them, like Cu_2S, AgS_2, Cu_3AsS_4, FeS_2, $CuFeS_2$ etc. included in the *Sulphide Class*; $CaCO_3$, $MgCO_3$, $FeCO_3$, $MnCO_3$, $ZnCO_3$ etc. included in the *Carbonate Class*.

Subclass (Family): On the basis of structural types. For example in fayalite and pyrope, each SiO_4 tetrahedron is isolated and not connected to another; while in talc and muscovite each SiO_4 tetrahedron is linked to three other tetrahedra to form a layered structure. Based on these structural attributes the first two are included in the family nesosilicate, while the other two belong to the family phyllosilicate.

In some classes, the subclasses are defined on the basis of chemical composition. Example: the two subclasses of the carbonate class, *hydrous* and *anhydrous*.

Group: Based on chemical/structural similarity to greater degree. For example Hematite group, Spinel group, Rutile group included within oxide class or Olivine group and Garnet group included within nesosilicate class.

Series: Based on solid solution relationship (vide Chapter 3). Example: albite and anorthite have a continuous solid solution relationship, and they constitute the plagioclase series.

Species: Based on individual mineral type.

It is to be noted here that the above definitions of hierarchical units may not be applicable in some cases, and different criteria may be used in the systematization of different classes of minerals. For example, in carbonate class the subclasses are defined on the basis of chemical composition (hydrous and anhydrous) while the groups are defined on the basis of structural resemblance (like the hexagonal calcite group, orthorhombic aragonite group etc.). In some classes one or more hierarchical units (like family or group or series) are omitted. So a strict adherence to the given scheme of classification may be impractical and it cannot be generalized.

Chemical Classification of Minerals

As already been mentioned, minerals have definite chemical composition. On that basis, some are pure metals or salts, while others have extremely complex chemical constituents. So depending on that widely variable range, minerals can be classified into following classes; the categorization being mainly on the basis of anion group.

Table 1.1: Classification of minerals on the basis of the anionic groups

Mineral group	*Anionic group*	*General properties*	*Sub-groups*
Native elements		Occur in nature, almost in pure form; can be metals, semi-metals or non-metals.	Native metal and semi-metal group;Native non-metal group
Sulphides	S	Origin is hydrothermal, magmatic, skarn and also sedimentary. Show high symmetry, have metallic properties like metallic lustre, high density, low hardness and electrical conductivity; derivatives of H_2S in the core.	*Tetrahedral sulphide group *Octahedral sulphide group;Other sulphides
Sulphosalts		General chemical formula is XYS, where X is Ag, Cu or Pb, Y is As, Sb or Bi and S is sulphur; have As and Sb in their metal site.	Sulphosalt group
Sulphates	SO_4	Origin is sedimentary and chemical sea and lake precipitates. Contain anionic $(SO_4)^{2-}$ group; occur as individual, unlinked groups.	Anhydrous sulphate group;Hydrous sulphate group
Oxides	O	Origin is magmatic, pegmatitic, hydrothermal and also as a result of exogenic processes in upper stratum of a lithosphere. Have strong ionic bonding, relatively hard, dense and refractory.	*Tetrahedral and octahedral oxide group; Spinels and other oxides with mixed or unusual combination
Hydroxides	OH	Hydroxides contain $(OH)^-$ group; have low bond strength, less hard and dense, formed at low temperature.	Hydroxide group
Halides	Cl, F, Br, I	Contain large-sized anions of halogens chlorine, bromine, iodine and fluorine; exhibit low hardness and medium to high melting point; poor conductor of heat and electricity in solid state, but good conductors in fused state.	Halide group
Carbonates	CO_3	Origin is sedimentary or metamorphic. Contain anionic complex $(CO_3)^2$; are strongly bonded, do not share oxygen atoms with one another; relatively soft except the uranyl and alkali carbonates.	Calcite group; Dolomite group; Aragonite group; Other carbonates

(Contd.)

(Contd.)

Nitrates	NO_3	Mainly of biogenic origin. Contain $(NO_3)^-$ group and resembles carbonate structure; have great strength.	Nitrate group
Borates	BO_3, B_4O_7, B_4O_{11}, B_5O_9	Origin is chemical laccustrine sediments or result of dissolution and redeposition of boron compounds. Contain boron-oxygen groups that are linked as double triangles, triple rings, sheets and chains; colour white, hardness and density low.	Anhydrous borate group;Hydrous borate group
Chromates	CrO_4, Cr_2O_7	Formed from the oxidation of copper-iron-lead sulphide ores, containing a small amount of chromium.	Chromate group
Phosphates, arsenates, vanadates	PO_4, AsO_4 VO_4	Mainly rare exogenic minerals, only monazite and apatite are magmatic; contains anionic $(PO_4)^{3-}$ group.	Phosphate group
Tungstates, molybdates	WO_4, MoO_4	Form a tetrahedral structure with four oxygen atoms with a centrally placed molybdenum or tungsten atom; do not share any oxygen atoms between them.	Tungstate group; Molybdate group
Silicates	Si_xO_y	Si:O ratio is 1:4	Nesosilicates
		Si:O ratio is 2:7	Sorosilicates
		Si:O ratio is 1:3	Inosilicates (single chain)
		Si:O ratio is 4:11	Inosilicates (double chain)
		Si:O ratio is 2:5	Phyllosilicates
		Si:O ratio is 1:2	Tectosilicates

* To be discussed in subsequent chapters

1.5 BIO-MINERALS

Many living organisms incorporate minerals into their body structures for support and protection. Humans and other invertebrate animals use calcium and phosphorus in their bones. But bones are not the only support and protection structures requiring minerals. Chickens and other birds have eggshells made of calcium carbonate, the same compound in chalk. Other animals that use calcium carbonate to produce protective coverings are coral and molluscs, such as clams, snails and chonches. Another group of organisms that produce hard coverings are diatoms, microscopic, unicellular algae. Seashells, egg shells, coral and chalk are composed of the same compounds, but the properties of each is very different.

Definition: Biominerals are a subset of the mineral kingdom, those created by living creatures. In spite of usually fine grain size and intimate association with organic materials, biominerals are readily identified as common mineral species. Iron hydroxides, oxyhydroxides, calcium carbonates and calcium phosphates from uni- and multi-cellular species are presented as examples of biominerals, and biomineralisation processes. Their special morphological and crystal chemical characteristics provide unique structural contributions to the life forms that create them.

Biomineralisation: The processes by which organisms form minerals are collectively called biomineralisation. The control by many organisms over mineral formation distinguishes this process from abiotic mineralisation.

Choice of Materials for Formation of Biominerals

Calcium is the cation of choice for most organisms comprising 50% of known biominerals—calcium phosphate, carbonate, oxalate and other mineral types.

- 25% of the minerals are non-crystalline.
- 25% of biogenic minerals are phosphates.
- Calcium carbonates are the most abundant biogenic minerals.

Table 1.2: List of biominerals

Minerals	*Types*
Calcium carbonate: Eight known polymorphs	Seven crystalline and one amorphous Three are pure calcium carbonate: aragonite, calcite and vaterite Two contain one water molecule per calcite:monohydrocalcite and amorphous calcite Calcium carbonate hexahydrate
Calcium phosphate	Bone Teeth
Celestite	Acantharians
Barite	Loxedes
Silica	Diatoms Radiolarians
Iron oxides	Magnetic bacteria Iron teeth-limpets and chitins

Uses

- Skeletons
- Gravity Sensors—otoliths
- Lenses—trilobites
- Vaterite—ascidians, some inner ears of fish, repair in molluscs
- Amorphous calcium carbonate—storage of Ca in plants
- Many molluscs use biominerals for protection, support and cellular processes.
- Organic compounds modify the properties and structure of $CaCO_3$ so that the biomineral properties and performance are different from those of the pure inorganic form of the minerals.

- Environment can influence the minerals utilized and crystal structures formed by minerals.
- Production and dissolution of biominerals have major influences on mineral cycling, including the oceans' capacity to absorb excess atmospheric carbon dioxide.

1.6 MINERALS IN METEORITES

Table 1.3: Minerals in meteorites

Mineral group	*Minerals*	*Characteristics*
Metallic iron	Kamacite	This is essentially metallic iron with up to 7.5% nickel in solid solution. In cut sections kamacite looks like metallic iron and the other important nickel-iron mineral, taenite. On etched and polished surfaces, sometimes it can be distinguished. Kamacite may have Neumann bands—lines that are created when meteorites are shocked. Kamacite occurs as rims around troilite or schreibersite—swathing kamacite. Sometimes on polished surfaces, kamacite may be distinguished from taenite because it is not as hard.
	Taenite	This is iron with more than 25 percent nickel in solid solution. The atoms of taenite arrange themselves in a cube form with an atom centred in the face. In non-meteoritic metals this mineral is called austenite.
	Plessite	This is not a mineral, but is a mixture of fine-grained kamacite and taenite.
Iron sulfide	Troilite (FeS)	Iron sulfide is similar to the earthly mineral pyrrhotite. Troilite is not magnetic. It is metallic with a slight bronze colour and slight greenish tint. It occurs as spherical or rounded nodules. Sometimes troilite occurs as plates (lamellae) parallel to the cubic structure of the cubic taenite structure of iron meteorites. With a few exceptions, troilite is present in most iron meteorites.
	Daubreelite ($FeCr_2S_4$)	This iron chrome sulfide is found as exsolution lamellae in troilite nodules or as particles in kamacite.
Iron carbides	Cohenite $(Fe, Ni, Co)_3C$	It is iron carbonate and closely related to cementite (Fe_3C), a mineral described in steel. It is brilliant silver. The crystal form is orthorhombic. Cohenite is extremely hard in nature. Cohenite is unstable at atmospheric temperature and pressure and over a long period of time will break down into kamacite and graphite.
	Haxonite $(Fe, Ni, Co)_{23}C_6$	It is a cubic iron carbide similar to minerals found in steel. It is hard, though not as hard as cohenite. It is bright silver. Like cohenite, it is unstable at atmospheric temperature and pressure and will gradually decompose to kamacite and graphite.

(Contd.)

Table 1.3 (*Contd.*)

Mineral group	*Minerals*	*Characteristics*
Phosphides $(Fe, Ni)_3P$	Schreibersite, Rhabdite	Iron-nickel phosphide. Schreibersite is yellowish metallic, very hard—though not as hard as Cohenite—brittle and magnetic. The crystal form appears as oriented plates in the matrix of kamacite and taenite. Plate shaped rhabdites are found in hexahedrites.
Oxides	Chromite $FeCrO_4$	This ore of chrome is common on earth and as a trace mineral in iron meteorites. It is black, with cubic crystal form and octahedral cleavage.
	Magnetite Fe_3O_4	A common ore of iron on earth is found in the crusts of most meteorites and in carbonaceous chondrites. Magnetite is submetallic black with cubic crystal form.
	Corundum Al_2O_3	This mineral is what rubies and sapphires are made of—it is also a trace mineral in meteorites. Don't look for red or blue. Gray is the probable colour.
Carbon	Graphite	The common mineral of pencil lead is mainly found as nodules in iron meteorites and in carbonaceous chondrites. It is black and very soft.
	Diamond	This is the same element as graphite. When graphite is subjected to the extraordinarily high temperature and pressure of meteorite impact, sometimes diamonds will form. Canyon diablo meteorites have some.
	Lonsdaleite	A hexagonal polymorph of diamond (diamond is cubic). Lonsdaleite is mixed with diamond at Canyon diablo.
Silicates*	Pyroxene	Enstatite, bronzite, hypersthene and orthopyroxene are common in meteorites. Clinopyroxenes (monoclinic crystal structure), diopside ($CaMg(SiO_3)_2$), hedenbergite ($CaFeSi_2O_6$), augite ($(Ca,Na)(Mg,Fe,Al)\ (Si,Al)_2O_6$) and pigeonite ($(Mg,Fe,Al)\ (Mg,Fe)\ Si_2O_6$) are also found in meteorites particularly achondrites.
	Olivine	It is a rock forming mineral on Earth that is believed to have been formed at great depth. The Earth's mantle may have significant olivine. In meteorites, olivine is important in some pallasites. The olivines are a solid solution series with forsterite containing the most magnesium and fayalite having the most iron. The olivine in most stony iron meteorites ranges in composition from 60 to 85 percent forsterite with the remainder being fayalite.
	Feldspars	These are calcium, sodium. aluminum silicates—the most important rock forming minerals in the crust of the Earth, but are relatively scarce in meteorites.
	Phyllosilicate	Clays are found in some meteorites.

* To be discussed in subsequent chapters.

Concluding Remarks

Minerals have formed the basis of many industries, researches and discoveries nowadays.

The initiation of mineralogy has started with the notion of understanding the mineralogical structure of the Earth. The only way to estimate the internal conditions of Earth is the seismological techniques which provide direct measurement of mantle properties. Phase diagrams of mineral systems support the basis of modelling and understanding the constitution of the mantle. This phase diagram is converted to velocity and density profiles for connecting seismology and mineral physics. And through various researches it has been possible to estimate the impacts of the elements on phase equillibria and crystal chemistry of minerals, as well as on physical properties of mineral assemblages and to compare them with seismic data. This has led to wide development in the field of researchers.

Depending upon the utility of various minerals, different modes of explorations are carried out for economically viable minerals. Profound use of minerals in industries leads to various types of extraction, processing and refining techniques to make the natural minerals suitable for industrial use.

Wide application of minerals in medical geology is prevalent in recent times. The use and variations of biogenic minerals have aided this process.

With the increasing requirements of different minerals, various synthetic minerals are manufactured from raw materials.

All these processes and techniques will be elaborated in subsequent chapters.

Think for a while

1. Citing few examples, show how the variation of anionic groups control the mineral characteristics.
2. What do you mean by the term *biominerals*? How it can be compared with the general mineral-types?
3. Minerals are also an important part in meteorites. How they can be related or compared to earth minerals?

FURTHER READING

Blackburn, W.H. and W.H. Dennen. Principles of Mineralogy, 2nd ed. William C. Brown, Dubuque, IA. 1994.

Deer, W.A., R.A. Howie and J. Zussman. Rock Forming Minerals. 5 vols. John Wiley and Sons, New York. 1962.

Klein, C. and C.S. Hurlbut, Jr. Manual of Mineralogy, 21st ed. John Wiley and Sons, New York. 1993.

Skinner, H.C.W. Biominerals. *Mineralogical Magazine*, October 2005, **69(5):** 621-641.

Wenk, Hans-Rudolf and Andrei Bulakh. Minerals. Cambridge University Press, 2006.

Internet Data Retrieved from:

Biomineralization. Session lectures. *J Biol Inorg Chem* (2007) **12** (Suppl 1): S207–S209.

Meteorite Minerals. from The Meteorite Market. www.meteoritemarket.com

CHAPTER

2

MINERAL CRYSTALS AND STRUCTURAL CLASSIFICATION

Crystallography is the study of the structure of crystals. It describes the structure of a crystal, the principles that govern the different types of possible structures, the crystal structures of specific substances, and the methods by which structures are determined. In numerous scientific fields, the knowledge of crystallography is essential and significantly important e.g. metallurgy, materials science, ceramics, physics and chemistry. Each of these fields has its own special requirements, but there is a basic core of knowledge common to them all. This consists of the description of the crystal structure in terms of the smallest unit representing it, the three-dimensional lattice with which the atom sites can be represented, and also the accepted method for determining planes and directions in crystal structures. Various aspects of crystallography e.g. symmetry, point groups and space groups, Miller indices, unit cell geometry, lattice, motif, planes and directions, crystallographic systems and forms, various crystal defects and symmetrical intergrowths of crystals called twinned crystals will be discussed to explain the crystal morphology and how it influences the various physical and chemical properties of minerals.

Atoms of elements combine in mineral formation to build specific geometric shapes on the smallest scale. These shapes combine to form crystal forms. So the science of crystallography deals with the study of crystals, which are defined as a rigid solid in which the particles are arranged in a repeating pattern. This pattern is governed by the symmetry elements present in the specific crystal form and can be categorized into one of the 32 possible crystal classes.

2.1 SYMMETRY

The key parameter to understand crystal structure is its symmetry. Symmetry is the order in arrangement and orientation of atoms in mineral crystals and the consequent development of various mineral properties.

Symmetry operation in crystallography can be defined as any motion or movement that brings any face, corner or edge into coincidence with similar face, edge or corner, elsewhere in the crystal.

Symmetry element: The geometrical locus, about which symmetry operation acts, is called symmetry element.

Table 2.1: Different symmetry operations used in crystallography

Symmetry operation	*Symmetry element*	*Symbols*
Rotation around an imaginary axis passing through centre	Rotation axis or symmetry axis	1, 2, 3, 4, 6
Reflection across a plane	Plane of symmetry or mirror	*m*
Inversion through a centre	Inversion point or centre of symmetry	*i* or *c*
Combined operation of rotation around an axis followed by inversion through the centre	Roto-inversion axis	A1, A2, A3, A4, A6

There are 32 possible combinations of symmetry operations that define the external symmetry of crystals. These 32 possible combinations result in the 32 crystal classes. These are often also referred to as the 32 point groups, which have been described later in this chapter.

Table 2.2: Additional operational symmetry used in crystallography

Elements	*Operation*	*Symbols*	*Comments*	*Figure*
Screw axis	Rotation × Translation; The symbol N_m indicates a rotation of 360/N° plus a translation m/N times the repeat (cell edge) in the direction of the rotation axis.	2_1,3_1,3_2, 4_1,4_2,4_3, 6_1,6_2,6_3, 6_4,6_5	The translational component of the operation is parallel to the axis of rotation. Its magnitude is a fraction of the unit translation in the direction of the rotation axis. There are (n – 1) possible screw axes for an n-fold rotation axis (e.g., one axis for 2-fold). The translational component is 1/n.	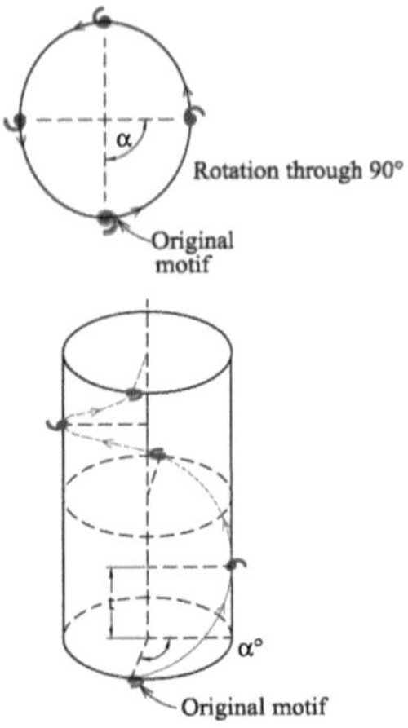

(*Contd.*)

(*Contd.*)

Glide planes	Reflection × Translation	*a*, *b*, *c*, *n*, *d*	The translational component is in the direction of the unit translations of the unit cell or along the diagonal of the cell. The component equals the unit translation divided by 2 (i.e., *a*, *b*, *c* glide = *a*/2, *b*/2, *c*/2). The component equals the diagonal divided by 2 {i.e., *n* glide = (*a*+*b*)/2, (*b*+*c*)/2), (*a*+*c*)/2}. The component equals the diagonal divided by 4 {i.e., *d* glide = (*a*+*b*)/4, (*b*+*c*)/4), (*a*+*c*)/4}.	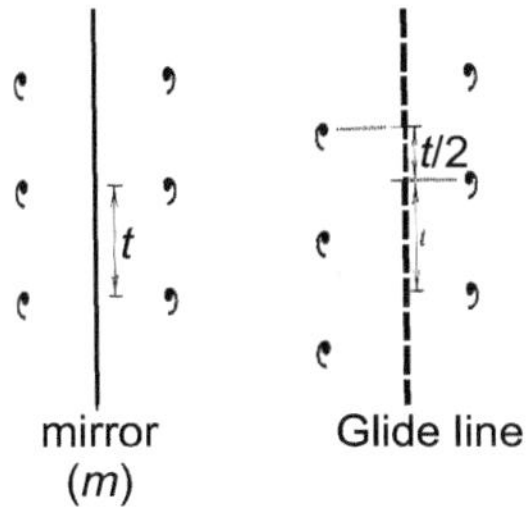

Table 2.3: Point groups and space groups

Group	*Symmetry operations*	*Number of groups*
Point groups: The mathematical group comprising the symmetry operations that leave at least one point unmoved and that leave the appearance of the crystal structure unchanged.	Reflection, rotation, inversion, improper rotation	There are 32 possible crystal classes. Each one can be classified into one of the seven crystal system.
Space groups: Composed of the translational symmetry operations in addition to the operations of the point group.	Pure translations which move a point along a vector; Screw axes, which rotate a point around an axis while translating parallel to the axis; Glide planes, which reflect a point through a plane while translating it parallel to the plane.	There are 230 distinct space groups.

Space groups control atom symmetry and point groups control crystal face symmetry.

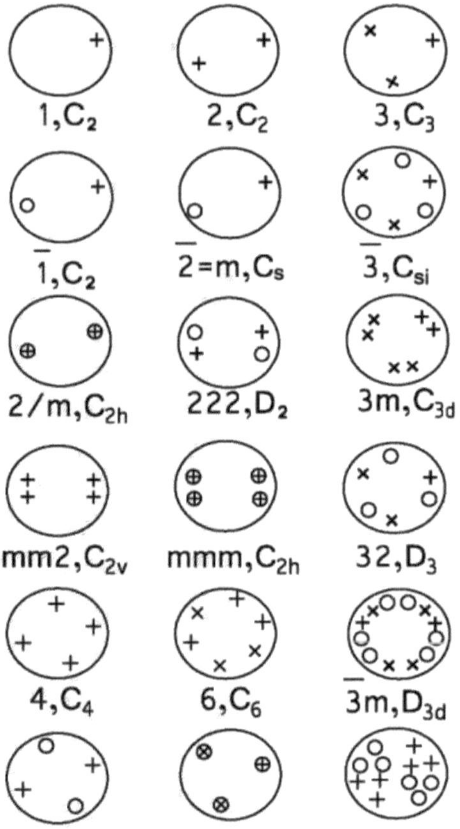

Fig. 2.1: The 32 point groups.

Crystallographic axes: These are imaginary reference lines constructed through crystals and in most instances coincide with symmetry axes or normals to symmetry planes. These axes aid in the orientation of crystals and help to explain other crystal concepts like unit cells and Miller indices (to be discussed later).

Crystallographic axes can also be defined as the imaginary lines in a crystal that can be used as a frame of reference for naming crystal faces. They are oriented parallel to their symmetries.

Interfacial angle: The angle between face-normals of a crystal.

The interfacial angle is controlled by the spacing between lattice points. From the imaginary 2-dimensional crystal lattice shown here the angle θ between the face that runs diagonally across the lattice and the horizontal face will depend on the spacing between the lattice points.

Since all crystals of the same substance will have the same spacing between lattice points (they have the same crystal structure), the *angles between corresponding faces of the same mineral will be the same.* This is known as the Law of constancy of interfacial angles.

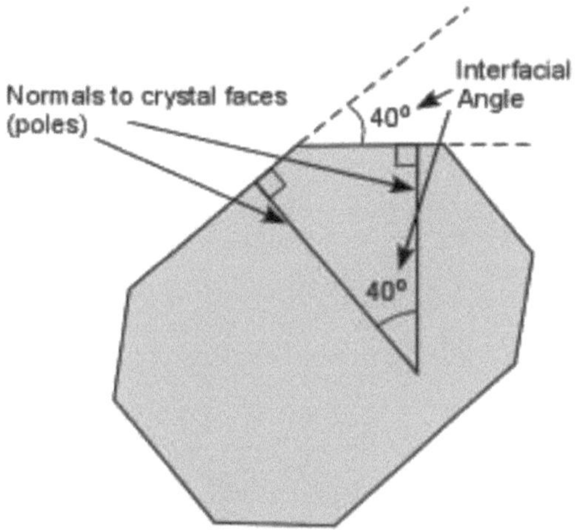

Fig. 2.2: Interfacial angle.

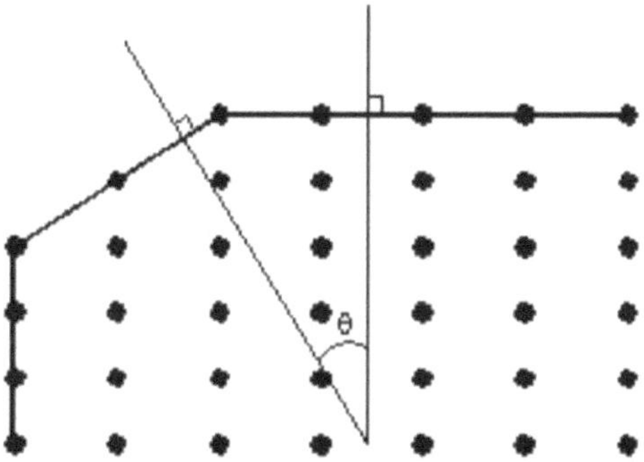

Fig. 2.3: Dependency of interfacial angles.

Axial ratios: Axial ratios express the relative lengths of the crystallographic axes. They are taken as relative to the length of the *b*-axis. Thus, an axial ratio is defined as follows:

$$\text{Axial Ratio} = a/b{:}b/b{:}c/b$$

where a = length of the (equivalent to x) crystallographic axis, b = length of the (equivalent to y) crystallographic axis, and c = length of the (equivalent to z) crystallographic axis.

Table 2.4: Variation of Miller indices with different crystallographic systems

Crystallographic systems	*Crystallographic axes*	*Miller indices*
Triclinic, monoclinic and orthorhombic	$a \neq b \neq c$	$a/b : 1 : c/b$ or $: 1 : c$
Tetragonal	$a = b \neq c$	$1 : 1 : c/b$ or $1 : c$
Isometric	$a = b = c$	$1 : 1 : 1$ or 1
Hexagonal	$a_1 = a_2 = a_3 \neq c$	$1 : 1 : 1 : c/a$ or $1 : c$

In recent days x-ray crystallography is used to determine the size of the unit cell, and hence, the absolute value of the crystallographic axes. For example, two specific cases have been shown here.

For *orthorhombic sulfur*, the unit cell dimensions as measured by x-rays are:

$$a = 10.47\text{Å}$$
$$b = 12.87\text{Å}$$
$$c = 24.39\text{Å}$$

Thus, the axial ratio for orthorhombic sulfur is:

$$10.47/12.87 : 12.87/12.87 : 24.39/12.87$$

or

$$0.813 : 1 : 1.903$$

For hexagonal quartz, the unit cell dimensions as determined by x-ray crystallography:

$$a_1 = a_2 = a_3 = 4.913\text{Å};$$
$$c = 5.405\text{Å}$$

where Å stands for Angstroms = 10^{-10} metre.

Thus the axial ratio for quartz is 1 : 1 : 1 : 5.405/4.913

or

$$1 : 1 : 1 : 1.1001$$

which simply says that the *c* axis is 1.1001 times longer than the *a* axis.

2.2 LATTICE, MOTIFS, UNIT CELLS, PLANES AND DIRECTION IN CRYSTALS

A crystal structure is a unique arrangement of atoms in a crystal. A crystal structure is composed of a *motif*, a set of atoms arranged in a particular way, and a *lattice*, an array of points repeating periodically in three dimensions. We specify the lattice using the vectors **a**, **b** and **c**.

The vector between any two lattice points (**r**) satisfies the relationship, $\mathbf{r} = n\mathbf{a} + p\mathbf{b} + q\mathbf{c}$, where n, p and q are integers. The simplest repeating unit in the arrangement of a crystal is called the *unit cell*. It is the smallest unit of a mineral that retains all of the physical, chemical and crystallographic properties of a mineral. The unit cell comprises the symmetry of each holohedral class of the crystal system. There is no fixed relationship between the positions of a unit cell and the lattice that is associated with it. The lattice only emphasizes on the size and shape of the unit cell, not position.

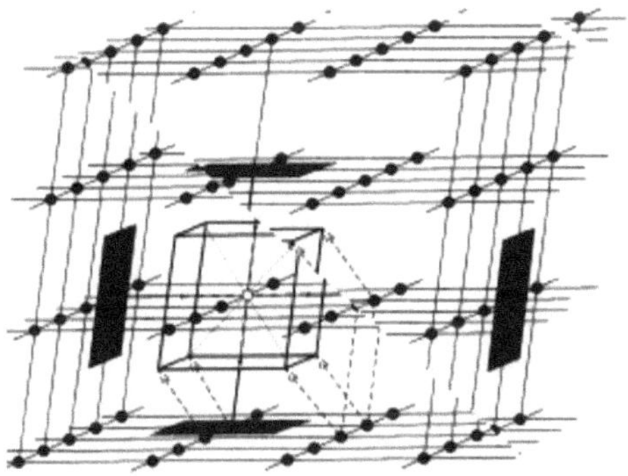

Fig. 2.4: Lattice and unit cell in 3D.

Bravais Lattice

In crystallography, a Bravais lattice is an infinite set of points generated by a set of discrete translation operations. A Bravais lattice looks exactly the same from whichever point one views it. The position vectors of a Bravais lattice in three dimensions are given by

$$\mathbf{R} = n_1\mathbf{a}_1 + n_2\mathbf{a}_2 + n_3\mathbf{a}_3$$

where n_1, n_2 and n_3 are integers and $\mathbf{a}_1$, $\mathbf{a}_2$ and $\mathbf{a}_3$ are three non-coplanar vectors, called primitive vectors.

It can also be described as a lattice system with the interaxial angles as *a*, *b* and gama, where these angles may or may not be the same. Where $\alpha = \beta = \gamma = 90°$, the cartisan coordinate system is achieved.

Lattice systems: These lattice systems are a grouping of crystal structures according to the axial system used to describe their lattice. Each lattice system consists of a set of three axes in a particular geometrical arrangement. There are seven lattice systems. They are similar to but not quite the same as the seven crystal systems and the six crystal families.

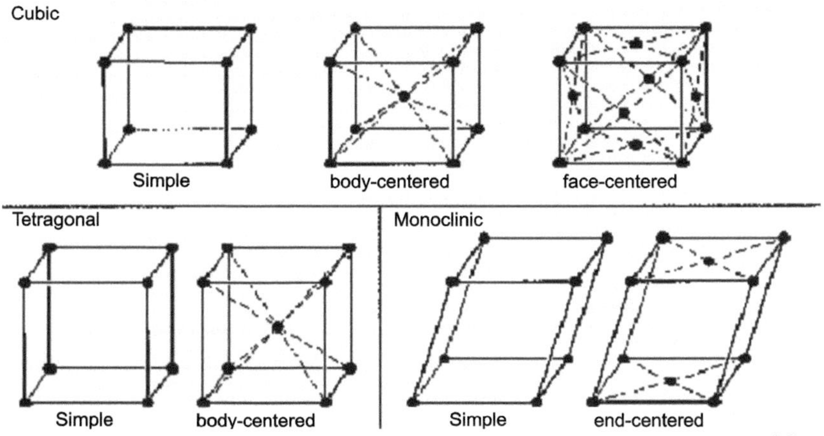

(Contd.)

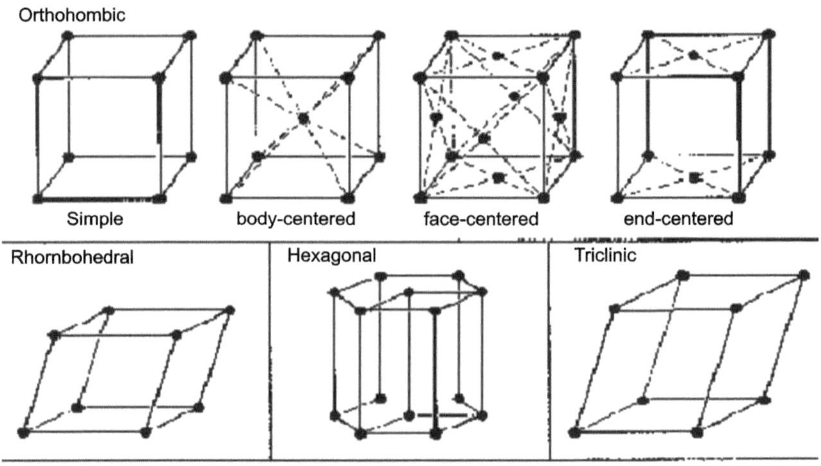

Fig. 2.5: Bravais lattices.

Planes and Directions in Crystals

The crystallographic directions are imaginery lines linking nodes (atoms, ions or molecules) of a crystal and the crystallographic planes are imaginery planes linking nodes.

Importance of identification of direction of planes and directions in a crystal

- Deformation under loading occurs on certain crystallographic planes and in certain directions; so for prediction of material failure, it is important to assure the probable mode of failure.
- Also mineral properties like electrical conductivity, thermal conductivity and elastic modulus vary in a crystal with orientation.

Labelling Convention

Simplest way is to use a 3-D system, where every location can be expressed using three numbers or indices—*a*, *b*, *c* and α, β, γ. These are called *lattice parameters*.

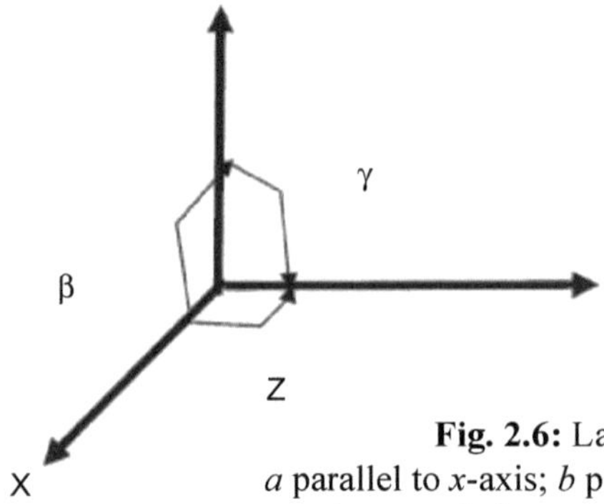

Fig. 2.6: Lattice parameters.
a parallel to *x*-axis; *b* parallel to *y*-axis; *c* parallel to *z*-axis; α angle between *y* and *z*; β angle between *z* and *x*; and γ angle between *x* and *y*.

Some directions and planes have a higher density of nodes. These high density planes have an influence on the behaviour of the crystal as follows:

- Optical properties: Refractive index is directly related to density (or periodic density fluctuations).
- Adsorption and reactivity: Physical adsorption and chemical reactions occur at or near surface atoms or molecules. These phenomena are thus sensitive to the density of nodes.
- Surface tension: The condensation of a material means that the atoms, ions or molecules are more stable if they are surrounded by other similar species. The surface tension of an interface thus varies according to the density on the surface.
- Microstructural defects: Pores and crystallites tend to have straight grain boundaries following higher density planes.
- Cleavage: This typically occurs preferentially parallel to higher density planes.
- Plastic deformation: Dislocation glide occurs preferentially parallel to higher density planes. The perturbation carried by the dislocation (Burgers vector) is along a dense direction. The shift of one node in a more dense direction requires a lesser distortion of the crystal lattice.

Linear and Planar Density
Linear density—Number of atoms per length whose centres lie on the direction vector for a specific crystallographic direction.

$$LD = \frac{\text{Number of atoms centered on a direction vector}}{\text{length of direction vector}}$$

- Planar density—Number of atoms per unit area that are centered on a particular crystallographic plane.

$$PD = \frac{\text{Number of atoms centered on a plane}}{\text{area of plane}}$$

Significance: Properties, in general, depend on linear and planar density.

Examples: Speed of sound along directions– Slip (deformation in metals) depends on linear and planar density. Slip occurs on planes that have the greatest density of atoms in direction with highest density (along closest packed directions on closest packed planes).

The unit cell configuration is mainly of two types:

1. *Primitive*: A cell with (nodes) atoms at all the corners and *not* at the centre of faces *nor* an atom at the very centre of the cell.

2. *Non primitive*: A cell with (nodes) atoms at all corners and also an atom on each of two opposite faces—if at top and bottom. *C-centered*: if on both side faces, *B-centered*: if on front and back faces, *A-centered*: or on all faces. *F- or face-centered*: the A, B and C lattices are symmetrically identical and can be converted to each other by an appropriate change of the crystallographic axes also. A cell with (nodes) atoms at the corners and one at the very centre of the unit cell is a *body-centered (I)* non-primitive cell.

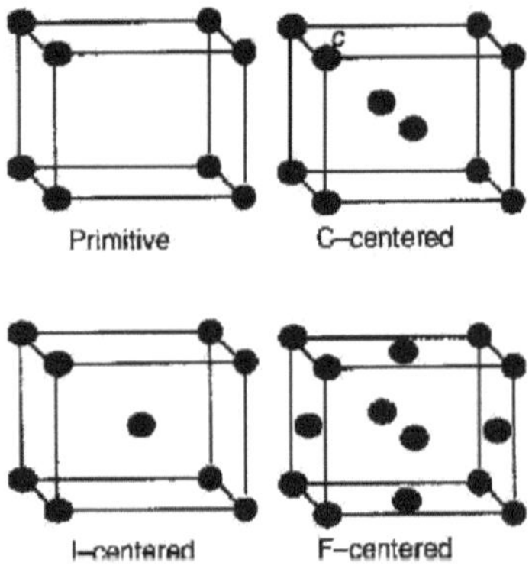

Fig. 2.7: Unit cell configuration.

Relation of Spacings to Cell Geometry

Calculation of interplanar angle:

$$\cos\varphi = \frac{h_1h_2 + k_1k_2 + l_1l_2}{(h_1 + k_1 + l_1)(h_2 + k_2 + l_2)}$$

The spacing d is the length of the perpendicular dropped from the origin to the plane of crystal. The specific values of $1/d_2$ for the crystal systems is given in the following table.

Table 2.5: Unit cell parameters relating to d_{hkl}

Crystal system	d^{-2}	
Isometric	$1/a^2 (h^2 + k^2 + l^2)$	
Tetragonal	$[(h^2 + k^2)/a^2] + 12/c^2$	
Orthorhombic	$(h^2/a^2) + (k^2/b^2) + (l^2/c^2)$	
Hexagonal	$(4/3a^2)(h^2 + hk + k^2) + l^2/c^2$	
Rhombohedral	$[1/a^2]\{(h^2 + k^2 + l^2)\sin^2\alpha + 2(hk + kl + lh)(\cos^2\alpha - \cos\alpha)\}/\{1 + 2\cos^3\alpha - 3\cos^2\alpha\}$	
Monoclinic	$\{h^2/a^2 + k^2/b^2 - 2\,hk\cos\gamma/ab\}/\sin^2\gamma + (l^2/c^2)$	(1st setting)
	$[(h^2/a^2) + (l^2/c^2) - (2\,hl\cos\beta/ac)] + (k^2/b^2)$	(2nd setting)
Triclinic	$[(h^2/a^2)\sin^2\alpha + (k^2/b^2)\sin 2\beta + (l^2/c^2)\sin^2\gamma + (2hk/ab)(\cos\alpha\cos\beta - \cos\gamma) + (2\,kl/bc)(\cos\beta\cos\gamma - \cos\alpha) + (2\,lh/ca)(\cos\gamma\cos\alpha - \cos\beta)]/(1 - \cos^2\alpha - \cos^2\beta - \cos^2\gamma + 2\cos\alpha\cos\beta\cos\gamma)$	

Reciprocal Lattice: a*, b* and c*

The reciprocal lattice generated has the same properties as the different pattern of the real (direct) lattice. The vector from the origin to the reciprocal lattice point (101) has a length reciprocal to the (101) interplanar spacing and a direction normal to (101) planes. Every reciprocal lattice point has this property.

a* : Perpendicular to the (100) planes of the real lattice and has a length equal to $1/d_{100}$.

b* : Perpendicular to the (010) planes of the real lattice and has a length equal to $1/d_{010}$.

c* : Perpendicular to the (001) planes of the real lattice and has a length equal to $1/d_{001}$.

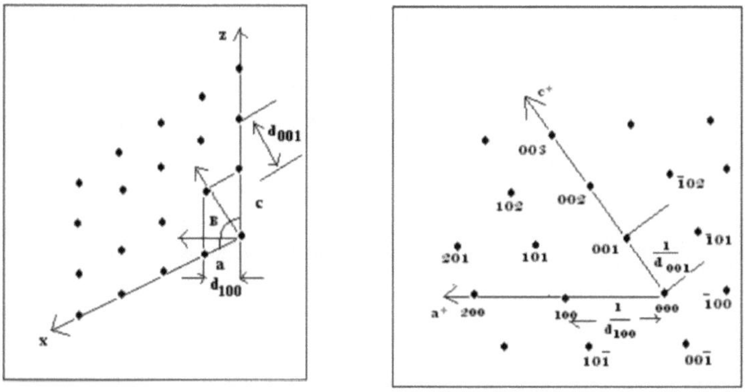

a b

Fig. 2.8: An example of a reciprocal lattice construction from a real lattice. (a) An *ac* section of a monoclinic lattice showing the *d* spacing of (100) and (001) planes. (b) The a*c* reciprocal lattice section in the same orientation as (a).

Choice of Unit Cell

There is always more than possible choice of unit cell. By convention the unit cell is chosen so that it is as small as possible while reflecting the full symmetry of the lattice. The best choice is one where: (a) the edges of the unit cell should coincide with the symmetry of the lattice; and (b) the edges of the unit cell should be related by the symmetry of the lattice. The smallest possible cell that contains all elements should be chosen.

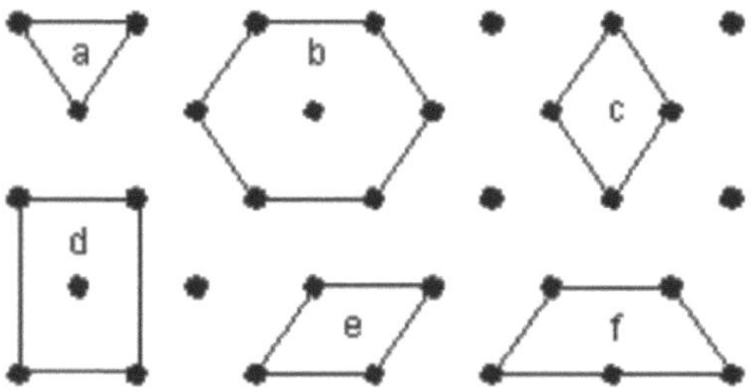

Fig. 2.9: Choice of unit cell in 2D.

For example, in the two-dimensional lattice shown here there are six possible choices to define the unit cell, labelled *a* through *f*. The lattice has two-fold rotational symmetry about an axis perpendicular to the page. Since the lattice itself does not have three-fold or six-fold rotational symmetry, choices *a* and *b* would not be wise choices for the unit cell. Choice *f* can be eliminated because it is really just half of cell *b*. The edges of *c* and *e* are not coincident or parallel to any two-fold axes that lie in the plane of the page. Thus our best choice would be cell *d*.

Miller index (hkl): A Miller index is of the form (*h k l*), where the integers *h*, *k* and *l* represent the *x*-, *y*-, and *z*-intercepts of the plane respectively. The Miller indices of a face consists of a series of whole numbers that have been derived from the intercept by their inversions and if necessary, the subsequent clearing of fractions.

Here the face [labelled (111)] that cuts all three axes at single unit length has the parameters 1a, 1b, 1c. Inverting these results in 1/1, 1/1, 1/1 to give the Miller index (111) (Fig. 2.10a).

Here for a complicated case, an orthorhombic crystal is taken. The small triangular face near the top that cuts all three axes had the parameters 1a, 1b, 1/3c. Inverting these becomes 1/1, 1/1, 3/1 to give the Miller index for this face as (113) (Fig. 2.10b).

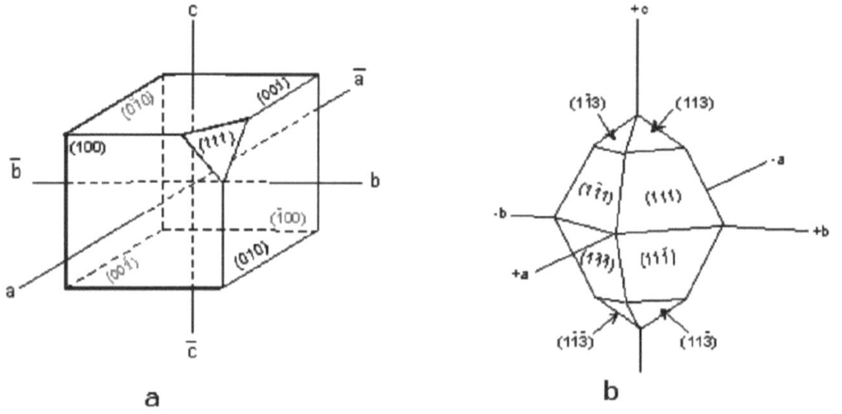

Fig. 2.10: (a) Miller index of a simpler case and (b) Miller index of a complex case.

2.3 CRYSTALLOGRAPHIC SYSTEMS

On the basis of the number of crystallographic axes, their relative length and the angular relationship existing between them, minerals can be categorized into six crystallographic systems. They are listed in Table 2.6 and shown in Fig. 2.11.

Table 2.6: Classes in different crystallographic systems

Crystal system	Crystal class	Symmetry	Name of class
Triclinic	1	none	Pedial
		i	Pinacoidal
Monoclinic	2	$1A_2$	Sphenoidal
	m	1m	Domatic
	2/m	i, $1A_2$, 1m	Prismatic
Orthorhombic	222	$3A_2$	Rhombic-disphenoidal
	mm2 (2mm)	$1A_2$, 2m	Rhombic-pyramidal
	2/m2/m2/m	i, $3A_2$, 3m	Rhombic-dipyramidal
Tetragonal	4	$1A_4$	Tetragonal- Pyramidal
		4	Tetragonal-disphenoidal
	4/m	i, $1A_4$, 1m	Tetragonal-dipyramidal
	422	$1A_4$, $4A_2$	Tetragonal-trapezohedral
	4mm	$1A_4$, 4m	Ditetragonal-pyramidal
	2m	1, $2A_2$, 2m	Tetragonal-scalenohedral
	4/m2/m2/m	i, $1A_4$, $4A_2$, 5m	Ditetragonal-dipyramidal
Hexagonal	3	$1A_3$	Trigonal-pyramidal
		1_3	Rhombohedral
	32	$1A_3$, $3A_2$	Trigonal-trapezohedral
	3m	$1A_3$, 3m	Ditrigonal-pyramidal
	2/m	1_3, $3A_2$, 3m	Hexagonal-scalenohedral
	6	$1A_6$	Hexagonal-pyramidal
		1_6	Trigonal-dipyramidal
	6/m	i, $1A_6$, 1m	Hexagonal-dipyramidal
	622	$1A_6$, $6A_2$	Hexagonal-trapezohedral
	6mm	$1A_6$, 6m	Dihexagonal-pyramidal
	m2	1_6, $3A_2$, 3m	Ditrigonal-dipyramidal
	6/m2/m2/m	i, $1A_6$, $6A_2$, 7m	Dihexagonal-dipyramidal
Isometric	23	$3A_2$, $4A_3$	Tetaroidal
	2/m	$3A_2$, 3m,$4A_3$	Diploidal
	432	$3A_4$, $4A_3$, $6A_2$	Gyroidal
	3m	$3A_4$, $4A_3$, 6m	Hextetrahedral
	4/m, 2/m	$3A_4$, $4A_3$, $6A_2$, 9m	Hexoctahedral

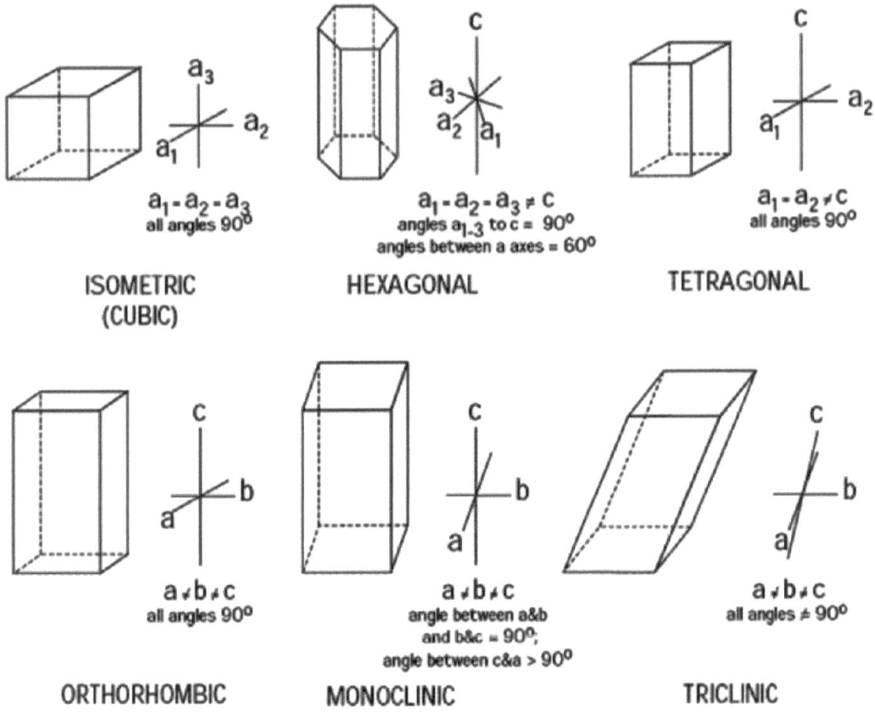

Fig. 2.11: Different crystallographic systems.

2.4 CRYSTAL MORPHOLOGY, STEREOGRAPHIC PROJECTIONS AND TWINNING

Crystal Morphology

The frequency with which a given face in a crystal is observed is proportional to the density of lattice nodes along that plane. Since the faces have direct relationship to the internal structure, they must have a direct and consistent angular relationship to each other. To understand it clearly we must know about form in crystallography.

Form in crystallography: A set of identical faces, related by a symmetry element is defined as a form in crystallography. A form consists of a group of crystal faces, all of which have the same relation to the symmetry elements and display the same chemical and physical properties because all are underlain by similar atoms in the same geometrical arrangement. Forms can be broadly divided into two types: open and close.

Open forms are those which do not completely enclose space. Close forms are those which enclose space. The open and close forms can be further divided into two types as shown in Table 2.7.

Table 2.7: Different types of forms in crystallography

Types	*Name of forms*	*Faces description*
Open	Pedions	Single face
	Pinacoids	Two parallel faces
	Prisms	3, 4, 6, 8 or 12 faces, all parallel to a common line
	Pyramids	3, 4, 6, 8 or 12 non-parallel faces, all intersect at a common point
	Domes	Two nonparallel faces related by a mirror
	Sphenoids	Two nonparallel faces related by a 2-fold axis
Close	Scalenohedrons	8-12 scalene triangle-shaped faces
	Trapezohedrons	6, 8 or 12 trapezoid-shaped faces
	Disphenoids	Four nonequilateral triangular faces
	Dipyramids	Two 3-, 4-, 6-, 8- or 12-sided pyramids related by a mirror
	Rhombohedrons	Six rhomb-shaped faces
	Tetrahedrons	Four equilateral triangular faces
	Cubes	Six square faces
	Octahedrons	Eight equilateral triangular-shaped faces
	Dodecahedrons	12 rhomb-shaped faces
	Pyritihedrons and tetaroids	12 five-sided faces
	Diploids and gyroids	24 five-sided faces

Hermann-Mauguin (International) Symbols

The use of Hermann-Mauguin symbols is inevitable in crystallography since these symbols are used to describe the crystal classes from the symmetry content. A procedure of deduction of these symbols is shown below.

Suppose a rectangular block has 32-fold rotation axes (A_2), three mirror planes (m), and a centre of symmetry (i). Write a number representing each of the unique rotation axes present. A unique rotation axis is one that exists by itself and is not produced by another symmetry operation. In this case, all three 2-fold axes are unique, because each is perpendicular to a different shaped face, so we write a 2 (for 2-fold) for each axis as **2 2 2**.

Next we write an "m" for each unique mirror plane. Again, a unique mirror plane is one that is not produced by any other symmetry operation. In this example, we can tell that each mirror is unique because each one cuts a different looking face. So, we write: 2 m 2 m 2 m.

If any of the axes are perpendicular to a mirror plane we put a slash (/) between the symbol for the axis and the symbol for the mirror plane. In this case, each of the 2-fold axes are perpendicular to mirror planes, so our symbol becomes: 2/m2/m2/m.

Stereographic Projection

Stereographic projection, introduced by William Hallowes Miller in 1839, is a most convenient tool for the representation of three-dimensional objects in two dimensions. Its primary importance is in depicting the angular relationship between the crystal phases.

To construct a stereographic projection, imagine that the object with a given symmetry or surface is at the centre of a sphere. Consider the sphere to have a polar axis that is bisected by an equatorial plane. Project features of interest on the object from the centre, out to the surface of the sphere.

Then project the points on the surface of the sphere through the equatorial plane to the point where the polar axis intersects the sphere in the opposite hemisphere. The stereographic projection is then given by the equatorial plane and all intersections of the plane by the projected points. If the projection point started in the northern hemisphere, its projection onto the equatorial plane is represented as a "plus". Points originating in the southern hemisphere are denoted with a "circle". Sometimes points generated by improper symmetry operations are also denoted with a comma to indicate opposite handedness.

The unit cell axis with highest symmetry is usually selected as the polar axis. Rotation axes not in the equatorial plane are drawn with the symbol representing the type of axis at the projection point on the equatorial plane. Rotation axes in the equatorial plane are drawn outside of the projection terminating in arrows. Mirror planes are drawn as thickened lines. Inversion centres are drawn as open circles in the centre of the polar axis.

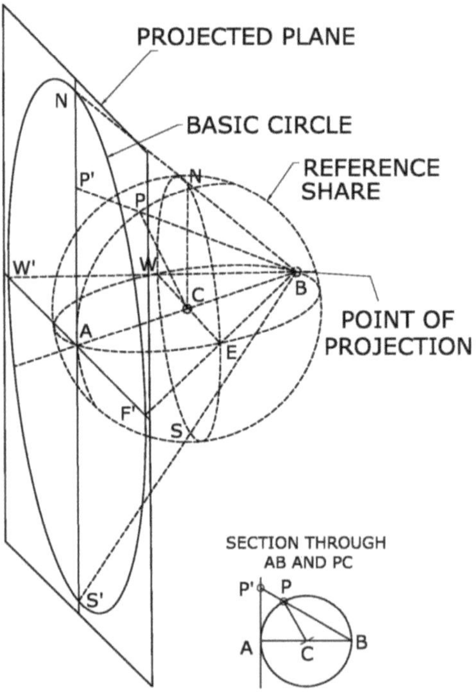

Fig. 2.12: Stereographic projection.

Points to remember while constructing stereographic projection for plotting crystal faces

- All crystal faces are plotted as poles (lines perpendicular to the crystal face. Thus, angles between crystal faces are really angles between poles to crystal faces.
- The **b** crystallographic axis is taken as the starting point. Such an axis will be perpendicular to the (010) crystal face in any crystal system. The [010] axis (note zone symbol) or (010) crystal face will therefore plot at $\phi = 0^o$ and $\rho = 90^o$.
- Positive ϕ angles will be measured clockwise on the stereonet, and negative ϕ angles will be measured counter-clockwise on the stereonet.
- Crystal faces that are on the top of the crystal ($\rho < 90^o$) will be plotted as open circles, and crystal faces on the bottom of the crystal ($\rho > 90^o$) will be plotted as "+" signs.
- Place a sheet of tracing paper on the stereonet and trace the outermost great circle. Make a reference mark on the right side of the circle (East).
- To plot a face, first measure the ϕ angle along the outermost great circle, and make a mark on your tracing paper. Next rotate the tracing paper so that the mark lies at the end of the E-W axis of the stereonet.
- Measure the ρ angle out from the centre of the stereonet along the E-W axis of the stereonet. Note that angles can only be measured along great circles. These include the primitive circle, and the E-W and N-S axes of the stereonet.
- Any two faces on the same great circle are in the same zone. Zones can be shown as lines running through the great circle containing faces in that zone. The zone axis can be found by setting two faces in the zone on the same great circle, and counting 90° away from the intersection of the great circle along the E-W axis.

Twinning

A twin is an intergrowth of two or more crystals, by means of a common point, line or plane and related by some symmetrical operations like reflection, translation or rotation.

Mechanism of Twinning: Three special symmetry elements used to describe types of twinning:

1. Reflection across a mirror plane referred to as the twin plane.
2. Rotation about a twin axis that is common to both of the twinned individuals.
3. Inversion about a point referred to as the twin centre.

On the basis of their origin, twinning can be classified as three types shown in Table 2.8.

Table 2.8: Different types of twinning

Types	*Description*	*Examples*
Growth twins	During crystal growth accidentally, a new crystal is added to the face of an already existing crystal, and the new crystal shares lattice points on the face of the existing crystal, but has an orientation different from the original crystal.	Contact twins, penetration twins.
Transformation twins	When a pre-existing crystal undergoes a transformation due to a change in pressure or temperature. Due to change of temperature or pressure when a new crystal structure and symmetry gets stable, different parts of the crystal become arranged in different symmetrical orientations, and thus form an intergrowth of one or more crystals.	Dauphiné and Brazil twinning in quartz. The combination of albite twinning and pericline twinning in alkali feldspar.
Gliding twins	During deformation, atoms can be pushed out of place. If this happens to produce a symmetrical arrangement, it produces gliding or deformation twins.	Polysynthetic twins on {012} of calcite.

Table 2.9: Types of twins

Twins	*Description*	*Examples*	*Figures*
Contact twin	A planar composition surface separating two individual crystals	Orthoclase	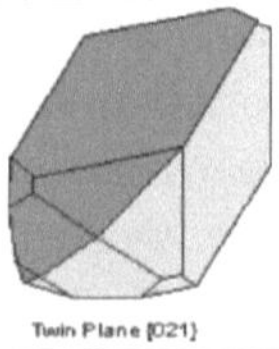
Penetration twin	An irregular composition surface separating two individual crystals	Staurolite, pyrite, fluorite	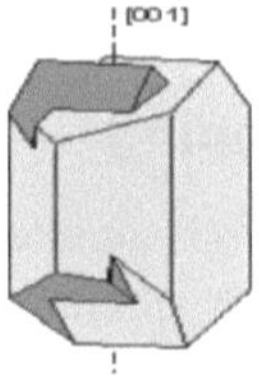
Polysynthetic twin	Compositions surfaces are parallel to one another	Plagioclase, calcite	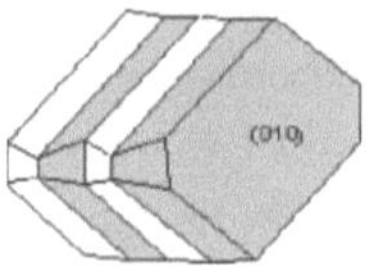
Cyclic twin	Composition surfaces are not parallel to one another but forms almost circular patterns	Crysoberyl, rutile, gypsum	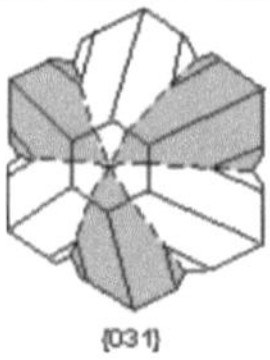

Table 2.10: Common twin laws of the crystal systems

Twin laws	*Description*	*Figures*
Triclinic system		
Albite law	{010} indicates that the twining occurs perpendicular to the **b** crystallographic axis.	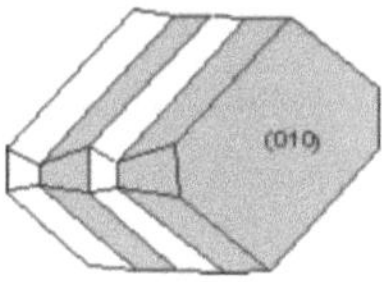
Pericline law	[010] is the twin axis; occurs as the result of monoclinic orthoclase or sanidine transforming to microcline. Combination of pericline and albite twinning produce a cross-hatched pattern, called *tartan twinning*, that is a distinguishing property of microcline.	
Monoclinic system		
Manebach law	{001} forms a contact twin commonly observed in the mineral orthoclase.	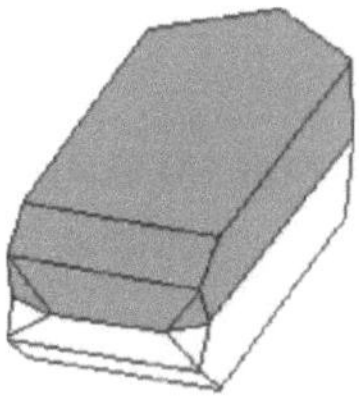 Twin Plane [001]
Carlsbad law	[001] forms a penetration twin in the mineral orthoclase. Crystals show two intergrown crystals, one rotated 180° from the other about the [001] axis.	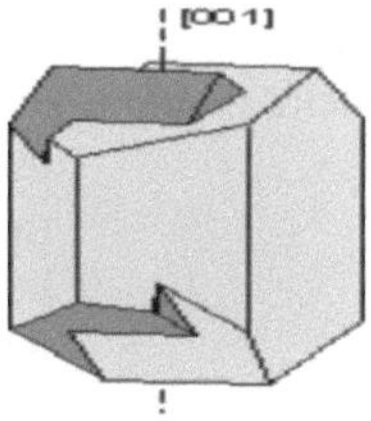
Baveno law	{021} forms a contact twin in the mineral orthoclase.	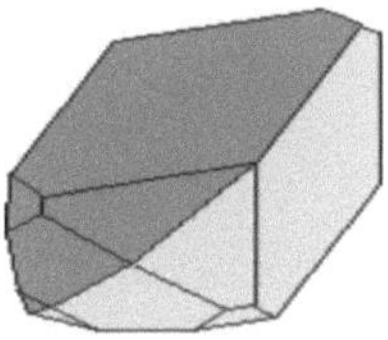 Twin Plane {021}
Swallow Tail Twins - {100}	Commonly observed in the mineral gypsum ($CaSO_4 \cdot 2H_2O$).	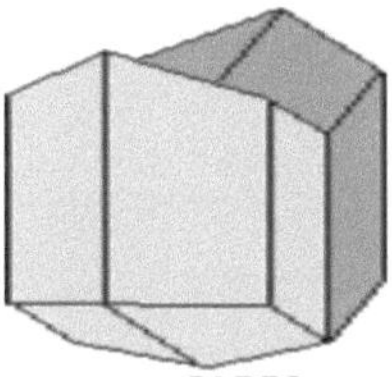 {100}

(*Contd.*)

Table 2.10 (*Contd.*)

Twin laws	*Description*	*Figures*
Orthorhombic system		
Cyclical twin law	Aragonite ($CaCO_3$), chrysoberyl ($BeAl_2O_4$) and cerrusite ($PbCO_3$) commonly develop cyclic twinning on {110}, which gives these minerals a pseudo-hexagonal appearance.	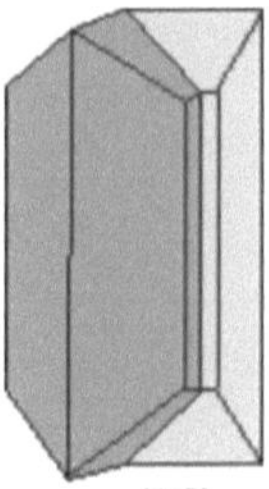
Staurolite twin law	Two types of interpenetration twins occur in staurolite, the {031} twins form a right-angled cross and the {231} twins form a cross at about 60°.	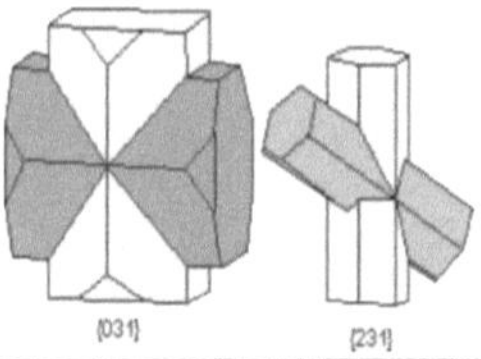
Tetragonal system		
Contact cyclical twin	Twinning occurs on {011}, the minerals showing this type of twinning are rutile (TiO_2) and cassiterite (SnO_2).	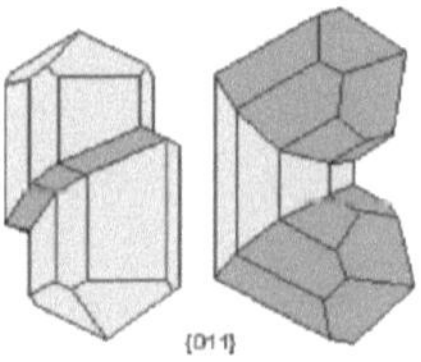
Hexagonal system		
Calcite twins	Two most common twin laws observed in calcite crystals are {0001} and the rhombohedron. Both are contact twins, but the {0112} twins can also occur as polysynthetic twins resulting from deformation.	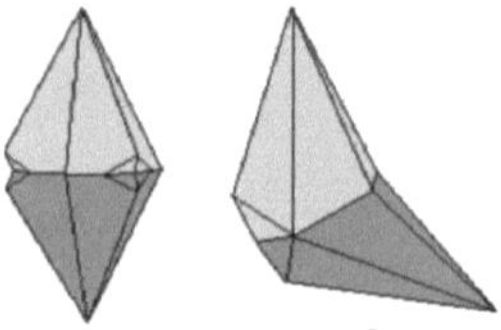
Brazil Law	{1120} is a penetration twin resulting from transformation.	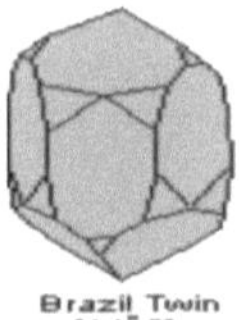
Dauphiné Law	[0001] is a penetration twin resulting from transformation.	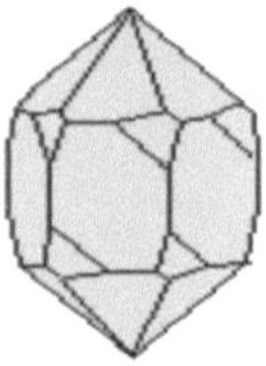

(*Contd.*)

(*Contd.*)

Japanese Law	{1122} is a contact twin resulting from accidents during growth.	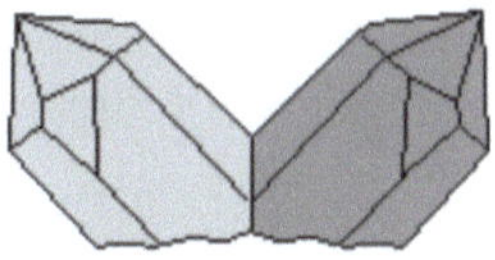 Japanese Twin $\{11\bar{2}2\}$
Isometric system		
Spinel Law	{111} is a twin plane, parallel to an octahedron. It occurs in spinel ($MgAl_2O_4$).	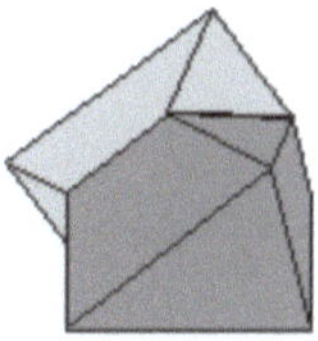 $\{\bar{1}\bar{1}1\}$
Iron Cross [001]	Pyrite (FeS_2) often shows the iron cross formed by interpenetration of two pyritohedrons. Since this occurs in the class 2/m, with no 4-fold rotation axes, the [001] twin axis gives the mineral apparent 4-fold symmetry about three perpendicular axes.	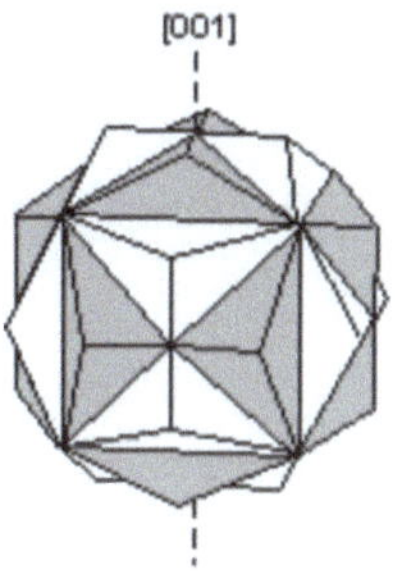

2.5 REFLECTION OF CRYSTALLOGRAPHIC FEATURES IN PHYSICAL PROPERTIES OF MINERALS

Relation between Crystal Structure and Morphology

The likelihood of developing a crystal face is related to the density of nodes within the lattice. Highest density of lattice points result in most prominent faces of a crystal. This is because low lattice node density faces grow the fastest (i.e., you need fewer atoms per unit length). The rapid growth rate of low density faces results in the development of high density faces (i.e., the low density faces grow themselves away, leaving behind the more stable high density faces).

The crystals with primitive lattice, the planes with decreasing lattice point densities and decreasing interplanar spacing d_{hkl} are {100}, {110}, {200}. In non-primitive lattice, the interplanar spacing d_{hkl} is not highest for planes with highest number of lattice points. In body-centered cubic crystal {110} has the highest lattice point density. In face-centered cubic crystal {111} has the highest lattice point density.

Figure 2.13 shows a square lattice. Along the line A-B, one atom is encountered each unit length. Along the line A-C, an atom is not encountered until a distance of 1.41 unit lengths is travelled. Note that as the angles between the vector **AB** and the vectors **AC** ≥ **AD** ≥ **AE** decrease, the distance needed to reach another lattice node increases. Think now why single crystal faces are often parallel or intersect at only a few similar angles.

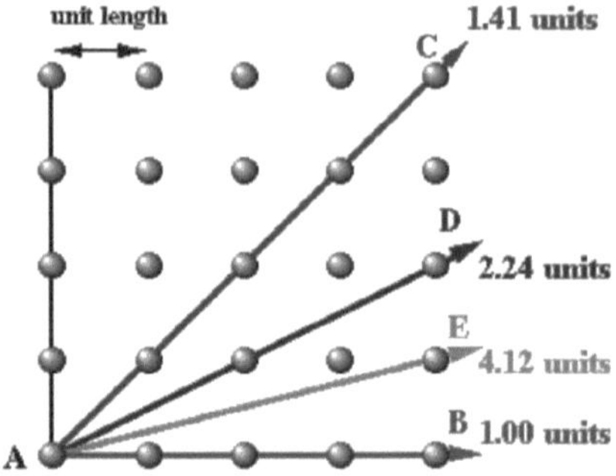

Fig. 2.13: Arrangement of atoms in a square lattice.

Due to various arrangements of atoms in various crystal structures, different forms are exhibited in the crystallographic systems (symmetry elements of each of the form and their corresponding classes which have been enumerated in Table 2.11.

Table 2.11: Different type of crystallographic systems

Forms	*Representative minerals*	*Figure*
Triclinic system		
Pinacoid	Microcline (K-feldspar), plagioclase, turquoise, amblygonite, polyhalite, chalcanthite, rhodonite, pectolite, ulexite and wollastonite	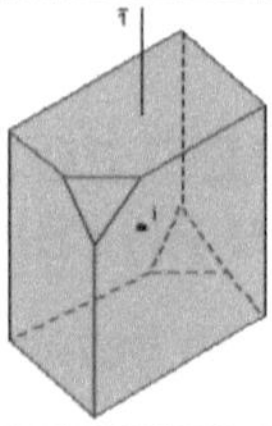
Monoclinic system		
Sphenoid	Halotrichite group	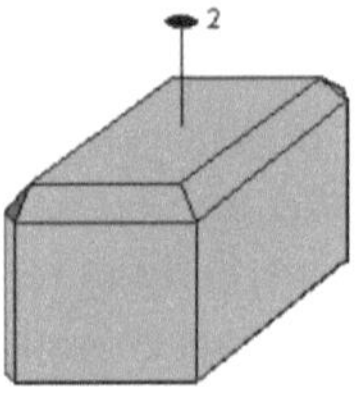
Dome	Two rare elements: Hilgardite $[Ca_2ClB_5O_8(OH)_2]$, Clinohedrite $[Ca_2Zn_2(OH)_2Si_2O_7.H_2O]$	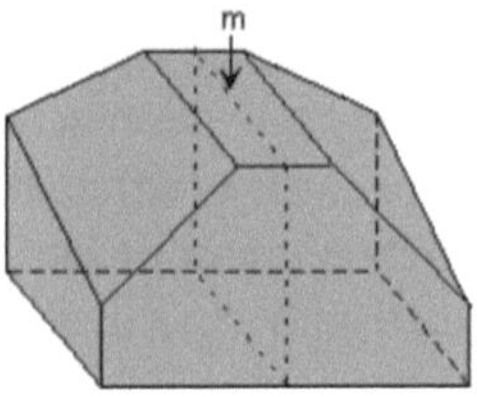

(*Contd.*)

(*Contd.*)

Prism	Micas (biotite and muscovite), azurite, chlorite, orpiment, clinopyroxenes, epidote, gypsum, malachite, kaolinite, orthoclase, spodumene, sphene, realgar and talc	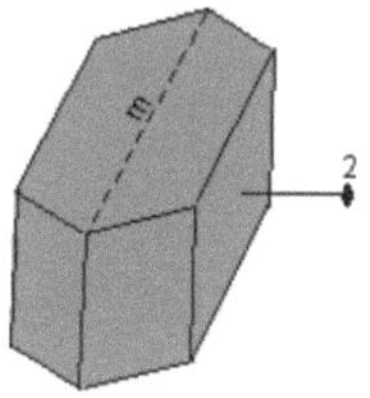
Orthorhombic system		
Rhombic disphenoid	Epsomite and olivenite	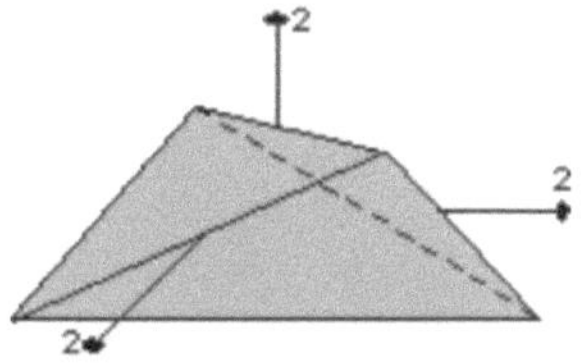
Rhombic pyramid	Hemimorphite $[Zn_4Si_2O_7(OH)_2.$ $H_2O]$Bertandite $[Be_4Si_2O_7(OH)_2]$	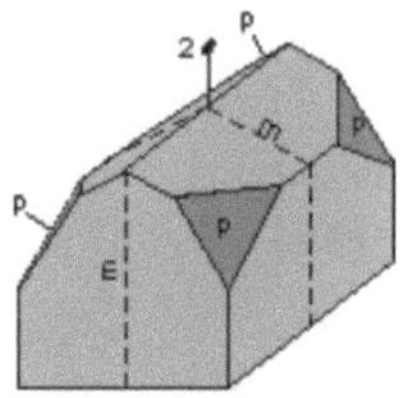
Rhombic dipyramid	Andalusite, anthophyllite, aragonite, barite, cordierite, olivine, sillimanite, stibnite, sulfur, marcasite, sillimanite, brookite, stibnite, goethite, lawsonite, and topaz	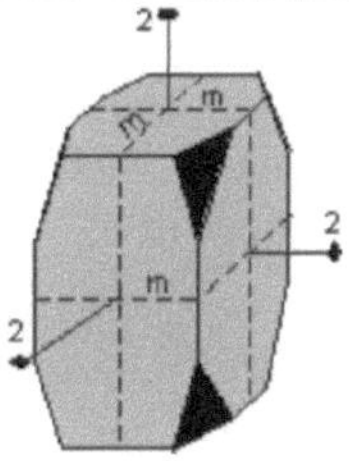
Tetragonal system		
Tetragonal pyramid	Wulfinite	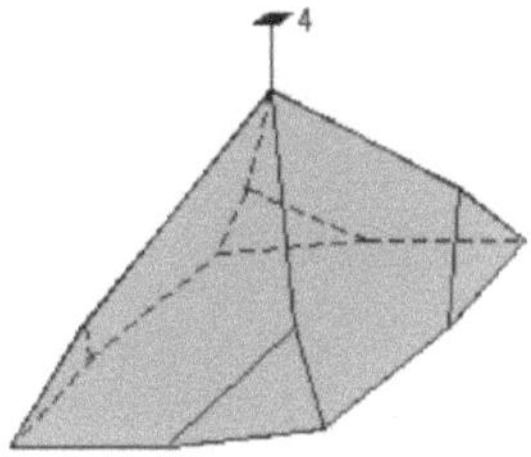
Tetragonal disphenoid	One rare mineral: Cahnite $[CaB(OH)_4AsO_4]$	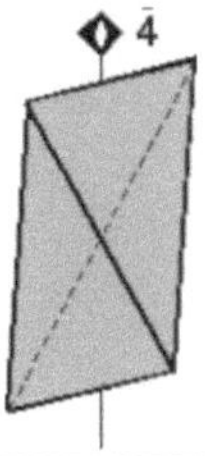

(*Contd.*)

Table 2.11 (*Contd.*)

Forms	*Representative minerals*	*Figure*
Tetragonal dipyramid	Scheelite, powellite and scapolite	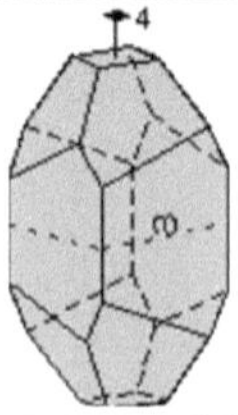
Tetragonal trapezohedron	Phosgenite $[Pb_2CO_3Cl_2]$	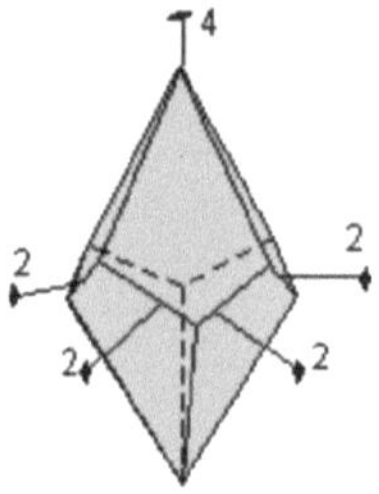
Ditetragonal pyramid	Diaboleite $[Pb_2Cu(OH)_4Cl_2]$	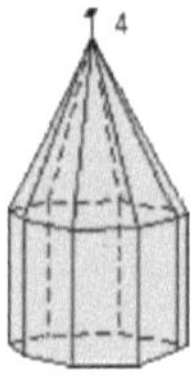
Tetragonal scalenohedron	Chalcopyrite and stannite	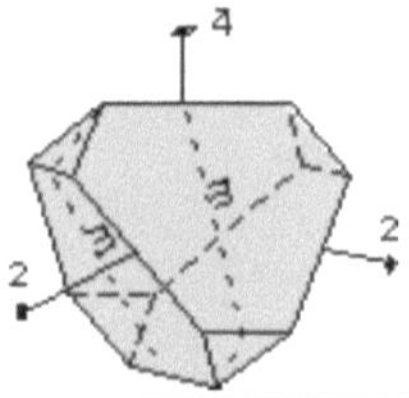
Ditetragonal dipyramid	Anatase, rutile, idocrase cassiterite, apophyllite, zircon and vesuvianite	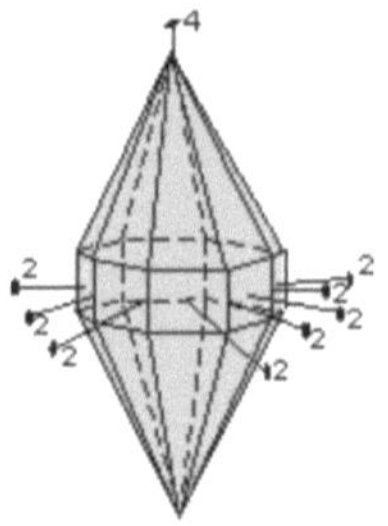
Hexagonal (Trigonal) system		
Trigonal pyramid	Gratonite $(Pb_4As_4S_{15})$	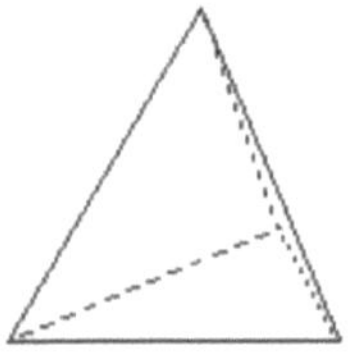

(*Contd.*)

(*Contd.*)

Rhombohedron	Dolomite, ilmenite and willemite	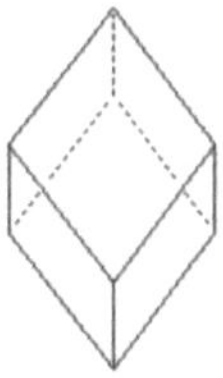
Trigonal trapezohedron	Cinnabar	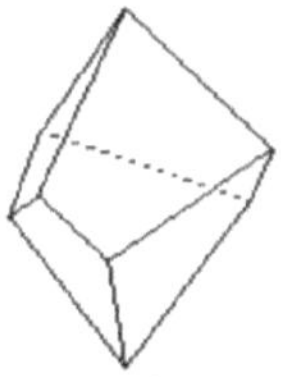
Ditrigonal pyramid	Tourmaline, pyrargyrite, prousite and alunite	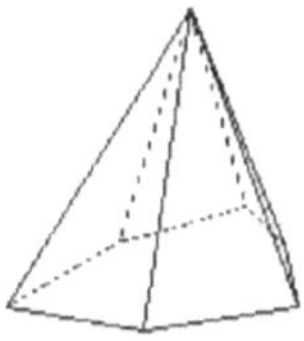
Hexagonal scalenohedron	Calcite, corundum, brucite, soda nitre, arsenic, millerite, antimony and bismuth	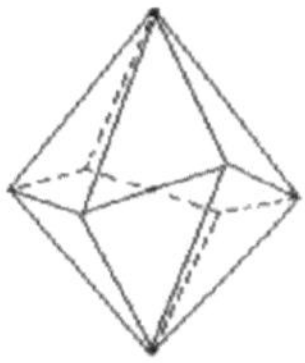
Hexagonal pyramid	Nepheline	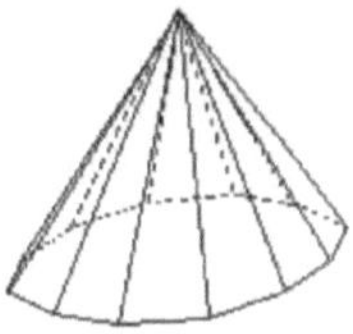
Trigonal dipyramid	No characteristic group	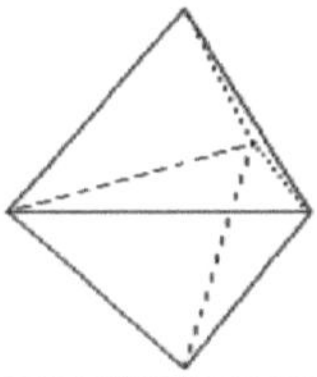
Hexagonal dipyramid	Apatite group	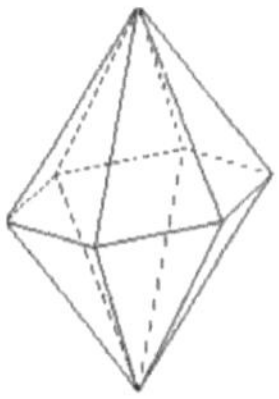

(*Contd.*)

Table 2.11 (*Contd.*)

Forms	*Representative minerals*	*Figure*
Hexagonal trapezohedron	High quartz and kalsilite	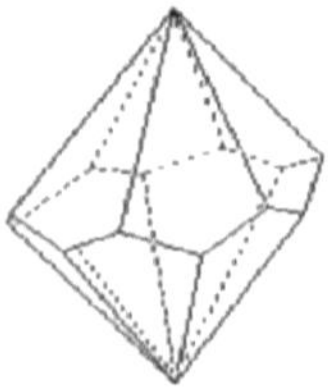
Dihexagonal pyramid	Wurtzite, greenockite and zincite	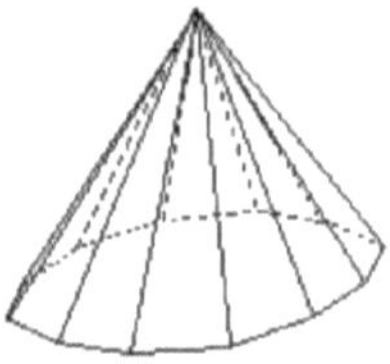
Ditrigonal dipyramid	Benitoite ($BaTiSi_3O_9$)	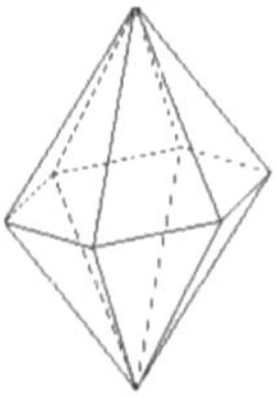
Dihexagonal dipyramid	Beryl, molybdenite, pyrrhotite and nicolite	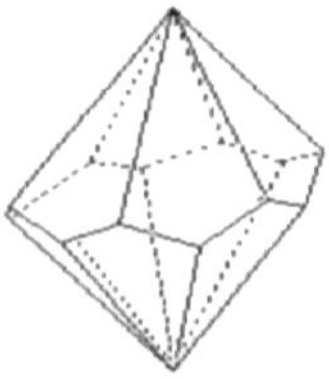
Isometric system		
Tetroidal	Cobaltite and ullmanite	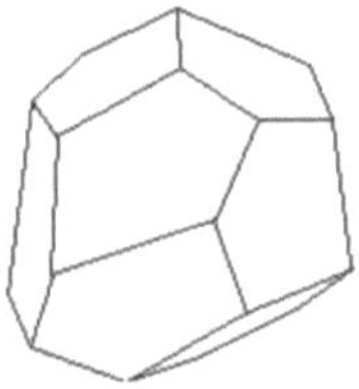
Diploidal	Pyrite, skutterudite, chloanthite and speerrylite	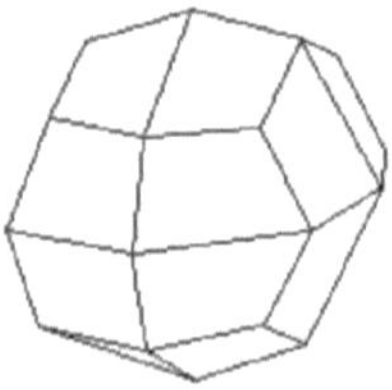

(*Contd.*)

(*Contd.*)

Gyroidal	No known minerals	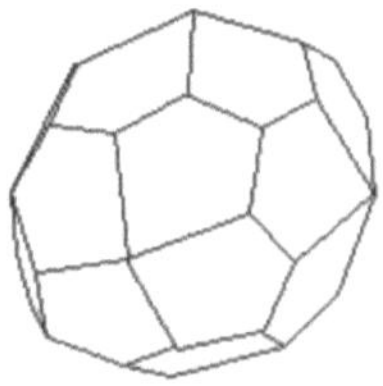
Hextetrahedral	Members of tetrahedrite-tennantite series	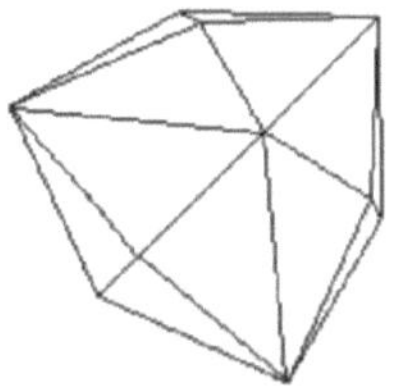
Hexoctahedral	Analcime, galena, leucite, garnet, silver, copper, gold, spinel group, cuprite, halite, sylvite, fluorite and uraninite	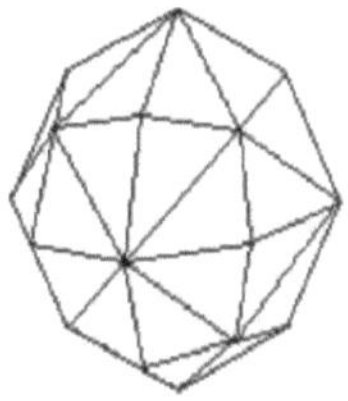

Piezoelectricity: Twenty of the 32 crystal classes are so-called piezoelectric, and crystals belonging to one of these classes (point groups) display piezoelectricity. All 21 piezoelectric classes lack a centre of symmetry.

Polarity: Any material develops a dielectric polarization when an electric field is applied, but a substance which has such a natural charge separation even in the absence of a field is called a polar material. Whether or not a material is polar is determined solely by its crystal structure. Only 10 of the 32 point groups are polar. All polar crystals are pyroelectric, so the 10 polar crystal classes are sometimes referred to as the pyroelectric classes.

Ferroelectricity: There are a few crystal structures, notably the perovskite structure, which exhibit ferroelectric behaviour. This is analogous to ferromagnetism, in that, in the absence of an electric field during production, the ferroelectric crystal does not exhibit a polarization. Upon the application of an electric field of sufficient magnitude, the crystal becomes permanently polarised. This polarization can be reversed by a sufficiently large counter-charge, in the same way that a ferromagnet can be reversed. However, it is important to note that, although they are called ferroelectrics, the effect is due to the crystal structure (not the presence of a ferrous metal).

2.6 DEFECTS IN CRYSTAL: POINT DEFECTS AND LINE DEFECTS

No crystal is actually pure or perfect. There is always some imperfections in the ideal arrangement of atoms in a crystal. So departure of a crystalline solid from a regular array of atoms or ions is termed as a crystallographic defect. It has been seen that the presence of relatively smaller amount of defects have a profound effect on the macroscopic properties of minerals. There are mainly three types of defects, observed in a crystal.

Table 2.12: Defects in crystal

Defects	*Description*
Point	Atoms missing or irregular places in the lattice (lattice vacancies, substitutional and interstitial impurities, self-interstitials).
Line	Groups of atoms in irregular positions (e.g. screw and edge dislocations).
Planar	The interfaces between homogeneous regions of the material (grain boundaries, stacking faults, external surfaces).

Now each of these defects along with their effects on crystals is to be discussed in details.

Point Defects

Vacancies: The higher is the temperature, more often atoms are jumping from one equilibrium position to another creating larger number of empty lattice positions, called vacancies in a crystal. The number of vacancies, N_v, increases exponentially with the absolute temperature, T, and can be estimated using the equation (Boltzmann Distribution):

$$N_v = N_s \exp(-E_v/k_b T)$$

where N_s is the number of regular lattice sites, k is the Boltzmann constant and E_v is the energy needed to form a vacant lattice site in a perfect crystal.

In order for atom to jump into a vacancy site, it needs to possess enough energy for example, thermal energy. The energy necessary for motion, E_m, is called *the activation energy for vacancy motion.* The average thermal energy of an atom is usually much smaller than the activation energy E_m and a large fluctuation in energy is needed for a jump. The probability of such fluctuation or frequency of jumps, R_j, depends exponentially from temperature and can be described by equation that is attributed to Swedish chemist Arrhenius:

$$R_j = R_0 \exp(-E_m/k_B T)$$

where R_0 is an attempt frequency proportional to the frequency of atomic vibrations.

Interstitials: Atoms that are squeezed in between regular lattice sites. If the interstitial atom is of the same species as the lattice atoms, it is called *self-interstitial*. They are generally high energy configurations. Small atoms in some crystals can occupy interstices without high energy, such as hydrogen in palladium.

Interstitial impurity: The atoms are often incorporated at a regular atomic site in the crystal structure. They introduce less distortion to the lattice and are more common in real materials and more mobile.

Complexes: Complexes can form between different kinds of point defects. For example, if a vacancy encounters an impurity, the two may bind together if the impurity is too large for the lattice. Interstitials can form 'split interstitial' or 'dumbbell' structures where two atoms effectively share an atomic site, resulting in neither atom actually occupying the site.

There are two special types of point defects:

Frenkel defect: It is a pair of cation (positive ion) vacancy and a cation interstitial.

Or it may also be an anion (negative ion) vacancy and anion interstitial. However anions are much larger than cations and it is not easy for an anion interstitial to form.

Schottky defect: It is a pair of anion and cation vacancies.

In both Frenkel and Schottky defects, the pair of point defects stay near each other because of strong coulombic attraction of their opposite charges.

Line Defects

Dislocations are caused by the termination of a plane of atoms in the middle of a crystal. In such a case, the surrounding planes are not straight, but instead bend around the edge of the terminating plane so that the crystal structure is perfectly ordered on either side. Dislocations can move if the atoms from one of the surrounding planes break their bonds and rebond with the atoms at the terminating edge. This area is called the dislocation core. Dislocations also create small elastic deformations of the lattice at large distances.

Planar Defects

Grain boundaries occur where the crystallographic direction of the lattice abruptly changes. This commonly occurs when two crystals begin growing separately and then meet.

Anti-phase boundaries occur in ordered alloys. In this case, the crystallographic direction remains the same and each side of the boundary has an opposite phase.

Stacking faults occur in a number of crystal structures, but the common example is in close packed structures. Face centered cubic (fcc) structures differ from hexagonal close packed (hcp) structures only in stacking order: both structures have close packed atomic planes with six-fold symmetry—the atoms form equilateral triangles. When stacking one of these layers on top of another, the atoms are not directly on top of one another. The first two layers are identical for hcp and fcc, and labelled AB. If the third layer is placed so that its atoms are directly above those of the first layer, the stacking will be ABA. This is the hcp structure, and it continues ABABABAB. However there is another location for the third layer, such that its atoms are not above the first layer. Instead, the fourth layer is placed so that its atoms are directly above the first layer. This produces the stacking ABCABCABC, and is actually a cubic arrangement of the atoms. A stacking fault is a one or two layer interruption in the stacking sequence, for example if the sequence ABCABABCAB were found in an fcc structure.

Twin boundaries are similar, and may be produced by deformation of materials. This gives rise to *shape memory metals*, which can recover their original shape if heated to a high temperature. Shape-memory alloys are twinned and when deformed they untwin. At high temperature the alloy returns back to the original twin configuration and restore the original shape.

Concluding Remarks

Crystallographic arrangement and growth favour the formation of different crystals in different minerals. Their variations lead to various properties in minerals which will be discussed in subsequent chapters.

Think for a while

1. Calculate the axial ratios for a triclinic crystal with dimension $a = 8.47$Å, $b = 10.87$Å and $c = 21.39$Å.
2. Write the symmetry of 4/mmm. How would you distinguish between {100} and {110} in crystals with this symmetry. Represent on a sketch stereogram the symmetry elements, crystal axes and {hkl} of this class and name two minerals belonging to it.
3. The crystal morphology has a profuse impact on the physical properties of the mineral. Explain with suitable examples.
4. How do some forms belong to more than one point group?

FURTHER READING

Blackburn, W.H. and W.H. Dennen. Principles of Mineralogy, 2nd ed. William C. Brown, Dubuque, IA. 1994.

Bloss, F.D. Crystallography and Crystal Chemistry: An Introduction. Mineralogical Society of America, New York. 1994.
Deer, W.A., R.A. Howie and J. Zussman. Rock Forming Minerals. 5 vols. John Wiley and Sons, New York. 1962.
Klein, C. and C.S. Hurlbut, Jr. Manual of Mineralogy, 21st ed. John Wiley and Sons, New York. 1993.
Philips, F.C. An Introduction to Crystallography, 3rd ed. John Wiley and Sons, New York. 1963.

Internet Data Retrieved from:

Texas Natural Science Center\The science of minerals.htm
www.tulaneuniversity.edu. Twinning, polymorphism, polytypism, pseudomorphism *by* Prof. Stephen A. Nelson.
Leonid V. Zhigilei, Department of Materials Science and Engineering, University of Virginia. *Crystal Defects*.
The Oklahoma Chemical Crystallography Lab.
Crystallographic Directions and Planes. Conchordia University.

CHAPTER

3

MINERAL CHEMISTRY

The Earth is composed of rocks. Rocks are aggregates of minerals. Minerals are composed of atoms. In order to understand rocks, we must first have an understanding of minerals. In order to understand minerals we must have some basic understanding of atoms—what they are and how they interact with one another to form minerals. A particular mineral cannot be formed unless the chemical ingredients necessary to make the mineral are present. Thus, the most common minerals are those that have a chemical composition made of the common elements found in their environment. The atomic bonds of minerals, rules governing the chemistry of the minerals, the different modes of their formation, inter-relationship between mineral species form the basis of mineral chemistry and shall be discussed in this chapter.

3.1 CHEMICAL COMPOSITION OF MINERALS AND UNIT CELL CONTENT

The minerals in the Earth have a wide and variable chemical composition, depending upon the chemical abundance of elements and compounds on the Earth. The major elements found in Earth crust in order of abundance are ahown in Table 3.1.

Table 3.1: Abundance of elements found in Earth crust

Elements	*Proportion*	*Elements*	*Proportion*
Oxygen	46.60%	Calcium	3.63%
Silicon	27.72%	Sodium	2.83%
Aluminium	8.13%	Potassium	2.59%
Iron	5.00%	Magnesium	2.09%

The chemical composition of minerals governs different mineral properties. But to understand such properties, alongwith chemical composition, the geometry of constituent atoms and ions and their binding forces are also important.

Calculation of Mineral Formulas

Chemical analyses are expressed in weight percentages. For determination of formula:

- These weight percentages are converted to atomic proportions by dividing the weight percentage of each oxide in the mineral by its formula weight.
  Weight % of XO_2/formula weight of XO_2
- The resulting mole numbers are normalized to a number of moles of oxides that has a mass of 100 grams.
- The mole number of each oxide is multiplied by the number of oxygen atoms in the oxide formula
  (Moles of XO_2)(oxygen units of XO_2/moles of XO_2)
- The sum of all the oxygen numbers is then obtained in terms of six oxygen units of mineral.
- The oxygen number of each oxide is multiplied by a normalization constant (equal to the number of oxygen atoms in the desired formula divided by the sum of oxygen numbers).
  Oxygen units of XO_2/six oxygen units of mineral
- The normalized oxygen numbers of each oxide is multiplied by the number of cations per oxygen in the oxide formula.
  (Oxygen units of XO_2/six oxygen units of mineral) (moles of X/oxygen units of XO_2)

The chemical formula thus can be anticipated from above.

An example for calculating the chemical formula, following the above procedure, has been enumerated below:

1. A chemical analysis is taken for calculation, where SiO_2 = 55.49%, MgO = 18.61% and CaO= 25.90%.
2. Weight % of SiO_2/Formula weight of SiO_2 = 55.49/60.085 = 0.92353
   Weight % of MgO/Formula weight of MgO = 18.61/40.311 = 0.46166
   Weight % of CaO/Formula weight of CaO = 25.90/56.079 = 0.46185
3. (Moles of SiO_2)(oxygen units of SiO_2/moles of SiO_2) = 0.92353 × 2 = 1.84705
   (Moles of MgO)(oxygen units of MgO/moles of MgO) = 0.46166 × 1 = 0.46166
   (Moles of CaO)(oxygen units of CaO/moles of CaO) = 0.46185 × 1 = 0.46185
4. Oxygen units of SiO_2/six oxygen units of mineral= 1.84705 × 2.16567= 4.00
   Oxygen units of MgO/six oxygen units of mineral = 0.46166 × 2.16567 = 1.00
   Oxygen units of CaO/six oxygen units of mineral = 0.46185 × 2.16567 = 1.00

5. (Oxygen units of CaO/six oxygen units of mineral) (moles of Ca/oxygen units of CaO) = 1.00 × 1 = 1.00
   (Oxygen units of MgO/six oxygen units of mineral) (moles of Mg/oxygen units of MgO) = 1.00 × 1= 1.00
   (Oxygen units of SiO_2/six oxygen units of mineral) (moles of Si/oxygen units of SiO_2) = 4.00 × 0.5 = 2.00

The resulting formula anticipated from above was $CaMgSi_2O_6$.

Unit cell content: Knowing the dimensions of the unit cell and the density of the substance, the unit cell content can be calculated. The volume (*V*) of unit cell of different crystals are shown in Table 3.2.

Table 3.2: Variation of unit cell volume with different crystallographic systems

Crystal system	*Unit cell volume*
Isometric	$V = a^3$
Tetragonal	$V = a^2c$
Hexagonal	$V = a^2c \sin(60°)$
Trigonal	$V = a^2c \sin(60°)$
Orthorhombic	$V = abc$
Monoclinic	$V = abc \sin(\beta)$
Triclinic	$V = abc\ (1 - \cos^2 \alpha - \cos^2 \beta - \cos^2 \gamma) + 2(\cos(\alpha) \cos(\beta) \cos(\gamma))^{1/2}$

The total weight (*M*) of atoms in the unit cell is expressed as $M = V \times G$ where *G* is the density. Dimension of unit cell is expressed in Ångstrom units (1 Å = 10^{-8} cm). So to obtain *M* in grams, the formula becomes $M = (V \times G \times 10^{-24})$ grams.

Unit Cell Contents and Crystal Densities

The relationship between crystal density (*D*), volume of formula unit in a cube and the unit cells can be expressed by the following formula:

$$D = \text{formula weight (FW)/molar volume} = \text{FW/ Volume of one formula unit} \times \text{N}$$

where N = Avogadro's number, 6.023×10^{23}.

When a unit cell contains more than one formula unit, supposing a unit cell of volume *V* contains *Z* formula units, then

$$V = \text{volume of one formula unit} \times Z$$

Therefore, $D = \text{FW} \times Z/V \times \text{N}$

For densities measured in g/cm^3, the *V* is to be multiplied by 10^{-24}, whereon

$$D = \text{FW} \times Z \times 1.66/\text{V (Å}^3\text{) g/cm}^3$$

By comparison of the experimental density of a material and the density calculated by X-ray diffraction, an information can be obtained on the presence of crystal defects such as vacancies as opposed to interstitials in the mechanism of solid solution formation.

3.2 PRINCIPLE OF CRYSTAL CHEMISTRY AND IONIC SIZES

Atom is the basic unit of all the crystal structures, which can also remain associated with other atoms in a group behaving as a single unit in the structure. Evaluation of the manner in which the various atoms become assembled into stable chemical compounds, both natural and synthetic, and their effective physical and chemical properties are the fundamental implications of science of crystal chemistry. So crystal chemistry encloses various aspects of nuclear and electronic structures and properties of atoms and the bonds that bind them.

Atom: The smallest sub-division of matter that retains the properties of the chemical element. An atom is composed of three different particles: *Protons*—positively charged, reside in the centre of the atom called the *nucleus*, *neutrons*—no charge, reside in the nucleus and *electrons*—negatively charged, orbit in a cloud around nucleus.

Bohr model defines the atom as:

- Electrons are arranged in discrete shells that they occupy around the nucleus corresponding to some energy levels.
  Shifting of an electron from one orbital to another requires the absorption or release of energy. Each shell is associated with a principal quantum number, n, where $n_K = 1$, $n_L = 2$, $n_M = 3$, $n_N = 4$, etc.
  The number of electrons in each shell is controlled by this principal quantum number by the following relationship:

$$\text{No. of electrons} = 2n^2$$

- *Valence electrons:* The electrons in the outermost shell of the atom, taking part in chemical reactions and thus determining the valency of the atom are called valence electrons.
- *Isotopes:* Atoms with same number of protons but different numbers of neutrons are known as isotopes. Some elements may have a single isotope, while others may have ten or more. For example carbon has three isotopes, C-12, C-13 and C-14. Among them, C-14 is naturally radioactive, so it is used to date rocks (to determine rock age), containing carbon. Uranium has got two isotopes, of which one is used as reactor fuel.

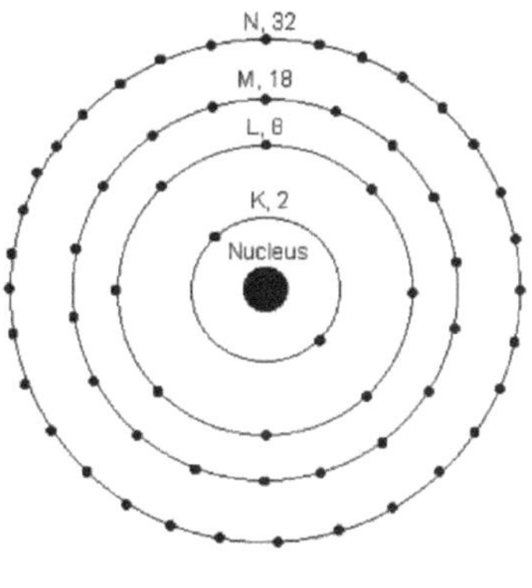

Fig. 3.1: Structure of an atom.

3.3 BONDING OF ATOMS AND THEIR EFFECTS ON PHYSICAL PROPERTIES

Bonding in Crystals

Many chemical and physical properties of crystals are decided by the forces that bind the atoms together in a crystal structure. These forces are known as chemical bonds. Bonding depends on the electronic structure of the atoms involved, in particular the valence electrons in the outermost shells, and on the size of the ion or atom.

Ionic Bond

When atoms become charged ions the force of attraction between oppositely charged ions results in an ionic bond.

Ionic bonds are non-directional in nature i.e., the attractive forces occur from all directions. Crystals made of ionically bonded atoms tend to have the following properties:

- Dissolve easily in polar solvents like water (H_2O is a polar solvent because the hydrogen ions occur on one side of the water molecule and give it a slight positive charge while the other side of the water molecule has a slight negative charge).
- Tend to form crystals with high symmetry.
- Moderate hardness and density.
- High melting temperatures.
- Generally poor conductors of heat and electricity (they are good materials for thermal and electrical insulation).

For example, Na has one electron in its outermost shell. It will tend to give up this electron to become Na^{+1} ion. Similarly, Cl has seven electrons in its outermost shell and would like to gain an electron to become Cl^{-1} ion. Once these atoms become Na^{+1} and Cl^{-1}, the force of attraction between the oppositely charged ions results in an ionic bond.

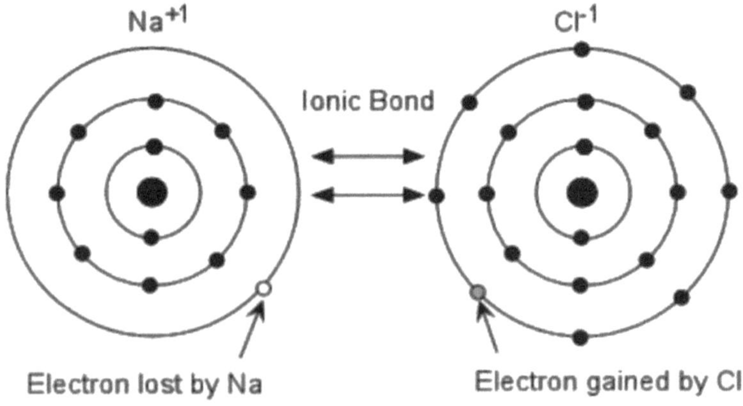

Fig. 3.2: Ionic bond in NaCl.

Covalent Bond

Covalent bonds develop when atoms can achieve a stable outer shell electron configuration by sharing electrons with another atom. Covalent bonds are strongly directional bonds that occur along the zone where the electrons are shared. The properties covalently bonded crystals are:

- Relatively insoluble in polar solvents like water.
- High melting temperatures.
- Generally form crystal structures of low symmetry.
- Tend to have high hardness.
- Generally poor conductors of heat and electricity.

For example, oxygen has six electrons in its outer shell, two are in s-orbitals and four are in p-orbitals. If it can gain two more electrons to fill its p-orbital shells, it would have the stable electronic configuration of a Noble gas. A second oxygen atom also needs two more electrons to achieve the stable configuration. If the two oxygen atoms each share two electrons with each other, then each oxygen has the stable electronic configuration. The covalent bond thus formed is a very strong bond. Thus we find oxygen gas is composed of O_2 molecules. Similarly, F_2 and Cl_2 gases are composed of covalently bonded F and Cl pairs.

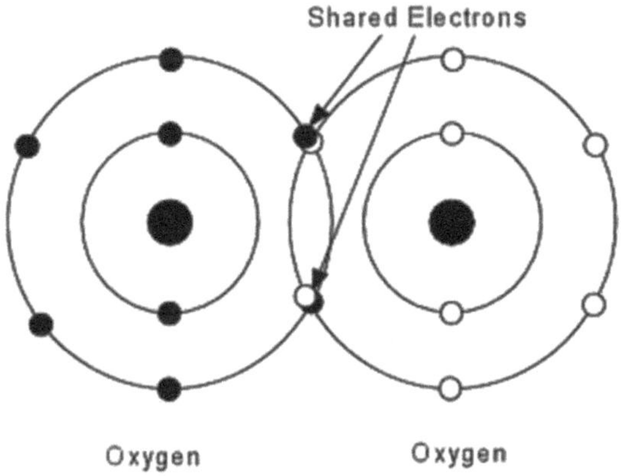

Fig. 3.3: Covalent bond in O_2.

Metallic Bonds

The type of bonds where positively charged atomic nuclei share electrons in their electron clouds freely. Pure metals appear to bind in this way. Crystals with metallic bonds have the following properties:

- Low to moderate hardness.
- Usually very malleable and ductile.
- Good thermal and electrical conductors.
- Soluble only in acids.
- Crystals with high symmetry.

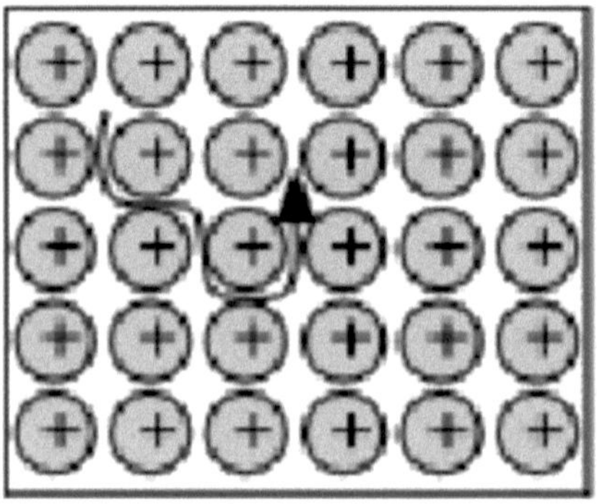

Fig. 3.4: Metallic bond.

Van der Waals Bonds

Bonds that result from polarization of atoms or molecules are called Van der Waals bonds. Bonding between sheets takes place as a result of the slight attraction between these residual charges from one sheet to another. The primary cause behind such type of bonding is small irregularities in the electron clouds around molecules, creating small forces of attraction.

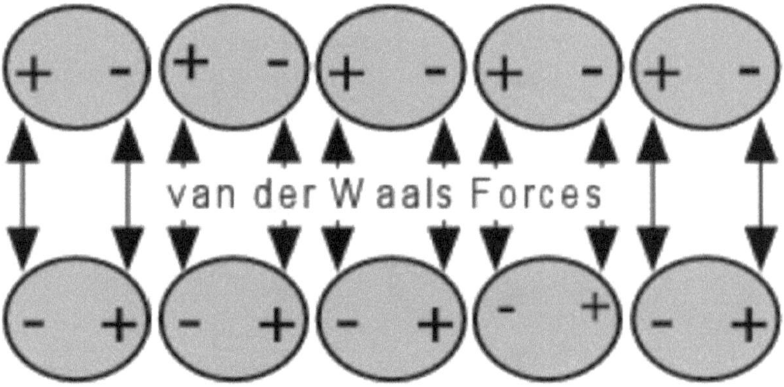

Fig. 3.5: Van der Waals bond.

This type of bond is anisotropic in nature, indicating their dependence on the relative orientation of the molecules.

Hydrogen Bonds

These occur in the special case of hydrogen, because H has only one electron. When hydrogen gives up this electron to become H^{+1} ion or shares its single electron with another atom in a covalent bond, the positively charged nucleus of the hydrogen atom is exposed, giving that end of the H ion a residual +1 charge. This is what causes the H_2O molecule to be a polar molecule seen here. Similarly, an OH^{-1} molecule, common in sheet silicate minerals like micas and clay minerals, although possessing a -1 charge will have exposed H nuclei that can bond to other negative residual charges forming a weak hydrogen bond. Layers of OH^{-1} molecules in the sheet silicates result in the easy cleavage along the {001} planes.

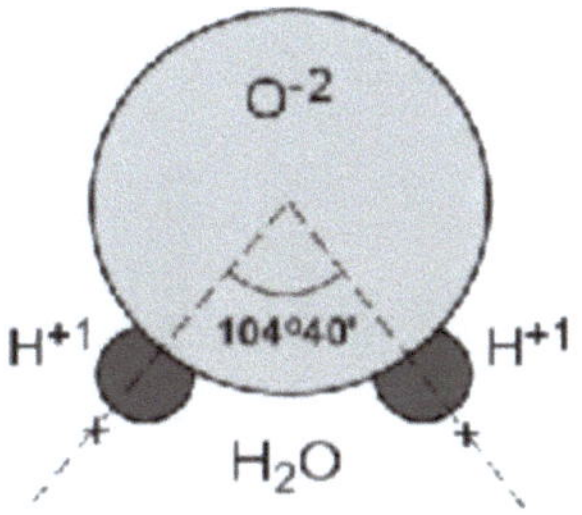

Fig. 3.6: Hydrogen bond.

Hydrogen bonding is actually a type of residual bond, that is not very common in minerals. So while highlighting various properties of bonds in Table 3.3, hydrogen bond is kept out of discussion.

Electronegativity

Electronegativity is the relative tendency of a bonded atom to attract electrons to itself.

Na-F: Na has an electronegativity of 0.9 and F has an electronegativity of 4.0. The difference in electronegativity is 4.0 – 0.9 = 3.1. Since 3.1 is greater than 1.7, the resulting bond is ionic.

F has such a strong attraction for electrons that it pulls the electron off the Na resulting in a negative charge for fluorine, F^-, and a positive charge for sodium, Na^+.

The bond between Na^+ and F^- is NOT covalent since the bonding electrons are not shared between the 'atoms'. Rather the bond is ionic since there has been a transfer of electrons from the least electronegative atom to the more electronegative atom resulting in two ions of opposite charge.

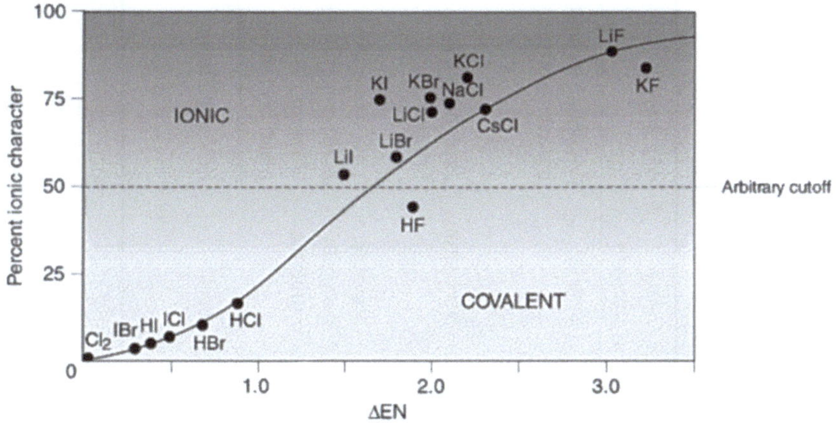

Fig. 3.7: Relation between electronegativity and ionic character of compounds.

Table 3.3: Effects of bonding on physical properties

Bonds	*Bond strength*	*Mechanical*	*Electrical*	*Thermal*	*Solubility*	*Structural*	*Examples*
Ionic	Strong	Moderate to high hardness; brittle	Poor conductor in solid state; conduct in fused state	Melting point (M.P.) moderate to high; low thermal coefficient	Soluble in polar solvents	Non-directed	Halite (NaCl), Calcite ($CaCO_3$) Fluorite (CaF_2)
Covalent	Very strong	High hardness; brittle	Insulators in solid and fused state	M.P. high; low coefficient	Very low	Highly directional	Diamond (C), Sphalerite (ZnS)
Metallic	Variable, generally moderate	Low to moderate hardness; gliding common; high plasticity; sectile, ductile, malleable	Good conductors	Variable M.P. and coefficient	Insoluble, other than in acids or alkalis	Non-directed	Copper, silver, gold, most metals
Van der Waals	weak	Soft and plastic	Insulators in solid and liquid state	Low M.P. and high coefficient	Soluble in organic solvents	Non-directed	Iodine, graphite, organic compounds

3.4 IONIC CRYSTALS AND IONIC RADII

Atoms with an electrical charge are called *ions*. Elements that tend to *loose* electrons and hence become positively charged are called *metals*, and the ions are known as *cations*. *Non-metals* tend to *gain* electrons and forms negatively charged ions, i.e. anion.

Ionic radii: The size of an atom/ion depends on the size of the nucleus and the number of electrons. Generally atoms with larger number of electrons have larger radii than those with smaller number of electrons. Atomic and ionic radii also depend on the type of bonding between the constituents, and on the coordination number. Thus, atomic and ionic radii will vary somewhat as a function of the environment in which the atoms or ions are found.

- Thus as the charge on the ion becomes more positive, there will be less electrons and the ion will have a smaller radius.
- As the charge on the ion becomes more negative, there will be more electrons and the ion will have a larger radius.
- As the atomic number increases in any given column of the Periodic Table, the number of protons and electrons increases and thus the size of the atom or ion increases.

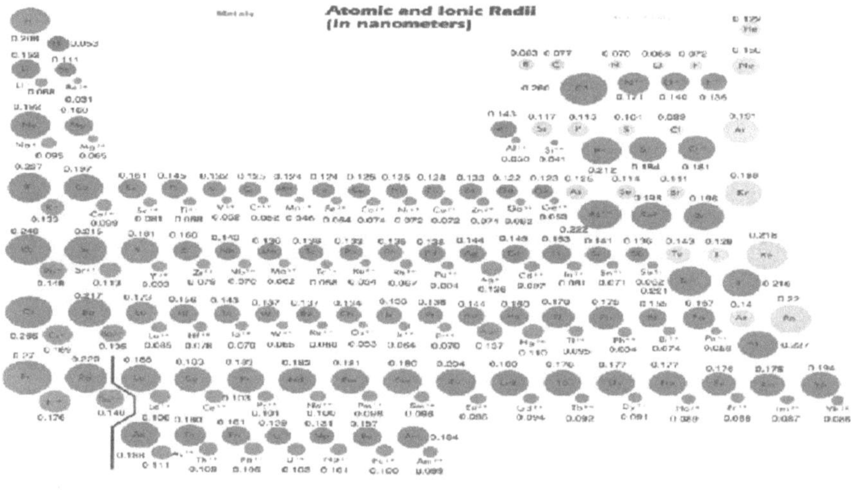

Fig. 3.8: Periodic Table.

3.5 COORDINATION NUMBER AND ATOMIC PACKING

Radius Ratio

In any given molecule or crystal structure each atom or ion will be surrounded by other atoms or ions. The number of ions or atoms that immediately surround an atom or ion of interest is called the *coordination number* (C.N.). Coordination number, C.N. depends on the relative size of the ions. This dependency is expressed by the radius ratio, which is radius of the cation/radius of the anion. The assemblage of anions around a cation forms a simple shape that resembles various forms, depending upon the radius ratio and coordination number.

For the elements that occur in common minerals in the Earth's crust, the most common coordinating anion is oxygen. Table 3.4 gives the ionic radius and coordination of these common metal cations coordinated with oxygen.

Table 3.4: Radius ratio and coordination number

Rx/Rz	C.N.	Type	
1.0	12	Hexagonal or Cubic Closest Packing	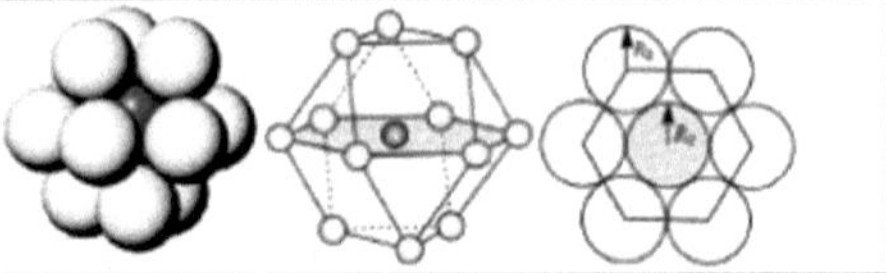
1.0 - 0.732	8	Cubic	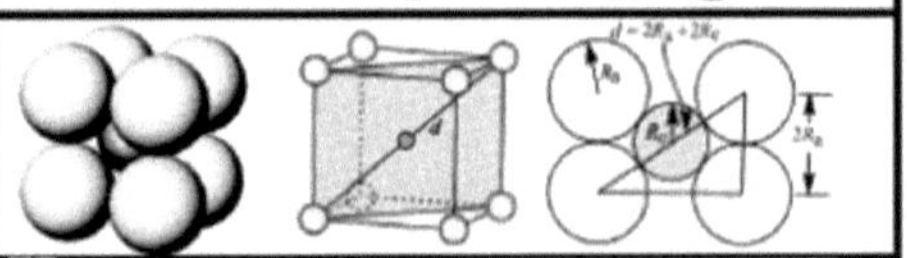
0.732 - 0.414	6	Octahedral	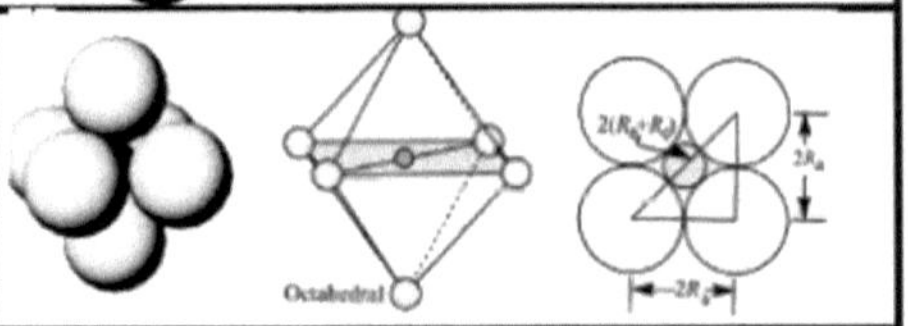
	4	Square Planar	
0.414 - 0.225	4	Tetrahedral	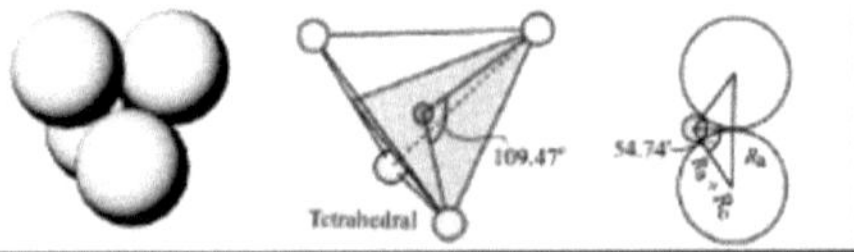
0.225 - 0.155	3	Triangular	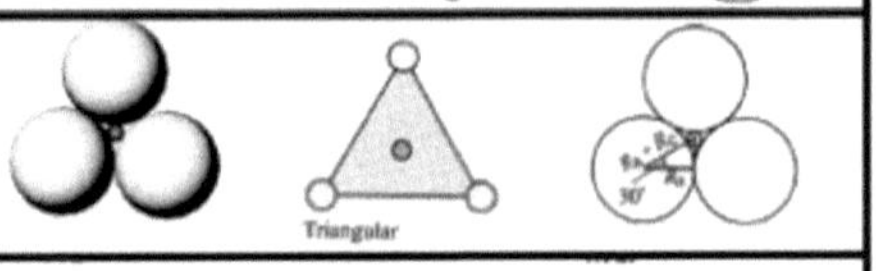
<0.155	2	Linear	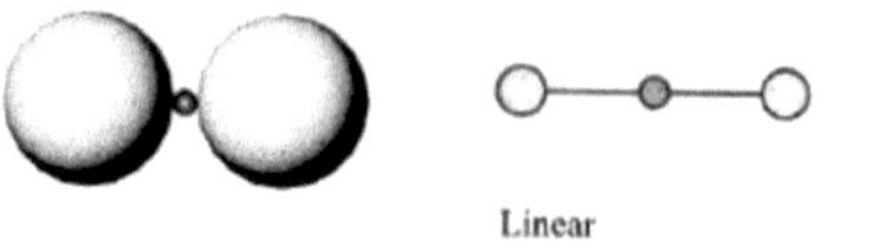

The coordination number can be calculated from the radius ratio, by principles of solid geometry. It is called the predicted coordination number. But in some cases the observed coordination numbers vary considerably from the predicted coordination numbers. The observed values deviate from the predicted values as shown in Table 3.6.

Table 3.5: Coordination number for different ions

Ion	*C.N. (with oxygen)*	*Coord. polyhedron*	*Ionic radius, Å*
K^+	8 - 12	Cubic to closest	1.51 (8) - 1.64 (12)
Na^+	8 - 6	Cubic to octahedral	1.18 (8) - 1.02 (6)
Ca^{+2}	8 - 6		1.12 (8) - 1.00 (6)
Mn^{+2}	6	Octahedral	0.83
Fe^{+2}	6		0.78
Mg^{+2}	6		0.72
Fe^{+3}	6		0.65
Ti^{+4}	6		0.61
Al^{+3}	6		0.54
Al^{+3}	4	Tetrahedral	0.39
Si^{+4}	4		0.26
P^{+5}	4		0.17
S^{+6}	4		0.12
C^{+4}	3	Triangular	0.08

Table 3.6: Deviations of C.N. from predicted values

	Illustrations	*Examples*
Radius ratio close to boundary limits	If the radius ratio of a cation with respect to an anion is very close the limiting value between two types of coordination, like Radius ratio = 0.225, 0.414 etc., then the cation can have both of those CN values. Such cations commonly form polymorphs and the cations have different CN in different polymorphs.	The radius ratio of Al^{3+} in Al-O bond is 0.36, close to the limiting value of 4 and 6 coordination. So both CN are possible for Al in sillimanite (Al_2SiO_5), the CN of one Al^{3+} is 4 and that of the other is 6.
Effect of high pressure	The ions in a crystal are squeezed together in high pressure environment, thus the ionic radius becomes effectively smaller. With increasing pressure, the decrease of ionic radius is more in O^{-2} than in the cations. So the effective radius ratio increases and consequently the CN also increases. Commonly the observed CN is greater than the predicted CN in minerals formed in high pressure.	The radius ratio of Si^{4+}: O^{2-} is 0.26, so the predicted Cn\N is 4. observed CN = predicted CN in quartz. But in stishovite which forms at pressure of nearly 80 kb (at normal temperature), CN of Si^{4+} is 6.
Covalent characters of the bonds	The geometric radius ratio rules are applicable only to the ionic compounds. If the bonding mechanism is not dominantly ionic, CN cannot be safely determined.	Radius ratio of Ca^{2+} : S^{2-} = 0.62 and Cd^{2+} : S^{2-} = 0.598. So the predicted CN for both cations should be 6. The CA-S bond is dominantly ionic, so the observed CN = predicted CN. But due to the covalent nature of Cd-S bond, the effective radius ratio is less and the CN of Cd is 4.

Packing of Atoms

In crystal structures, atoms or ions behave as spheres, which remain in contact with each other to maintain the stability of the structures. The mode of packing in crystal structure is expressed by packing index, which is formulated as:

$$\text{Packing index} = (\text{Volume of ions/volume of unit cell}) \times 10$$

In a crystal structure, same ions, in same number may exhibit different modes of packing. In this respect, the volume in a crystal structure, occupied by atoms can also be determined with the help of atomic packing factor (APF), which is expressed as:

$$\text{APF} = N_{\text{atoms}} \times V_{\text{atom}}/V_{\text{unit cell}}$$

where N_{atoms} = no of atoms in unit cell, V_{atom} = volume of an atom and $V_{\text{unit cell}}$ = volume occupied by the unit cell.

Table 3.7: Different types of atomic packing

Crystal structures	*APF*
Simple cubic: There is one lattice point at each of the eight corners of a cube. There is one host atom wholly inside the cube, because each of the eight corner atoms contributes one eighth of an atom to the cell interior. The total volume of the cell which is occupied by the host atoms is $V_s = {}^4/_3\pi r^3 \times Z$. The packing efficiency is about 52%.	0.52
Body-centered cubic: One host atom (lattice point) at each corner of the cube and one host atom in the centre of the cube: $Z = 2$. Each corner atom touches the central atom along the body diagonal of the cube, and the unit cell edge is an irrational number, about $2.3r$. The corner atoms do not touch one another. The packing efficiency is about 68%.	0.68
Hexagonal close-packing: The interplanar spacing between adjacent planes is proportional to r. The proportionality constant is an irrational number called the *Closest Packed Interlayer Spacing*, CPIS, and its value is about 1.6. Thus, the interlayer spacing is about $1.6 \times r$ (compared to $2 \times r$ for simple hexagonal stacking). Thus, the "c" unit cell edge (in the stacking direction) has a length $c = 2 \times \text{CPIS} \times r$, the ratio $c{:}a = \text{CPIS}$, and it can be shown that this lattice has a packing efficiency which is identical to the FCC lattice.	0.74
Face-centered cubic: There is one host atom at each corner and one host atom in each face. Since each corner atom contributes one eighth and each face atom contributes one half of its volume to the cell interior (and there are six faces), then $Z = {}^1/_8 \times 8 + {}^1/_2 \times 6 = 4$. The corner and face atoms touch along the face diagonal, and the cube edge (a) is about $2.8r$. The packing efficiency is about 74%.	0.74

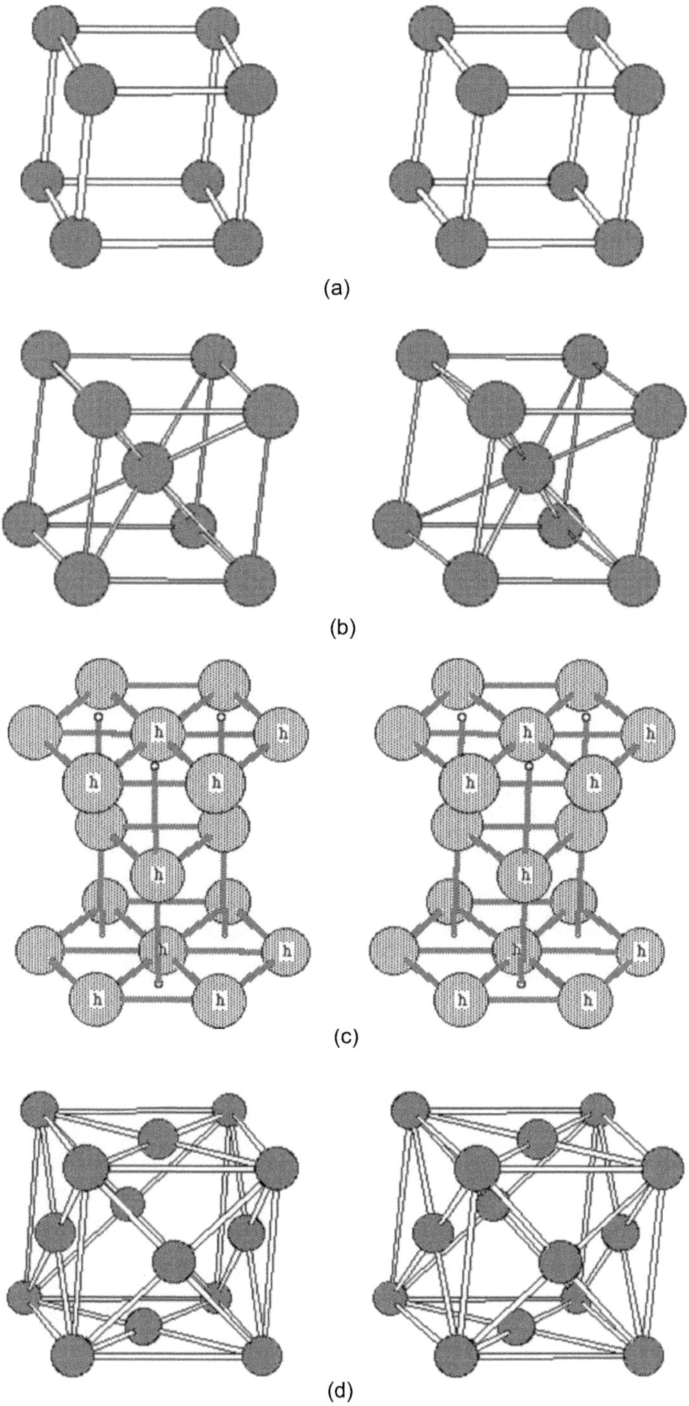

Fig. 3.9: (a) Simple cubic; (b) Body-centered cubic; (c) Hexagonal close-packing; and (d) Face-centered cubic.

3.6 PAULING'S RULES

Linus Pauling enunciated certain rules that explain the way ions combine in a crystal structure and henceforth they are known as Pauling's rules.

Rule 1: Around every cation, a coordination polyhedron of anions forms, in which the cation-anion distance is determined by the radius sums and the coordination number is determined by the radius ratio. It states that the different types of coordination polyhedra are determined by the radius ratio, Rx/Rz, of the cation to the anion.

Rule 2: The Electrostatic Valency Principle: An ionic structure will be stable to the extent that the sum of the strengths of the electrostatic bonds that reach an ion equal the charge on that ion.

Electrostatic valency (e.v.) = Charge on the ion/C.N.

Based on this principle three types of bonding can take place in compounds.

Isodesmic Bond

In NaCl, each Cl ion is also surrounded by six Na ions in octahedral coordination. So, again, the 1/6 of a positive charge from each Na reaches the Cl ion and thus the Cl ion sees 6 × 1/6 = 1 positive charge, which exactly balances the –1 charge on the Cl. Hence in the case of NaCl the charge is exactly balanced on both the cations and anions. In such a case, we say that the bonds are of equal strength from all directions. When this occurs the bonds are said to be *isodesmic*.

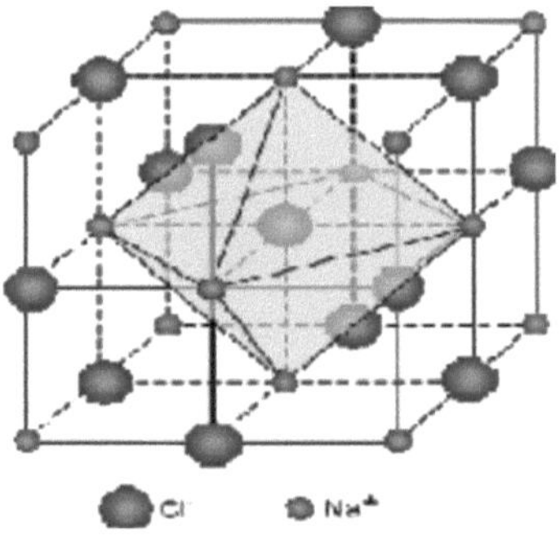

Fig. 3.10: Isodesmic bond.

Mesodesmic Bond

When the e.v. reaching the cation is exactly 1/2 the charge of the anion like for Si^{+4} in tetrahedral coordination with O^{-2}, it will form mesodesmic bond. Here, the e.v. reaching the Si is 4/4 = 1. This leaves each oxygen with a –1 charge that it has not shared. Since this –1 is exactly 1/2 the original charge on O^{-2}, the oxygens in the SiO_4^{-4} group can be just as tightly bound to ions outside the group as to the centrally coordinated Si. In this case the bonding is said to be mesodesmic.

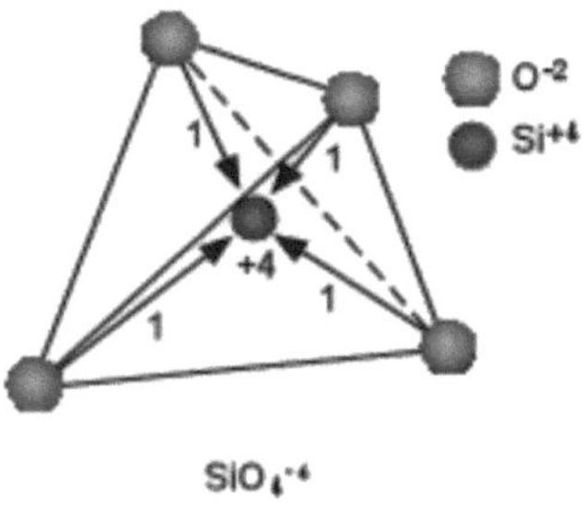

Fig. 3.11: Mesodesmic bond.

Anisodesmic Bond

In case for C^{+4} ion in triangular coordination with O^{-2}, e.v. = 4/3 (C has a charge of +4 and is coordinated by three oxygens). Thus, of the three oxygens each contribute 4/3 charge to the carbon ion, and the charge on the carbon is balanced. But, each oxygen still has 2/3 of a charge that it has not used. Thus, a carbonate structural group is formed $–CO_3^{-2}$. In cases like this, where the electrostatic valency is greater than 1/2 the charge on the anion (4/3 > ½ × 2), the anion will be more strongly bound to the central coordinating cation than it can be bonded to other structural groups. When this occurs the bonding is said to be anisodesmic.

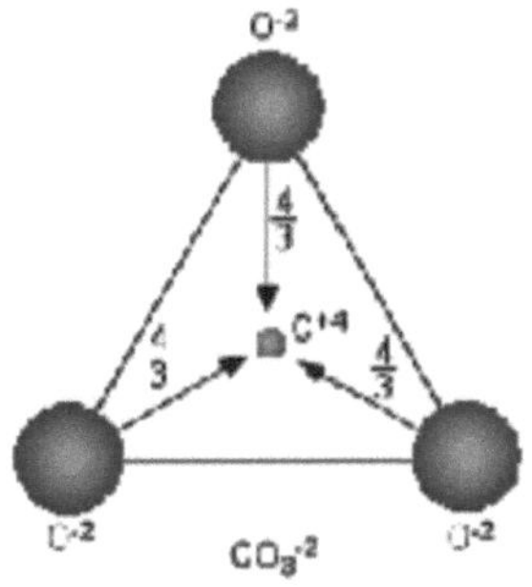

Fig. 3.12: Anisodesmic bond.

Rule 3: Shared edges, and particularly faces of two anion polyhedra in a crystal structure decreases its stability.

Polyhedra are usually joined at their corners; joining at their edges is much less common, whereas joining by sharing faces is rare. Sharing of faces among polyhedra generally decreases the stability of the structure, especially if the centres of these polyhedra are occupied by cations with high valency and small coordination numbers. When the bonding in a crystal is largely ionic, the crystal is likely to have a structure that keeps the cations as far apart as possible. However, because anions commonly have more than one cation bonded to them, cations are never far apart. Cations will be farthest apart if they share only one anion, i.e. their coordination polyhedra share only corners. If two anions are shared, the coordination polyhedra have an edge in common and the cations are closer together. If three anions are held in common, the coordination polyhedra of the two cations share faces and the cations are unreasonably closer. This effect is especially large when the radius ratio approaches the lower stability limit of the polyhedron.

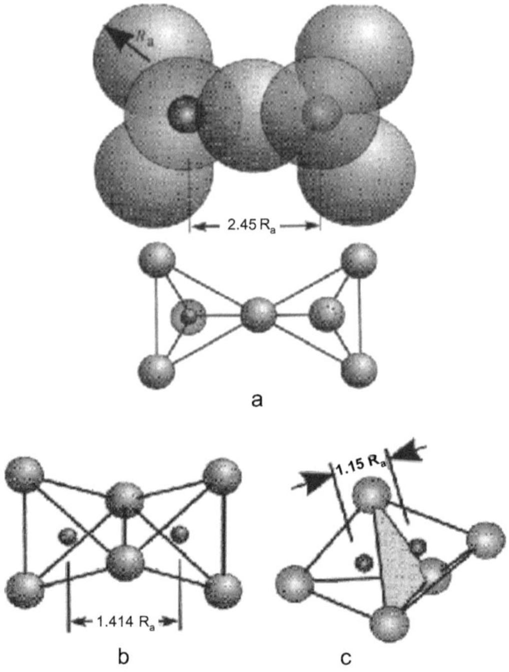

Fig. 3.13: (a) Corner sharing (most stable), (b) Edge sharing and (c) Face sharing (least stable).

Rule 4: In a crystal structure containing several cations, those of high valency and small coordination number tend *not* to share polyhedral elements.

Sharing of polyhedral elements for cations of high charge will place cations close enough together that they may repel one another. Thus, if they do not share polyhedral elements they can be better shielded from the effects of other positive charges in the crystal structure.

Rule 5: The Principle of Parsimony: The number of different kinds of constituents in a crystal tends to be small.

This means that there are only a few different types of cation and anion sites in a crystal. Even though a crystal may have tetrahedral sites, octahedral sites and cubic sites, most crystals will be limited to this small number of sites, although different elements may occupy similar sites.

Table 3.8: Applications of Pauling's Rules in silicate minerals

Seven types of polymerization in silicates

- The linked tetrahedra are usually SiO_4, but can also involve AlO_4 tetrahedra, Al^{3+} is larger than Si^{4+}, but not too much larger, so can substitute Al for Si.
- Oxygens in a tetrahedron NOT shared with another tetrahedron are Non-Bridging Oxygens (NBO).
- The NBO/T (T = Al+Si) ratio determines the type of silicate mineral.

	Si : O	
Island Silicates Ex: Olivine	1:4	$(SiO_4)^{4-}$ tetrahedra form isolated units No linkage NBO T = 4
Sorosilicates Ex: Epidote	2:7	Two tetrahedra share a corner oxygen $(Si_2O_7)^{6-}$ unit NBO/T = 3
Cyclosilicates	1:3	Tetrahedra share two corners to form rings having a Si:O ratio = 3:1 NBO/T = 2
(Ring Silicates) Ex: Beryl		The rings can be 3-membered, as in the rare mineral benitoite $BaTiSi_3O_9$ In the more common 3-membered ring, the unit is $(Si_3O_9)^{6-}$ or the rings can be six-membered, as in the mineral beryl $Be_3Al_2Si_6O_{18}$ (emerald). The unit is $(Si_6O_{18})^{12}$.
Single-Chain Silicates Ex: Pyroxene	1:3	Tetrahedra share two corner oxygens to form chains; Same NBO/T as cyclosilicates; the unit is $(SiO_3)^{2-}$ or $(Si_2O_6)^{4-}$.
Double-Chain Silicates Ex: Amphibole	4:11	Some tetrahedra share two corners and some three corners, so on average NBO/T=1.5; Unit is $(Si_4O_{11})^{6-}$.
Phyllo-Silicates (Sheet Silicates) Ex: Mica group	2:5	Tetrahedra share three corners to form sheets infinite in two directions. Sheets are also linked to a layer of AlO_6 or MgO_6 octahedra, forming a t-o or a t-o-t supersheet. Bonding between supersheets is either VdW or ionic NBO/T = 1 Unit is $(Al,Si)_2O_5$.
Tecto-Silicates (Framework Silicates) Ex: Feldspar group	1:2	Complete sharing of corner oxygens to form a three dimensional framework NBO/T = 0, unit is $(Al,Si)O_2$.

3.7 ATOMIC SUBSTITUTION AND SOLID SOLUTIONS

Atomic Substitution

Homogeneous crystalline solids of variable chemical composition: Many minerals vary in their composition. Elements are readily substituted (atomic substitution) for one another in many crystal structures (when ionic radius and charge permits).

- The difference in valences between the substituting ion and substituted ion should not be more than 1, as that between Na^{+1} and Ca^{+2}.
- At room temperature difference in the size of substituting ions must be <15%.
- Atomic substitution is greater at higher temperature (crystal lattices are more open) and can accommodate greater ionic radius deviation (than 15%). For ex: Na^{+1} IR = 0.97, K^{+1} IR = 1.33, Ca^{+2} IR = 0.99.

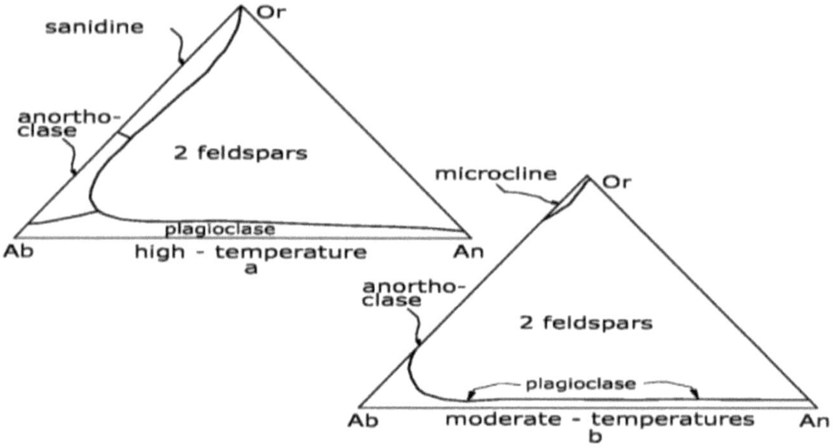

Fig. 3.14: Phase relation at Or-Ab-An system: (a) at high temperature and (b) at moderate temperature.

Solid Solutions

A solid solution is formed when two metals are completely soluble in liquid and solid state. In other words, when homogeneous mixtures of two or more kinds of atoms (of metals) occur in the solid state, they are known as solid solutions. The more abundant atomic form is referred as solvent and the less abundant atomic form is referred as solute. In this phenomenon substitution of one element by another takes place. It can also be referred as a homogeneous crystalline solid of variable composition.

For example, in olivine structure a solid solution exists between Mg_2SiO_4 (forsterite) and Fe_2SiO_4 (fayalite).

Factors affecting the Solid Solution

- Ionic size of the substituting ion (differ by less than 15-30%).
- Ionic charge: Must have electrostatic neutrality.
- Temperature: Atomic sites are more accommodating at higher temperatures.

Types of Solid Solution

Substitutional solid solution: Defined by the process in which different elements occupy same lattice position in a crystal, i.e. Mg, Fe, Mn in the structure of dolomite. Here the atoms of the solvent or parent material are replaced in the crystal lattice by atoms of the solute material.

Dependency factors:

- Crystal structure
- Relative size
- Chemical affinity
- Relative valency

Examples:

- Olivines Fe_2SiO_4 and Mg_2SiO_4. Fe^{+2} and Mg^{+2} are about the same size, thus they can substitute for one another in the crystal structure and olivine thus can have a range of compositions expressed as the formula $(Mg, Fe)_2SiO_4$.
- Alkali feldspars: $KAlSi_3O_8$ (orthoclase) and $NaAlSi_3O_8$ (albite). K^{+1} can substitute for Na^{+1}.
- Plagioclase feldspars: $NaAlSi_3O_8$ (albite) and $CaAl_2Si_2O_8$ (anorthite). $NaSi^{+5}$ can substitute for $CaAl^{+5}$ (a complex solid solution).

Coupled substitution: Here a solid solution occurs between two ions with different charges, other substitutions must occur in order to maintain charge neutrality. In apatite, there are several known substitutions that can couple with the Ca^{2+} La^{3+} substitution to maintain charge balance. These include Na^{1+}, Ca^{2+}, Si^{4+}, P^{5+}, O^{2-} and F.

3.8 INTERSTITIAL AND DEFECT SOLID SOLUTIONS

Foreign ions fill up the interstices of the lattice, i.e. there are interstices in the amphibole to accommodate a sodium ion.

In this system the carbon (solute) atom occupies an interstitial position between iron (solvent) atoms. Normally, atoms which have atomic radii less than one angstrom are likely to form interstitial solid solutions.

Examples are atoms of carbon (0.77 Å), nitrogen (0.71 Å), hydrogen (0.46 Å), oxygen (0.60 Å) etc.

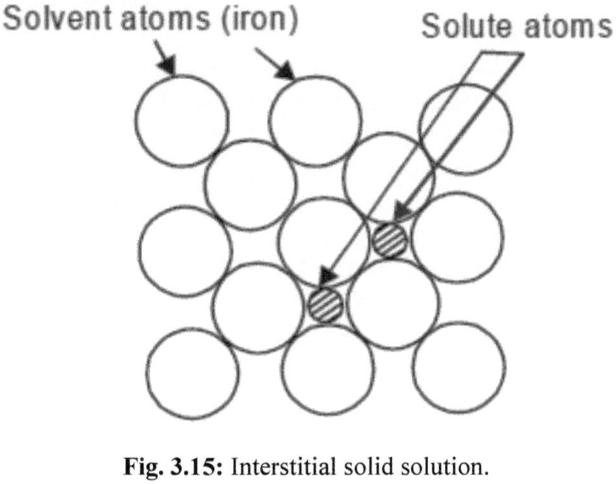

Fig. 3.15: Interstitial solid solution.

Omission solid solution: A more highly charged cation replaces lower charged ions, creating vacancies, i.e. in microcline ($KAlSi_3O_8$), one Pb^{+2} can replace two K^{+1} ions, resulting in vacancy. Also exchange of single higher charge cation for two or more lower charged cations which creates a vacancy (e.g. pyrrhotite – $Fe_{(1-x)}S$) with x ranging 0-0.2; where Fe^{+2} absent from some octohedral sites, some iron probably Fe^{+3} to restore electrical neutrality.

3.9 ISOMORPHISM, POLYMORPHISM AND PSEUDOMORPHISM

Isomorphism

The phenomenon in which two substances may exist with analogous formulas but have relatively similar sizes of cations and anions and therefore have comparable crystal structures, is called isomorphism and the substances are said to be isomorphous.

- Anions and cations of isomorphous minerals have the same relative size, the same coordination.
- Crystallize in the same crystal structure.
- Share similarity of crystal structure but not (necessarily) chemical behaviour.

The phenomena of isomorphism is very widespread in minerals. So, the elemental composition of a mineral of wolframite is mapped by the formula (Fe, Mn) $[WO_4]$. It represents an isomorphic intermixture where atoms of manganese replace atoms of iron. The mineral olivine $(Mg,Fe)_2$ $[SiO_4]$ also represents an isomorphic intermixture where atoms of magnesium in structure are replaced with atoms of iron.

Isomorphous Replacement

The concept of isomorphous replacement helps to understand how a single family of minerals can offer such a wide variety of colours. There are several mineral groups that offer isomorphous replacement series of formations. Among them are the topaz family, the tourmaline family and the garnet family.

Along with simple cases there can be the difficult isomorphic displacement of the whole complexes in crystalline structures. A classical example of such difficult displacement are minerals from group of feldspars-plagioclases. Plagioclases represent the continuous number of minerals where pair Ca^{2+} and Al^{3+} are replaced respectively by Na^+ and Si^{4+} (CaAl↔NaSi). End members of this series are termed as anorthite $Ca[Al_2Si_2O_8]$ and albite $Na[AlSi_3O_8]$. According to composition change physical properties, optical properties, density, etc. of plagioclases also change.

Types of Isomorphism

Isovalence isomorphism: When the ions having identical valence are crossly replaced.

Heterovalence isomorphism: When there is displacement of ions of different valencies.

Polymorphism

The phenomenon in which an element or compound exists with different crystalline structures, but with same chemical combination is called polymorphism. In mineralogy it means that a single chemical composition can exist with two or more different crystal structures. If a crystal is subjected to different pressures and temperatures, the arrangement of atoms changes, as the sizes of the atoms which control the atomic arrangement change. When the crystal structure changes to the more stable structure a different mineral will form.

Types of Polymorphism

Displacive polymorphism
- Involves minor reassembly of the crystal structure.
- Bonds remain intact with slight variation of bond angles.
- Changes are instantaneous and reversible.

Example: Transformation of α-quartz to β-quartz below 580° C.

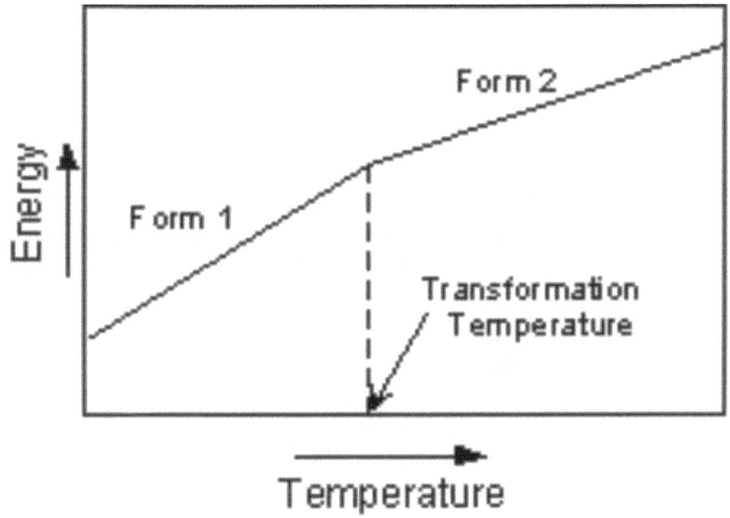

Fig. 3.16: Example of displacive transformation.

Reconstructive polymorphism

- Involves vivid rearrangement of the crystal structure.
- Breaking of chemical bonds.
- Regeneration of a new crystal structure by reorienting the atoms.

- Activated by change of temperature and pressure.
- Requires high energy.

Example: Transformation of diamond to graphite.

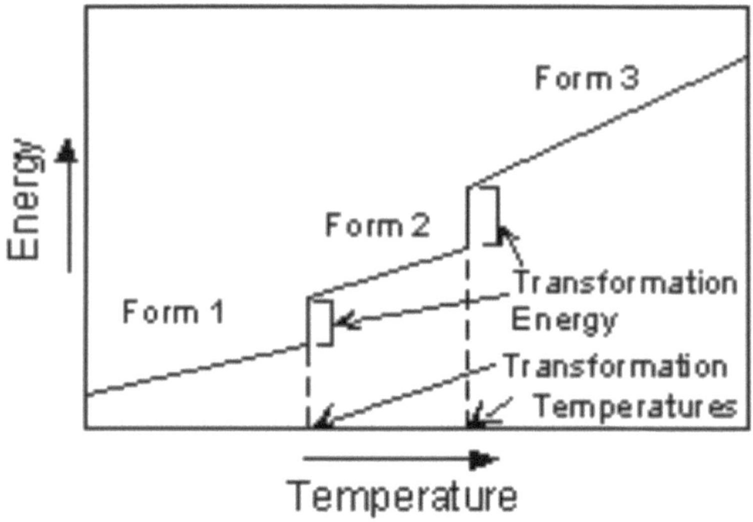

Fig. 3.17: Example of reconstructive transformation.

Order-Disorder Polymorphism

- Decrease of regularity of a crystal structure at higher temperature, resulting in high-temperature disordered forms.
- No temperature can be fixed in this case, since the change is gradual.

Example: Transformation of sanidine-disordered high temperature form to orthoclase-semiordered low temperature form to microcline-ordered low temperature form.

Table 3.9: Important polymorphs

Examples	*Polymorph*
Al_2SiO_5	Kyanite, Andalusite, Sillimanite
C	Diamond and Graphite
SiO_2	Cristobalite, Tridymite, Stishovite, Coesite, α-quartz, β-quartz

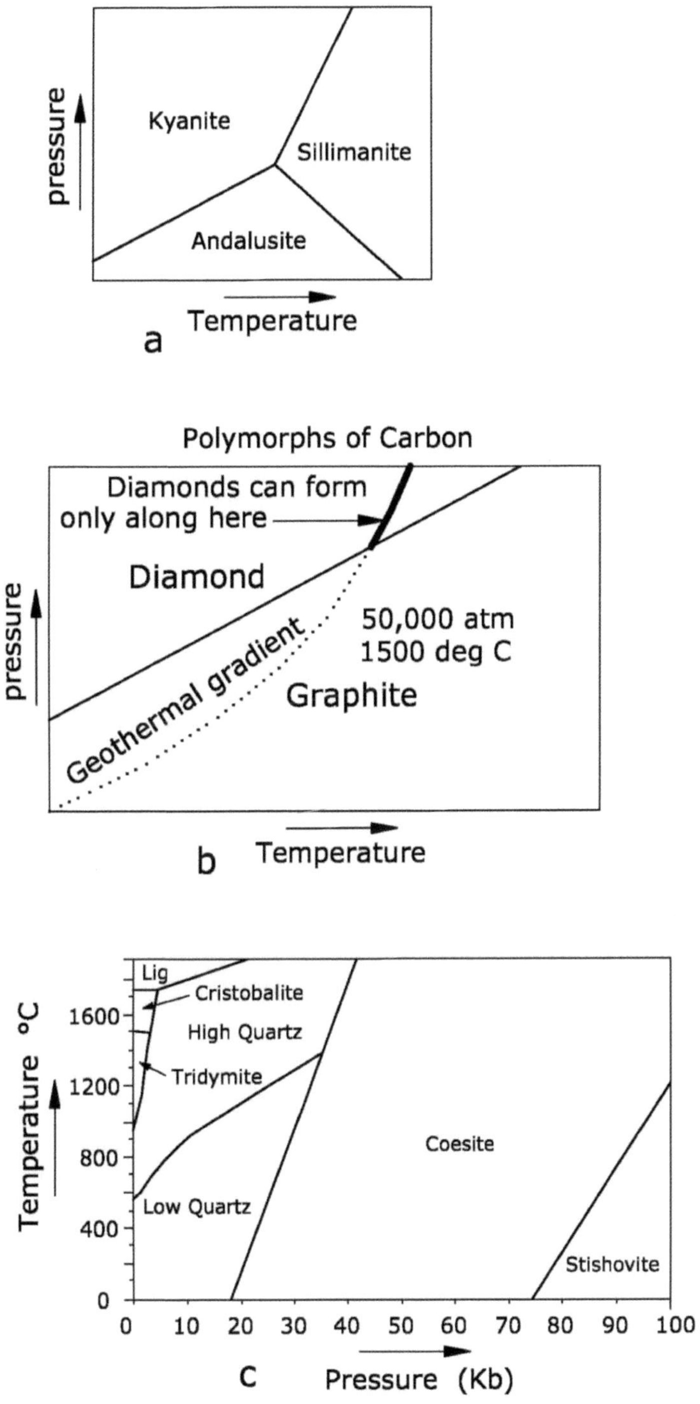

Fig. 3.18: Polymorphism of (a) Al_2SiO_5, (b) carbon and (c) SiO_2.

Pseudomorphism

The phenomenon in which the existing mineral has the appearance of another mineral is termed pseudomorphism. It happens when the internal structure and chemical composition are altered, but external form remains preserved.

Mechanisms of Pseudomorphism

Substitution: Replacement of existing chemical constituents by different ones, as in pseudomorphism of quartz with fluorite.

Encrustation: Formation and preservation of a new mineral as a crust on a pre-existing mineral, which gets subsequently removed.

Alteration: Partial removal of the existing mineral and partial replacement by new mineral, as in pseudomorphism of limonite after pyrite, anglesite after galena, or anhydrite after gypsum.

3.10 NON-CRYSTALLINE MINERALS

Apart from the crystalline minerals, there are some minerals, that are non-crystallized in nature. They can be sub-divided into two groups as shown in Table 3.10.

Table 3.10: Types of non-crystalline minerals

Non-crystalline minerals	*Properties*	*Mode of origin*	*Examples*
Metamict	Noncrystalline pseudomorphs after originally crystalline material; isotropic; unable to diffract x-rays; glassy or pitchy in appearance; conchoidal fracture; radioactive due to presence of uranium (<1%).	Breakdown of the crystal structure through bombardment by alpha particles, releasing from disintegrating radio-active elements.	Zircon ($ZrSiO_4$) often has U and Th atoms substituting for Zr in the crystals structure. Since U and Th have radioactive isotopes, zircon is often seen to occur in various stages of metamictization.
Amorphous	Sometimes known as super-cooled liquids; absence of periodicity in atom location; isotropic.	Rapid cooling from the molten state or by slow hardening of gelatinous material.	Opal in glasses and gels.

Concluding Remarks

So far, the geometry, classification and crystal chemistry have been covered. It has been seen that minerals exhibit a highly variable range of chemical composition and geometrical pattern. Chemical differences in minerals are the result of various processes like solid solution and exsolution, while geometrical variability is due to phenomenon like polymorphism, isomorphism or pseudomorphism. After covering all these, our next focus will be on the different physical properties, which are controlled by these geometric and chemical variations in minerals.

Think for a while

1. Explain the various types of atomic packing to construct crystals of various morphology.
2. How to explain the order-disorder polymorphism with the example of K-feldspar?
3. Write down the names of the polymorphs of Al_2SiO_5. Mention the P-T conditions and crystallographic system for each of the polymorphs of Al_2SiO_5.

FURTHER READING

Blackburn, W.H. and W.H. Dennen. Principles of Mineralogy, 2nd ed. William C. Brown, Dubuque, IA. 1994.

Bloss, F. Donald. Crystallography and Crystal Chemistry. Holt Rinehart, Winston, New York. 1971.

Deer, W.A., R.A. Howie and J. Zussman. Rock Forming Minerals. 5 vols. John Wiley and Sons, New York. 1962.

Henderson, Paul. Inorganic Geochemistry. Pergamon, New York. 1982.

Klein, C. and C.S. Hurlbut, Jr. Manual of Mineralogy, 21st ed. John Wiley and Sons, New York. 1993.

Pauling, Linus. The Nature of the Chemical Bond, 3rd Edition. Cornell Univ. Press, Ithaca, NY. 1960.

Smyth, J.R. and D.L. Bish. Crystal Structures and Cation Sites of the Rock Forming Minerals. Allen and Unwin, London. 1988.

CHAPTER

4

PHYSICAL PROPERTIES

Both the physical properties of minerals and gemstones and their beauty are dependent on their chemical composition and atomic structure. Physical properties can be useful in distinguishing between both mineral species and individual minerals within a group or series.

Minerals are identified by their distinguishing physical properties which are discussed in details in this chapter. The physical properties are colour (which depends on the crystal structure, symmetry, different bondings in the crystals, mode of crystallization, defects/impurities in the mineral structure, absorption capacity of different wavelengths of the light spectrum), lustre (depending on type of atomic bonds, symmetry, refraction, absorption capacity of different wavelengths of the light spectrum), crystal habit (symmetry), cleavage (symmetry), parting (symmetry), tenacity (types of bonds and symmetry), hardness (types of bonds, symmetry and packing of atoms), specific gravity (types of bonds, symmetry and packing of atoms), thermal properties (heat flow, atomic bonds), electrical property (types of bonds, surface conditions), magnetic properties (chemical composition, atomic bonding), and radioactivity (chemical composition).

4.1 PROPERTIES CONTROLLED BY SYMMETRY

Colour

- Colour is one of the physical properties most commonly used to describe a mineral.
- Colour is caused due to absorption, or lack of absorption of different wavelengths of light by the mineral/material.

- Atoms present in the minerals when exposed to white light, absorb certain wavelengths and emit another wavelength to get rid of the extra energy, provided by the absorbed wavelength. The different types of bonding decides the energy state of the participating electrons and hence produce different colours.
- Some minerals are nearly always of the same colour like azurite (blue) and sulfur (yellow) and they are known as *idiochromatic* minerals.
- Many minerals come in a variety of colours—the changes are caused by slight chemical impurities or through exposure to heat. Example of such minerals are quartz, diamond, beryl, corundum etc. and they are known as *allochromatic* minerals.
- Ions of certain elements are highly absorptive of selected wavelengths of light. Such elements are called *chromophores*; they possess strong pigmenting capabilities. The elements vanadium (V), chromium (Cr), manganese (Mn), iron (Fe), cobalt (Co), nickel (Ni), and copper (Cu) are chromophores. A mineral whose chemical formula stipulates the presence of one or more of these elements may possess a vivid and distinctive colour.
- Colour-change occurs when the surface is exposed to moisture and air—it tarnishes or oxidizes. Colour also varies with the change of oxidation state, for example, beryl (aquamarine) = Fe^{++}, is blue; beryl (heliodor) = Fe^{+++}, is yellow (the oxidation state of impurities present in beryl).
- Some minerals have common names (varietal names) that describe a specimen with a certain colour e.g. quartz—rock crystal (colourless), smoky quartz (brown), citrine (yellow), amethyst (violet), rose quartz (pink).
- Colour can be described as metallic or non-metallic and is often described along with lustre.

Colours may also be caused due to the following reasons:

- Charge transfer can only occur in compounds that have at least two elements in different and variable oxidation states. Charge transfer can produce very intense colours in gems and minerals. The term *charge transfer* refers to the process where electrons are swapped between elements.
  Examples of elements that can participate in charge transfer are: Fe^{2+} and Fe^{3+}; Ti^{3+} and Ti^{4+}; Mn^{2+}, Mn^{3+} and Mn^{4+} etc.
- *Colour centres* are imperfections in crystals that cause colour [defects that cause colour by absorption of light). Colour centres (also known as F-centres or farbe (German for colour) centres] are created when atoms are oxidized or removed. This is usually done by radiation. In most cases the hole left behind is occupied by an electron trying to proxy for the missing atom. This electron comes from a neighbouring atom and the unpaired electron left behind is prone to absorb light energy and thereby create colours. The most familiar examples of minerals coloured by colour centres are amethyst and smoky quartz but fluorite, green diamonds, and brown topaz are also good examples. Colour centres are one of the few colouring mechanisms that can be removed by heating or exposing the mineral to strong light. They are most often caused due to radiation damage, e.g., damage due to exposure to gamma rays. This irradiation may be from both natural (U, Th, K in minerals)

Table 4.1: Colour classification of minerals

Colour Grouping	*Rock and mineral example*
Metallic	
Golden, golden yellow	Gold, pyrite
Bronze	Chalcopyrite, nickel ore
Copper, copper-red	Copper
Silver, silvery-yellow, silvery-gray, bright silver, dark silver, black	Antimony, galena, manganite, silver
Peacock feathers, rainbow	Bornite, peacock ore
Non-metallic	
Colourless	Barite, quartz, rock crystal, selenite
White, beige, creamy, dirty-white, snow white	Calcite, gypsum, muscovite mica, quartz, stilbite, talc
Yellow, orange, brown, browny-beige	Barite rose, calcite, cancrinite, celestite, jasper, siderite, sphalerite, sulfur
Green, turquoise green, moss green, lime green	Amazonite feldspar, apatite, bloodstone, emerald, epidote, fluorite, grossular garnet, jade, malachite, turquoise
Blue, sky blue, turquoise blue, pale blue, steel blue, deep blue	Azurite, celestite, kyanite, labradorite, lapis lazuli, sodalite, turquoise
Red, purple, maroon, violet, mauve, pink, burgundy, reddish-brown	Almandine garnet, amethyst, apatite, dolomite, feldspar, fluorite, halite, lepidolite, rhodonite, rose quartz, ruby
Black, brown-black, dark gray, gray	Biotite mica, diopside, fluororichterite, hornblende, titanite

or artificial sources. In rare cases, UV light can produce colour centres. Examples include missing F^- ions in fluorite defects (see Fig. 4.1) which makes purple colour. Coupled substitution of Al^{3+} and Na^+ for Si commonly occurs in trace levels in the quartz structure. Irradiation makes electron in tetrahedrally coordinated Al^{3+} susceptible to relocation and trapping elsewhere in the crystal. This produces smokey quartz.

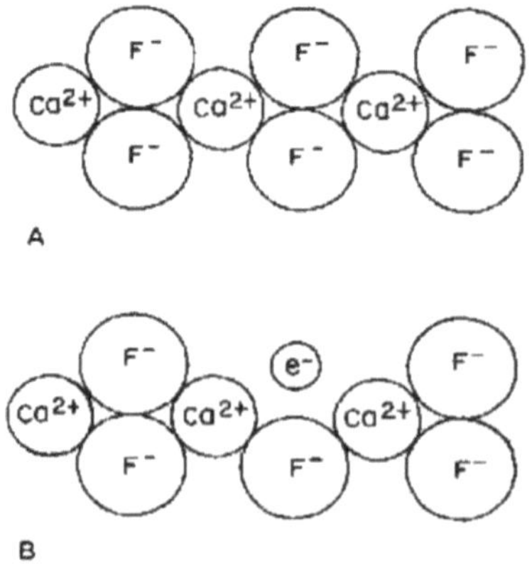

Fig. 4.1: Fluorite structure (schematic): (A) normal, and (B) containing an "F-centre" where a fluorine ion has been replaced by an electron.

Since the crystal must remain electrically neutral, an electron usually occupies the empty position to produce the F-centre or "electron colour centre" as shown in figure. This unpaired electron can now exist in excited states, the energy of which is controlled by the same crystal field factors described previously; colour and fluorescence can now occur by the same types of transitions as described above.

- Interference of light at the surface of a mineral as the angle of incidence changes is known as a *play of colours*. Light will diffract as it passes through regularly spaced mineral boundaries. The wavelength of light that is diffracted from a white light source is related to angle of incidence and the spacing between the diffraction planes.

Examples: Opal, which is composed of hexagonally closest packed hydrated-silica spheres, shows play of colours. When light passes over many tiny sharp edges or between many repeated points of differently refracting media an interference-like phenomenon occurs; light is spread out into specific colours. This principle is used in the diffraction grating spectroscope. This is what causes the play of colour in opals, light being bent and diffracted as it passes innumerable regular stacked layers of minuscule silica gel spheres.

Labradorite which is composed of alternating lamelae of different composition plagioclase felspars also show this property. Closely spaced fracture planes, cleavage planes and twinning planes may also cause a play of colours.

Streak

Streak is closely related to colour but is a different property because the colour of the mineral may be different than the colour of the streak. Streak is actually *the colour of the powder of a mineral.* It is called streak because the proper way to test for streak is to rub a mineral across a tile of white unglazed porcelain and to examine the color of the "streak" left behind. It has proven to be a powerful property because it is generally very consistent from specimen to specimen for a given mineral. Some minerals have a different colour powder than their actual colour. Every mineral has an inherent streak no matter what colour it is.

Two minerals that have similar outward colour may have different colours when powdered. For instance, the minerals hematite and galena can be confused as both have a gray colour. However, hematite's streak is cherry-red, while galena's streak is lead gray. Hematite is probably the most well known example of streak with its completely surprising streak colour.

For example, calcite occurs in many different colours, shapes and varieties. But every single variety of calcite has a white streak. Streak is useful to distinguish two minerals that have the same colour, but a different streak. A fine example is distinguishing gold, which has a yellow streak, and pyrite, which has a black streak. Most light coloured, nonmetallic minerals have a white or colourless streak, as do most silicates, carbonates, and most transparent minerals. The streak test is most useful for identifying dark coloured minerals, especially metals. Most mineral references don't make a distinction between a white or colourless streak, since the difference is minimal. A mineral with a white or colourless streak will not leave visibly coloured powder.

Lustre

Lustre refers to the general appearance of a mineral surface in reflected light. It can be described as, when light is incident upon an opaque mineral, no portion of the light is transmitted through it, due to its high absorption index and almost 20% to 50% of it is reflected. But in case of transparent minerals, the amount of reflected light is as little as 5%. These relative differences in opacity and transparency are termed as lustre. Lustre is the relative difference of brightness of mineral surfaces. The minerals with less degree of transparency have greater reflectivity therefore their brightness is also greater. The wavelength of light reflected from a mineral surface determines its colour, and amplitude of the reflected light determines its lustre.

Table 4.2: Classification of lustre

Type	*Description*	*Example*
Dull/earthy	Very dull, mainly in minerals that are porous	Kaolinite, orthoclase
Waxy	Like the surface of a candle	Opal, chalcedony
Greasy/oily	Like an oily surface	Nepheline
Pearly	Like a pearl, play of colours, light	Talc, muscovite
Silky	Has a shiny surface like a piece of silk cloth	Some varieties of gypsum, kernite, ulexite and in fibrous minerals
Glassy/vitreous	Looks like glass	Quartz, many rock-forming minerals, obsidian— "nature's glass"
Resinous	Looks like freshly-broken shellac, usually yellow-brown	Sphalerite
Adamantine	High lustre, almost brilliant, "diamond-like"	Sphalerite
Sub-metallic	Silvery or metallic lustre but mineral is transparent or translucent	Hematite
Metallic	Very shiny, like processed metals, highly reflective opaque minerals	Pyrite, gold, silver

As a whole, lustre of a mineral is determined by:

- *Type of atomic bond present in mineral:* Minerals possessing metallic lustre indicate the presence of metallic bonding within the crystal lattice of the material. Examples of minerals which exhibit metallic lustre are native copper, gold, silver, galena, pyrite, and chalcopyrite. Vitreous lustre occurs in minerals with predominant ionic bonding. Covalent bonding results in adamantine lustre, like in diamond.

- *Absorption coefficient:* Minerals possessing a high absorptive index exhibit metallic lustre, while with low absorptive index show dull lustre.
- *Refraction of the mineral:* The ranges of R.Is for various groups of minerals are: metallic (>3), sub-metallic (2.6-3), and non-metallic (<2.6) lustre. Minerals with high R.I. generally show metallic lustre like galena, pyrite etc. while the non-metallic lustre display a wide range of R.I. values. Minerals showing resinous lustre have a refractive index greater than 2.0, ex: sphalerite (ZnS). Minerals with vitreous luster have refractive index of 1.5 to 2.0, ex: quartz and tourmaline. Minerals showing adamantine lustre have high R.I. of 1.9-2.6.
- *Amount of dispersion from crystal lattice:* Silky lustre results when light is dispersed from an aggregate of fine parallel fibres, as in malachite and serpentine. Dull or earthy lustre is seen in minerals, composed of an aggregate of tiny grains.
- *Texture of the exposed mineral surface:* A light-scattering surface which is slightly rough, such as that of nepheline, may exhibit greasy lustre.
- *Cleavage present*: Pearly lustre is typically exhibited by mineral surfaces which are parallel to planes of perfect cleavage. Layer silicates such as talc often demonstrate a pearly lustre on cleavage surfaces. Another such example is feldspar.

Crystal Habit

In nature perfect crystals are rare. The faces that develop on a crystal depend on the space available for the crystals to grow. If crystals grow into one another or in a restricted environment, it is possible that no well-formed crystal faces will be developed. However, crystals sometimes develop certain forms more commonly than others, although the symmetry may not be readily apparent from these common forms. The term used to describe general shape of a crystal is *habit*.

Cleavage

Crystals often contain planes of atoms along which the bonding between the atoms is weaker than along other planes. In such a case, if the mineral is struck with a hard object, it will tend to break along these planes. This property of breaking along specific planes, which are parallel to possible crystal faces is termed cleavage.

- If a mineral has cleavage along {100} it will break easily along planes parallel to the (100) crystal face, and any other planes that are related to it by symmetry. Thus, if the mineral belongs to the tetragonal crystal system it should also cleave along faces parallel to (010), because (100) and (010) are symmetrically related by the 4-fold rotation axis. The mineral will be said to have two sets of cleavage.
- Cleavage can mainly be of the following types: perfect, imperfect, good, distinct, indistinct and poor.

- Cleavage is said to be *basal* when it occurs perpendicular to the major axis of the mineral and *prismatic* when it occurs parallel to the major axis.
- Different forms of cleavage exist on different minerals, depending on the mode of a mineral's crystallization.

Table 4.3: Different habits of individual and group of crystals

Type	*Description*	*Example*	*Figure*
Aciculur	Crystals that grow in fine needles	Natrolite	
Bladed	Like a wedge or knife blade	Kyanite	
Blocky	Rectangular and bow-like	Oligoclase	
Equants	Any three perpendicular axes through the crystal are equal	Fluorite	
Fibrous	Look like fibers, threads, parallel lines	Okenite	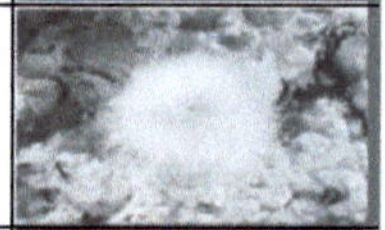
Platy	Flattened and thin crystals	Wulfenite	
Prismatic	Abundance of prism faces	Indicolite	
Stubby	Slightly elongated	Topaz	
Tabular	Divide easily into thin plates or sheets, a stack is know as a "book"	Wulfenite	
Groups of Distinct Crystals			
Dendritic	Tree-like growth	Sal ammoniac	
Rossette	Petal-like crystals arranged in flattened radial habit	Barite	

Table 4.4: Forms of cleavage

Cleavage	*Description*	*Example*
Basal cleavage	Cleavage exhibited on a horizontal plane of the mineral by way of its base. Minerals with basal cleavage can sometimes be "peeled".	Minerals of mica group
Cubic cleavage	Cleavage exhibited on minerals of the isometric crystal system crystallized as cubes; small cubes evenly break off of an existing cube.	Galena
Octahedral cleavage	Cleavage exhibited on minerals of the isometric crystal system crystallized as octahedrons; flat, triangular "wedges" peel off of an existing octahedron.	Fluorite
Prismatic cleavage	Cleavage exhibited on some prismatic minerals in which a crystal cleaves by breaking off thin, vertical, prismatic crystals off the original prism.	Aegirine
Pinacoidal cleavage	Cleavage exhibited on some prismatic and tabular minerals in which a crystal cleaves on the pinicoidal plane, which is the third dimension aside from the basal and prismatic sides.	Barite
Rhombohedral cleavage	Cleavage exhibited on minerals crystallizing in the hexagonal crystal system as rhombohedrons, in which small rhombohedrons break off of the existing rhombohedron.	Calcite

Parting

Parting is characteristically similar to cleavage. It is easily confused with cleavage, and is often present on minerals that do not exhibit any cleavage. There are two causes of parting:

1. Two separate pressures pushed toward the centre of a crystal after its formation, causing the crystal interior to evenly dislodge on a flat, smooth plane.
2. Twin crystals that separated from one another, leaving a flat, smooth plane.

Table 4.5: Examples of partings and habits

Parting	*Crystal habit*	*Example*
Fibrous parting	Occurs in fibrous aggregates of crystals	Asbestos
Basal parting	Crystals stacked like segments in a column	Corundum
Flaky parting	Occurs in aggregates of small flaky crystals	Graphite, chlorite and talc

Fracture

If the mineral contains no planes of weakness, it will break along random directions called fracture. Several different kinds of fracture patterns are observed. Fracture can be confused with cleavage. Cleavage and fracture differ in that cleavage is the break of a crystal face where a new face (resulting in a smooth plane) is formed, whereas fracture is the "chipping" of a mineral. All minerals exhibit a fracture, even those that exhibit cleavage. If a mineral with cleavage is chipped a certain way, it will fracture rather than cleave.

Mechanism of Fracture

- Occur at stress, required to break atomic bonds, not due to inherent flaws.
- Large-scale applied stress due to seismic activity result in dynamic fracture.
- Excess of resistance to plastic shear than the cohesive strength of cleavage planes result in cleavage fracture.
- Combined stress for cleavage fracture and that necessary to cause fracturing along grain boundaries result in brittle intergranular fractures.
- When the general plasticity is >10%, low temperature ductile fracture results.
- Due to dislocation creep processes, transgranular creep fracture results at specified temperature.
- At temperature greater than that for previous type of fracture and at low applied stress, intergranular creep fracture results.
- Creep rupture results under localized condition of deformation and at reduction of cross-sectional area to zero. [For detailed study of these fractures, consult any structural geology text book.]

Table 4.6: Different types of fractures

Fracture	*Description*	*Example*
Conchoidal	Breaks along smooth curved surfaces concave surfaces like conch-shells	Quartz, opal, chert, flint
Sub-conchoidal	Smooth with irregular rounded corners	
Fibrous and splintery	Similar to the way wood breaks	Serpentine, chrysotile
Hackly	Jagged fractures with sharp edges	Silver
Uneven or Irregular	Rough irregular surfaces	Hematite, limonite, magnetite, pyrite, olivine
Crumbly	Minerals crumble when broken	Howlite
Even	Forms a smooth surface	Kyanite
Earthy	Resembles broken soil	Kaolinite

Tenacity

Tenacity is how tough a mineral is, how easily a mineral will break, split, crumble or change shape. Terms used to describe this trait are listed in Table 4.7.

Table 4.7: Types of tenacity in minerals

Tenacity	*Description*	*Example*
Elastic	Can be bent and when let go, they resume their previous shape	Mica
Ductile	Can be pulled to make very thin threads	Gold
Flexible	May be bent but does not return to original form	Talc and chlorite
Fragile	Break into pieces easily	Selenite
Friable	Crumbles easily	Kaolinite
Sectile	Deforms plastically and can be cut with a blade to make shavings	Gypsum
Malleable	Flatten out into thin sheets without breaking	Gold

Hardness

Hardness of minerals is their resistance against cracking and it is determined by the presence of weak bond in the mineral. Though integral, but hardness is a relative property of a mineral and thus its estimation is dependant upon its mode of determination.

Determination of Hardness

Hardness is determined mainly by static and dynamic indentation, scratching, grinding, drilling and the measurement of the recoil height of an impinging hammer.

In mineralogy the most widely used methods are scratching and indentation methods. Hardness is determined on the basis of Moh's relative scale of hardness exhibited by some common minerals. This scale is not linear, the difference in hardness between the first standard minerals being much less than that between the last ones. These minerals are listed in Table 4.8, along with the hardness of some common objects.

Table 4.8: Variation of hardness of minerals in Moh's scale

Hardness	*Mineral*	*Common objects*
1	Talc	
2	Gypsum	Fingernail (2+)
3	Calcite	Copper Penny (3+)
4	Fluorite	
5	Apatite	Steel knife blade (5+), Window glass (5.5)
6	Orthoclase	Steel file
7	Quartz	
8	Topaz	
9	Corundum	
10	Diamond	

Dependency of Hardness

1. Is dependant on direction, since hardness is a vectorial property. When there is significant difference in hardness in different directions, it can be a very diagnostic property of the mineral. It is thus wise to perform the hardness test by attempting to scratch the mineral in different directions. A single mineral has different hardness depending on different crystallographic directions:
   - Kyanite has a hardness of 5 parallel to the length of the crystal, and a hardness of 7 when scratched along a direction perpendicular to the length.
   - Calcite has a hardness of 3 for all surfaces except the {0001} plane. On {0001} it has a hardness of 2.
2. Varies inversely to the crystal interatomic spacing.
3. Increases with the increase of cation valency.
4. Increases discontinuosly with the increase in coordination number.
5. Increases with higher density of atomic packing.
6. Increases at transition from the ionic to the covalent bond.
7. Decreases in the presence of electron gas shell.
8. Decreases due to the presence of hydroxyl or water molecules.

Specific Gravity

Specific Gravity (SG) indicates how many times more the mineral weighs compared to an equal amount of water. In cgs units density is grams per cm^3, and since water has a density of 1 g/cm^3, specific gravity would have the same numerical value as density, but no units (units would cancel).

- In case of isostructural minerals, if a mineral has higher atomic number cations it has a higher specific gravity. For example in carbonates the following observation listed in Table 4.9 has been made.

Table 4.9: Cation dependence of specific gravity in ISO structure minerals

Mineral	*Composition*	*Atomic number of cation*	*SG*
Aragonite	$CaCO_3$	40.08	2.94
Strontianite	$SrCO_3$	87.82	3.78
Witherite	$BaCO_3$	137.34	4.31
Cerussite	$PbCO_3$	207.19	6.58

- Those with high specific gravities usually feel heavier. Most common silicate minerals have a specific gravity between about 2.5 and 3.0. These would feel light compared to minerals with high specific gravities.
- Specific gravity has been measured by weighing a mineral specimen on a balance scale while it is submerged first in air and then in water. The difference between the two measurements is the weight of the volume of

water which was displaced by the sample. The specific gravity of the mineral specimen is thus:

$$G = m_{air}/[m_{air} - m_{water}]$$

Because the density of water at 4° celsius is 1.00 g/cm^3, the density of a mineral in units of grams per centimetre cube (g/cm^3) is equal to its (unitless) specific gravity.

- Native elements, which contain only one type of atom and whose molecular structure is that of cubic or hexagonal closest packing, are relatively dense.
- Minerals whose chemical composition contains heavy metals—atoms of greater atomic number than iron (Fe, atomic number 26)—are more dense than atoms whose chemical composition does not include such elements.
- Minerals formed at the high pressures deep within the earth's crust are in general more dense than minerals formed at lower pressures and shallower depths. Thus for example, coesite and stishovite which are formed under higher pressure at greater depth have higher specific gravity than the quartz, formed at lower pressure and shallow depth.
- Though colour and SG are independent properties, such a relation is observed in certain solid solution series like fosterite-fayalite or enstatite-ferrosilite. In these cases, increase of iron content in the solid solution series brings about darkening of colour as well as increasing of SG. A general trend relating colour to density is also prevalent; this trend states that dark-coloured minerals are often fairly heavy whereas light-coloured ones are frequently relatively light. But there is exception. Graphite is dark coloured but of low density (C; 2.23 g/cm^3) while barite is light in colour but unexpectedly heavy ($BaSO_4$; 4.5 g/cm^3).
- The average rock you would pick up has an SG of about 2.75 because most of the Earth's crust is made up of quartz, calcite and feldspar.

Table 4.10: Range of specific gravity

Description	*SG*	*Mineral examples*
Very light	< 2	Borax
Light	2-2.5	Gypsum, halite, selenite, ulexite
Average	2-3	Calcite, dolomite, feldspar, muscovite mica, quartz, talc, turquoise
Above average/slightly heavy	3-4	Biotite mica
Heavy	4-5	Almandine garnet, apatite, barite, celestite, chalcopyrite, fluorite
Very heavy	5-10	Galena, hematite, magnetite, nickel-iron, pyrite
Extremely heavy even for a metallic mineral	>10	Gold, silver
Super heavy	20+	Platinum

4.2 THERMAL PROPERTIES: HEAT FLOW IN MINERAL CRYSTALS

The thermal properties of minerals hold great significance in the theoretical aspects of mineralogical reactions. These properties are of two types: those related to transfer of heat and those related to changes in heat content occurring during mineralogical reactions.

Table 4.11: Thermal properties in minerals

Properties	*Denotation*	*Description*
Specific heat	c [J $kg^{-1}K^{-1}$]	Heat required to raise the temperature of unit mass of material by one degree
Thermal conductivity	k or λ [W $m^{-1}K^{-1}$]	Heat flow per unit temperature gradient
Thermal diffusivity	K or α [m^2/s]	Governs rate of spread of temperature disturbances
Latent heat, coefficient of thermal expansion		Change in volume per unit volume for a temperature change of 1°C

Thermal Conductivity

Thermal conductivity is the amount of heat transmitted through a unit thickness in a direction normal to unit area surface, due to a unit temperature gradient under steady state conditions.

- Is a linear function of density for constant mean atomic weight.
- Is controlled by silicon-oxygen ratio in case of silicates.
- Decreases with the increase of mean atomic weight for a group of minerals of similar crystal structure and bonding.
- Minimum at an intermediate composition for a series of minerals, forming binary solid solution.
- Is related linearly to elastic wave velocities in case of silicates.

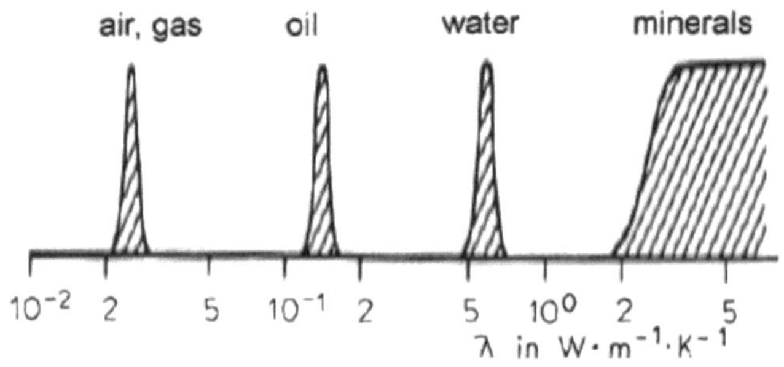

Fig. 4.2: Variation of thermal conductivity.

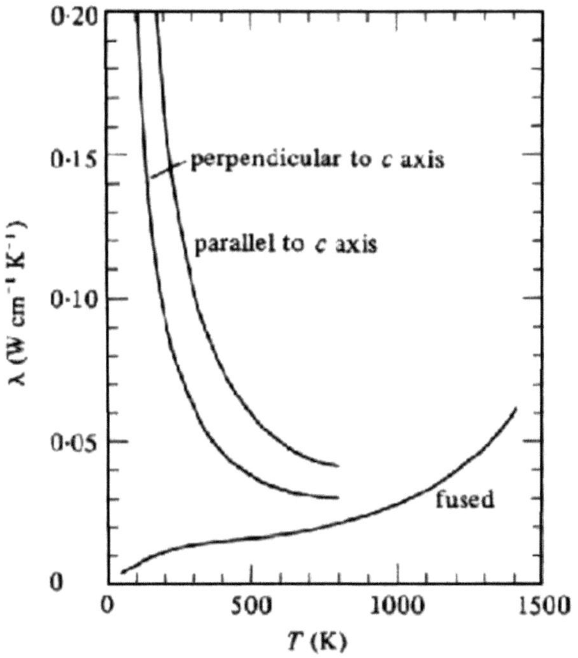

Fig. 4.3: Thermal conductivity in quartz.

Table 4.12: Thermal properties of some rock-forming minerals

Minerals	*Specific heat* λ *in* $W\ m^{-1}\ K^{-1}$	*Thermal conductivity* *c in* $kJ\ kg^{-1}\ K^{-1}$
Hematite	11.2-14.7	0.61
Magnetite	4.7-5.3	0.6
Pyrite	19.7	0.5
Fluorite	9.5	0.85
Chlorite	4.2-5.14	0.6
Halite	5.3-10	0.79-0.84
Apatite	1.37-1.4	0.7
Calcite	3.25-3.9	0.79-0.83
Dolomite	3.9-5.5	0.86-0.93
Anhydrite	4.6-5.75	0.52-0.62
Quartz	6.6-13	0.7-0.74
Vitreous silica	1.36	0.7
Olivines	3.16-5.06	0.55
Muscovite	1.7-2.32	0.76
Biotite	1.17-1.73	0.77
Clay min. (mean)	2.9	
Feldspar	2.3	
Orthoclase	2.31-3.2	0.67-0.75
Plagioclase (mean)	2.31	

4.3 MAGNETIC PROPERTY

There are many minerals that remain more or less affected by the magnetic field. Magnetism occurs when there is an imbalance in the structural arrangement of the iron ions. Iron occurs in two principal ionic states called ferrous and ferric ions. The ferrous ion has charge of (+2); the ferric ion has a charge of (+3). The two ions have different atomic radii. This fact can lead to the different ions being placed in separate positions in a crystal structure. Electrons that move from the ferrous to the higher positively charged ferric ions create a slight magnetic field. On the basis of magnetic property, minerals can be classified as in Table 4.13.

Table 4.13: Types of magnetism

Types	*Description*	*Examples*
Diamagnetic	Slightly repelled by a strong magnetic field	Calcite, zircon
Paramagnetic	Slightly attracted by a strong magnetic field	Pyrite, chalcopyrite (i.e. all Fe-bearing minerals), beryl, diopside
Ferromagnetic	Strongly attracted by simple bar magnets or horse-shoe magnet	Magnetite, pyrrhotite, kamacite, native iron, tacnite

4.4 ELECTRICAL PROPERTIES

Minerals can be grouped under different classes, depending on the electrical properties. The sub-divisions can be presented with the help of a chart for easy understanding of the student.

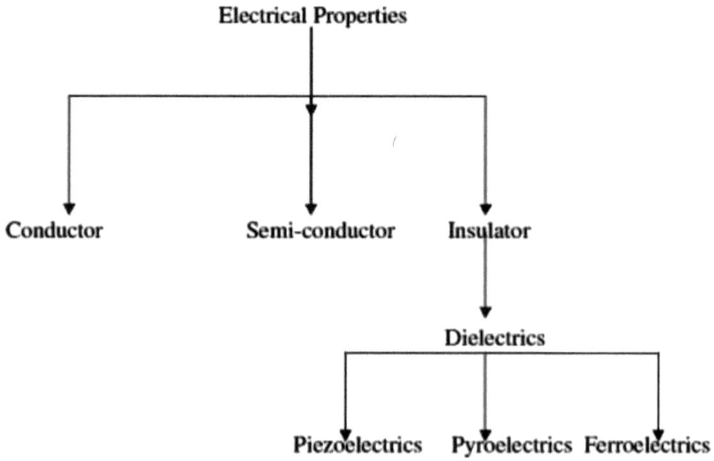

Fig. 4.4: Different electrical properties observed in minerals.

Conductor

Conduction in mineral terms is the ability of a mineral to conduct electricity. Only a very small number of minerals are good conductors; they are the metallic elements and the mineral graphite. This property mainly depends on the presence of free-migrating charge carriers within the material. Such carriers are electrons, which can freely migrate only in metallic and semimetallic bonds in case of perfect crystals. While in case of defective structures, relatively small amount of charge may be transferred. Conduction is an important property that can distinguish true metals from metallic looking sulfides and oxides.

Classification of minerals into semiconductors and dielectrics is possible, except for the native metals (conductors).

Semiconductor: Solids that conduct electricity poorly are called semiconductors. Mobility appears in two charge carriers; electrons in a normally unfilled conduction band and holes in a normally filled valence band in the presence of external electric and magnetic fields and temperature gradients. The specific conductance for semiconductors is denoted by ó. The main minerals to fall within the group of semiconductors are pyrite, arsenopyrite, chalcopyritre, galena and molybdenite.

Dielectricity

Dielectric constants: Dielectric constant (ε) is a number relating the ability of a material to carry alternating current to the ability of vacuum to carry alternating current. The capacitance created by the presence of the material is directly related to the dielectric constant of the material.

Factors influencing dielectricity

- Effect of the frequency of the radar beam**:** Expressed as function of frequency. Hydrothermal minerals have higher dispersion curves, while minerals with stable crystallographic structure and composition have lower dispersion.
- Effect of soil moisture: Water a dipolar liquid with high dielectric constant, i.e. at 23° C and $\lambda = 3.2$ cm, $\varepsilon'' = 61.5$ and $\varepsilon'' = 31.4$ (where ε' being the real part and ε'' the imaginary part). So presence of water in minerals has profound variation of dielectric constant in those minerals.
- Effect of chemical substitutions: Types of cation, anions, the ionic radii, the ionic polarizability directly control the polarization intensity and thereby the dielectric constant of a mineral.
- Effect of density: According to Olhoeft (1981), a linear relation between ε' and the density d of Hawaii basalt has been deduced as

$$\ln \varepsilon' = 0.839 + 0.524\, d$$

Piezoelectricity

A number of dielectric crystals has the property of acquiring polarization, i.e. dipole moments under externally applied mechanical stress. Electrifiction by stress is called a piezoelectric phenomenon. Piezoelectric solids are a subset of dielectric materials, i.e. all piezoelectrics are dielectrics but all dielectrics are not piezoelectrics. Examples of piezoelectric solids or crystals are crystalline quartz, Rochelle salt, barium-strontium titanate, lead-zirconium titanate, tourmaline etc. Crystals having a centre of symmetry do not show any piezoelectric property, e.g. crystals belonging to cubic system. Among 32 classes of crystals only 20 classes show piezoelectric effect. When pressure is applied to crystal it is deformed resulting in an altered position of the centre of the mass; however, the centre of charge distribution is unaffected. If the electrically neutral crystal has a centre of symmetry no polarization occurs due to applied stress. In general if a crystal is polarized due to application of an external electrical field, a mechanical strain is generated which is known as *electrostriction* (examples are ionic solids). The converse is not true in general. In a piezoelectric crystal stress develops polarization and vice-versa. Crystals without centre of symmetry and assymmetric charge distribution are piezoelectric. Piezoelectric effect varies from one direction to another in a crystal and in some specific directions it may not show any piezoelectric effect at all. But in some other directions it may be very pronounced. In a metal-oxygen tetrahedral complex if pressure is applied along metal-oxygen bond, polarization occurs, whereas no polarization occurs when pressure is applied perpendicular to the tetrahedral edge.

Pyroelectricity

Crystals which show spontaneous polarization in their natural primitive cells have non-vanishing dipole moments and are called pyroelectric. When such crystals are heated they become positively charged on one side and negatively on the other. The situation is reversed during cooling. Examples are tourmaline and crystalline quartz. Pyroelectric crystals are a subset of piezoelectric crystals, i.e. all pyroelectrics are piezoelectric but all piezoelectrics are not pyroelectric. Pyroelectric crystals possess permanant dipole moment in ordinary state which, however, is not manifested due to neutralization by surface charges under external air exposure. When these crystals are heated some of the surface charges dissappear manifesting polarization. The dipole moment may change due to other reasons. Out of 32 crystal classes only 11 crystal classes show pyroelectricity.

Table 4.14: Symmetry groups in which pyroelectricity may be possible

Point group	*Polar direction*
1	Every direction
2	[010]
m	All directions in the (010) plane
mm2, 3, 3m, 4, 4mm, 6, 6mm	[001]

Ferroelectricity: The relative permittivity of normal crystals (or dielectrics) is low. Some crystals exhibit relative permittivity several orders of magnitude larger than ordinary dielectrics, e.g. $BaTiO_3$ in one polymorph has $\varepsilon \sim 20{,}000$, Rochelle salt $\sim 5 \times 10^3$ etc. By analogy with magnetic properties, this behaviour is called ferroelectricity and the materials are called ferroelectrics (it has nothing to do with iron).

Ferroelectric materials possess spontaneous polarization in absence of electric field or mechanical stress. Ferroelectrics are a subset of pyroelectrics, i.e. all ferroelectric materials are pyroelectric and hence piezoelectric but all pyroelectric or piezoelectric materials are not ferroelectric. The feature that distinguishes a ferroelectric crystal from a pyroelectric is that the direction of spontaneous polarization can be changed under an applied field. Thus ferroelectric crystals exhibit hysteresis in an applied field. Spontaneous polarization and hysteresis are essential characteristics of a ferroelectric material.

Ferroelectrics have wide applicability in microelectronic devices.

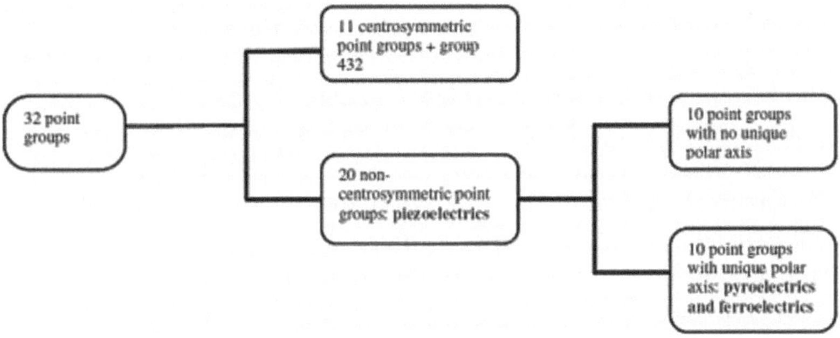

Fig. 4.5: Relationship between point groups and piezoelectric and pyroelectric properties.

4.5 RADIOACTIVE PROPERTIES

Mineral radioactivity is due to alpha, beta and gamma radiation from the unstable isotopes in the composition.

Alpha decay is due to the ejection of a helium nucleus (two protons and two neutrons) from the parent isotope. This alpha particle may or may not be accompanied by gamma radiation and a daughter isotope which is two protons and two neutrons lighter than the parent isotope. Beta decay is due to the ejection of an electron from a neutron in the parent nucleus. This particle may also be accompanied by gamma radiation and a daughter isotope which is one proton heavier and one neutron lighter than the parent isotope. Electron Capture (EC) decay is the result of the nucleus capturing one of the atom's orbital electrons. This decay is accompanied by gamma radiation and a daughter isotope which is one neutron heavier and one proton lighter than the parent isotope.

Radioactive minerals are discrete minerals of uranium and/or thorium as exemplified by uraninite (of U) and thorite (of Th). Besides they include those in which either U or Th or both occur in notable quantity like the minerals of Rare Metals (Nb-Ta, Li, Be, Sn, W etc.) and Rare Earth Elements (REE: La to Lu, Y).

Radioactive minerals are broadly divided into primary and secondary discrete minerals of U and Th, and nearly 200 such minerals are known. The primary minerals are those formed directly from magmas, hydrothermal solutions and groundwater. Secondary minerals are formed due to remobilization of elements from primary minerals, their transportation in solution and later precipitation due to over-saturation in oxidizing or supergene environment.

The most common primary minerals of uranium are uraninite (pitchblende, if microcrystalline) (oxide), coffinite (silicate) and brannerite (complex oxide). In these, U occurs mostly in U^{4+} state, besides some U^{6+} in pitchblende. Secondary minerals of U^{6+} uranium occurring in supergene conditions are many (~180) in which U occurs entirely in the U^{6+}. This includes various oxides and hydrated oxides, silicates, vanadates, carbonates, sulphates, molybdates, phosphates and arsenates, and their complex derivatives. The most common primary minerals of thorium are thorianite, thorouraninite (oxides) and thorite/uranothorite (silicates). A few secondary (supergene) minerals of Th are known, the most common being thorogummite. Thorium occurs in several minerals, the most common being the rare earth-thorium-phosphate mineral, monazite, which contains up to about 12% thorium oxide.

Radioactive minerals are unstable, meaning the elements in their structure continually break down. This destroys the mineral crystal lattice, causing it to lose its crystal shape. When this happens, its crystal edges become rounded, and the mineral eventually becomes amorphous. During this process, it also becomes opaque and develops a pitchy lustre. Minerals that have gone through this breakdown process are known as metamict.

Half-life: Half-life is defined as *the time required for one-half of the radioactive (parent) isotopes in a sample to decay to radiogenic (daughter) isotopes.* Mathematically, the half-life can be represented by an exponential function.

Occurrence of Radioactive Minerals

Radioactive minerals occur in very low content in different rock types, viz., in (i) magmatic or igneous rocks (~2 ppm in the crust), (ii) sedimentary rocks and (iii) metamorphic rocks of low grade like phyllite and schist. In the magmatic rocks, they usually concentrate in acidic, plutonic rocks which include granitoids-pegmatite and volcanic rocks like rhyollites and acid tuffs. Radioactive minerals like monazite and zircon also occur in placer derived from such magmatic rocks. Amongst the sedimentary rocks, sandstones, pyrite-bearing quartz-pebble conglomerates and phosphatic and carbonaceous rocks are good hosts for uranium minerals. Low grade metamorphic rocks like phyllite and schist subjected to metasomatic alterations involving addition of volatiles within the structurally weak zones are loci for uranium minerals. Thorium minerals occur generally in acidic igneous rocks and high-grade metamorphic rocks and placers derived from these rocks.

Radioactive Clocks

Uranium decays through several stages, all of which are radioactive, to lead, giving off helium in the process. Radioactivity is a constant process, repeatedly halving its value in a period, aptly called the 'half-life', that can be accurately determined in the laboratory. It can therefore provide a very accurate 'atomic clock'.

Two such processes that are particularly important to geology are radioactive isotopes of potassium (K) and rubidium (Rb).

The half-life of potassium-40 is 4.5 billion years, very similar to the 4.6 billion years estimated age of the Earth. During radioactive decay, a potassium-40 nucleus captures an orbital electron, converting a proton into a neutron, thereby changing to argon-40. Determining the ratio of argon-40 to potassium-40 in a mineral sample can provide a very accurate clock for dating ancient rocks.

Concluding remarks

Depending upon the kinds of atoms, their arrangement and the nature of the chemical bonding present, the physical properties of minerals vary widely. Apart from their implication in mineral identification, physical properties are significantly used for various applications and academic studies.

Think for a while

1. Hardness varies with change in crystallographic direction within the same mineral. Justify the statement citing suitable example.
2. How heat flow can be controlled within mineral crystals?
3. How will you distinguish between following minerals with the help of physical properties:
   A. Calcite and Aragonite
   B. Hematite and Magnetite
   C. Pyrite and Chalcopyrite
   D. Galena and Sphalerite
4. Enumerate the factors that control the development of cleavage in a crystalline substance with particular reference to pyroxene and mica.

FURTHER READING

Dixon, D.W. Radiation hazards to collectors of geological specimens containing natural radioactivity. *National Radiological Protection Board Report* 131 (UK): 23 pp. 1983

Klein, C. and C.S. Hurlbut, Jr. Manual of Mineralogy, 21st ed. John Wiley and Sons, New York. 1993.

Perkins, D. *Mineralogy*, 2nd ed. Prentice Hall of India. 2002.

Nassau, K. The origin of colour in minerals. *American Mineralogist*, **63,** 219-229. 1978.

Zussman, J. Physical Methods in Determinative Mineralogy. Academic Press, New York. 1977.

Internet Data Retrieved from:

Examples of Fracture and Cleavage - Treasure Hunting Wiki.htm

GeologyData_Info- Info Portal of Geology with special reference to Rajasthan, India.htm

CHAPTER

5

OPTICAL MINERALOGY AND ITS USES

Optical mineralogy is the study of minerals by measuring their optical properties. Most commonly, rock and mineral samples are prepared as thin sections or grain mounts for study in the laboratory with a petrographic microscope. Optical mineralogy is used to identify the mineralogical composition of geological materials in order to help reveal their origin and evolution. Some of the properties and techniques used include: refractive index, birefringence, Michel-Lévy colour chart, pleochroism, extinction angle, conoscopic interference pattern and Becke line test—to be covered in this chapter.

Optical mineralogy is the optical microscopic study of minerals for identification and also for study of properties. This identification of mineralogical compositions of different rock-types and study of texture, structure etc. in turn are useful in deciphering the information about their origin and evolution. The investigation in optical mineralogy is done under polarizing microscope, by transmitting light.

5.1 POLARIZED LIGHT AND POLARIZING MICROSCOPE

Light

Light is electromagnetic radiation that has properties of waves. The electromagnetic spectrum can be divided into several bands based on the wavelength. Visible light represents a narrow group of wavelengths between about 380 nm and 730 nm. Light waves are described in terms of velocity, frequency and wavelength.

The velocity (V) and the wavelength are related in the following equation,

$$F = V/\lambda$$

where F = frequency, V = velocity of the wave, λ = wavelength and A = amplitude as shown in Fig. 5.1.

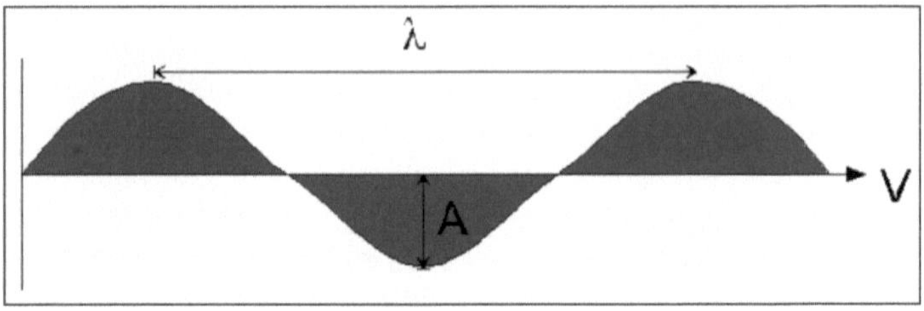

Fig. 5.1: Relationship between frequency, velocity and wavelength.

Reflection and Refraction

When light strikes an interface between two substances with different refractive indices, two things occur. An incident ray of light striking the interface at an angle, i, measured between a line perpendicular to the interface and the propagation direction of the incident ray, will be reflected off the interface at the same angle, i. In other words the angle of reflection is equal to the angle of incidence. If the second substance is transparent to light, a ray of light will enter the substance and will be refracted, or bent, at an angle r, the angle of refraction, and thus bending will depend on the refractive index n of the medium.

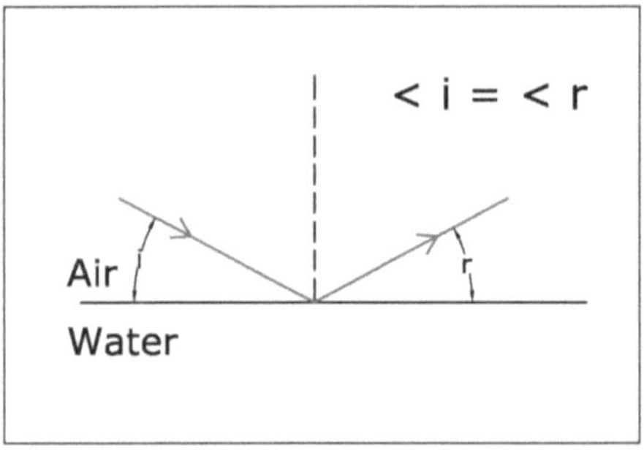

Fig. 5.2a: Reflection.

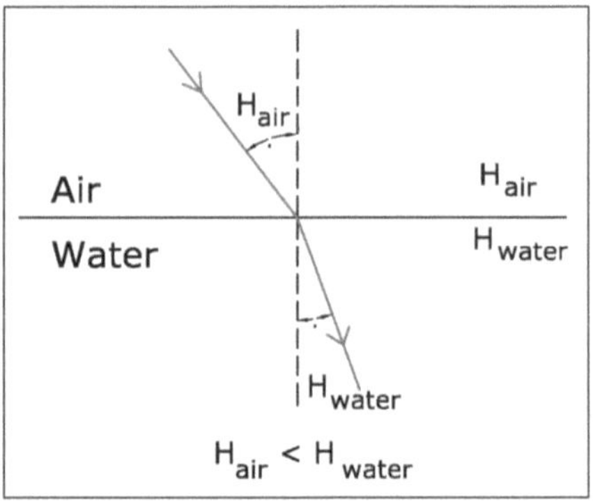

Fig. 5.2b: Refraction.

The **Index of Refraction** (n), is the effectiveness of bending the light ray by a mineral/material where $n = V_v/V_m$, V_v = velocity of light in vacuum and V_m = velocity of light in the material.

Snell's Law defines the amount of bending that takes place when a light ray strikes a refractive boundary, e.g., an air-glass interface, at an angle other than 90°.

Snell's Law states that

$$\sin \theta_1 / \sin \theta_2 = n_2/n_1$$

where n_1 is the index of refraction of the medium in which the incident ray travels, θ_1 is the angle, with respect to the normal at the refractive boundary, at which the incident ray strikes the boundary, n_2 is the index of refraction of the medium in which the refracted ray travels, and θ_2 is the angle, with respect to the normal at the refractive boundary, at which the refracted ray travels. The incident ray and refracted ray travel in the same plane, on opposite sides of the normal at the point of incidence.

Snell's law can be used to calculate the amount of bending of incident light while travelling into the new medium.

As in reflection, the angles from the normal to the surface, at the point of contact can be measured. The constants n are the indices of refraction for the corresponding media.

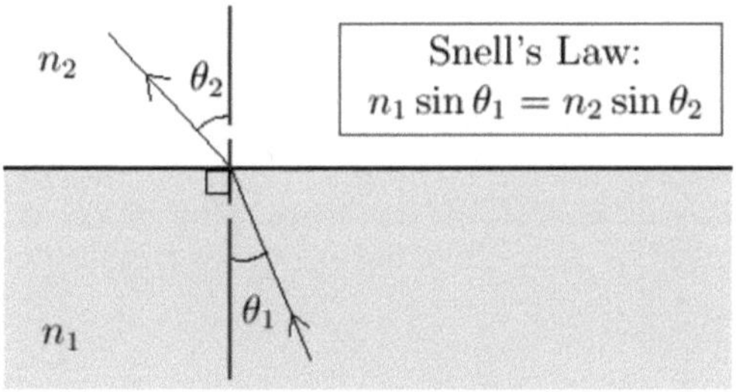

Fig. 5.3: Snell's Law.

To uderstand the law better, a table of refractive index for the common media is given.

Table 5.1: n for light of wavelength 600 nm

Substance	*Refractive Index, n*
Air (1 atmosphere pressure, 0 degree C)	1.00029
Water (20 degrees C)	1.33
Crown glass	1.52
Flint glass	1.66

In our simple example above, suppose we shine a light of wavelength 600 nm from water into air, so that it makes a 30° angle with the normal of the boundary. If we wish to find the angle x that the outgoing ray makes with the boundary, Snell's Law gives

$$1.33 \sin 30^\circ = 1.00029 \sin x$$
$$x = 41^\circ$$

Moreover in case of a more complex case, Snell's Law can be illustrated in the following way. If you stand behind a window made of uniform glass, then you know by now that the images of the things on the other side of the window have been refracted. Assuming that the air on both sides of your window has the same refractive indices, we have the situation as explained in Fig. 5.4.

We find that the incoming and outgoing light beams are actually parallel.

Rearranging Snell's Law, with *i* and *r* being the incident and refracted angles,

$$n_1\sin(i) = n_2\sin(r)$$
$$(n_1/n_2)\sin(i) = \sin(r)$$

a qualitative description of refraction becomes clear. When we are travelling from an area of higher index to an area of lower index, the ratio n_1/n_2 is greater than one, so that the angle *r* will be greater than the angle *i*, i.e. the refracted

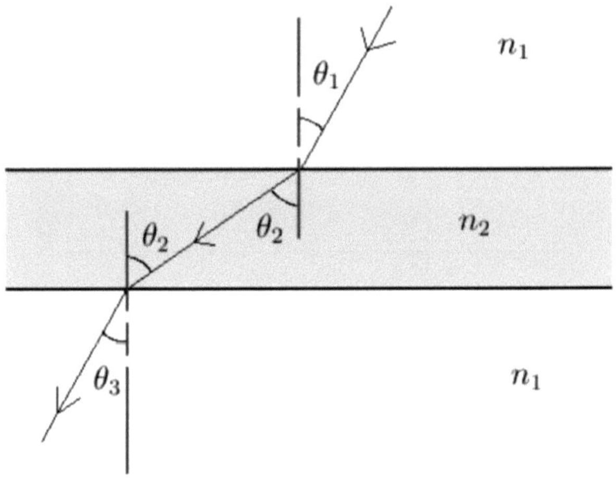

$$\begin{aligned}
\theta_2 &= \sin^{-1}\left(\frac{n_1}{n_2}\sin\theta_1\right) \\
\theta_3 &= \sin^{-1}\left(\frac{n_2}{n_1}\sin\theta_2\right) \\
&= \sin^{-1}\left(\frac{n_2}{n_1}\sin\left(\sin^{-1}\left(\frac{n_1}{n_2}\sin\theta_1\right)\right)\right) \\
&= \sin^{-1}(\sin\theta_1) \\
&= \theta_1
\end{aligned}$$

Fig. 5.4: Complex illustration of Snell's Law.

ray is bent away from the normal. When light travels from an area of lower index to an area of higher index, the ratio is less than one, and the refracted ray is smaller than the incident one; hence the incident ray is bent toward the normal as it hits the boundary.

There is also an angle, *ic*, called the ***critical angle for total internal reflection*** where the refracted ray travels along the interface between the two substances.

In the diagram beside, imagine that we are trying to send a beam of light from a region with refractive index n_1 to a region with index n_2 and that $n_2 <$

n_1. If i and r are the angles made with the normal for the incident and refracted rays, then Snell's Law yields

$$i = \sin^{-1} (n_1/n_2 \sin r)$$

This occurs when the angle $r = 90°$. In this case, applying Snell's law: $n_i \sin (ic) = n_r \sin (90°) = n_r$ [since $\sin (90°) = 1$]

$$\sin (ic) = n_r/n_i$$

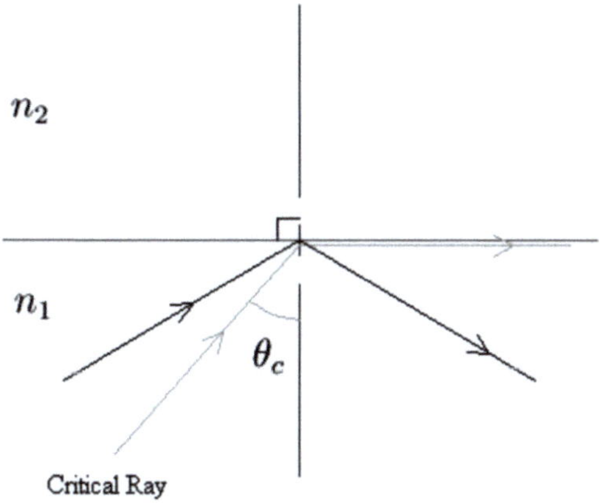

Fig. 5.5: Illustration of Snell's Law.

Polarized Light

Light waves vibrate perpendicularly to its propagation direction. Unpolarized light waves vibrate in different directions in a plane perpendicular to the propagation direction. Polarisation makes these light waves to vibrate in one direction, parallel to a particular plane. In other words, the electromagnetic radiation theory of light implies that light consists of electric and magnetic components which vibrate at right angles to the direction of propagation. In optical mineralogy only the electric component, referred to as the electric vector,

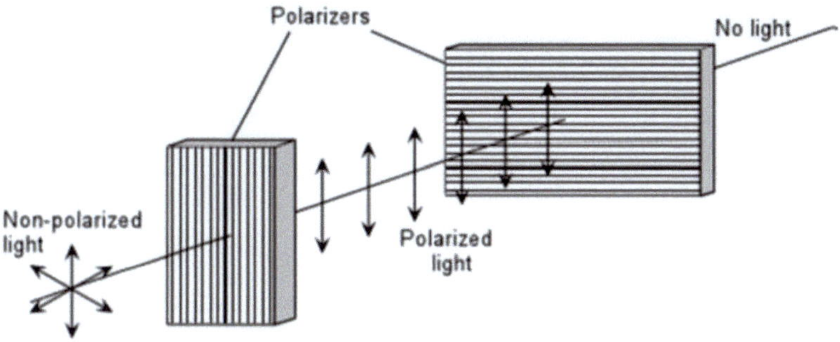

Fig. 5.6: Polarization of light.

Optical classification of minerals

- **Isotropic** (velocity of light is same in all directions)
- **Anisotropic** (velocity of light is different in different directions)
  - **Uniaxial** (has a single optic axis)
  - **Biaxial** (has two optic axes)

is considered and is referred to as the **vibration direction** of the light ray. The vibration direction of the electric vector is perpendicular to the direction in which the light is propagating.

Polarizing Microscope

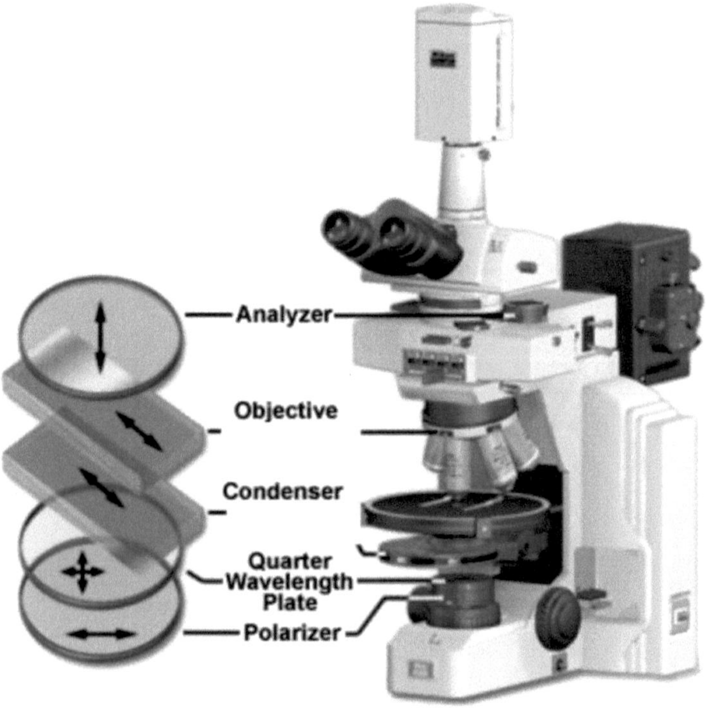

Fig. 5.7: Optical microscope.

Table 5.2: Parts of an optical microscope

Parts	*Functions*
Microscopic stage	A rotating stage, beneath which there is a polarizer
Objective lenses	Objective lenses of magnification, ranging from about 2× to 50× set above the stage
Condenser	A part above the polarizer, which is used to condense parallel beams of light to a particular point
Upper polarizer	The upper polarizer, also known as the analyzer, is set above the objectives and is oriented at 90° to the lower polarizer, which can be inserted or removed from the path of the light beam. Without the upper polarizer, the mineral is said to be viewed under plane polarized light and with the upper polarizer, the mineral is viewed under crossed polars. The optical properties of minerals are different for these two above-mentioned conditions.
Accessory plate	Above the analyzer, there is space for insertion of accessory plates. The most common kind of accessory plates are quartz plate, gypsum plate and mica plate.
Bertrand lens	Above the accessory plates, generally a diaphragm is provided, which enables to view minerals in conoscopic illumination.

Mechanisms by which light waves are constrained within a microscope are shown through Figs 5.8a-5.8d.

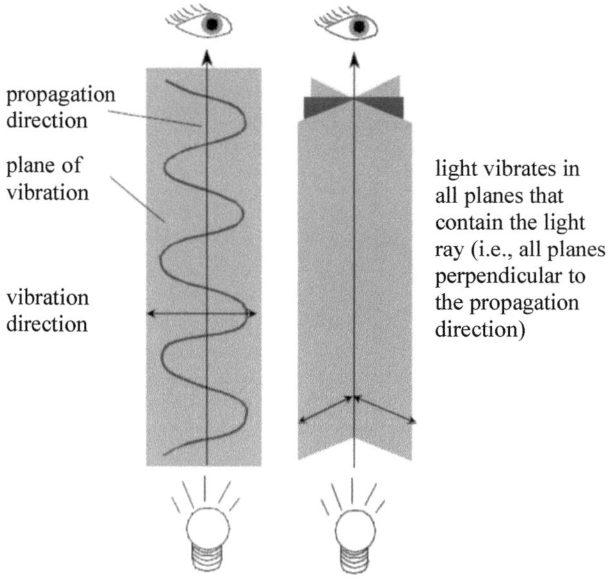

Fig. 5.8a: Behaviour of light.

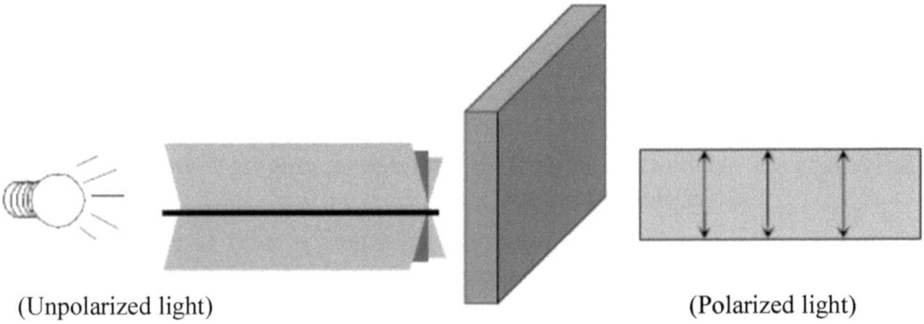

Fig. 5.8b: Light passes through lower polarizer.

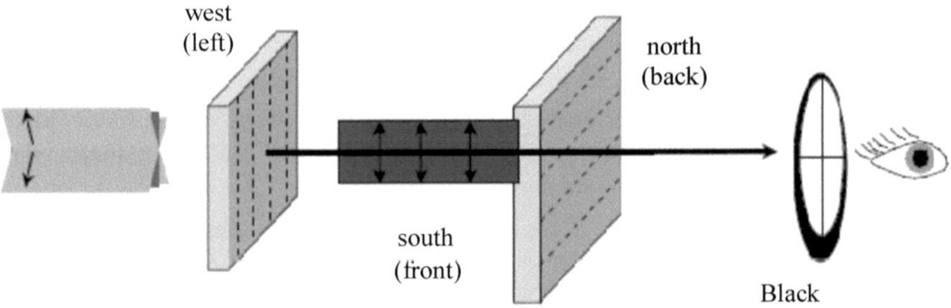

Fig. 5.8c: Upper polarizer is inserted.

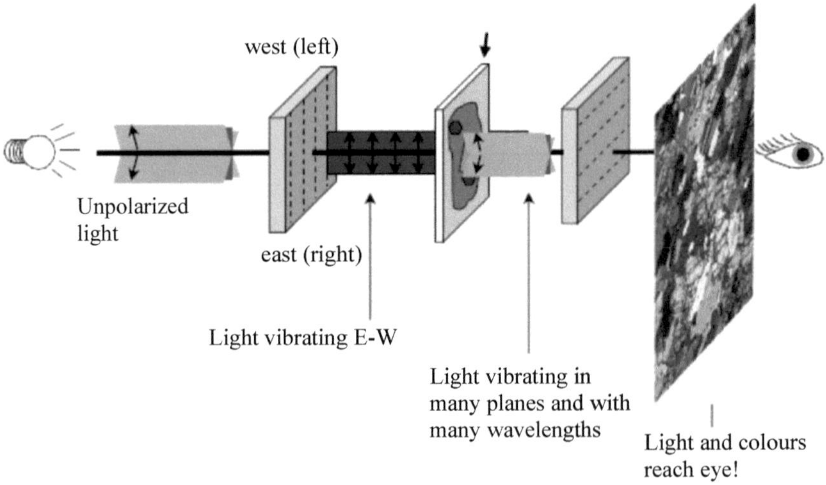

Fig. 5.8d: A thin section of rock is inserted.

Minerals can be subdivided, based on the interaction of the light ray travelling through the mineral and the nature of the chemical bonds holding the mineral together, into two classes: **Isotropic minerals** and **Anisotropic minerals**.

Isotropic materials show the same velocity of light in all directions because the chemical bonds holding the minerals together are the same in all directions, so light travels at the same velocity in all directions. Examples of isotropic material are volcanic glass and isometric minerals (cubic): fluorite, garnet and halite. In isotropic materials the wave normal and light ray are parallel.

Anisotropic minerals have a different velocity for light, depending on the direction the light is travelling through the mineral. The chemical bonds holding the mineral together will differ depending on the direction the light ray travels through the mineral. Anisotropic minerals belong to tetragonal, hexagonal, orthorhombic, monoclinic and triclinic systems. In anisotropic minerals the wave normal and light ray are not parallel.

5.2 OPTICAL PROPERTIES

Optical Properties Studied under Plane Polarized Light

Colour: The minerals display a variety of colours under microscope under transmitted light. Some are colourless (quartz, calcite, felspar, muscovite, etc.). Some minerals are yellow or brown (rutile, tourmaline, biotite), some are green (diopside, hornblende, chlorite), blue (glaucophane), pink (garnet), etc. The colour of minerals are found to vary with the orientation in which they are cut in respect to the crystallographic directions.

Pleochroism: The phenomenon of selective absorption rate of a mineral with change of its vibration direction is called pleochroism and the minerals exhibiting this property are called pleochroic, like biotite, tourmaline, clinopyroxene, staurolite etc. Due to selective absorption of different wavelengths at different vibration directions, the mineral changes its colour when the stage is rotated slowly in PPL.

Shape and size: Grain shapes or outline and their generalized size (fine, medium or coarse-grained) can be identified.

Cleavage: If the mineral has one or more good cleavages they will be indicated by systems of cracks, like that of biotite, muscovite, hornblende etc.

Refractive index: The refractive index can be identified by the appearance of the rough section, with well-defined borders if they have a much stronger refraction than the medium in which they are mounted. Refractive index can be low (borax), medium (spinel, garnet) or high (olivine, zircon). It may be defined as the ratio of the velocity of light in vacuum to that in the medium and is given by $n = c/v$.

- R.I can be explained by the assumption that light is expressed by two vectors—electric vector E and the magnetic vector B; both being normal to the direction of propagation of light wave.

- E generates electric dipoles in a crystal as a result of distortion of internal charge distribution, which in turn developes a net dipole moment per unit volume, termed the electric polarization P. The relationship between E and P is shown as

$$(\varepsilon - \varepsilon_0)\, E = P$$

where ε = permittivity of the medium and ε_0 = permittivity of the vacuum.

- Electric dipoles results in an additional electric field component, resulting in an electric flux density D within the medium, given by $D = \varepsilon\, E$.
- Relative permittivity of the medium $\varepsilon/\varepsilon_0$ is known as dielectric constant K of the medium, that is $D = \varepsilon_0 KE$.
- K is useful to describe the optical properties in terms of variation of R.I with direction.
- R.I of a non-magnetic material is equal to ½K.

Table 5.3: Symmetry control on the variation of refractive index

System	*Variation of refractive index direction*
Isometric	Sphere; one value of R.I needed
Tetragonal, hexagonal, trigonal	Uniaxial Ellipsoid; two principal values of R.I needed
Orthorhombic	Biaxial ellipsoid; three principal values of R.I needed
Monoclinic	One axis of indicatrix parallel to y-axis and other two lie on xz-plane
Triclinic	No constraints given

Becke Line Test

In order to determine whether the index of refraction of a mineral is greater than or less than the mounting material the Becke Line Method is used. A band or rim of light visible along the grain boundary in plane light when the grain mount is slightly out of focus.

Becke line may lie inside or outside the mineral grain depending on how the microscope is focussed. The direction of movement of the Becke line is determined by lowering the stage, with the Becke Line always moving into the material with the higher refractive index. The Becke Line can be considered to form from a cone of light that extends upwards from the edge of the mineral grain. By changing focus, the movement of the Becke line can be observed. If focus is sharp, such that the grain boundaries are clear, the Becke line will coincide with the grain boundary. Becke line will always move towards the material of higher RI upon lowering the stage.

The Becke lines observed are interpreted to be produced as a result of the lens effect and/or internal reflection effect.

Inclusions: The presence of any secondary or foreign particles, like glass, fluids including gases can also be identified.

Optical Properties Studied under Crossed Polars

Double refraction: In the phenomenon of double refraction a ray of unpolarized light is split into two rays at right-angles to each other. For uniaxial minerals, the ordinary ray or O ray (ω) and the extraordinary ray or E ray (ε).

Retardation represents the distance that one ray lags behind another. Retardation is measured in nanometres, 1 nm = 10^{-7}cm, or the number of wavelengths by which a wave lags behind another light wave.

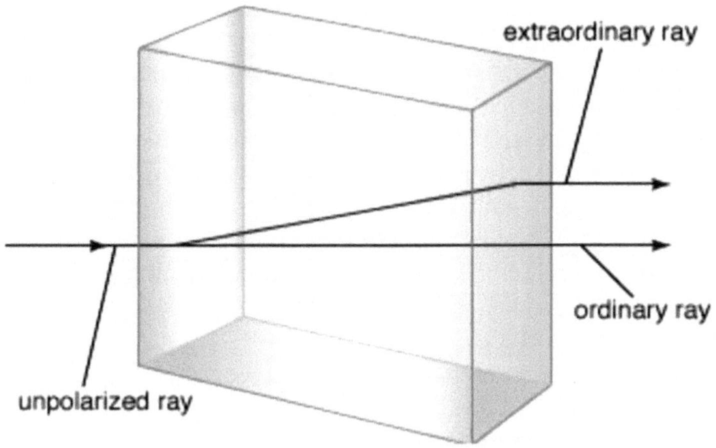

Fig. 5.9: Phenomenon of double refraction.

Monochromatic ray, of plane polarized light, upon entering an anisotropic mineral is split into two rays, the *fast* and *slow* rays, which vibrate at right angles to each other. Due to differences in velocity the slow ray lags behind the fast ray, and the distance represented by this lagging after both rays have exited the crystal is the retardation *D*.

The magnitude of the retardation depends on the thickness (*d*) of the mineral and the differences in the velocity of the slow (V_s) and fast (V_f) rays. The time it takes the slow ray to pass through the mineral is given by:

$$t_s = d/V_s \tag{5.1}$$

During this same interval of time the fast ray has already passed through the mineral and has travelled an additional distance = retardation.

$$t_s = d/V_f + \Delta/V \tag{5.2}$$

Substituting (5.1) in (5.2), yields

$$d/V_s = d/V_f + \Delta/V$$

Rearranging

$$\Delta = d\,(V/V_s - V/V_f)$$

Or, $$\Delta = d(n_s - n_f) \text{ since, } V/V_s = n_s \text{ and } V/V_f = n_f$$

The relationship $(n_s - n_f)$ is called birefringence. It represents the difference in the indices of refraction for the slow and fast rays.

Birefringence: After understanding the retardation property, birefringence can be illustrated in the following way. The two rays are known as slow ray and fast ray, since they travel with differential velocities, unless they are travelling parallel to an optic axis. Hence the refractive indices of the two rays are also different. The difference in their refractive indices is called birefringence. The property called birefringence has to do with anisotropy in the binding forces between the atoms forming a crystal, so it can be visualized as the atoms having stronger 'springs' holding them together in some crystalline directions. A useful situation with mineral crystals occurs when there are two distinct indices of refraction, and they are called birefringent materials. This is associated with uniaxial crystals, which belong to the hexagonal, tetragonal, and trigonal crystal systems. In a uniaxial crystal, there is one direction such that any light in that direction in the crystal has the same speed, regardless of its state of polarization. This direction is called the optic axis. The remaining crystal systems (orthorhombic, monoclinic, and triclinic) have two optic axes and are said to be biaxial.

Two indices of refraction characterize birefringent materials. Such materials form uniaxial crystals, and may be characterized as positive or negative uniaxial crystals based on the comparison of the indices.

Table 5.4: Minerals with their corresponding indices for O-ray and E-ray

Crystal	N_o	N_e
Tourmaline	1.669	1.638
Calcite	1.6584	1.4864
Quartz	1.5443	1.5534
Sodium nitrate	1.5854	1.3369
Ice	1.309	1.313
Rutile (TiO_2)	2.616	2.903

Calcite: An Important Birefringent Mineral

Calcite is a crystalline form of calcium carbonate, $CaCO_3$, and the crystal is a lattice of positively charged calcium and negatively charged carbonate ions. The very large optical effects shown by calcite arise because the flat carbonate ions are arranged in parallel sheets. Light travelling perpendicular or parallel to them has very different electronic interactions. The crystal is anisotropic and its optical properties are direction dependent.

Light entering the crystal splits into two polarised components because the differently oriented electric vectors of the radiation interact differently with the induced dipoles of the carbonate oxygen.

One ray, called the ordinary or O ray, behaves in the familiar way of light passing through glass or water (as predicted by Snell's law of refraction). The second—the extraordinary or E ray—is indeed extraordinary, it is refracted through a different angle to the O ray. Worse, Snell's Law does not predict the refraction because its extent, given by the refractive index, is direction dependent. The mineral calcite, also known as Iceland spar, is a widely used material in optics because of its birefringence. Its birefringence is so large that a calcite crystal placed over a dot on a page will reveal two distinct images of the dot. One image will remain fixed as the crystal is rotated, and that ray through the crystal is called the 'ordinary ray' since it behaves just as a ray through glass. However, the other image will rotate with the crystal, tracing out a small circle around the ordinary image. This ray is called the 'extraordinary ray'.

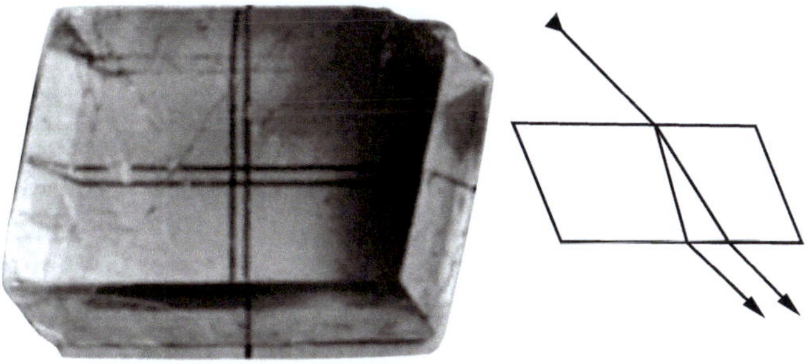

Fig. 5.10: Double refraction of calcite.

The indices of refraction for the O- and E-rays are 1.6584 and 1.4864 respectively. This gives total internal reflection critical angles of 37.08° for the O- and 42.28° for the E-rays when in contact with air. This means that for any angle between these two values, the O-ray will be totally reflected but the E-ray will be partially transmitted. This gives linear polarization since only the E-ray emerges.

Calcite is used in polarizing prisms such as the Nicol prism, the Glan-Foucault prism, and the Wollaston prism.

Applications of Birefringence

Birefringence is widely used in

- Optical devices, such as liquid crystal displays, light modulators, colour filters, wave plates, optical axis gratings, etc.
- It also plays an important role in second harmonic generation and many other nonlinear processes.

- It is also utilized in medical diagnostics: needle aspiration of fluid from a gouty joint will reveal negatively birefringent urate crystals.
- *Elastic birefringence* is observed in anisotropic elastic materials. In these materials, shear waves split. The study of birefringent shear waves in the earth is a part of seismology.
- Birefringence is also used in optical mineralogy to determine the chemical composition, and history of minerals and rocks.
- Birefringent materials are used in many devices which manipulate the polarization of light, such as wave plates, polarizing prisms, and Lyot filters.

Table 5.5: Electromagnetic waves in an anisotropic material

Propagation direction	*Ordinary ray*		*Extraordinary ray*	
	Polarization	n_{eff}	*Polarization*	n_{eff}
z	xy-plane	n_o	*n/a*	*n/a*
xy-plane	xy-plane	n_o	z	n_e
xz-plane	y	n_o	xz-plane	$n_e < n < n_o$
Other	Analogous to xz-plane			

Extinction: The phenomenon where the anisotropic mineral goes to complete darkness on rotation of the stage, after every 90°, is called extinction. It refers to that position where the vibration direction of the mineral section happens to coincide with the vibration direction of the polarizers.

Types of Extinction

1. Parallel Extinction

The mineral grain is extinct when the cleavage or length is aligned with one of the crosshairs.

The extinction angle (EA) = 0°, e.g., orthopyroxene or biotite.

2. Inclined Extinction

The mineral is extinct when the cleavage is at an angle to the crosshairs, i.e., EA > 0°, e.g., clinopyroxene or hornblende.

3. Symmetrical Extinction

The mineral grain displays two cleavages or two distinct crystal faces. It is possible to measure two extinction angles between each cleavage or face and the vibration directions. If the two angles are equal, then symmetrical extinction exists.

$$EA_1 = EA_2 \text{ e.g. amphibole or calcite.}$$

4. No Cleavage

Minerals which are not elongated or do not exhibit a prominent cleavage will still go extinct every 90° of rotation, but there is no cleavage or elongation direction from which to measure the extinction angle, e.g., quartz or olivine.

Undulatory Extinction

Deformation of minerals can result in strain of the crystal structure, which causes different parts of the same mineral to have different crystallographic and therefore optical orientations. When this occurs, the parts of the crystal with different orientations will go extinct at different rotational positions. This is referred to as undulatory extinction. It is common in quartz found in metamorphic rocks.

Interference colours: Apart from the extinction positions, the anisotropic minerals display various colours under crossed polars. These colours are known as interference colours, which result from the retardation of different wavelengths, which in turn depends on the orientation, birefringence and thickness of a crystal. Interference colours can be of different order for different minerals, depending on the amount of retardation.

Order	*Colour*	*Retardation*
First order	Grey to yellow	200-550 nm
Second order	Violet to red	> 550 nm
Third order	Violet to red again	After every 550 nm

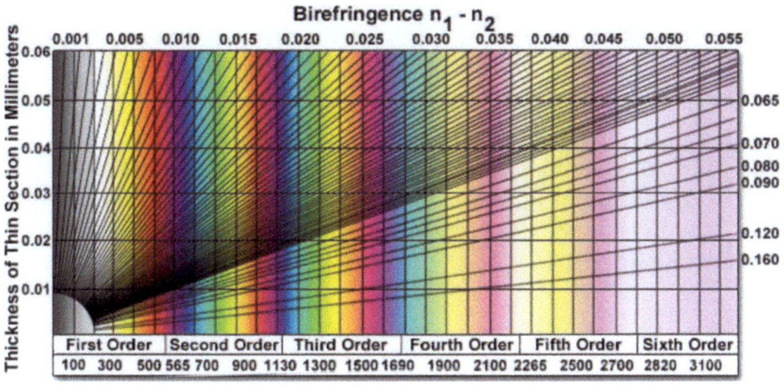

Fig. 5.11: Michel–Levy Birefringence chart.

A relationship between interference colour and retardation can be graphically illustrated in the classical Michel-Levy interference colour chart, presented above. This graph plots retardation on the abscissa and specimen thickness on the ordinate. The Michel-Levy chart is utilized by comparing the highest-order interference colours displayed by the specimen in the microscope to those contained on the chart. Once the appropriate colour has been located, the nearest vertical line along the interference colour is followed to the nearest horizontal line representing the known thickness. Birefringence is determined by selecting the diagonal line crossing the ordinate at the intersection of the specimen interference colour and thickness value.

Twinning: Twinning occurs when two separate crystals share some of the same crystal lattice points in a symmetrical manner. The result is an intergrowth of two separate crystals in a variety of specific configurations.

Table 5.6: Types of twins

Twins	*Description*	*Examples*
Contact twin	Have a planar composition surface separating two individual crystals	Orthoclase
Penetration twin	Have an irregular composition surface separating two individual crystals	Staurolite, pyrite, fluorite
Polysynthetic twin	The compositions surfaces are parallel to one another	Plagioclase, calcite
Cyclic twin	Composition surfaces are not parallel to one another but forms almost circular patterns	Crysoberyl, rutile, gypsum

5.3 ANISOTROPY AND CRYSTAL STRUCTURE

Anisotropy is most easily observed in single crystals, in which atoms, ions or molecules are arranged in regular lattices. Thus this property is mainly attributed to its crystal structure, rather than its symmetry.

Table 5.7: Dependency of anisotropy on crystal structure, observed in some minerals

Minerals	*Cause*	*Effect*	
Graphite	C-C interatomic distance between layers 3.35 Å and interatomic distance within layers is 1.42 Å.	Normal to layers	Electrical resistivity 5×10^{-3} Ωm; Semiconductor; Thermal expansion coefficient at 600 K, 28.3×10^{-6} K^{-1};
		Parallel to layers	Electrical resistivity 5×10^{-6} Ωm; Metallic conductor; Thermal expansion coefficient at 600 K, almost zero
Calcite	Ca atoms in between CO_3 groups. Internal electric field is distorted more within the planes than normal to the planes	Normal to layers	R.I is 1.486
		Parallel to layers	R.I is 1.658

5.4 DETERMINATION OF EXTINCTION ANGLE AND SIGN OF ELONGATION

Determination of Vibration Direction of Polarizer

In general, the lower polarizer is set to allow light to pass only in the E-W direction (i.e., the vibration direction is constrained to this orientation, thus producing plane polarized light). The simplest way to check this is to use a thin section of the mineral biotite. Biotite is a phyllosilicate mineral with one excellent direction of cleavage. In addition, biotite strongly absorbs light that vibrates parallel to its cleavage direction.

Thus in plane polarized light, biotite should be darkest (generally dark brown) when the cleavage is oriented parallel to the vibration direction of the lower polarizer and lightest (very light brown to yellow) when the cleavage is at right angles (90°) to the vibration direction of the lower polarizer.

Determination of Extinction Angle

The angle between a cleavage (or direction of elongation) and extinction is called extinction angle.

To measure the extinction angle, the microscopic stage is rotated so that length or cleavage is N-S.

The angle is recorded from the stage goniometer. The stage is rotated until mineral is extinct, it means that either the fast or slow ray is parallel to N-S. If already extinct, no more rotation is needed. The stage is rotated clockwise to extinction and angle is recorded. The extinction angle is the difference between the two. If measured with anticlockwise rotation, the extinction angle is 90°-measured value. Smaller value is the one reported.

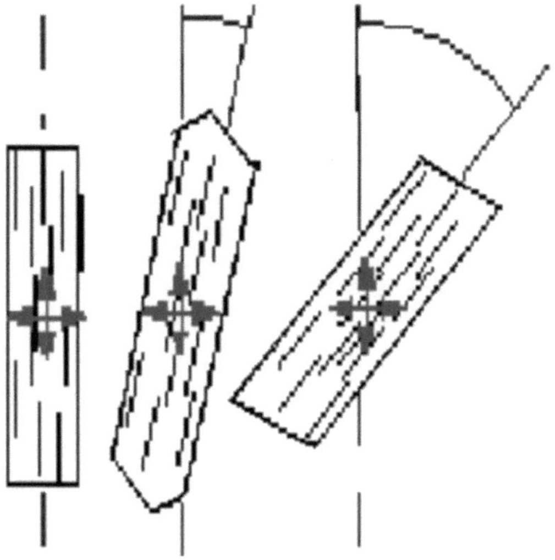

Fig. 5.12: Determination of extinction angle.

Determination of Sign of Elongation

Sign of elongation describes the relations between the principal vibration directions and the length of an anisotropic, elongate mineral. Length slow means that the slow ray vibrates more or less parallel to the length of an elongate mineral. Length fast means that the fast ray vibrates more or less parallel to the length of an elongate mineral. Length slow is also known as positive elongation. Length fast is known as negative elongation.

A mineral has a sign of elongation only if:

- It is anisotropic;
- It has an elongate habit; and
- The principal vibration directions are nearly parallel to the length (parallel extinction).

Use an accessory plate to determine fast ray and slow ray.

To determine the sign of elongation, it has to be found out whether the length direction is fast or slow. This can be done by using one of the accessory plates. The accessory plates consist of pieces of quartz, muscovite, or gypsum mounted on a holder so that their vibration directions are known. The gypsum plate is most commonly used to determine fast ray and slow ray. In the following procedure, the convention is to have the slow ray of the accessory plate oriented NE-SW.

1. Use extinction to locate the vibration direction, which should be nearly parallel to the length direction of the grain;
2. Align the length direction with the fast ray direction of the gypsum plate by rotating the stage counter clockwise 45°;
3. Determine the retardation (interference colour) before and after inserting the plate.

If retardation cancels out (colour decreases), the length direction is the slow ray. If retardation adds up (colour increases), the length direction is the fast ray.

5.5 INDICATRIX

Indicatrix is a geometric form used to illustrate the refractive indices and the vibration direction of light passing through a mineral in any direction.

Uniaxial Indicatrix

The uniaxial indicatrix is constructed by first orienting a crystal with its c-axis vertical. Since the c-axis is also the optic axis in uniaxial crystals, light travelling along the c-axis will vibrate perpendicular to the c-axis and parallel to the ω refractive index direction. Light vibrating perpendicular to the c-axis is associated with the O-ray. Thus, if vectors are drawn with lengths proportional to the refractive index for light vibrating in that direction, such vectors would define a circle with radius ω. This circle is referred to as the ***circular section*** of the uniaxial indicatrix.

Light propagating along directions perpendicular to the c-axis or optic axis is broken into two rays that vibrate perpendicular to each other. One of these rays, the e-ray vibrates parallel to the c-axis or optic axis and thus vibrates parallel to the ε refractive index. Thus, a vector with length proportional to the ε refractive index will be larger than or smaller than the vectors drawn perpendicular to the optic axis, and will define one axis of an ellipse. Such an ellipse with the ε direction as one of its axes and the ε direction as its other axis is called the ***principal section*** of the uniaxial indicatrix.

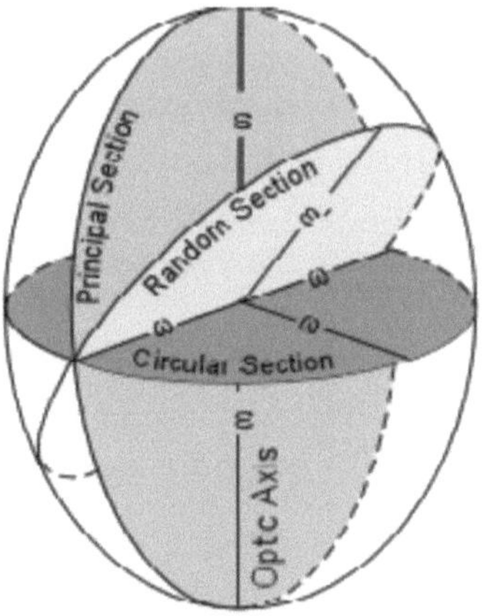

Fig: 5.13: Uniaxial indicatrix.

Light vibrating parallel to any direction associated with a refractive index intermediate between ε and ω will have vector lengths intermediate between those of ε and ω and are referred to as ε′ directions.

Thus, the uniaxial indicatrix is seen to be an ellipsoid of revolution. Such an ellipsoid of revolution would be swept out by rotating the ellipse of the principal section by 180°. Note that there are an infinite number of principal sections that would cut the indicatrix vertically. Light propagating along one of the ε′ directions is broken into two rays, one vibrates parallel to a ε′ direction and the other vibrates parallel to the ω direction. An ellipse that has a ε′ direction and a ω direction as its axes is referred to as a random section of the indicatrix.

Optic Sign

Recall that uniaxial minerals can be divided into two classes based on the optic sign of the mineral.

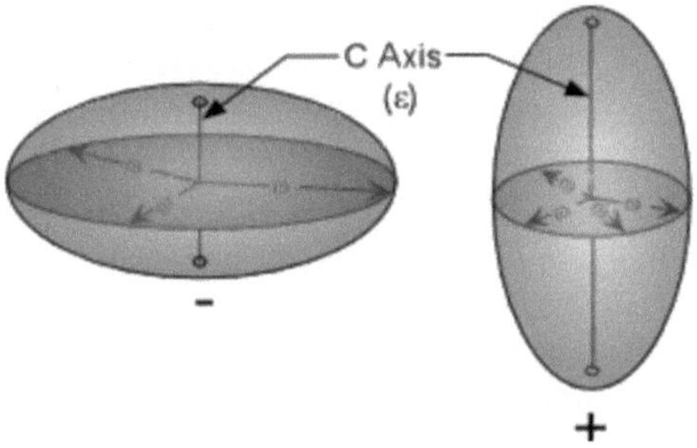

Fig. 5.14: Optic sign determination of uniaxial minerals.

If $\omega > \varepsilon$, the optic sign is negative and the uniaxial indicatrix would take the form of an oblate spheroid. Note that such an indicatrix is elongated in the direction of the stroke of a minus sign.

If $\varepsilon > \omega$, the optic sign is positive and the uniaxial indicatrix would take the form of a prolate spheroid. Note that such an indicatrix is elongated in the direction of the vertical stroke of a plus sign.

Table 5.8: List of uniaxial minerals

Minerals	*Optical sign*	α	β
Quartz	Positive	1.544	1.553
Calcite	Negative	1.658	1.486
Nepheline	Negative	1.540	1.536
Apatite	Positive	1.633	1.630
Zircon	Positive	1.99	1.93
Tourmaline	Negative	1.645-1.670	1.625-1.640

Biaxial Minerals

All minerals that crystallize in the orthorhombic, monoclinic or triclinic crystal systems are biaxial. Biaxial crystals have two optic axes, and this distinguishes biaxial crystals from uniaxial crystals. Like uniaxial crystals, biaxial crystals have refractive indices that vary between two extremes, but also have a unique intermediate refractive index. Biaxial refractive indices are as follows:

- The smallest refractive index is given the symbol α (or X).
- The intermediate refractive index is given the symbol β (or Y).
- The largest refractive index is given the symbol γ (or Z).

All biaxial minerals have optical symmetry equivalent to $2/m^2/m^2/m$. But, in each of the crystal systems, the optical directions have different orientations corresponding to the crystallographic directions.

- In orthorhombic crystals the optical directions correspond to the crystallographic axes, i.e. the X direction and its corresponding refractive index, α can be either the *a*, *b* or *c* crystallographic axes, the Y direction and β can be parallel to either *a*, *b* or *c*, and the Z direction or γ, can be parallel to either *a*, *b* or *c*.
- In monoclinic crystals, one of the X (α), Y (β), or Z (γ) directions or indices is parallel to the *b* crystallographic axis, and the other two do not coincide with crystallographic directions.
- In triclinic crystals none of the optical directions or indices coincide with crystallographic directions, although in some rare case one of the indices might coincide with one of the crystallographic directions.

The Biaxial Indicatrix

The biaxial indicatrix, like the isotropic and uniaxial indicatrices, diagrammatically illustrates the refractive index for vibration directions of light. It is shown in the diagram below.

The biaxial indicatrix has three principal axes, labeled α, β and γ.Directions that have refractive indices between α and β, are referred to as α′. Directions with refractive indices between γ and β are referred to as γ′. Note that the β direction also must occur in the plane that includes α and γ. Similarly, if we

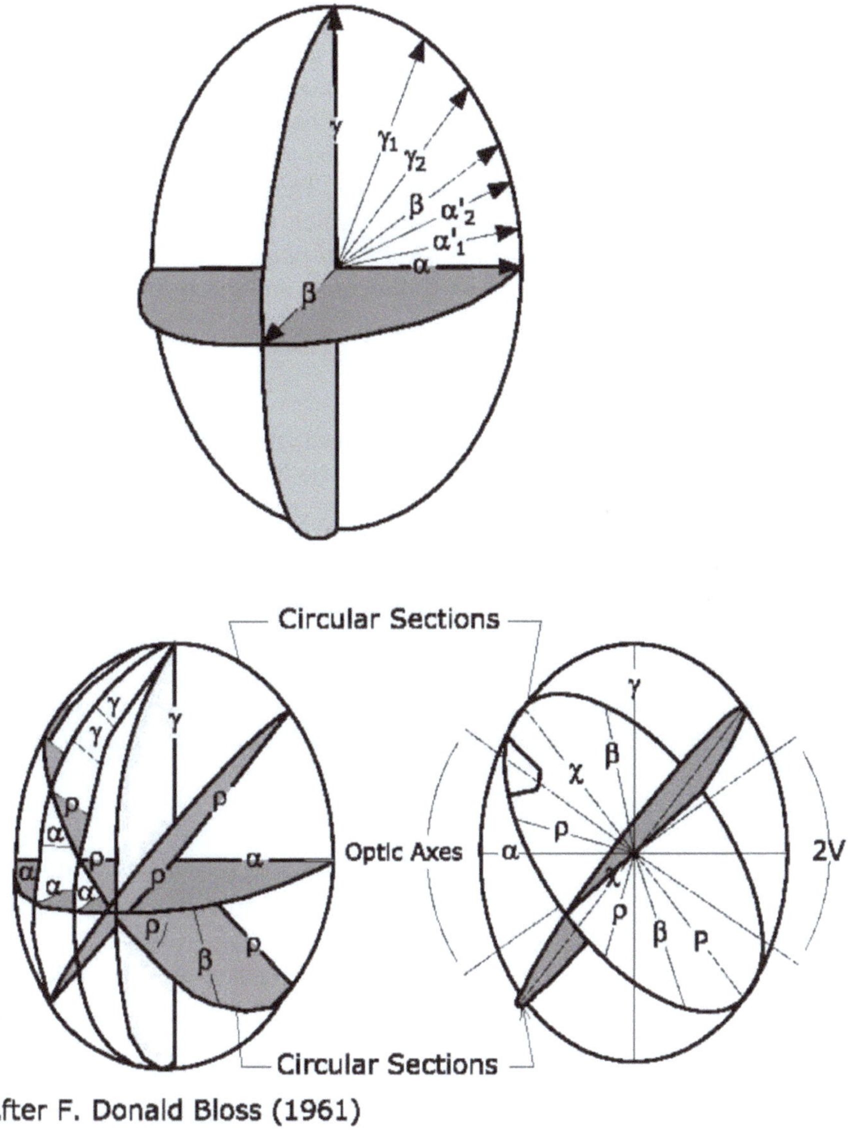

Fig. 5.15: Biaxial indicatrix.

were to draw all other possible planes that include the γ direction, β would have to occur in each of these as well. This results in two sections that would be circular with a radius equivalent to the β refractive index.

These two sections are referred to as the circular sections. In the diagrams we see the two circular sections, each having a radius equal to the β refractive index. In the left-hand diagram some of the other possible planes that include γ are shown. In the righthand diagram these planes are removed to show only the circular sections. Lines drawn perpendicular to the circular sections is the optic axes. This is why minerals that exhibit these optical properties are called biaxial.

The acute angle between the optic axes is called the $2V$ angle.

Just like in uniaxial minerals, if one is looking down one of the optic axes, light travelling along the optic axis will be vibrating in the β direction, and thus the mineral would be extinct for all rotation positions. The three principal planes of the biaxial indicatrix are shown here. The plane containing the α and γ directions also contains the optic axes, which are perpendicular to the β directions. This plane is called the *optic axial plane* or *OAP*. The other two principal planes contain the γ and β directions and the α and β directions, respectively.

Optic Sign of Biaxial Minerals

The optic sign of biaxial minerals depends on whether the β refractive index is closer to that of α or to γ. There are several ways that this can be stated, so we will look at all of them.

Biaxial Positive

A mineral is biaxial positive if β is closer to α than to γ. In this case the acute angle, $2V$, between the optic axes is bisected by the γ refractive index direction. Thus we say that γ is the ***acute bisectrix*** (***BXA***), because it bisects this angle. In the case of a biaxial positive mineral, $2V\gamma$ is the acute bisectrix, while $2V\alpha$ bisects the obtuse angle between the optic axes (called the ***obtuse bisectrix*** (***BXO***).

Biaxial Negative

A mineral is biaxial negative if β is closer to γ than to α. In this case the acute angle, 2V, between the optic axes is bisected by the α refractive index direction. Thus we say that α is the acute bisectrix.

In the case of a biaxial negative mineral, $2V\alpha$ is the acute bisectrix, while $2V\gamma$ is the obtuse bisectrix.

Note that $2V\alpha + 2V\gamma = 180°$.

If $2V = 90°$ the mineral has no optic sign.

If $2V = 0$ the mineral is uniaxial.

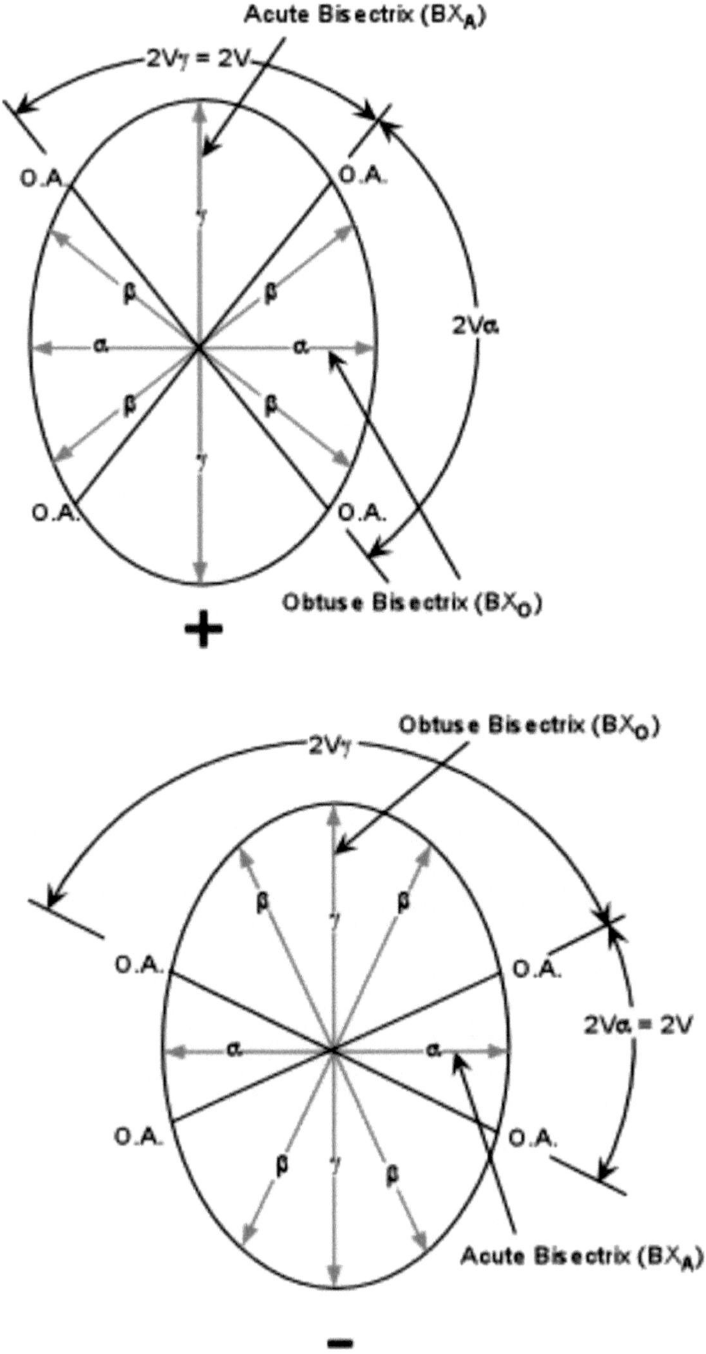

Fig. 5.16: Optic sign determination of biaxial minerals.

Table 5.9: List of biaxial minerals

Olivine	$(Mg,Fe)_2SiO_4$
Orthopyroxene	$(Mg,Fe)_2Si_2O_6$
Clinopyroxene	$(Ca,Mg,Fe,Al)_2Si_2O_6$
Hornblende	$Ca_2(Mg,Fe,Al)_5Si_8O_{22}(OH)_2$
Plagioclase	$(Na,Ca)Al(Al,Si)Si_2O_8$
Alkali feldspar	$KAlSi_3O_8$
Biotite	$K_2(Mg,Fe)_3AlSi_3O_{10}(OH,F,O)_2$
Muscovite	$KAl_2(AlSi_3O_{10})(OH)_2$
Chlorite	$(Mg,Al,Fe)_3(Si,Al)_4O_{10}(OH)_2{*}(Mg,Al,Fe)_3(OH)_6$
Tremolite	$Ca_2Mg_5Si_8O_{22}(OH)_2$
Actinolite	$Ca_2Fe_5Si_8O_{22}(OH)_2$

5.6 INTERFERENCE FIGURE

An interference figure forms under conoscopic illumination, which consists of strongly converging light. The figure is a combination of extinction and interference colours produced by rays going through the sample at different angles. To generate conoscopic illumination, you need to insert the auxiliary condensing lens. To observe the interference figure, you can either insert the Bertrand lens or remove the ocular, and you need to use the 40× objective lens.

All of the work done so far in lab has been with the microscope arranged in mode called orthoscope. In the orthoscope mode, light incident on and passing through the crystals on the microscope stage is perpendicular to the lower and upper surfaces of the crystals. In conoscope mode, a lens—called a condensing lens—is inserted between the light source and the crystals, and is focussed such that the incident light is no longer perpendicular to the top and bottom of the crystal, but crosses within the crystal i.e. focussed within the crystal and then emerges divergently. In addition, another lens, called the **Bertrand Lens**, is inserted between the objective and ocular lenses.

To put the microscope in the conoscope mode you must do the following:

1. Switch to the highest power objective lens.
2. Switch in the condensing lens located beneath the stage, and raise it to a position so that the top is just below the stage.
3. Switch in the Bertrand lens, located just below the ocular lenses.
4. Initially make sure that the compensator plate is not in (i.e. the compensator plate should be at the intermediate position).
5. Put the analyzer in.

Now imagine that we have a grain of some uniaxial mineral on the stage oriented in a position such that the C axis (optic axis) is perpendicular to the stage.

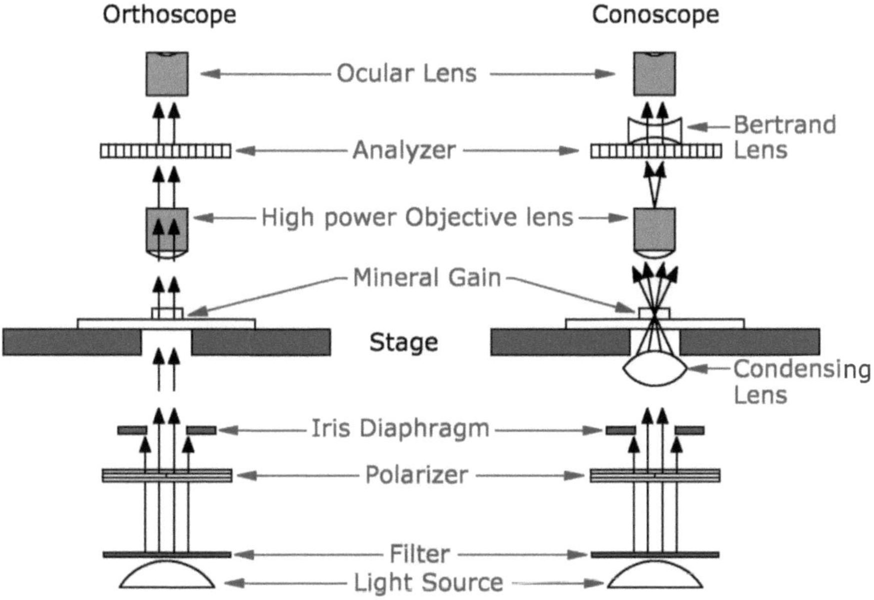

Fig. 5.17: Orthoscopic and Conoscopic illumination settings in polarizing microscope.

- It is a grain that shows only the ω refractive index,
- one that shows no relief change upon rotation, and
- one that would remain extinct on rotation if the analyzer is inserted in orthoscope mode.

Interference figures are used:

- To determine whether a mineral is uniaxial or biaxial
- To determine the optic sign
- To measure the $2V$ angle in case of biaxial minerals.

Interference figures can be studied from two types of principal sections:

- A section cut perpendicular to optic axis, i.e. Basal section
- A section cut parallal to optic axis, i.e. Prismatic section

Following types of interference figures are found in uniaxial and biaxial minerals.

Centred Uniaxial Interference Figure

A centred uniaxial interference figure is shown in the Fig. 5.18. It consists of dark bands that cross the field of view to form a black cross called **isogyres**. Depending on the thickness and birefringence of the mineral, one may also see bands of colour that form circles that are concentric with the centre of the cross. These colour bands are called **isochromes**.

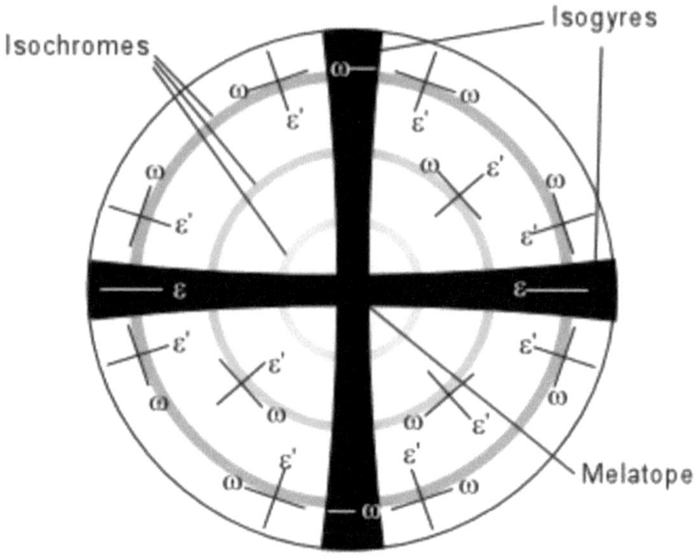

Fig. 5.18: Centered uniaxial interference figure.

At the intersection of the bars of the cross is the **melatope** where light that has travelled along the optic axis exits in the field of view. For a perfectly centred uniaxial interference figure, that is where the optic axis is exactly perpendicular to the microscope stage, the cross will not change position or orientation during a full 360° rotation of the stage.

Before discussing other types of interference figures, formation of isogyres and isochromes should be understood clearly.

Formation of Isochromes

For formation of isochromes, convergent light is required which is produced by the auxiliary condenser and is focussed on the sample.

- Light which travels along the optic axis is not split into two rays, $n_{epsilon}' = n_{omega}$, and exits the mineral to form the melatope. No retardation 'between' rays.
- Light following paths 2 and 4 experience moderate retardation $n_{epsilon}' < n_{omega} \sim 550$ nm
- Light following paths 3 and 5 experience moderate retardation $n_{epsilon}' << n_{omega} \sim 1100$ nm because light makes a larger angle with optic axis and must take a longer path through the sample.

Optic axis is vertical and optical properties vary symmetrically about the optic axis. Rings of equal retardation are produced around the melatope, which are known as isochromes. Number of isochromes depends on retardation and the thickness of the sample.

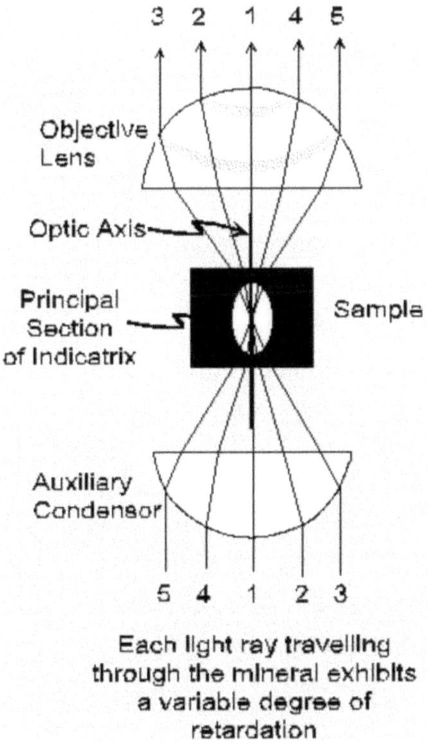

Fig. 5.19: Formation of isochromes.

Formation of Isogyres

Isogyres form when the vibration directions in the interference figure parallel the vibration directions of the polars. These are areas of extinction.

In the uniaxial indicatrix the ordinary rays vibrate perpendicular to the optic axis and are analogous to lines of latitude on the surface of the indicatrix. They vibrate as tangents to the circular isochromes.

Extraordinary rays vibrate parallel to the optic axis and are analogous to lines of longitude on the indicatrix surface and vibrate along radial lines from melatope outwards.

Off-centred Uniaxial Interference Figure

Finding a grain with the optic axis oriented exactly perpendicular to the stage will sometimes be very difficult. It would be much more common to find one wherein the optic axis is at a slight angle to being perpendicular to the microscope stage. Such a grain will exhibit the following properties:

- It is a grain that shows ω refractive index and an ε′ refractive index that is close to the ω refractive index, thus, it will show only a small change in relief change upon rotation.

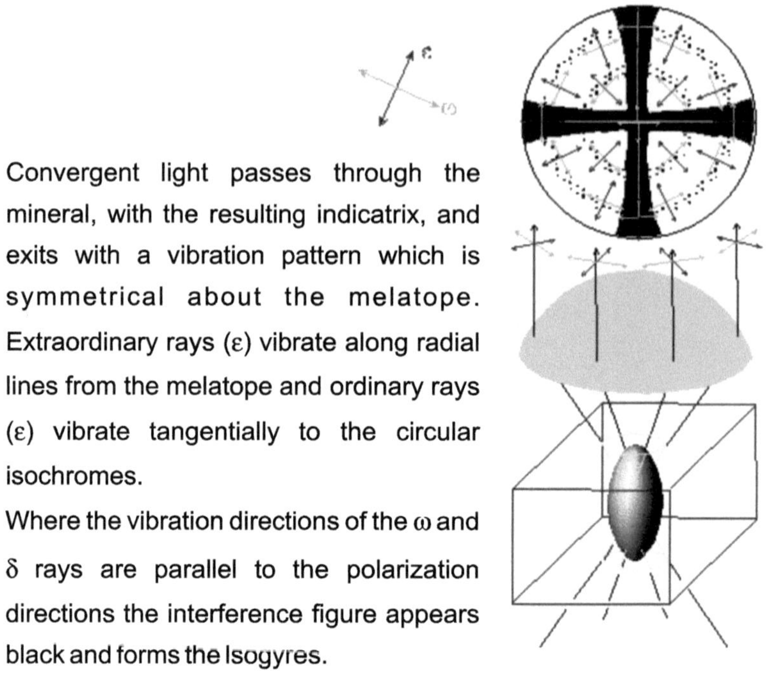

Fig. 5.20: Formation of isogyres.

- It would also show very low order (first order grey) interference colours between extinction positions if the analyzer is inserted in orthoscope mode.

Such a grain would show an off centred optic axis figure as shown here. On rotation of the stage, the melatope would rotate in a circle around the perimeter of the field of view, and the bars of the isogyres would remain oriented E-W and N-S.

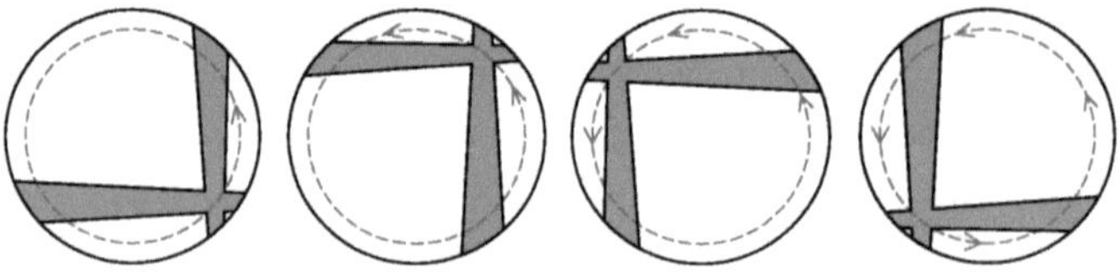

Fig. 5.21: Off-centred uniaxial interference figure.

Biaxial Interference Figures

Four primary types of biaxial interference are seen. Only two of these are commonly used, but it is essential to discuss all four so that you can recognize each.

Acute Bisectrix Figure (BXA)

Looking down the acute bisectrix (the γ direction perpendicular to the stage if the crystal is optically positive, or the α direction perpendicular to the stage if the crystal is negative), at 45° off extinction in conoscope mode, one would see the interference figure shown in the left-hand diagram below.

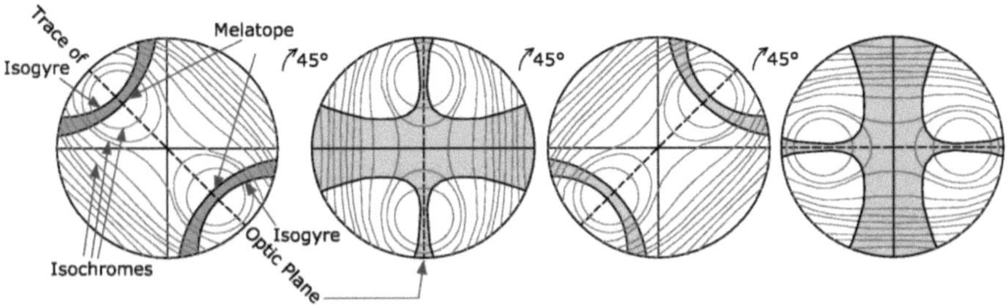

Fig. 5.22: Acute bisectrix figure (BXA).

The dark isogyres mark the positions where light vibrating parallel to the polarizer has passed through the crystal. At the points of maximum curvature of the isogyres are the two melatopes that mark the positions where rays that travelled along the optic axis emerge from the field of view.

Note that the distance between the two melatopes is proportional to the angle $2V$ between the optic axes. Also seen are isochromes, which show increasing interference colours in all directions away from the melatopes. The number of isochromes and maximum order of the interference colours seen will increase with increasing thickness and absolute birefringence of the crystal.

Shown in the figure is the trace of the optic axial plane which includes the two optic axes.

As the stage is rotated 45° from this initial position, the isogyres will close to produce a cross.

In this position the crystal would be extinct in orthoscope mode. The melatopes will be rotated so that both lie along the N-S crosshair. Rotation by an additional 45° will result in the isogyres, then separating again to show the interference figure in the third diagram. Another 45° rotation will again cause the isogyres to close into a cross, this time with the OAP lying parallel to the polarizing direction of the microscope. The crystal would again be extinct in orthoscope mode. Another 45° rotation would return the view to the first diagram in the series.

Optic Axis Figure (OA)

If one of the optic axes is oriented perpendicular to the stage, placing the microscope in conoscope mode, it will result in an optic axis interference figure. This is similar to the BXA figure, except that one of the isogyres and melatopes will be outside the field of view (unless the $2V$ angle is very small).

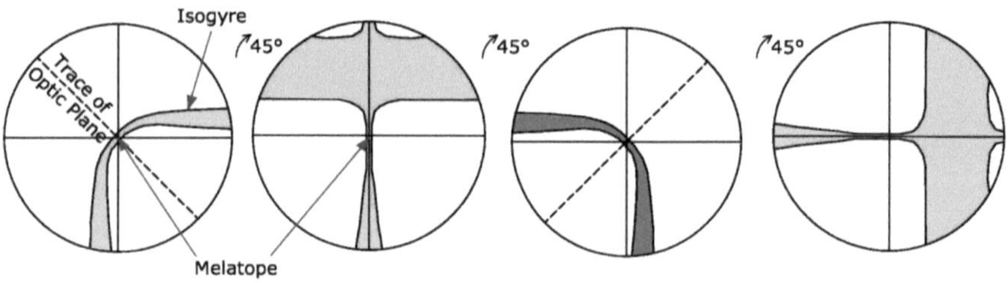

Fig. 5.23: Optic axis figure (OA).

During rotation of the stage, the melatope will remain at the cross-hair intersection and the isogyres will close to form an off-centred cross and then separate to show the curved isogyre in the adjacent quadrant of the field of view. OA figures are easiest to find among randomly oriented grains, because a grain that shows such a figure will show no change in relief on a 360° rotation (analyzer out), and will remain extinct through a 360° rotation (analyzer inserted).

Obtuse Bisectrix Figure (BXO)

A BXO figure will be similar to the BXA figure, except that the melatopes will be outside the field of view most of the time during a 360° rotation. Still, every 90° the broad cross will form as the OAP becomes parallel to one of the crosshairs.

Optic Normal Figure (O.N.)

If the principal β direction of the indicatrix is oriented perpendicular to the stage, such that the crystal's privileged directions are α and γ, then changing to conoscope mode will produce an optic normal figure, also called a **flash figure**. In this figure, when one of the two privileged directions lines up with the polarizer, a broad cross covering almost the entire field of view will be observed. This cross however will quickly disappear after just a slight rotation of the stage (this is why it is often called a **flash figure**).

Optic Sign Determination

Once the interference figure has been obtained and identified as to whether it is uniaxial or biaxial, the optic sign of the mineral can be determined using an accessory plate, either gypsum, quartz or mica.

The optic sign tells us whether the ordinary ray corresponds to the fast or slow ray.

Omega = Fast	Optically Positive	epsilon = Slow
Omega = Slow	Optically Negative	epsilon = Fast

To determine optic sign of a uniaxial mineral:

1. Obtain an optic axis interference figure; one that is centred in field of view.
2. Insert accessory plate into the light path.
3. Observe the interference colours:
    - in two quadrants the colours increase, move to the right,
    - in other two quadrants the colours decrease, move to the left.
4. Look at the NE quadrant of the interference figure.

In the centred uniaxial optic axis interference figure, remember

- omega vibrates parallel to isochromes
- epsilon vibrates radially from centre.

The accessory plate vibration direction is NE-SW, and corresponds to slow direction of plate. *It is parallel to extraordinary ray vibration direction in NE Quadrant of the interference figure.*

Examining the NE quadrant of the interference figure, two possibilities may occur:

1. The interference colours will increase, move to the right on the colour chart, when the accessory plate is inserted. This tells us that the extraordinary ray, of the mineral, must be the slow ray and therefore the mineral is optically positive.
2. The interferecne colours will decrease, move to the left on the colour chart, when the accessory plate is inserted. This tells us that the extraordinary ray, of the mineral, must be the fast ray and therefore the mineral is optically negative.

The SW quadrant of the interference figure will exhibit the same colour changes, observed in the NE quadrant because omega and epsilon vibration directions are the same. The NW and SE quadrants exhibit the reverse colour changes. Gypsum plate is used to determine the optic sign, provided not too many isochromes are present.

Optic Sign Using the Gypsum Plate

Under crossed polars, without the gypsum plate, a first order grey interference colour has a retardation of approximately 200 nm.

This first order grey colour, on inserting the gypsum plate, will either:

1. Increase to second order blue-green, the colour shown on the left below, (200 + 550 = 750 nm)
   giving a total retardation = 750 nm
   or
2. Decrease to first order yellow, the colour shown on the right below, (200 – 550 |–350| nm)
   giving a total retardation = 350 nm.

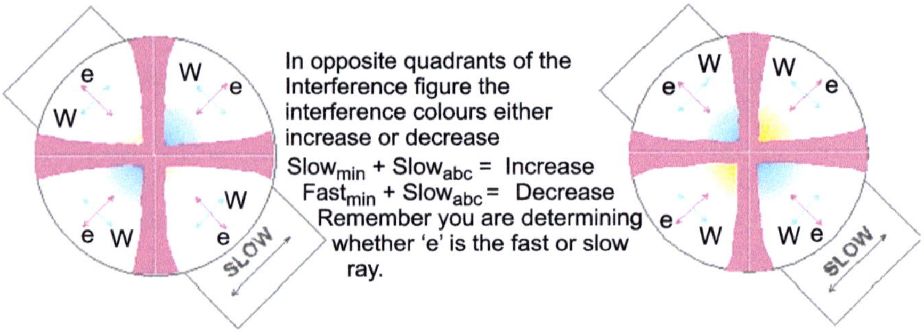

Fig. 5.24: Optic sign determination using a gypsum plate.

The blue or green colour results from the addition of the slow vibration direction of plate to the slow vibration direction of mineral. The yellow colour results from the subtraction of the slow vibration direction of plate from the fast vibration direction of mineral.

Optic Sign Using the Quartz Wedge

If the interference figure displays numerous isochromes colour changes produced with the gypsum plate become difficult to detect. In this case the quartz wedge is used.

Inserting quartz wedge results in the movement of the isochromes about the isogyres. In quadrants where the colours subtract, i.e. where the fast ray of the mineral is parallel to slow ray direction of the quartz wedge, the isochromes move outward as lower order colours form near the melatope and displace

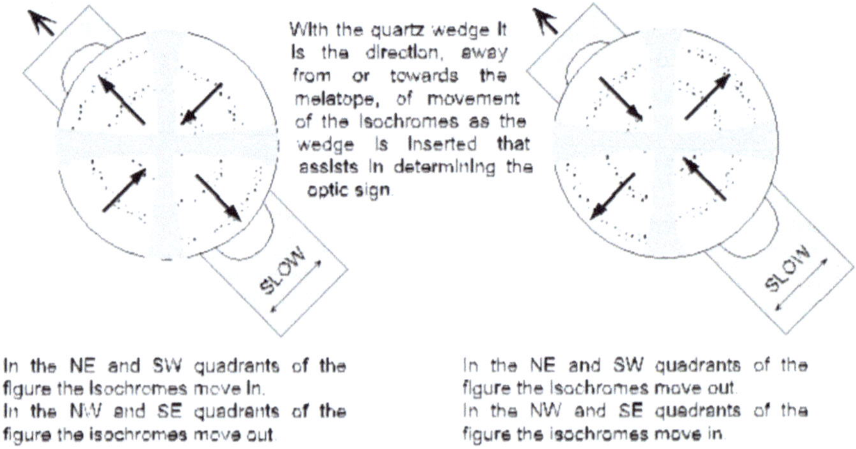

Fig. 5.25: Optic axis determination using a quartz plate.

higher order colours. In quadrants where the colours add, and the slow ray of the mineral is parallel to the slow ray of the quartz wedge, the isochromes move inwards, towards the melatope. The isogyre, on insertion of the accessory, adopts the interference colour corresponding to the retardation of the accessory.

5.7 USES OF OPTICAL MINERALOGY

1. Uses of Petrographic Microscope

- Polarizing microscopy exploits the interference of the split light rays, as they are re-united along the same optical path to extract information about anisotropic materials.
- Polarized light microscopy is perhaps best known for its applications in the geological sciences, which focus primarily on the study of minerals in rock thin sections.
- Determination of texture and structure.
- Determination of rock-type.
- Determination of crystallization sequence.
- Document deformation history.
- Observe frozen-in reaction.
- Constrain P-T history.
- Note weathering/alteration.
- Determination of micro-structures.
- A wide variety of other materials can readily be examined in polarized light, including both natural and industrial minerals, cement composites, ceramics, mineral fibres, polymers, starch, wood, urea, and a host of biological macromolecules and structural assemblies.

2. Determination of Plagioclase Compositions Using Michel-Levy Method

The composition of a plagioclase can be determined by measuring the symmetrical extinction angles of albite twins on sections at right angles to the a-axis (sections normal to {010}). In this orientation twin lamellae appear as very sharp, dark and light stripes or bands under crossed polarisers and display maximum extinction angles. When the twin lamellae are parallel to the vertical cross hair of the microscope they merge into a single grey tone across the entire mineral. In the correct orientation, rotating the microscope stage to the left and right of this vertical position should produce similar extinction angles of the albite twin lamellae (Fig. 5.26). If the extinction angles are within 5°, they can be averaged and the composition of the plagioclase from the Michel-Levy chart (Fig. 5.27).

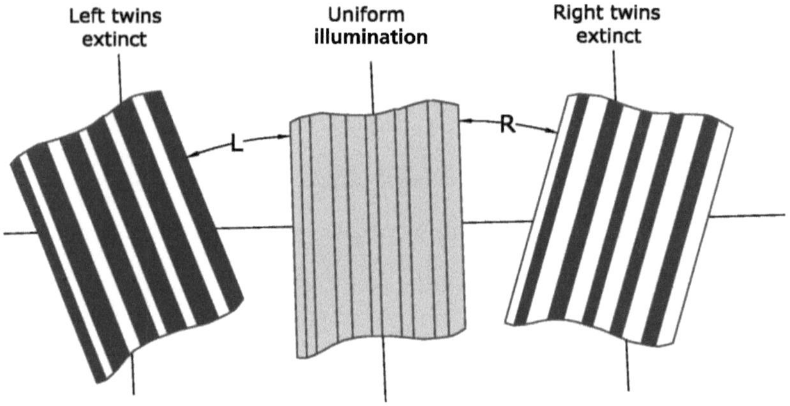

Fig. 5.26: The method of determining the extinction angles in albite twins in plagioclase in sections normal to (010).

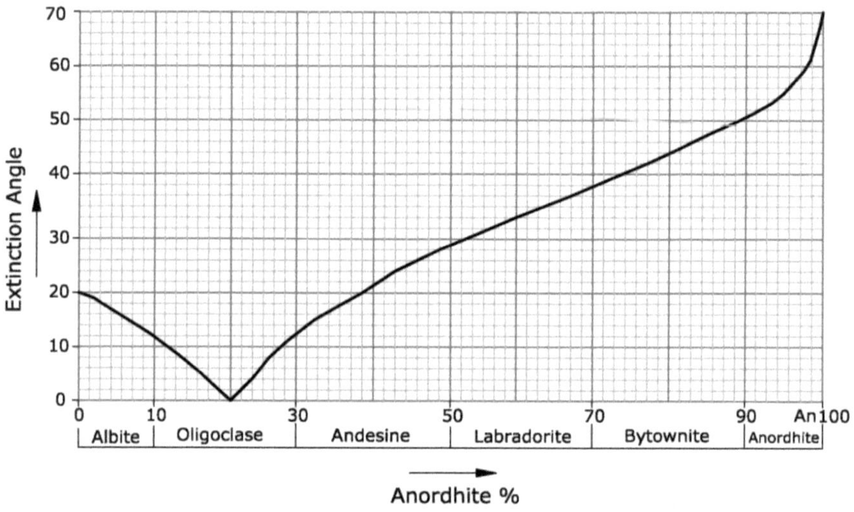

Fig. 5.27: Curve showing the maximum extinction angle of albite twins in plagioclase sections normal to (010) (Michel-Levy method).

Concluding Remarks

In this chapter, the optical properties, i.e. the properties of the minerals that can be studied under petrographic microscope have been covered in details. The properties and their variations in minerals have different implications, which will help the readers to understand this subject more clearly in forthcoming chapters.

Think for a while

1. How is R.I of a mineral related to the density of minerals? Explain with the help of suitable examples.
2. How a uniaxial and biaxial mineral can be distinguished? Explain all the possible processes, starting from examining of crystal form, observation under microscope to study of interference figures to distinguish between the two.
3. (a) A thin plate of calcite does not twinkle when seen in plane polarised light. What does it tell about the orientation of the plate?
   (b) A thick plate of quartz cut perpendicular to the optic axis does not extinguish between cross polars. What does it tell about behaviour of light within the plate?
   (c) A thin plate of augite shows first order red colour when seen in white light between cross polars. What does it tell about double refraction in the plate?

FURTHER READING

Blackburn, W.H. and W.H. Dennen. Principles of Mineralogy, 2nd ed. William C. Brown, Dubuque, IA. 1994.

Bloss, F.D. An Introduction to the Methods of Optical Crystallography. Holt, Rinehart and Winston, New York. 1961.

Deer, W.A., R.A. Howie and J. Zussman. Rock Forming Minerals. 5 vols. John Wiley and Sons, New York. 1962.

Kerr, P.F. Optical Mineralogy, 4th ed. McGraw Hill, New York. 1977.

Klein, C. and C.S. Hurlbut, Jr. Manual of Mineralogy, 21st ed. John Wiley and Sons, New York. 1993.

Nesse, W.D. Introduction to Optical Mineralogy, 2nd ed. Oxford University Press, New York. 1991.

Perkins, D. Mineralogy, 2nd ed. Prentice Hall of India. 2002.

CHAPTER

6

SPECIAL MINERAL PROPERTIES AND RELATED STRUCTURES

The specific properties detected in definite minerals are the function of mineral structures. The properties are useful in identification of these minerals and have important applications in various fields.

The special properties that will be discussed in this chapter are magnetism, fluorescence, phosphorescence, absorption and adsorption.

6.1 MAGNETISM

A few minerals exhibit this property and it becomes one of the most diagnostic property for identification.

Magnetism is generated in minerals due to imbalance in structural arrangement of iron ions. The transfer of electrons from ferrous (Fe^{2+}) state to ferric (Fe^{3+}) state creates a slight magnetic field. The movement of electrons in the crystalline structure produces magnetism in minerals and they get attracted by magnets.

Dependency of Magnetism

- Amount of iron, cobalt or nickel in chemical composition.
- Amount of alignment which takes place.

The magnetic strength of a mineral or rock is measured by a parameter called *magnetic susceptibility*. In case of non-magnetic substance, it is zero and in case of highly magnetic substance like magnetite, it is almost 20. This value is negative in case of diamagnetic minerals and is about 10^{-6} for paramagnetic minerals.

Table 6.1: Types of magnetism observed in minerals

Magnetism	*Property*	*Representative minerals*
Anti-ferromagnetic	There is an antiparallel alignment of spins in two interpenetrating structures, resulting in a lacking of overall magnetization.	
Ferromagnetic	Strongly attracted to a magnet; there is synchronized arrangement of magnetic atoms, producing a bulk magnetic moment and large magnetization.	Magnetite, maghemite
Magnetic	Prominently attracted to magnet.	Iron, nickel
Magnetism due to heating	On heating, magnetism is generated.	Ilmenite
Moderately magnetic	Attracted to magnet.	Pyrrhotite
Weakly magnetic	Weakly attracted to magnet.	Babbingtonite, chromite, columbite, tantalite, ferberite, franklinite
Diamagnetic	Slightly repelled by strong magnetic field.	Quartz
Paramagnetic	Weakly attracted to very strong magnetic fields.	Montmorillonite

Table 6.2: Magnetic susceptibility of different minerals

Minerals	*Magnetic susceptibility*
Pyrite	0.0001 – 0.005
Hematite	0.001 – 0.0001
Pyrrhotite	0.001 – 1.0
Chromite	0.0075 – 1.5
Magnetite	0.1 – 20.0

Curie temperature: A temperature below which a magnetic mineral may loose its magnetism is called Curie temperature. In case of ferromagnets, thermal energy overcomes the electronic exchange forces and produces a randomizing effect at Curie temperature.

Magnetism and Anisotropy

Magnetic properties of minerals depend on the atomic arrangement and directions of the crystal lattice. In case of ferromagnetics, there is an axis of easiest magnetization and one axis of its most difficult magnetization. This character of minerals is known as magnetocrystalline anisotropy.

Table 6.3: Magnetism and crystal structure

Material	*Crystal structure*	*Axis of easiest magnetization*	*Axis of difficult magnetization*
Fe	BCC	[100]	[111]
Ni	FCC	[111]	[100]
Co	HCP	[001]	[1010]
Fe_3O_4	Cubic	[111]	[100]

Table 6.4: Some important types of magnetism

Type of magnetism	*Properties*	*Example of minerals*
Diamagnetism	All the orbital shells are filled and there are no unpaired electrons; when exposed to a field, a negative magnetization is created, producing negative susceptibility and the susceptibility is temperature independent.	Quartz (SiO_2), calcite ($CaCO_3$) water
Paramagnetism	Most commonly contains iron. Some of the atoms or ions in the material have a net magnetic moment due to unpaired electrons in partially filled orbitals. In the presence of a field, a partial alignment of the atomic magnetic moments in the direction of the field, resulting in a net positive magnetization and positive susceptibility; the efficiency of the field in aligning the moments opposed by the randomizing effects of temperature. This results in a temperature dependent susceptibility, known as the Curie Law.	Montmorillonite (clay), nontronite (Fe-rich clay), biotite (silicate), siderite (carbonate), pyrite (sulfide)
Ferromagnetism	The atomic moments exhibit very strong interactions. These interactions are produced by electronic exchange forces and result in a parallel or antiparallel alignment of atomic moments. Spontaneous magnetization (net magnetization that exists inside a uniformly magnetized microscopic volume in the absence of a field, depending upon spin magnetic moments of electrons) and the existence of magnetic ordering temperature (Below the Curie temperature, the ferromagnet is ordered and above it, disordered.).	Magnetite, pyrrhotite, kamacite, native iron, tacnite.

Magnetite—The Most Essential Ferromagnetic in Nature

Magnetite is chemically an iron-oxide ($Fe^{+2}Fe^{+3}{}_2O_4$), representing spinel group of minerals. The mineral contains about 72.36% of iron and about 27.64% of oxygen. According to crystallographic groups, magnetite belongs to isometric group with octahedral and dodecahedral forms.

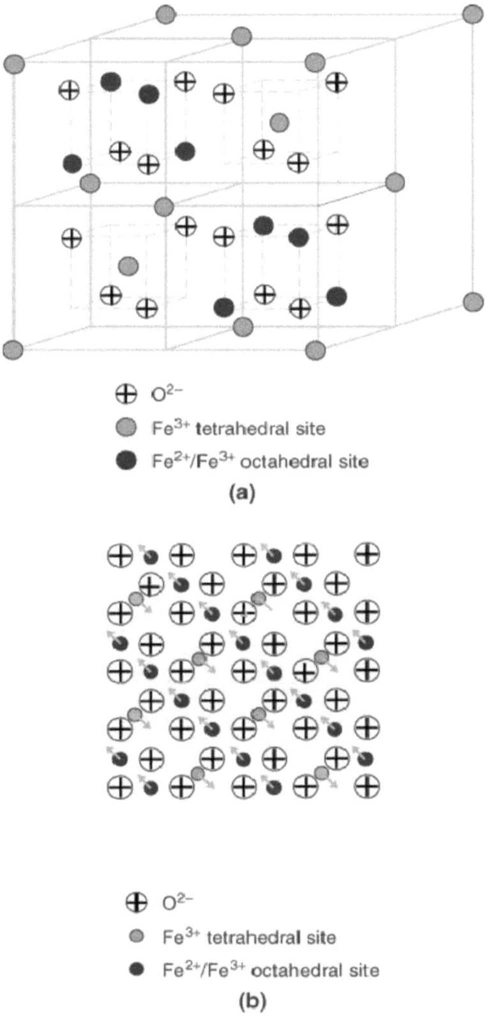

Fig. 6.1: (a) Structure of magnetite and (b) crystal lattice of magnetite.

The most distinctive property of magnetite, that separates it from other minerals, is its magnetic property. This magnetism originates due to the arrangement and spin of the constituting electrons. These electrons are localized and oriented by quantum shells, i.e., l, n, m and s. The quantum number, s, indicates the electron spin, which is the cause of magnetic properties. According to Hund's Rule, the orbitals fill up one electron at a time, so in some cases there are unpaired electrons that do not have any opposing magnetic fields. These unpaired electrons produce the magnetic fields in minerals. In magnetite the ferrous ion produces the magnetic field. However magnetite is considered as a ferromagnetic, because the magnetic moments of the electrons are antiparallel so they cannot pull in any single direction. But in some special instances, when magnetite is formed by cooling from a liquid phase of matter under the influence of earth's magnetic field, it behaves as a natural magnet and is known as lodestone. The Curie temperature for magnetite is at 580°C.

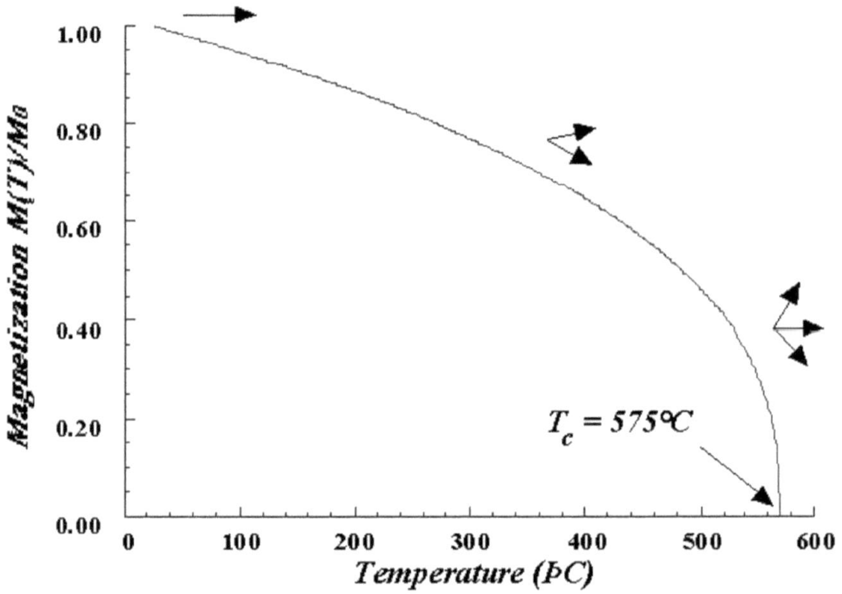

Fig. 6.2: Plot of magnetization vs temperature for magnetite.

Uses of Magnetite

- Magnetite is used in the mining of iron and other products.
- To study palaeomagnetism and theory of plate tectonics magnetite plays a key role.
- Ferrofluid is made of small particles (~10 nm) of magnetite, surrounded by a surfactant such as tetramethylammonium hydroxide. The surfactant is needed to keep the particles of magnetite from agglomerating (clumping together) due to magnetic and van der Waals interactions. It produces a slippery skin around the small particle of magnetite. A group of these prepared particles is like a solution that acts like a medium density liquid which is affected by magnetic fields. When a magnet is brought near it, the liquid splits up and starts to group itself into spikes or hairs along the magnetic field lines. It is used to seal rotating shafts, and in speakers to help dampen the vibrations of the speaker coil, and help cool the coil.

6.2 LUMINESCENCE

Emission of light from nonincandescent materials is called luminescence.

Occurrence of Luminescence

The cause of luminescence in solid material is an "excitant" that raises the energy level of the material and the extra energy is given off as visible light. These excitants include parts of the electromagnetic spectrum like infra-red, visible, ultra-violet and X-ray radiations, as well as cathode and radium emanations, chemical reactions, heat, friction and many other factors. An activator, a small amount of chemical element or impurity which is foreign to the pure mineral, is considered a necessary factor for a mineral to exhibit luminescence. In some cases it has been shown that if the proportion of activator is greater or less than a certain optimum range, luminescence cannot be produced (without change in temperature).

Characteristic Properties of Luminescence

(i) Practically no pure elements and very few simple compounds are luminescent.
(ii) Most substances with a strong luminescence are white or only slightly coloured.
(iii) Most substances become luminescent or their luminescence is strongly increased when they are cooled down to liquid air temperature. Heating to a certain upper limit which usually lies between 100°C and 400°C destroys the fluorescence in almost every case.
(iv) Liquid solutions are seldom phosphorescent.
(v) The light-emissive capacity of many materials which are luminescent under normal conditions is reduced or totally suppressed by the presence of some other substance (poisons).
(vi) The number of impurity molecules able to activate phosphors is very large. One condition seems to be that the ionic (or atomic?) radius of the activator must be smaller than the radius of the cation of the basic lattice.
(ix) Activating impurities do not occupy analogous positions in the lattice of all phosphors. In some cases the activator ions are interstitial in the lattice and in others mix crystals are formed. In the second case there is a modification in the size of the crystal lattice and the luminescence may be dependent either on the formation of the mix crystal or the introduction of activators into the modified lattice.

Among the types of luminescence, fluorescence and phosphorescence are most important, in terms of significance and presence, so they have been studied in details in forthcoming discussion.

Types of Luminescence

On the basis of their characteristics, different types of luminescence are described in Table 6.5 with their respective description and example.

Table 6.5: Different types of luminescence

Types	*Description*	*Example*
Fluorescence	The spontaneous transfer of a molecule from an excited state and a high energy level to a lower energy level. The mean life is very short and practically independent of temperature.	Scheelite
Phosphorescence	The mineral continues to glow after the light has been removed.	Pectolite
Triboluminescence	Property of some minerals to glow when they are crushed, struck, scratched or even rubbed in some cases.	Amblygonite, calcite, feldspar, fluorite, lepidolite, micas, pectolite, quartz and sphalerite
Thermoluminescence	Property of some minerals to glow when they are heated.	Tourmaline, apatite, calcite, some feldspar, fluorite and lepidolite.

6.3 FLUORESCENCE

Fluorescence is a form of luminescence in which a substance is irradiated with light of a certain wavelength and, as a result, emits light of a longer wavelength. The light from the ultraviolet lamps reacts with the chemicals of a mineral and

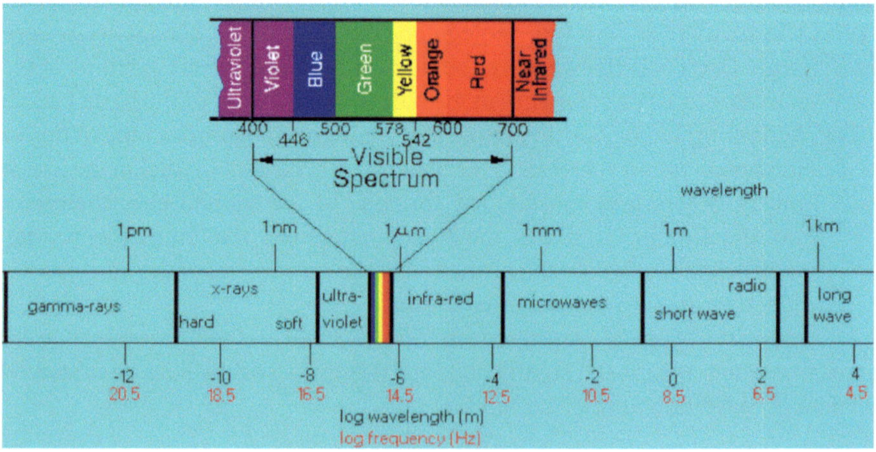

Fig. 6.3: Electromagnetic spectrum.

causes the mineral to glow. Absorption of photons leads to near immediate re-emission of photons with a (in most cases) longer wavelength. This emission ends almost immediately after the exciting radiation is cut off. Fluorescent minerals emit visible light when activated by invisible ultraviolet light (UV), X-rays and/or electron beams.

Fluorescence in minerals generally occur under two UV wavelengths: short wavelength (SW) and long-wavelength (LW). SW lamps produce UV radiation at 253.7 nm, while radiation from LW lamps has two different wavelengths one that peaks about 351 nm (called long wave three fifty, LW350) and the other at about 368 nm (called long wave three seventy, LW370). Some minerals fluoresce only under SW or one of the LW wavelengths, but not both, SW and LW350 or LW370. For example, most scheelite fluoresces only under SW light, while most ruby corundum fluoresces only under one of the LW wavelengths (it usually is brighter under LW370). Minerals will fluoresce under both SW and LW, usually in the same colour or nearly so, and a very high percentage of them fluoresce brighter under SW, like calcite, sodalite hackmanite etc.

Conditional Factors

- The exciting radiation has to be photonic.
- The emitted radiation has to lie within the visual range of electromagnetic spectrum.
- Increasing the temperature of a mineral reduces temporarily any fluorescence which may be present, and reducing the temperature temporarily increases the intensity of fluorescence if present.

Types of fluorescent substances

Clay particles

Scheelite

Organic compounds like humic acids, petroleum and droplets of brine.

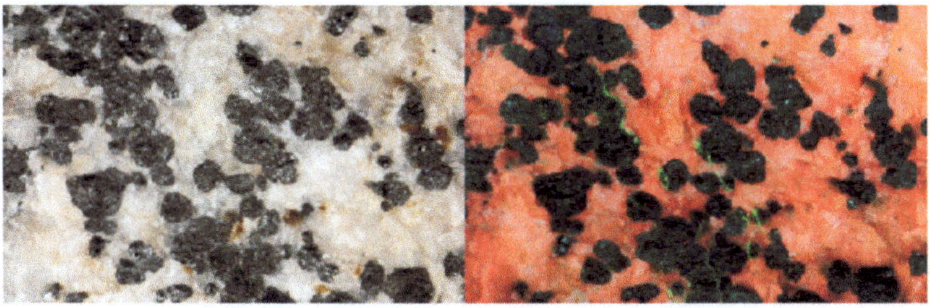

Fig. 6.4: Calcite (a) white light illuminated and (b) UV illuminated.

Occurrence of Fluorescence

The phenomenon of fluorescence is attributed to electron excitation in sub-atomic level. The electrons occurring in the inner shells carry lesser energy than the electrons of the outer shells. When certain atoms are exposed to ultraviolet (UV) light, a photon (particle of light energy) of UV will cause an electron residing in a lower-energy inner electron shell to be temporarily boosted to a higher-energy outer shell and the electron is said to be excited. It will then drop back to its original inner electron shell, releasing its extra energy in the form of a photon of visible light. This visible light is the fluorescent colour.

Most minerals do not fluoresce when pure. It takes certain impurities in certain quantities to make the mineral fluoresce. Such impurities are called "activators". Different activators can make the same mineral fluoresce in different colours. Different minerals require different activators, and in different quantities. There are also impurities, notably ferrous iron, that can prevent fluorescence despite the presence of an activator. Because the activators and deactivators may or may not be present in any given specimen, different specimens of the some mineral (especially from different locales) may vary in colour and degree of fluorescence. A common fluorescent mineral is calcite. It comes in just about all the observed fluorescent colours due to different activators. Red and pink fluorescent calcites are often activated by a combination of lead and manganese.

There are a few minerals that will fluoresce when pure. These are called "self-activated" minerals, and include scheelite, powellite and several uranium minerals. Others suspected of being self-activated include benitoite, cerussite, anglesite and perhaps many other lead minerals. Scheelite, a major ore of tungsten, is often found by its brilliant sky-blue fluorescence. If it has a little molybdenum in it (which makes it troublesome to extract the tungsten), this colour is modified to white or yellow, providing a quick way to assess the commercial value of a find.

Several secondary uranium minerals, such as autunite, are also characteristically fluorescent and show a bright yellowish green colour. This is due to the presence of uranyl ion; this ion is so prone to fluorescence that trace amounts of it cause yellowish-green fluorescence in a very large number of minerals, including adamite, apophyllite, aragonite, calcite, quartz and opal. Any yellowish-green fluorescence other than willemite is likely to be due to the uranyl ion.

Controlling Factors

- **Intrinsic fluorescence:** Some minerals fluoresce as a result of their composition. There is no contamination with foreign substances, the fluorescence is solely due to the chemical and physical arrangement.

  Example: Scheelite and most of the fluorescing uranyl minerals belong to this category. Most fluorescing borate minerals and a number of lead minerals are intrinsic fluorescents.

- **Activators and co-activators (primers):** A number of ions in a given crystal lattice may be replaced by foreign ions of comparable size and valence called *activators*, causing fluorescence. The presence of activators helps to explain why a single mineral, especially when found in different locations, can fluoresce in different colours.
  Example: Manganese can easily replace calcium in minerals like calcite. Manganese is capable of activating the green fluorescence of willemite in which it occupies a site with tetrahedral coordination in respect to the surrounding oxygen atoms. Sometimes Mn replaces zinc atoms in Zinc-bearing minerals and is capable of absorbing a fairly broad spectrum of ultraviolet wavelengths. In calcite, as a carbonate, manganese is locked in a distorted octahedral site where it is not capable of directly absorbing UV and transforming it into fluorescent light. In calcite, when the replacing ion present is only manganese, it will not fluoresce. However, lead, when present in small amounts, can absorb the UV and cause the crystal lattice of the calcite to vibrate. These vibrations constantly "flex" the manganese ions and the bonds that connect them to their neighbours just enough to create new transition-possibilities for the electrons. As a result the calcite fluoresces red and these atoms that 'pre-ignite' fluorescence are called primers or co-activators. The other examples include aluminum, antimony, arsenic, bismuth, copper, gallium, indium, lead, manganese, thallium, tin and various RE elements.

Table 6.6: Specific activators for different minerals

Activators	*Minerals*
Manganese	Calcite and willemite
Lead	Anthophyllite, apatite, axinite, benstonite, calcite, clinohedrite, halite, margarosanite, pectolite, sphalerite, svabite, tirodite, tremolite, walstromite and wollastonite, as well as yellow fluorescence to esperite and green fluorescence to willemite
Uranium	Green fluorescence to adamite, apophyllite, aragonite, calcite, gyrolite, hyalite, and quartz, var. chalcedony
Rare earth like europium	Blue to violet fluorescence to apatite, calcite, danburite, microcline and especially fluorite
Chromium	Deep red LW fluorescence to corundum, var. ruby, spinel, uvarovite and kyanite
Disulphide ion S^{2-}	Yellow fluorescence to scapolite and orange fluorescence to sodalite
Mercury	Orange fluorescence to calomel

- **Defects in the crystal lattice:** Missing or dislocated atoms lead to crystal defects. These defects disturb the electromagnetic balance between the atoms in their vicinity. Ions of which the electron mantle is deformed by electromagnetic pressure of any kind may become fluorescent.

These defects include:

⇒ Frenkel defects, which are ions in abnormal (ions located in sites, specified for other type of ions) locations in the crystal structure
⇒ Schottky defects, which are vacancies due to missing ions
⇒ Interstitial defects, which are impurity ions in crystal lattice sites not normally occupied.

- Such defects change the normal absorption and dissipation of light energy within and near the structurally imperfect parts of the crystal. Such point defects, commonly causes of daylight colour in minerals that would otherwise be colourless or white, are also associated with yellow-green fluorescence in diamond, blue fluorescence in sphalerite, many cases of white fluorescence in calcite, and with fluorescence in some fluorite and in various sulfate minerals.
- Foreign ions can cause crystal imperfections due to their different ionic radius from that of the replaced ion. Strontium, for example, usually replaces about 5% of calcium or barium in gypsum and barite. Defects like this may cause fluorescence by exerting electromagnetic pressure on the atoms and groups of atoms near them. These foreign ions are called 'promoter ions'. They do not fluoresce themselves but rather 'promote' fluorescence centres within the crystals.

Fluorescence and Related Structures

Atomic character: The relaxation of exited atoms to their ground states through a number of discrete jumps explains Stokes' law which states that the wavelength of fluorescent emission is always greater than that of the exciting radiation (there are some exceptions: anti-Stokes' law). Sir George G. Stokes is generally recognized as the discoverer of fluorescence due to ultraviolet light in 1852. He observed this phenomenon in the famous green Weardale (England) fluorite which fluoresces blue in daylight. The term 'fluorescence' is thus derived from the mineral fluorite. When atoms are chemically bound to each other, the complexity of the equations that govern the shape of the electron orbitals increases abruptly. The number of possible transitions of the electrons beneath the valence electrons becomes significantly larger. That is why fluorescence is usually confined to the realm of chemical compounds, both organic and inorganic and is only rarely observed in unbound elements.

Fluorite: The fluorescent colours are generally blue, but red, purple, yellow, green and white have also been recorded. The yellow colour occurs due to the impurities like yttrium or organic matter in the crystal lattice and the blue colour is due to the europium ions. Upon cooling, a relationship became apparent between the fluorescence and the twinning, the pink fluorescence becoming intensified at the intersections of the crystals.

Cassiterite: The initial fluorescence showed some relationship to the fractured surfaces of the crystals, but at lowered temperature the emitted light came from the entire crystal surface. The lead (Pb^{2+}) ion causes the yellow fluorescence in cerussite as well as in anglesite; such minerals are said to be "self-activated."

Barite in dolomite: The only fluorescence noted occurred at low temperature, but it was concentrated in the dolomite around the base of the barite crystal.

Packing index: The packing index applies only to minerals with an ionic bonding. The fluorescent or occasionally fluorescent minerals usually have an index which lies somewhere close to 5.0, with some higher and some lower values. In some cases, lower indices would probably be increased by the addition of an activator, and in others equivalent changes might reduce higher values.

Table 6.7: List of some fluorescent minerals

Adamite, Allophane, Analime, Anhydrite, Aragonite, Austinite, Austinite
Barite, Basaluminate, Bussamite, Bayleyite, Becquerelite, Beryl, Blodite, Boltwoodite, Brucite
Calcite, Calomel, Cassiterite, Celestine, Cerussite, Chabazite, Chlorapaitite, Chrysoberyl, Clinohedrite, Colemanite, Cookeite, Corumdum, Cotunnite, Cowiesite, Cristobalite
Dolomite, Dickite, Diopside, Dumontite, Dumortrerite
Edenite, Elbaite, Epsomite, Ettringite, Eucryotite
Ferrierit, E Fluorapatite, Fluorite
Gearksutit, E Glauberite, Gmelinite, Gonnardite, Greenockite, Grossular, Gypsum, Gyrolite
Halite, Harmotome, Hawleyite, Hectorite, Helvite, Hemimorphite, Hevlandite, Hunite, Hydrocerussite, Hydroxyherdrite, Hydromagnesite, Hydrozincite
Junitoite, Kutnahorite, Lamontite, Lepidolite, Leucite, Levyne, Liebigite
Magnesite, Mammothite, Manganoaxinite, Margarite, Marialite, Matlockite, Mesolite, Meta-Autuntic, Metatorbernite, Meta-Zeunerite Minum, Morssanite, Montmorillonite, Mordenite
Natroalunite, Nepheline, Opal, Orthoclase
Pectolite, Phlogopite, Phosgenite, Plagioclase, Powellite, Prehnite, Pyromorphite, Pyrophyllite
Quartz, Realgear, Rhodochrosite
Sabugalite, Sanidine, Scheclite, Schoepite, Schrockingerit, Sepiolite, Sodium-Zippeite, Sphalerite, Spinel, Stevensite, Stilbite, Stolzite, Strontianite, Sulfur, Swartzite
Talmessite, Thaumasite, Thenardite, Thomsonite, Thorite, Tilasite, Titanite, Tobermorite, Topaz, Torbernite, Termolite, Tridymite
Uranociricite, Uranophan, E Uranospinate, Uvorvite, Vandanite
Wickenburgite, Willemite, Witherite, Wollastonite, Wulfenite, Wurtzite
Xonotime, Xonolite, Zuenerite, Zincite, Zircon, Zunyite

Table 6.8: Fluorescence of different minerals under different wavelengths

Mineral	*Wave length*	*Number*	*Red*	*Orange*	*Yellow*	*Green*	*Blue*	*Violet*	*White*
Apatite	LW	3	•	YES	YES	•	L	•	•
	SW	2	•	YES	YES	•	•	•	•
Aragonite	LW	2	•	•	•	YES	•	•	YES
	SW	3	•	•	S	YES	•	•	YES
Barite	LW	5	YES	L	YES	•	•	YES	YES
	SW	4	YES	•	YES	•	•	YES	YES
Calcite	LW	7	YES	YES	YES	YES	YES	YES	YES
	SW	6	YES	•	YES	YES	YES	YES	YES
Celestite	LW	3	•	•	YES	•	YES	•	YES
	SW	3	•	•	YES	•	YES	•	YES
Cerussite	LW	3	•	YES	YES	•	•	•	L
	SW	2	•	YES	YES	•	•	•	•
Fluorite	LW	3	•	•	YES	•	YES	•	YES
	SW	3	•	•	YES	•	YES	•	YES
Gypsum	LW	3	•	•	YES	•	YES	•	YES
	SW	3	•	•	YES	•	YES	•	YES
Scapolite	LW	3	•	L	YES	•	L	•	•
	SW	3	S	S	YES	•	•	•	S
Scheelite	LW	2	•	L	YES	•	•	•	•
	SW	3	•	•	YES	•	S	•	YES
Sphalerite	LW	4	L	YES	L	•	L	•	•
	SW	1	•	YES	•	•	•	•	•
Witherite	LW	3	•	•	YES	•	YES	•	YES
	SW	3	•	•	YES	•	YES	•	YES
Wollastonite	LW	4	•	YES	YES	•	YES	•	YES
	SW	4	•	YES	YES	•	YES	•	YES
Zircon	LW	2	•	YES	YES	•	•	•	•
	SW	2	•	YES	YES	•	•	•	•

Yes = Fluorescent under both LW and SW Ultraviolet
L = Fluorescent under Long Wave UV light sources
S = Fluorescent under Short Wave UV light sources

Applications

- Used to make mercury vapour street lights.
- Gemstones, minerals, fibres and many other materials, which may be encountered in forensics or with a relationship to various collectibles, may have a distinctive fluorescence or may fluoresce differently under short-wave ultraviolet, long-wave ultra violet, or X-rays.
- Used to recognise clarity in minerals.

- Crude oil (petroleum) fluoresces in a range of colours, from dull brown for heavy oils and tars through to bright yellowish and bluish white for very light oils and condensates. This phenomenon is used in oil exploration drilling to identify very small amounts of oil in drill cuttings and core sample.

6.4 PHOSPHORESCENCE

The ability of a mineral to emit light even after the initial activating ultraviolet light, X-rays or cathode radiation is removed is called phosphorescence. The intensity of light gradually fades with time and after certain span (few seconds or minutes), it completely disappears.

Causes of Phosphorescence

Photonic energy sometimes unbinds the electrons from the original atoms. These free electrons remain sheltered in vacancies within the crystal. They cannot return or move to any other locations, until energised by some external force. This force produce phosphorescence, which may be a photon or crystal vibration. The phenomenon is believed to happen due to the presence of traces of organic materials or cations within the atomic structure of minerals, crystal defects, missing atoms, or substitution of foreign "impurity" atoms in the crystal of the mineral.

Properties of Phosphorescence

- The light produced by phosphorescent minerals consists commonly of rays less refrangible than those of the exciting light.
- Thus the ultraviolet portion of the spectrum is usually the most efficient in exciting rays belonging to the visible part of the spectrum.
- The phosphorescence of calcium sulphide and other phosphorite depends on the presence of minute quantities of other substances, such as copper, bismuth and manganese.
- Impurity level controls the intensity of phosphorescence. It is maximum at a certain definite proportion of impurity and with the increase of the impurity phosphorescent intensity level decreases.

Applications

- Used in the manufacture of long afterglow pigments of red, blue and yellow-green luminescent colours. These pigments are then added to translucent dyes or daylight pigments.
- Used in the manufacture of plastic products with strong afterglow effect.
- Phosphorescence quenching helps in pO_2 (pressure of oxygen) analysis in blood and optic nerves.
- Used in the manufacture of phosphorescent inks.

6.5 ABSORPTION

Absorption is the property of selecting particular wavelengths of light or the fraction of incident electromagnetic radiation over a range of frequencies. Every chemical entity has absorption lines at definite wavelengths, corresponding to the differences between the energy levels of its atomic orbitals.

Atoms and molecules may change states when they absorb specific amounts of energy. An electron in some orbital may be transferred to a more energetic orbital by absorbing exactly one photon whose energy is equal to the energy difference of the two orbitals.

Molecular states are defined by the molecule's modes of vibration and rotation. These vibrational and rotational modes are quantised, similar to the atomic orbitals, and may be excited by absorbing single *photons*.

After some random amount of time, the atoms and molecules return back to their original, lower energy state. In atoms, the excited electron returns to a lower orbital, emitting a photon. In molecules, the vibrational or rotational mode decays, also emitting a photon.

When this decay occurs, the photon produced is not necessarily emitted in the same direction as the original photon. The most common angle of this is assumed to be about 45° of the original photon. This applies to any situation where gases lie between a light source and an observer: the observer will see

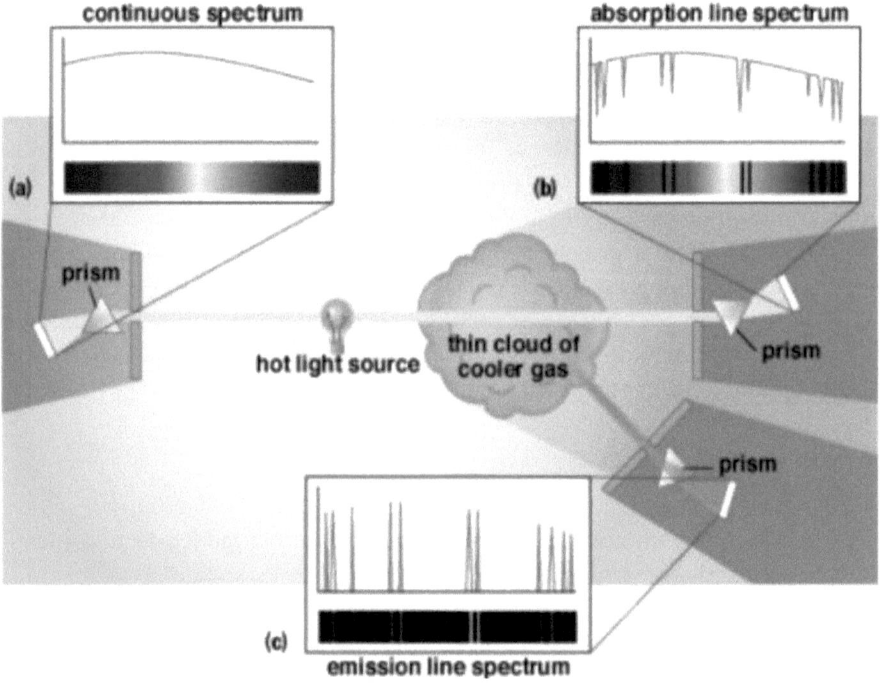

Fig. 6.5: Different types of absorption spectrum.

gaps in the spectrum of the light corresponding to the wavelengths of the photons which were absorbed. These gaps occur despite the re-emission of photons because the re-emitted photons are equally likely to travel in all directions, and it is statistically unlikely to travel along the original path to the observer. These gaps appear as black lines in an image of the spectrum.

Uses

- Used in deducing the presence of elements in stars and other gaseous elements, which cannot be measured directly.
- The methyl blue dye absorption by clay minerals is used to determine their cation-exchange capacities and their surface areas.
- Absorption spectrum studies of clay minerals help to understand the presence (semi-quantitative), the site of substitution, and the valence state of the iron in the lattice of clay minerals.

6.6 ADSORPTION

Adsorption is the phenomenon of accumulation of large number of molecular species at the surface of liquid or solid phase in comparison to the bulk.

Occurrence of Adsorption

Adsorption is a surface phenomenon, caused due to presence of unbalanced or residual forces at the surface of liquid or solid phase. These unbalanced residual forces have tendency to attract and retain the molecular species with which it comes in contact with the surface. Adsorption process involves two components: Adsorbent and Adsorbate. Adsorbent is the substance on the surface of which adsorption takes place. Adsorbate is the substance which is being adsorbed on the surface of adsorbent.

In mineral species, due to cleavage of a big crystal into smaller unit, residual forces or vacancies gets generated on the surface of the solid. Occupancy of these vacancies by some other molecular species results into adsorption.

Properties of adsorption

- Surface phenomenon
- Spontaneous process
- Exothermic reaction

Controlling Factors

- **Temperature:** Increases at low temperature conditions.

$$A + B \Longleftrightarrow AB + \text{Heat}$$

- **Pressure:** With the increase in pressure, adsorption increases up to a certain extent till saturation level is achieved. After saturation level is achieved no more adsorption takes place, no matter how high the pressure is applied.
- **Surface area:** Increases with increase in surface area.
- **Activation of adsorbent:** Activation of adsorbent surface is done so as to provide more number of vacant sites on surface of adsorbent.
- **Surface area of adsorbent:** Regulates the rate of adsorption positively.

Types of Adsorption

Forces of attraction exist between adsorbate and adsorbent. These forces of attraction can be due to weak Vander Waal forces of attraction or due to chemical bond which are strong forces of attraction. On the basis of type of forces of attraction existing between adsorbate and adsorbent, adsorption can be classified into two types: Physical adsorption or Chemical adsorption.

Table 6.9: Types of adsorption

Types	*Properties*
Physical	The forces of attraction between adsorbate and adsorbent are weak Vander Waal forces of attraction; formation of multilayer of adsorbate on adsorbent; low enthalpy of adsorption i.e. $\Delta H_{adsorption}$ is 20-40 kJ/mol; takes place at low temperature. As temperature increases, it decreases.
Chemical	The forces of attraction between adsorbate and adsorbent are chemical forces of attraction or chemical bond; formation of unilayer of adsorbate on adsorbent, high enthalpy of adsorption, i.e. 200-400 kJ/mole, takes place at all temperature. With the increases in temperature, it first increases and then decreases.

Uses

- Silica and alumina gels are used as adsorbents for removing moisture and for controlling humidity of rooms.
- In case of zeolites, it is used in catalytic processes in the production of lead-free high octane gasoline.
- Used to increase the oxygen content of water and air due to preferential adsorption of nitrogen.
- Used to make desiccant silica gel by adsorption of water vapour into internal pores.
- Used widely in the chemical industries for separation of impurities in aromatics, olefins-paraffins, fructose-glucose, oxygen-nitrogen, normal-cyclic paraffins and also for many novel separations in fine chemical industry.
- Adsorption of dyes from solutions has been extensively used for the determination of surface areas and examining the accessibility of porous structures to dye molecules and has been extensively used in the qualitative evaluation of active carbons in industry.

Table 6.10: Adsorbent minerals

Materials	*Bulk density lb/ft³*	*Heat capacity BTU/lb°F*	*Pore volume cm³/g*	*Surface area m²/g*	*Average pore diameter Å*	*Regeneration temperature °C*	*Maximum allowable Temperature °C*
Carbon	22-34	0.27-0.36	0.56-1.20	600-1600	15-25	100-140	150
Alumina		0.21-0.25	0.29-0.37	210-360	18-48	200-250	500
(granules)	38-42						
(pellets)	54-58						
Silica gel	44-56	0.22-0.26	·0.37	750	22	120-250	400
Zeolite							
anhydrous sodium alumino-silicate	44	0.19			4	200-300	600
Anhydrous calcium alumino-silicate	44	0.19			4	200-300	600
Anhydrous alumino-silicate	38				13	200-300	600

Concluding Remarks

By studying the different special properties significant application of minerals in different sectors, especially industries can be well understood. So a brief idea about each of the properties has been enumerated here.

Think for a while

1. Some minerals exhibit magnetism while others do not. Why? On the basis of this statement classify the minerals on the basis of magnetic property.
2. Fluorescent minerals have many special applications in modern-day industries. Illustrate with suitable examples.
3. How the phenomenon of absorption can be related to atomic structure of minerals?
4. Is there any effect of absorption on any other properties of minerals? Answer with appropriate examples.

FURTHER READING

Blackburn, W.H. and W.H. Dennen. Principles of Mineralogy, 2nd ed. William C. Brown, Dubuque, IA. 1994.

Klein, C. and C.S. Hurlbut, Jr. Manual of Mineralogy, 21st ed. John Wiley and Sons, New York. 1993.

Pham Till Hang and G.W. Brindley. Methylene blue absorption by clay minerals. Determination of surface: Areas and cation exchange capacities (clay-organic studies xviii). *In:* Clays and Clay Minerals, 1970, Vol. 18, pp. 203-212. Pergamon Press.

Internet Data Retrieved from:

The Internet Encyclopedia of Science. Spectra and spectroscopy.

Bruce H. Fine. Experiencing Fluorescence, at the February 14, 2003 meeting of the Calgary Rock and Lapidary Club.

Mineral Gallery - The Fluorescent Minerals.htm

Alan Wilkins M.D. Ultraviolet Light and its Use with Fluorescent Minerals: FMS member #894, Coto de Caza, CA.

CHAPTER

7

DESCRIPTIVE MINERALOGY

Descriptive Mineralogy should include first of all a description of the crystallographic, general physical and chemical characters of each mineral species, and should further give an account of its mode of occurrence and characteristic associations. In the case of minerals possessing an economic value, a brief statement of their uses is of interest. The order in which these various items are under each mineral in this chapter is as follows:

1. Chemical composition
2. Occurrence
3. Crystallization
4. Structure
5. General physical properties
6. Application

Descriptive mineralogy should also point out the chemical and physical relationships existing between the different mineral species. It will be noted that many minerals fall into definite groups, the members of which have chemical and crystallographic features in common. The most scientific classification of minerals recognises these facts and places the minerals having analogous chemical compositions together, and further groups them according to crystallographic and physical similarities.

7.1 NATIVE ELEMENTS

Apart from the free gases in the Earth's atmosphere, some 20 elements occur in nature in a pure or nearly pure form. Known as the native elements, they are classified into three families: metals, semi-metals, and non-metals. The most common native metals, which are characterized by simple crystal structures, make up three groups: the gold group, consisting of gold, silver, copper and lead; the platinum group, composed of platinum, palladium, iridium and osmium; and the iron group, containing iron and nickel-iron. Mercury, tantalum, tin and zinc are other metals that have been found in the native state. The native semi-metals are divided into two isostructural groups (those whose members share a common structure type): (1) antimony, arsenic and bismuth, with the latter two being more common in nature, and (2) the rather uncommon selenium and tellurium. Carbon, in the form of diamond and graphite, and sulfur are the most important native non-metals.

Metals

Silver (Ag)
Gold (Au)
Copper (Cu)
Platinum (Pt)

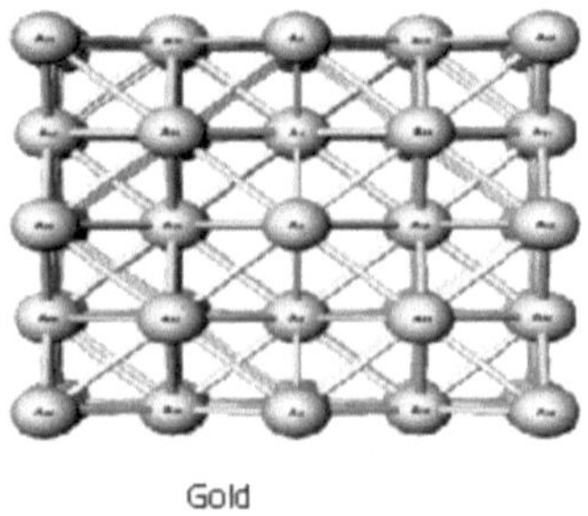

Fig. 7.1. Structure of gold.

Semi-metals

Arsenic(As)
Antimony (Sb)
Bismuth (Bi)

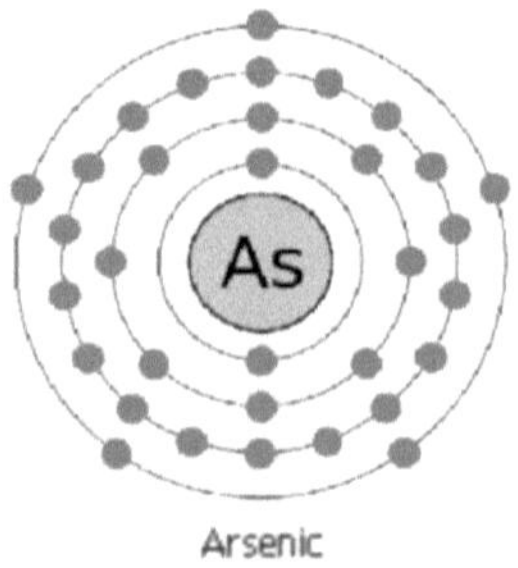

Fig. 7.2. Structure of arsenic.

Non-metals

Diamond (C)
Graphite (C)
Sulphur (S)

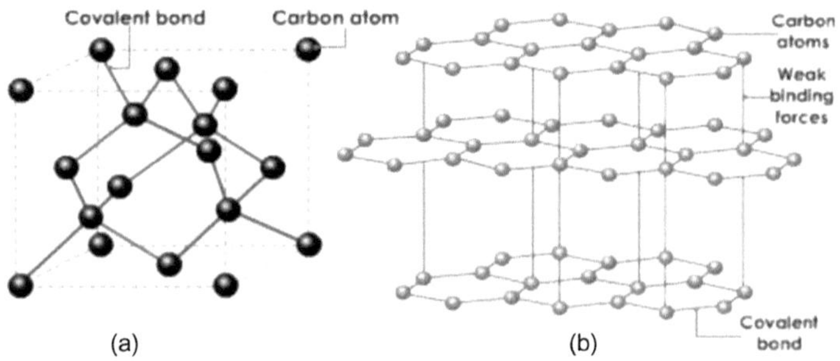

Fig. 7.3. (a) Structure of diamond and (b) Structure of graphite.

Table 7.1: Mineralogical description of native elements

	Mode of occurrence	*Properties*	*Applications*
Metals			
Silver (Ag)	Hydrothermal, oxidized zones of ore deposits Association: Acanthite, prousite, cobaltite, copper, zeolite and quartz.	Crystallography: Isometric 4/m 32/m; Crystal habit: Cubic, dendritic aggregates, wires, massive; Colour: Silver-white; Lustre: Metallic; Transparency: Opaque; Streak: silver-white; Cleavage: None; Fracture: Hackly; Hardness: 2.5-3; Sp. gravity: 10.1-10.5.	Widely used in photographic industry, jewellery and electronics
Gold (Au)	Mainly of hydrothermal origin; contact metamorphic deposits and pegmatites; placer deposits; Association: Pyrite, quartz, nagyagite ($AuPbSb_{0.75}Bi_{0.25}Te_{2.5}S_6$), calaverite, sylvanite and other sulphides.	Crystallography: Isometric 4/m 32/m; Crystal habit: Cubic and octahedral, skeletal and dendritic, leaves, nuggets; Colour: Gold-yellow; Lustre: Metallic; Transparency: Opaque; Streak: Gold-yellow; Cleavage: None; Fracture: Hackly; Hardness: 2.5-3; Sp. gravity: 15.6-19.3.	Used in jewellery, medicines and electronics
Copper (Cu)	Hydrothermal deposits; formed as a result of supergene enrichment. Association: Silver, calcite, malachite, azurite etc.	Crystallography: Isometric 4/m 32/m; Crystal habit: Cubic, dendritic aggregates, sheets, slabs, massive; Colour: Copper-red; Lustre: Metallic; Transparency: Opaque; Streak: Copper-red; Cleavage: none; Fracture: Hackly; Hardness: 2.5-3; Sp. gravity: 8.7-8.9.	Electronics, electrical enginnering, ingredient in gold alloys.
Platinum (Pt)	Magmatic segregation; in ultrabasic rocks, placer deposits. Association: Chromite, olivine, magnetite.	Crystallography: Isometric 4/m 32/m; Crystal habit: Cubic, nuggets, grains, scales; Colour: White grey to silver grey; Lustre: Metallic; Transparency: Opaque; Streak: White grey to silver grey; Cleavage: None; Fracture: Hackly; Hardness: 4-4.5; Sp. gravity: 21.47.	Used in catalytic convertors, chemical and rocket industry
Semi-metals			
Arsenic (As)	Primarily hydrothermal; Association: Silver, dyscrasite, barite, cinnabar and nickeline.	Crystallography: Trigonal 32/m; Crystal habit: Rhombohedral crystals, botryroidal aggregates, granular, massive; Colour: Tin-white; Lustre: Metallic; Transparency: Opaque; Streak: black; Cleavage: Perfect; Fracture: Uneven; Hardness: 3.5; Sp. gravity: 5.8.	Used in wood preservation, in pigments, medical applications, agriculture and as animal feed

(*Contd.*)

Table 7.1 (*Contd.*)

	Mode of occurrence	*Properties*	*Applications*
Antimony (Sb)	Primarily hydrothermal pegmatites; Association: Silver, stibnite, sphalerite.	Crystallography: Trigonal 32/m; Crystal habit: Rhombohedral crystals, botryroidal aggregates, massive; Colour: tin-white; Lustre: Metallic; Transparency: Opaque; Streak: Black; Cleavage: Perfect Fracture: uneven; Hardness: 3.5; Sp. gravity: 6.7.	Used in the semiconductor industry to produce diodes, infrared detectors, and Hall-effect devices. Used as an alloy; a hardener in lead for storage batteries
Bismuth (Bi)	Hydrothermal veins, greisens and pegmatites; Association: Chalcopyrite, arsenopyrite, nickeline and many other minerals.	Crystallography: Trigonal 32/m; Crystal habit: Rhombohedral crystals, granular, massive; Colour: Silver-white; Lustre: Metallic; Transparency: Opaque; Streak: Silver white; Cleavage: perfect; Fracture: uneven; Hardness: 2-2.5; Sp. gravity: 9.7-9.8.	Used as ore
Non-metals			
Diamond (C)	Mainly in kimberlite pipes as magmatic deposits and also as placers.	Crystallography: Isometric 4/m 32/m; Crystal habit: Octahedral and cubic crystals; Colour: Colourless, yellow, brown, blue, green, white, pink, black, red; Lustre: adamantine; Transparency: Transparent to translucent; Streak: White; Cleavage: Perfect; Fracture: Conchoidal; Hardness: 10; Sp. Gravity: 3.5; Special property: Occasionally fluorescent and phosphorescent.	Used as gemstrone; carbonado and bort used in abrasives
Graphite (C)	Primarily of metamorphic origin; also exhibit magmatic lines.	Crystallography: Hexagonal 6/m 2/m 2/m; Crystal habit: Hexagonal tabular crystals, massive; Colour: Black to steel grey; Lustre: Metallic to dull, earthy; Transparency: Opaque; Streak: Black to steel grey; Cleavage: Perfect; Fracture: Flaky; Hardness: 1-2; Sp. Gravity: 2.3.	Metallurgy, nuclear industry, used in lubricants

(*Contd.*)

(*Contd.*)

Sulphur (S)	Primarily the product of hydrothermal fumaroles; associated with celestite, calcite, aragonite, gypsum.	Crystallography: Orthorhombic 2/m 2/m 2/m; Crystal habit: Dipyramidal, disphenodic, thick tabular crystals, botryroidal and stalactitic aggregates; Colour: Strong yellow, yellow-browm, yellowish grey; Lustre: resineous to greasy; Transparency: Transparent to translucent; Streak: Yellow; Cleavage: Imperfect; Fracture: Conchoidal; Hardness: 1.5-2.5; Sp. Gravity: 2.1.Hackly; Hardness: 4-4.5; Sp. gravity: 21.47.	Used to make sulphur compounds; used to produce matches, vulcanized rubber, dyes, and gunpowder; an important fuel in pyrotechnic mixtures; used as fungicide in plants

7.2 SULPHIDES AND SULFOSALTS

This important class includes most of the ore minerals. The similar but rarer sulfarsenides are grouped here as well. Sulfide minerals consist of one or more metals combined with sulfur; sulfarsenides contain arsenic replacing some of the sulfur.

Sulfides are generally opaque and exhibit distinguishing colours and streaks. The non-opaque varieties (e.g., cinnabar, realgar and orpiment) possess high refractive indices, transmitting light only on the thin edges of a specimen.

Few broad generalisations can be made about the structures of sulfides, although these minerals can be classified into small groups according to similarities in structure. Ionic and covalent bonding are found in many sulfides, while metallic bonding is apparent in others as evidenced by their metal properties. The simplest and most symmetric sulfide structure is based on the architecture of the sodium chloride structure. A common sulfide mineral that crystallises in this manner is the ore mineral of lead, galena. Its highly symmetric form consists of cubes modified by octahedral faces at their corners. The structure of the common sulfide pyrite (FeS_2) also is modelled after the sodium chloride type; a disulfide grouping is located in a position of coordination with six surrounding ferrous iron atoms. The high symmetry of this structure is reflected in the external morphology of pyrite. In another sulfide structure, sphalerite (ZnS), each zinc atom is surrounded by four sulfur atoms in a tetrahedral coordinating arrangement. In a derivative of this structure type, the chalcopyrite ($CuFeS_2$) structure, copper and iron ions can be thought of as having been regularly substituted in the zinc positions of the original sphalerite atomic arrangement.

Arsenopyrite (FeAsS) is a common sulfarsenide that occurs in many ore deposits. It is the chief source of the element arsenic.

Sulfosalts: There are approximately 100 species constituting the rather large and very diverse sulfosalt class of minerals. Some common examples are listed here. The sulfosalts differ notably from the sulfides and sulfarsenides with regard to the role of semi-metals, such as arsenic (As) and antimony (Sb), in their structures. In the sulfarsenides, the semi-metals substitute for some of the sulfur in the structure, while in the sulfosalts they are found instead in the metal site. For example, in the sulfarsenide arsenopyrite (FeAsS), the arsenic replaces sulfur in a marcasite- (FeS_2-) type structure. In contrast, the sulfosalt enargite (Cu_3AsS_4) contains arsenic in the metal position, coordinated to four sulfur atoms. A sulfosalt such as Cu_3AsS_4 may also be thought of as a double sulfide, $3Cu_2S - As_2S_5$

Tetrahedral sulphide group	
Sphalerite	(ZnS)
Chalcopyrite	($CuFeS_2$)
Bornite	Cu_5FeS_4
Enargite	Cu_3AsS_4
Octahedral sulphide group	
Galena	(PbS)
Pyrrhoetite	$Fe_{1-X}S$
Niccolite	NiAs
Other sulphides	
Pentlandite	$(Ni,Fe)_9S_8$
Molybdenite	MoS_2
Millerite	NiS
Cinnabar	(HgS)
Covellite	(CuS)
Chalcocite	Cu_2S
Argentite	Ag_2S
Pyrite	(FeS_2)
Cobaltite	(CoAsS)
Marcasite	(FeS_2)
Arsenopyrite	(As_2S_3)
Skutterudite	$(Co, Ni)As_{3-x}$
Stibnite	(Sb_2S_3)
Orpiment	(As_2S_3)
Realgar	(AsS)
Sulfosalts	
Tetrahedrite	$(Cu,Fe)_{12}(Sb,As)_4S_{13}$
Pyrargyrite	Ag_3SbS_3

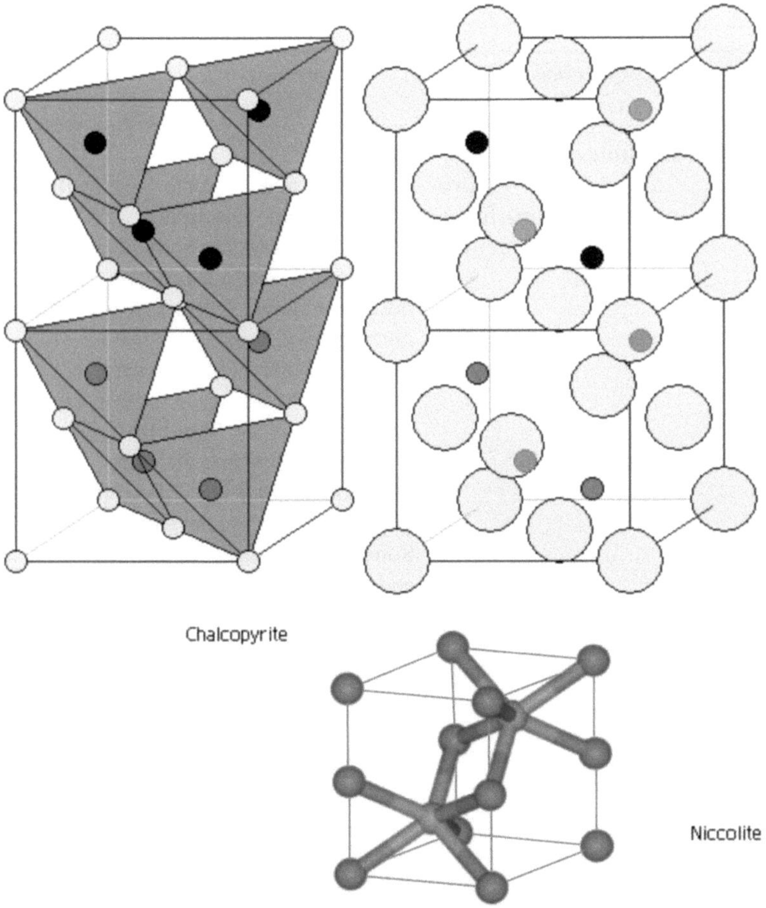

Fig. 7.4: Structures of chalcopyrite and niccolite.

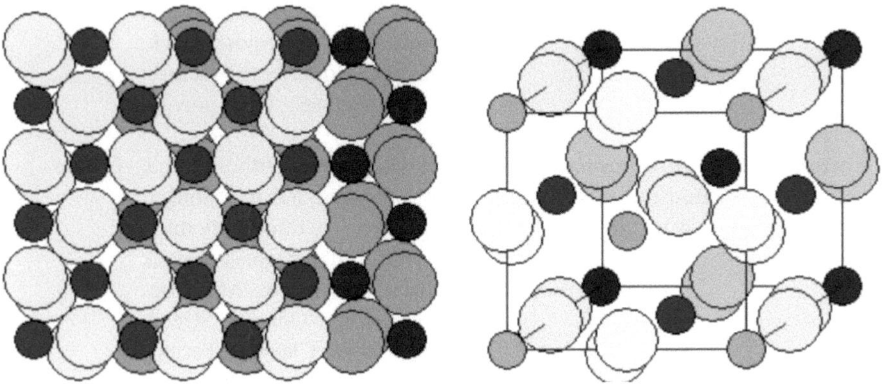

Fig. 7.5: Structure of pyrite.

Table 7.2: Mineralogical descriptions of sulphide minerals

	Occurrence and association	*Properties*	*Applications*
Tetrahedral sulphide group			
Sphalerite ZnS	Magmatic, hydrothermal. Skarns, hydrothermal sedimentary deposits.	Crystallography: Isometric 43/m; Crystal habit: Tetrahedral and dodecahedral crystals, botryoidal, fibrous, stalactite aggregates, massive; Colour: Yellow, orange, green, brown; Lustre: Resineous to adamantine; Transparency: Transparent to translucent; Streak: Yellow or light brown.; Cleavage: Perfect; Fracture: Conchoidal; Hardness: 3.5-4; Sp. gravity: 3.9-4.1; Special property: Sometimes exhibit iridescence and fluorescence.	Important Zn ore
Chalcopyrite $CuFeS_2$	Magmatic, hydrothermal, sedimentary. Association: Galena, pyrite, sphalerite, tetrahedrite etc.	Crystallography: Tetragonal 42/m; Crystal habit: Tetrahedral crystals, botryoidal aggregates, massive; Colour: Brass-yellow; Lustre: Metallic; Transparency: Opaque; Streak: green-black; Cleavage: Imperfect; Fracture: uneven; Hardness: 3.5-4; Sp. gravity: 4.3.	Important Cu ore
Bornite Cu_5FeS_4	Magmatic, hydrothermal, sedimentary, in skarns and pegmatites	Crystallography: Tetragonal 42/m; Crystal habit: Pseudo-cubic, octahedral crystals, massive; Colour: Copper-red; Lustre: Metallic; Transparency: Opaque; Streak: greyish black; Cleavage: Imperfect; Fracture: Uneven; Hardness: 3; Sp. gravity: 5-6.	Important Cu ore
Enargite Cu_3AsS_4	Hydrothermal in medium to low-temperature.	Crystallography: Orthorhombic mm^2; Crystal habit: Tabular, prismatic, striated crystals, massive; Colour: Greyish black to black; Lustre: Metallic; Transparency: Opaque; Streak: Greyish-black; Cleavage: Perfect, good, poor; Fracture: Uneven; Hardness: 3; Sp. gravity: 4.5.	Cu ore

(*Contd.*)

(*Contd.*)

Octahedral sulphide group			
Galena(PbS)	Magmatic, hydro-thermal, metamorphic; associated with pyrite, sphalerite, chalco-pyrite, quartz etc.	Crystallography: Isometric; 4/m 32/m; Crystal habit: Cubic crystals, tabular, massive; Colour: Lead-grey; Lustre: Metallic; Transparency: Opaque; Streak: Lead-grey; Cleavage: Perfect; Fracture: Sub-conchoidal; Hardness: 2.5; Sp. gravity: 7.6.	Important Pb ore
Pyrehotite $Fe_{1-X}S$	Magmatic, pegmatitic, hydro-thermal, sedimentary, metamorphic. Associated with pyrite and pentlandite.	Crystallography: Hexagonal 62/m; Crystal habit: Tabular, pyramidal, prismatic, massive; Colour: Bronze-yellow to brown; Lustre: Metallic; Transparency: Opaque; Streak: Dark grey-black; Cleavage: Poor; Fracture: Uneven; Hardness: 4; Sp. gravity: 4.6.	Important Fe ore
Niccolite NiAs	Hydrothermal, magmatic.	Crystallography: Hexagonal 6/m 2/m 2/m; Crystal habit: Striated, botryoidal, granular, massive; Colour: Light copper-red; Lustre: Metallic; Transparency: Opaque; Streak: Light brown black; Cleavage: Poor; Fracture: Uneven; Hardness: 5-5.5; Sp. gravity: 4.6.	Primary source of nickel
Other sulphides			
Pentlandite $(Ni,Fe)_9S_8$	Typically magmatic. Association: chalcopyrite, pyrrhotite.	Crystallography: Isometric 4/m 32/m; Crystal habit: Massive; Colour: Light bronze to yellow bronze; Lustre: Metallic; Transparency: Opaque; Streak: Light bronze to brown; Cleavage: Perfect; Fracture: Uneven; Hardness: 3.5-4; Sp. gravity: 5.0.	Important Ni ore
Molybdenite MoS_2	Pegmatitic, hydrothermal. Metamorphic, contact metamorphic, porphyric.	Crystallography: Hexagonal 6/m 2/m 2/m; Crystal habit: Tabular and prismatic crystals, scaly aggregates; Colour: Silver, lead-grey; Lustre: Metallic; Transparency: Opaque; Streak: Green-grey; Cleavage: Perfect; Fracture: Flexible; Hardness: 1-1.5; Sp. gravity: 4.7.	Mo ore

(*Contd.*)

Table 7.2 (*Contd.*)

	Occurrence and association	*Properties*	*Applications*
Millerites NiS	Low-temperature hydrothermal and product of oxidation of Ni sulphides. Association: Ankerite, pyrrhotite, barite.	Crystallography: Hexagonal 3/m; Crystal habit: Aciculur crystals, cleavable masses; Colour: Brass yellow; Lustre: Metallic; Transparency: Opaque; Streak: Greenish black; Cleavage: Perfect; Fracture: Uneven; Hardness: 3-3.5; Sp. gravity: 5.5.	Sometimes used as a nickel ore
Cinnabar HgS	Low-temperature hydrothermal. Association: Realgar, mercury, pyrite, marcasite etc.	Crystallography: Trigonal; Crystal habit: Rhombohedral, thick, tabular, prismatic crystals, massive; Colour: Red to brownish red; Lustre: Adamantine to metallic; Transparency: Opaque; Streak: crimson; Cleavage: Perfect; Fracture: sub-conchoidal; Hardness: 2-2.5; Sp. gravity: 8.2.	Used in Chinese carved lacquerware; has historically been used in traditional Chinese medicine.
Covellite CuS	Found in oxidation zone; associated with chalcopyrite, chalcocite, bornite etc.	Crystallography: Hexagonal 6/m 2/m 2/m; Crystal habit: Hexagonal tabular crystals, massive; Colour: Indigo-blue; Lustre: Submetallic to resineous; Transparency: Opaque; Streak: Lead-grey; Cleavage: Perfect; Fracture: Uneven; Hardness: 1.5-2; Sp. gravity: 4.7.	Used as a natural super-conductor.
Chalcocite Cu_2S	Hydrothermal, sedimentary, metamorphic and secondary oxidation deposits. Association: Pyrite, chalcopyrite, bornite etc.	Crystallography: Monoclinic 2/m; Crystal habit: Prismatic to tabular crystals, granular, massive; Colour: Blue-white, lead-grey; Lustre: Metallic; Transparency: Opaque; Streak: Greyish black; Cleavage: Poor; Fracture: Conchoidal; Hardness: 2.5-3; Sp. gravity: 5.8.	Cu ore
Argentite Ag_2S	Hydrothermal in low temperature. Association: Galena, silver and Ag sulfosalts.	Crystallography: Isometric 4/m 32/m; Crystal habit: Octahedral and cubic crystals, dendritic, massive; Colour: Lead-grey to black; Lustre: Metallic; Transparency: Opaque; Streak: Black; Cleavage: Poor; Fracture: Sub-conchoidal; Hardness: 2-2.5 Sp. gravity: 7.1.	Ag ore

(*Contd.*)

(*Contd.*)

Pyrite FeS_2	Magmatic segregation, sedimentary, metamorphic, hydro-thermal. Association: pentlandite, pyrrhotite.	Crystallography: Isometric 2/m 3/m; Crystal habit: Cubic crystals, striated, stalactitic, massive; Colour: Light brass-yellow; Lustre: Metallic; Transparency: Opaque; Streak: Green black to brown black; Cleavage: Indistinct; Fracture: Conchoidal to uneven; Hardness: 6-6.5; Sp. gravity: 4.8-5.	Potential source of sulphur; production of sulfuric acid; applied to calcareous soils, resulting in the increased availability of certain trace elements, an increase in nutritive parameters, and an increase in dry matter production.
Cobaltite CoAsS	Hydrothermal and metamorphic.	Crystallography: Orthorhombic 2 mm; Crystal habit: Granular and massive; Colour: Silver-white, greyish black; Lustre: Metallic, adamantine to dull; Transparency: Opaque; Streak: Grey-black; Cleavage: Perfect; Fracture: Uneven; Hardness: 5.5; Sp. gravity: 6-6.4.	Thermo-electric; used as Co ore.
Marcasite FeS_2	Sedimentary and hydrothermal. Association: Pyrite, pyrrhotite, galena, sphalerite etc.	Crystallography: Orthorhombic 2/m 2/m 2/m; Crystal habit: Tabular, prismatic, pyramidal, often twinned in the form of cockscomb-like aggregates, stalactitic, massive; Colour: Tin-white; Lustre: Metallic; Transparency: Opaque; Streak: Grey-black; Cleavage: Poor; Fracture: Uneven; Hardness: 6-6.5; Sp. Gravity: 4.9.	Production of sulfuric acid.
Arsenopyrite As_2S_3	Pegmatitic, hydro-thermal. Metamorphic, contact metamorphic skarns.	Crystallography: Monoclinic 2/m; Crystal habit: Thick tabular to prismatic striated crystals, granular, massive; Colour: Silver-white; Lustre: Metallic; Transparency: Opaque; Streak: Black; Cleavage: Good; Fracture: Uneven; Hardness: 5.5-6; Sp. gravity: 6.1.	As ore

(*Contd.*)

Table 7.2 (*Contd.*)

	Occurrence and association	*Properties*	*Applications*
Skutterudite $(Co, Ni)As_{3-x}$	Hydrothermal in medium to high temperature.	Crystallography: Isometric 2/m3; Crystal habit: Cubic crystals, skeletal aggregates, granular, massive; Colour: Tin-white to silver-grey; Lustre: Metallic; Transparency: Opaque; Streak: Black; Fracture: Uneven; Hardness: 5.5-6; Sp. gravity: 6.1-6.8.	It is a minor ore of cobalt and nickel.
Stibnite Sb_2S_3	Hydrothermal in medium and low temperature. Association: Realgar, orpiment, galena, marcasite, chalcedonic quartz etc.	Crystallography: Orthorhombic 2/m 2/m 2/m; Crystal habit: Thick to thin prismatic, needles, massive; Colour: Steel-grey; Lustre: Metallic; Transparency: Opaque; Streak: Lead grey; Cleavage: Perfect; Fracture: Conchoidal to uneven; Hardness: 2; Sp. gravity: 4.6.	Sb ore
Orpiment As_2S_3	Low-temperature, hydrothermal, in hot springs and fumaroles. Association: realgar, calcite, stibnite etc.	Crystallography: Monoclinic 2/m; Crystal habit: Prismatic, foliated and fibrous aggregates; Colour: Lemon-yellow to orange; Lustre: Resinous; Transparency: Translucent; Streak: Pale yellow to yellow; Cleavage: Perfect; Fracture: Even or sectile; Hardness: 1.5-2; Sp. gravity: 3.49.	As ore, pigment.
Realgar AsS	Low temperature, hydrothermal, sublimation product of volcanic gases, in hot springs and sediments. Association: As-Sb minerals.	Crystallography: Monoclinic 2/m; Crystal habit: Prismatic, striated, granular, massive; Colour: Red to orange; Lustre: Resinous; Transparency: Trans-parent to translucent; Streak: Red to orange; Cleavage: Good; Fracture: Conchoidal; Hardness: 1.5-2; Sp. gravity: 3.56.	As ore
Sulfosalts			
Tetrahedrite $(Cu,Fe)_{12}(Sb,As)_4S_{13}$	Hydrothermal in medium and low temperature, contact metamorphic. Association: Galena, pyrite, chalcopyrite, bornite etc.	Crystallography: Isometric 43/m; Crystal habit: Tetrahedral crystals, granular, massive; Colour: Silver to greyish black; Lustre: Metallic; Transparency: Opaque; Streak: Brown to black; Cleavage: None; Fracture: Subconchoidal; Hardness: 3-4.5; Sp. gravity: 4.5-5.1.	Used as an ore of copper.

(*Contd.*)

(*Contd.*)

Pyrargyrite Ag_3SbS_3	Hydrothermal and secondary oxidation deposits. Association: Acanthite, silver, sulfosalts, calcite, quartz.	Crystallography: Trigonal 3/m; Ag ore Crystal habit: Prismatic, rhombohedral to scalenohedral crystals, granular, massive; Colour: Ruby red; Lustre: Adamantine; Transparency: Translucent; Streak: Red to purple; Cleavage: Good; Fracture: Subconchoidal; Hardness: 2; Sp. gravity: 5.85.

7.3 SULPHATES

Anhydrous sulphate group

Barite	$Ba(SO_4)$
Celestine	$Sr(SO_4)$
Anglesite	$Pb(SO_4)$
Anhydrite	$Ca(SO_4)$

Hydrous sulphate group

Gypsum	$Ca(SO_4).\ 2H_2O$
Melanterite	$Fe(SO_4).\ 7H_2O$
Epsomite	$Mg(SO_4).\ 7H_2O$
Jarosite	$KFe_3(SO_4)_2(OH)_6$
Alunite	$KAl_3(SO_4)_2(OH)_6$

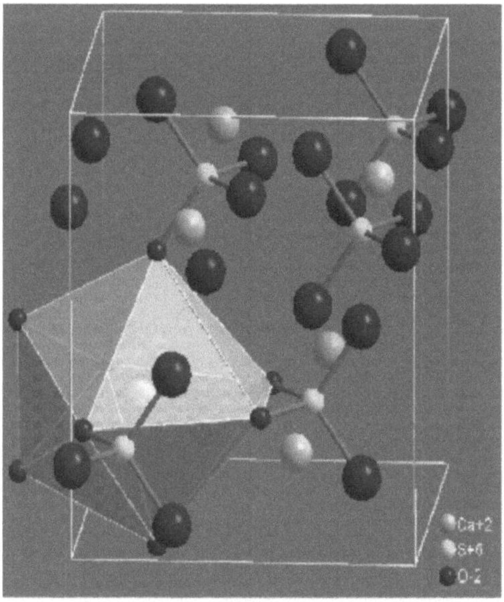

Fig. 7.6: Structure of anhydrite.

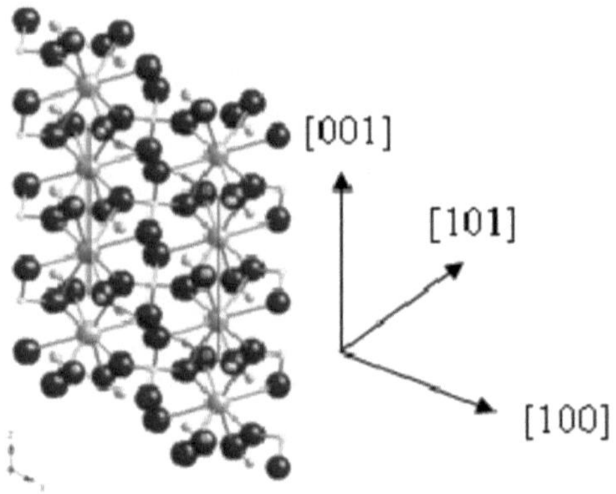

Fig. 7.7: Structure of gypsum.

Table 7.3: Mineralogical description of sulphate minerals

	Occurrence and association	*Properties*	*Applications*
Anhydrous sulphate group			
Barite $Ba(SO_4)$	Hydrothermal and sedimentary. Association: Fluorite, calcite, cinnabar etc.	Crystallography: Orthorhombic 2/m 2/m 2/m; Crystal habit: Crystals, rosettes or concretions; Colour: Colourless to white; Lustre: Vitreous to resineous; Transparency: Transparent to translucent; Streak: White; Cleavage: Perfect and good; Fracture: Uneven; Hardness: 3-3.5; Sp. gravity: 4.5.	Used as a weighing agent in drilling muds; as a pigment in paints and as a weighed filler for paper and cloth; primary ore for barium.
Celestine $Sr(SO_4)$	Commonly sedimentary, rarely hydrothermal. Association: Gypsum, anhydrite.	Crystallography: Orthorhombic 2/m 2/m 2/m; Crystal habit: Tabular, fibrous, granular, nodular; Colour: Colourless to pale blue or white; Lustre: Vitreous to pearly; Transparency: Transparent to translucent; Streak: White; Cleavage: Perfect and good; Fracture: Uneven; Hardness: 3-3.5; Sp. gravity: 4.0.	Has several interesting metaphysical characteristics; Used to aid mental activities and processes, integrating conciousness with instinct; as a healing stone.

(*Contd.*)

(*Contd.*)

Anglesite $Pb(SO_4)$	Secondary. Association: Cerrusite and other minerals.	Crystallography: Orthorhombic 2/m 2/m 2/m; Crystal habit: Massive, also granular to compact; Colour: Colourless to white; Lustre: Subadamantine to resinous; Transparency: Transparent to opaque; Streak: White; Cleavage: Good; Fracture: Conchoidal; Hardness: 2.5-3; Sp. gravity: 6.4.	Used as a secondary lead ore.
Anhydrite $Ca(SO_4)$	Commonly sedimentary, rarely pegmatitic and hydrothermal. Association: Gypsum, calcite and other minerals.	Crystallography: Orthorhombic 2/m 2/m 2/m; Crystal habit: Massive, fine granular to fibrous; Colour: Colourless to blue-white; Lustre: Vitreous to pearly; Transparency: Transparent; Streak: White; Cleavage: Perfect and good; Fracture: Uneven to splintery; Hardness: 3.5; Sp. gravity: 3.0.	Used as soil conditioner; mixed with Portland cement to control the rate of curing; manufacture of gypsum wallboard; mineral filler in joint compounds, plastics and paints; a source of sulfur.
Hydrous sulphate group			
Gypsum $Ca(SO_4).2H_2O$	Rarely hydrothermal, mostly secondary and weathering product. Association: Anhydrite, halite etc.	Crystallography: Monoclinic 2/m; Crystal habit: Massive, foliated; Colour: Colourless, white, grey, yellow, brown, orange or black; Lustre: Vitreous; Transparency: Transparent to translucent; Streak: White; Cleavage: Distinct and good; Fracture: Absent; Hardness: 2; Sp. gravity: 2.3.	A component of Portland cement used to prevent flash setting of concrete; plaster ingredient; fertilizer and soil conditioner; Plaster of Paris (surgical splints; casting moulds; modelling); adding hardness to water used for homebrewing.

(*Contd.*)

Table 7.3 (*Contd.*)

	Occurrence and association	*Properties*	*Applications*
Melanterite $Fe(SO_4).7H_2O$	Secondary. Association: Found in coal and lignites.	Crystallography: Monoclinic 2/m; Crystal habit: Stalactitic, fibrous, encrusting; Colour: Green, green-blue, greenish-blue; Lustre: Vitreous; Transparency: Translucent; Streak: Coloured; Cleavage: Perfect and distinct; Fracture: Conchoidal; Hardness: 2; Sp. gravity: 1.9.	
Epsomite $Mg(SO_4).7H_2O$	Secondary. Association: Halite	Crystallography: Orthorhombic 222; Crystal habit: Fibrous, woolly or colloform; Colour: Colourless, white, pink or green; Lustre: Vitreous to silky; Transparency: Transparent to translucent; Streak: White; Cleavage: Perfect and distinct; Fracture: Conchoidal; Hardness: 2-2.5; Sp. gravity: 1.7.	Used in the preparation of pharmaceutical products, as well as in dying material; used in the production of paper and sugar; used for the production of Epsom salts.
Jarosite $KFe_3(SO_4)_2(OH)_6$	Secondary. Association: Natrojarosite and other sulphates.	Crystallography: Hexagonal 3/m; Crystal habit: As fibrous, nodular; Colour: White to grey; Lustre: Pearly to vitreous; Transparency: Transparent to translucent; Streak: White; Cleavage: Imperfect to poor; Fracture: Absent; Hardness: 3.5-4; Sp. gravity: 2.6-2.9.	Used as substitute for gypsum.
Alunite $KAl_3(SO_4)_2(OH)_6$	Secondary. Association: quartz, kaolinite, allophane, diaspore.	Crystallography: Hexagonal 3/m; Crystal habit: Fibrous columnar; Colour: Ocherous yellow, brown; Lustre: Adamantine to vitreous; Transparency: Translucent; Streak: Pale yellow; Cleavage: Distinct; Fracture: Uneven to conchoidal; Hardness: 2.5-3.5; Sp. gravity: 2.9-3.3.	Used as a good source of alum. Sometimes, it is also used to recover potassium and aluminum. Commercially it is used as alloys, tools, equipment.

7.4 HALIDES

The halides consist of about 80 chemically related minerals with diverse structures and widely varied origins. The most common are halite (NaCl), sylvite (KCl), chlorargyrite (AgCl), cryolite (Na_3AlF_6), fluorite (CaF_2), and atacamite. The structure of sodium chloride is illustrated in Fig. 7.8. By the arrangement of the ions, it is evident that no molecules are present in the structure. Each cation and anion is in octahedral coordination with its six closest neighbours. The NaCl structure is found in the crystals of many *XZ*-type halides, including sylvite (KCl) and chlorargyrite (AgCl). Some sulfides and oxides of *XZ* type crystallize in this structure type as well—for example, galena (PbS), alabandite (MnS), and periclase (MgO).

Several XZ_2 halides have the same structure as fluorite (CaF_2). In fluorite, calcium cations are positioned at the corners and face centres of cubic unit cells. (A unit cell is the smallest group of atoms, ions, or molecules from which the entire crystal structure can be generated by its repetition.) Each fluorine anion is in tetrahedral coordination with four calcium ions, while each calcium cation is in eight-fold coordination with eight fluorine ions that form the corners of a cube around it.

Uraninite (UO_2) and thorianite (ThO_2) are two of the several oxides that have a fluorite-type structure.

Halite	(NaCl)
Fluorite	(CaF_2)
Chlorargyrite	(AgCl)

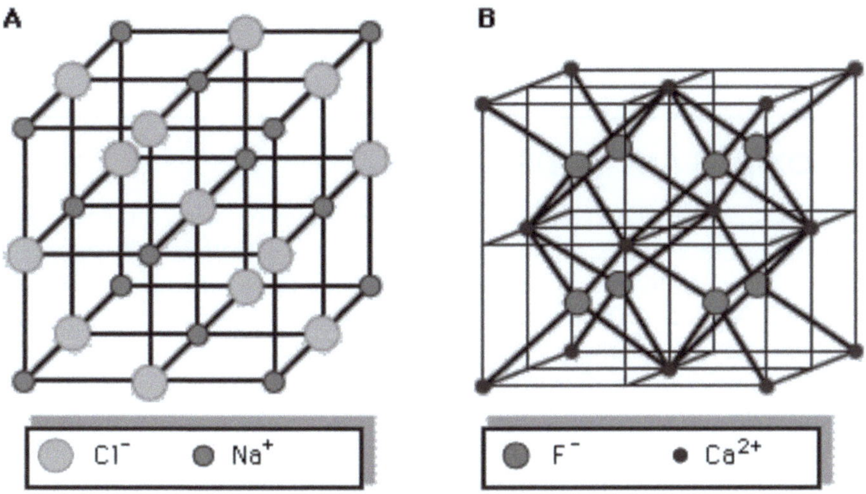

Fig. 7.8: Structures of halite and fluorite.

Table 7.4: Mineralogical descriptions of halides

	Occurrence and association	*Properties*	*Applications*
Halite $NaCl$	High temperature fumaroles, sedimentary. Association: Sylvite, camallite and other minerals.	Crystallography: Isometric 4/m 32/m; Crystal habit: Perfect cubic, granular, massive; Colour: Colourless, grey, white, red, blue; Lustre: Vitreous; Transparency: Transparent to translucent; Streak: White; Cleavage: Perfect cubic; Fracture: Conchoidal; Hardness: 2; Sp. gravity: 2.2.	Food and chemical industries.
Fluorite CaF_2	Rarely magmatic, mainly hydrothermal and metasomatic. Association: Tourmaline, quartz, galena, sphalerite, calcite, barite.	Crystallography: Isometric 4/m 32/m; Crystal habit: Cubic, granular, massive; Colour: Colourless, grey, white, red, blue, brown, black; Lustre: Vitreous; Transparency: Transparent to translucent; Streak: White; Cleavage: Perfect; Fracture: Conchoidal to splintery; Hardness: 4; Sp. gravity: 3.2.	Metallurgy, chemical industry, special optics.
Chlorargyrite $AgCl$	Secondary origin. Association: Native silver, cerussite, galena, limonite, calcite, barite, jarosite and wulfenite.	Crystallography: Isometric 4/m 32/m; Crystal habit: Columnar, massive; Colour: Colourless, pearl-grey to grey-green; Lustre: Resinous; Transparency: Transparent to translucent; Streak: White; Cleavage: Absent; Fracture: Poor conchoidal; Hardness: 2.3; Sp. gravity: 6-6.5.	

7.5 OXIDES AND HYDROXIDES

These classes consist of oxygen-bearing minerals; the oxides combine oxygen with one or more metals, while the hydroxides are characterized by hydroxyl $(OH)^-$ groups. Unlike the minerals of the sulfide class, which exhibit ionic, covalent, and metallic bonding, oxide minerals generally display strong ionic bonding. They are relatively hard, dense and refractory.

These minerals generally occur in small amounts in igneous and metamorphic rocks and also as preexisting grains in sedimentary rocks. Several oxides have great economic value, including the principal ores of iron (hematite and magnetite), chromium (chromite), manganese (pyrolusite, as well as the hydroxides, manganite and romanechite), tin (cassiterite) and uranium (uraninite).

Members of the hematite group are of the X_2O_3 type and have structures based on hexagonal closest packing of the oxygen atoms with octahedrally coordinated (surrounded by and bonded to six atoms) cations between them. Corundum and hematite share a common hexagonal architecture. In the ilmenite structure, iron and titanium occupy alternate Fe-O and Ti-O layers.

The XO_2-type oxides are divided into two groups. The first structure type, exemplified by rutile, contains cations in octahedral coordination with oxygen. The second resembles fluorite (CaF_2); each oxygen is bonded to four cations located at the corners of a fairly regular tetrahedron, and each cation lies within a cube at whose corners are eight oxygen atoms. This latter structure is exhibited by uranium, thorium and cerium oxides, whose considerable importance arises from their roles in nuclear chemistry.

The spinel-group minerals have type XY_2O_4 and contain oxygen atoms in approximate cubic closest packing. The cations located within the oxygen framework are octahedrally (six-fold) and tetrahedrally (four-fold) coordinated with oxygen.

The $(OH)^-$ group of the hydroxides generally results in structures with lower bond strengths than in the oxide minerals. The hydroxide minerals tend to be less dense than the oxides and also are not as hard. All hydroxides form at low temperatures and are found predominantly as weathering products, as, for example, from alteration in hydrothermal veins. Some common hydroxides are brucite [$Mg(OH)_2$], manganite [MnO(OH)], diaspore [AlO(OH)] and goethite [FeO(OH)]. The ore of aluminum, bauxite, consists of a mixture of diaspore, boehmite [AlO(OH)]—a polymorph of diaspore), and gibbsite [$Al(OH)_3$], plus iron oxides. Goethite is a common alteration product of iron-rich occurrences and is an iron ore in some localities.

Tetrahedral and octahedral oxide group

Corundum	(α-Al_2O_3)
Hematite	(Fe_2O_3)
Rutile	(TiO_2)
Pyrolusite	(MnO_2)
Cassiterite	(SnO_2)

Spinels and other oxides with mixed or unusual combination

Spinel	($MgAl_2O_4$)
Magnetite	($FeFe_2O_4$)
Chromite	($FeCr_2O_4$)
Franklinite	$ZnFe_2O_4$
Chrysoberyl	$BeAl_2O_4$
Uraninite	UO_2
Cuprite	Cu_2O

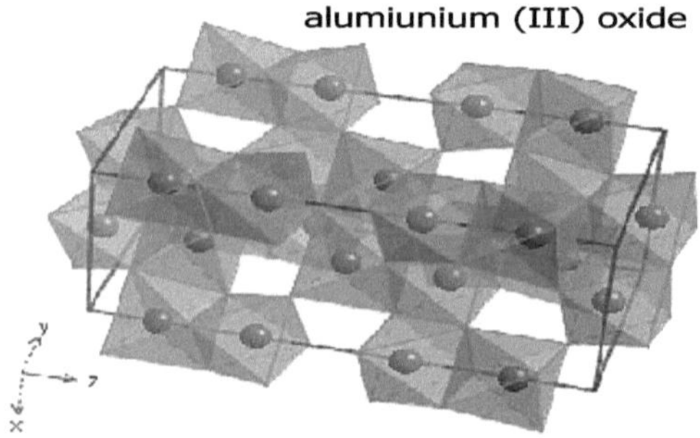

Fig. 7.9: Structure of corundum.

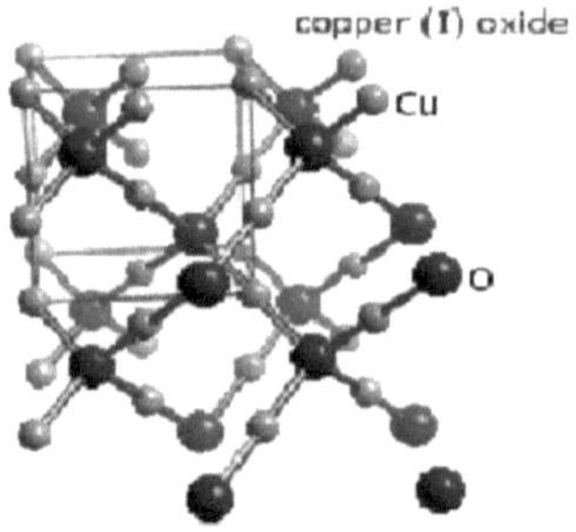

Fig. 7.10: Structure of cuprite.

Table 7.5: Mineralogical descriptions of oxide minerals

	Occurrence and association	*Properties*	*Applications*
Tetrahedral and octahedral oxide group			
Corundum α-Al_2O_3	Magmatic in andesites, pegmatites; metamor-phic and placers. Association: Andalusite, topaz, spinel.	Crystallography: Hexagonal 32/m; Crystal habit: Found as crystals, massive, granular, grey; Colour: Blue or pink; Lustre: Vitreous; Transparency: Transparent to translucent; Cleavage: Absent; Fracture: Uneven to conchoidal; Hardness: 9; Sp. gravity: 4.0.	Emery as abrasive material, sapphire and emerald as gemstones.
Hematite Fe_2O_3	Magmatic, hydrothermal, sedimentary, metamorphic.	Crystallography: Hexagonal 32/m; Crystal habit: As rosettes, columnar, granular, colloform, lamellar, compact, micaceous; Colour: Reddish brown to black; Lustre: Metallic to earthy; Transparency: Translucent to opaque; Streak: Cherry-red; Cleavage: Absent; Fracture: Uneven to subconchoidal; Hardness: 5-6.5; Sp. gravity: 5-6.5.	Important Fe ore

(*Contd.*)

(*Contd.*)

Rutile TiO_2	Magmatic, metamorphic and also in placers. Association: Monazite, topaz, beryl, quartz.	Crystallography: Tetragonal 4/m 2/m 2/m; Crystal habit: Well-formed crystals, also compact massive; Colour: Reddish brown to black, red; Lustre: Adamantine, splendant to submetallic; Transparency: Transparent to opaque; Streak: Pale brown, yellowish; Cleavage: Distinct and poor; Fracture: Uneven to subconchoidal; Hardness: 6-6.5; Sp. gravity: 4.2.	Main uses for rutile are the manufacture of refractory ceramic, as a pigment, and for the production of titanium metal; widely used as a welding electrode covering.
Pyrolusite MnO_2	Secondary and hydro-thermal; Association: Oxides of iron and manganese.	Crystallography: Tetragonal 4/m; 2/m 2/m; Crystal habit: Massive, columnar, colloform, granular and powdery; Colour: Light to dark steel grey, metallic; Transparency: Opaque; Streak: Black; Cleavage: Perfect; Fracture: Uneven; Hardness: 6-6.5; Sp. gravity: 5.0.	Important Mn ore
Cassiterite SnO_2	Magmatic in pegmatites, hydrothermal, metamorphic and in placers. Association: Wolframite, topaz.	Crystallography: Tetragonal 4/m; 2/m 2/m; Crystal habit: Colloform with radiating fibres, massive and granular; Colour: Brown or black; Lustre: Subadamantine to submetallic; Transparency: Transparent to opaque; Streak: White to brown; Cleavage: Imperfect; Fracture: Uneven to subconchoidal; Hardness: 6-7; Sp. gravity: 6.8-7.1.	Important Sn ore
Spinels and other oxides with mixed or unusual combination			
Spinel $MgAl_2O_4$	Magmatic, meta-morphic and placers. Association: Corundum, sillimanite and other minerals.	Crystallography: Isometric 4/m 32/m; Crystal habit: As crystals, also massive, granular and compact; Colour: Variously coloured; Lustre: Vitreous; Transparency: Transparent; Streak: White; Cleavage: Absent; Fracture: Conchoidal to splintery; Hardness: 7.5-8; Sp. gravity: 3.6.	Gemstone

(*Contd.*)

Table 7.5 (*Contd.*)

	Occurrence and association	*Properties*	*Applications*
Magnetite $FeFe_2O_4$	Magmatic, metamorphic an also sedimentary. Association: Franklinite.	Crystallography: Isometric 4/m 32/m; Crystal habit: Massive, granular and as crystals; Colour: Black; Lustre: Submetallic; Transparency: Opaque; Streak: Black; Cleavage: Absent; Fracture: Uneven to subconchoidal; Hardness: 5.5-6.5; Sp. gravity: 5.2.	Fe ore
Chromite $FeCr_2O_4$	Magmatic. Association: Magnetite, uvarovite.	Crystallography: Isometric 4/m 32/m; Crystal habit: Granular to compact masses; Colour: Black; Lustre: submetallic; Transparency: Opaque; Streak: Brown; Cleavage: Absent; Fracture: Uneven; Hardness: 5.5; Sp. gravity: 4.6.	Important Cr ore; used as a refractory material.
Franklinite $ZnFe_2O_4$	Metamorphic. Association: Willemite, zincite.	Crystallography: Isometric 4/m 32/m; Crystal habit: Granular, massive; Colour: Black; Lustre: Metallic; Transparency: Opaque; Streak: Dark brownish; Cleavage: Absent; Fracture: Uneven to conchoidal; Hardness: 6; Sp. gravity: 5.1-5.5.	Zn ore
Chrysoberyl $BeAl_2O_4$	Magmatic, metamorphic. Association: Schorl, phenakite.	Crystallography: Orthorhombic 2/m 2/m 2/m; Crystal habit: Thin to thick tabular crystals; Colour: Yellow, green, blue-green; Lustre: Vitreous; Transparency: Transparent to translucent; Streak: White; Cleavage: Good; Fracture: Uneven to conchoidal; Hardness: 8.5; Sp. gravity: 3.8.	Gemstone
Uraninite UO_2	Mainly hydrothermal in ore veins, skarns, magmatic, sedimentary and in placers. Association: Other U-bearing minerals.	Crystallography: Isometric 4/m 32/m; Crystal habit: Massive, often colloform; Colour: Black; Lustre: Submetallic to pitch-like to dull; Transparency: Opaque; Streak: Brownish black; Cleavage: Absent; Fracture: Uneven to conchoidal; Hardness: 5.5- 6; Sp. gravity: 10.8.	It has a small amount of radio-activity; hence used in: nuclear Navy ships and submarines, weapons, nuclear reactors in power plants, atomic bombs; also used on the outsides of army tanks, bullets and missiles. Uranium is used in ceramic glazes, fertilizers or chemicals used in photography.

(*Contd.*)

(*Contd.*)

Cuprite Cu_2O	Secondary, as a result of oxidation. Association: Limonite, native copper, malachite, azurite.	Crystallography: Isometric 4/m 32/m; Crystal habit: As crystals or aggregates, also massive, granular or earthy; Colour: Shades of red; Lustre: Adamantine, splendant to metallic to earthy; Transparency: Subtransparent to subtranslucent; Streak: Brown red; Cleavage: Poor; Fracture: Uneven to conchoidal; Hardness: 3.5-4; Sp. gravity: 6.1.	Ore of copper.

Hydroxides

Gibbsite	γ-$Al(OH)_2$, Al_2O_3, $3H_2O$
Brucite	$Mg(OH)_2$
Goethite	α-$FeO(OH)$
Diaspore	α-$AlO(OH)$, Al_2O_3, H_2O
Romanechite (Psilomelane)	$(Ba, H_2O)(Mn^{4+},Mn^{3+})_5O_{10}$

Table 7.6: Mineralogical description of hydroxide minerals

	Occurrence and association	*Properties*	*Applications*
Gibbsite γ-$Al(OH)_2$, Al_2O_3, $3H_2O$	Hydrothermal, secondary, metamorphic. Association: Goetite.	Crystallography: Monoclinic 2/m; Crystal habit: Tiny crystals with pseudohexagonal outlines; Colour: White; Lustre: Pearly; Transparency: Transparent; Cleavage: Perfect; Fracture: Uneven to conchoidal; Hardness: 2.5-3.5; Sp. gravity: 2.4.	Ore of aluminium.
Brucite $Mg(OH)_2$	Hydrothermal and rarely metamorphic. Association: Serpentine, chlorite, dolomite.	Crystallography: Hexagonal 32/m; Crystal habit: Foliated masses and fibrous aggregates; Colour: White to pale green, grey, brown or blue; Lustre: Pearly; Transparency: Transparent; Streak: White; Cleavage: Perfect; Fracture: Sectile; Hardness: 2.5; Sp. gravity: 2.4.	Used as a flame retardant and also constitutes a significant source of magnesium for industry.

(*Contd.*)

Table 7.6 (*Contd.*)

	Occurrence and association	*Properties*	*Applications*
Goethite α-FeO(OH)	Secondary, hydrothermal. Association: Pyrite and Fe-sulphides.	Crystallography: Orthorhombic 2/m 2/m 2/m; Crystal habit: Massive, colloform, stalactitic; Colour: Shades of brown; Lustre: Submetallic to dull; Transparency: Transparent to translucent; Streak: Brown yellow; Cleavage: Perfect; Fracture: Uneven; Hardness: 5-5.5; Sp. gravity: 3.3-4.3.	Ore of aluminium.
Diaspore α-AlO(OH)	Hydrothermal. Association: Andalusite, pyrophyllite, corundum.	Crystallography: Orthorhombic 2/m 2/m 2/m; Crystal habit: Foliated, massive, scaly; Colour: White or colourless; Lustre: Brilliant to vitreous; Transparency: Transparent; Cleavage: Perfect; Fracture: Conchoidal; Hardness: 6.5-7; Sp. gravity: 3.2-3.5.	Ore of aluminium.
Romanechite (Psilomelane) Ba, H_2O $(Mn^{4+},Mn^{3+})_5$ O_{10}	Secondary. Association: Pyrolusite, goethite and calcite.	Crystallography: Monoclinic 2/m; Crystal habit: Massive, colloform and earthy; Colour: Black to grey; Lustre: Submetallic to dull; Transparency: Opaque; Streak: Infusible; Hardness: 5-6; Sp. gravity: 4.7.	Is a valuable ore of manganese, which is used in steel making.

7.6 CARBONATES

The carbonate minerals contain the anionic complex $(CO_3)^{2-}$, which is triangular in its coordination—i.e., with a carbon atom at the centre and an oxygen atom at each of the corners of an equilateral triangle. These anionic groups are strongly bonded, individual units and do not share oxygen atoms with one another. The triangular carbonate groups are the basic building units of all carbonate minerals and are largely responsible for the properties particular to the class.

Carbonates are frequently identified using the effervescence test with acid. The reaction that results in the characteristic fizz, $2H^+ + CO\ ^2/_3\ (H_2O) + CO_2$, makes use of the fact that the carbon-oxygen bonds of the CO_3 groups are not quite as strong as the corresponding carbon-oxygen bonds in carbon dioxide.

The common anhydrous carbonates are divided into three groups that differ in structure type: calcite, aragonite and dolomite. The copper carbonates azurite and malachite are the only notable hydrous varieties.

The members of the calcite group share a common structure type. It can be considered as a derivative of the NaCl structure in which CO_3 groups substitute for the chlorine ions and calcium cations replace the sodium cations. As a result of the triangular shape of the CO_3 groups, the structure is rhombohedral instead of isometric as in NaCl. The CO_3 groups are in planes perpendicular to the three-fold *c* axis, and the calcium ions occupy alternate planes and are bonded to six oxygen atoms of the CO_3 groups.

Members of the calcite group exhibit perfect rhombohedral cleavage. The composition $CaCO_3$ most commonly occurs in two different polymorphs: rhombohedral calcite with calcium surrounded by six closest oxygen atoms and orthorhombic aragonite with calcium surrounded by nine closest oxygen atoms.

When CO_3 groups are combined with large divalent cations (generally with ionic radii greater than 1.0 Å), orthorhombic structures result. This is known as the aragonite structure type. Members of this group include those with large cations: $BaCO_3$, $SrCO_3$ and $PbCO_3$. Each cation is surrounded by nine closest oxygen atoms.

The aragonite group displays more limited solid solution than the calcite group. The type of cation present in aragonite minerals is largely responsible for the differences in physical properties among the members of the group. Specific gravity, for example, is roughly proportional to the atomic weight of the metal ions.

Dolomite [$CaMg(CO_3)_2$], kutnahorite [$CaMn(CO_3)_2$] and ankerite [$CaFe(CO_3)_2$] are three isostructural members of the dolomite group. The dolomite structure can be considered as a calcite-type structure in which magnesium and calcium cations occupy the metal sites in alternate layers. The calcium (Ca^{2+}) and magnesium (Mg^{2+}) ions differ in size by 33 percent, and this produces cation ordering with the two cations occupying specific and separate levels in the structure. Dolomite has a calcium-to-magnesium ratio of approximately 1: 1, which gives it a composition intermediate between $CaCO_3$ and $MgCO_3$.

Calcite group	
Calcite	$CaCO_3$
Magnesite	$Mg(CO_3)$
Siderite	$Fe(CO_3)$
Rhodochrosite	$Mn(CO_3)$
Smithsonite	$Zn(CO_3)$
Dolomite group	
Dolomite	$CaMg(CO_3)_2$
Ankerite	$CaFe(CO_3)_2$
Aragonite group	
Aragonite	$Ca(CO_3)$
Witherite	$Ba(CO_3)$
Strontianite	$Sr(CO_3)$
Cerussite	$Pb(CO_3)$

Other carbonates

Malachite	$Cu_2(CO_3)(OH)_2$
Azurite	$Cu_3(CO_3)(OH)_2$

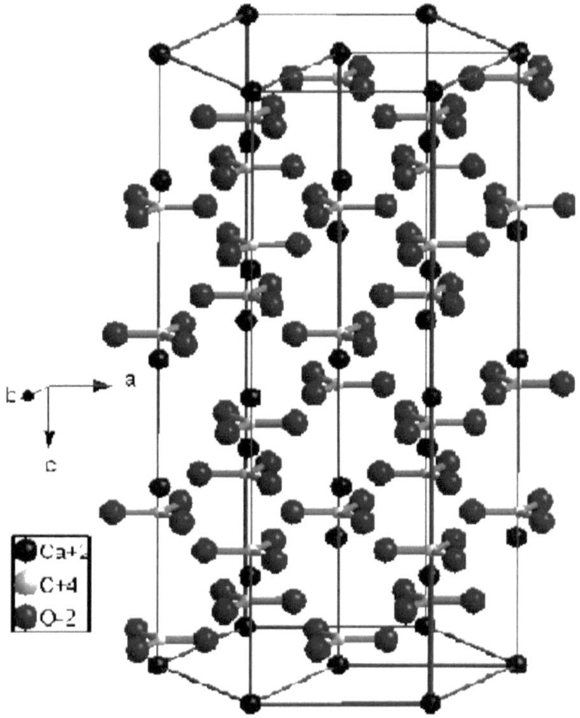

Fig. 7.11: Structure of calcite.

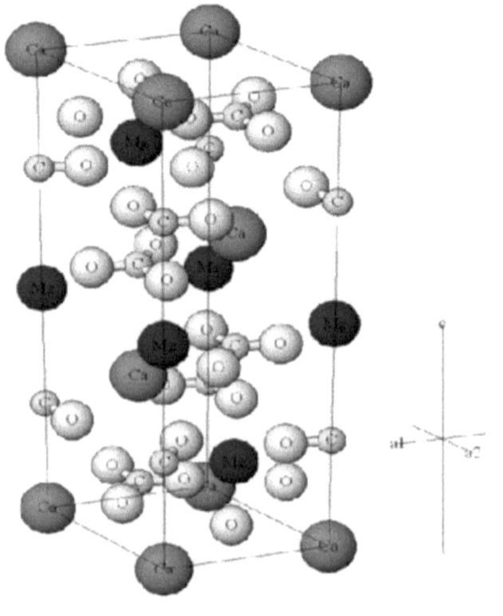

Fig. 7.12: Structure of malachite.

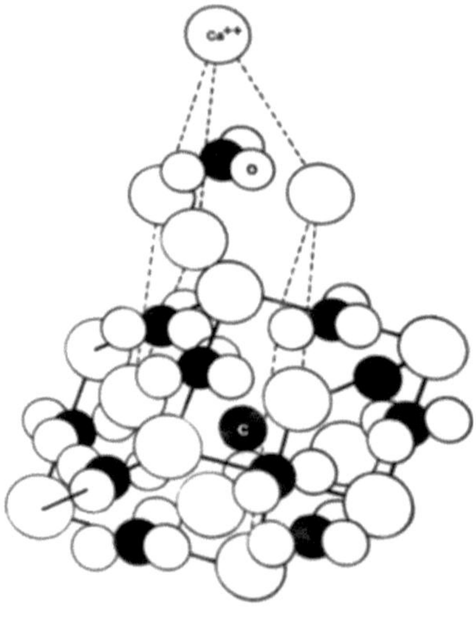

Fig. 7.13: Structure of aragonite.

Table 7.7: Mineralogical description of carbonate minerals

	Occurrence and association	*Properties*	*Applications*
Calcite group			
Calcite $CaCO_3$	Magmatic, hydrothermal, metamorphic, sedimentary and secondary. Association: Sulphides, barite, quartz, fluorite, dolomite and siderite; nepheline, orthoclase and Nb-Ta minerals.	Crystallography: Hexagonal 32/m; Crystal habit: Coarse granular, stalactitic, nodular, encrusting; Colour: Colourless to white; Lustre: vitreous; Transparency: Transparent; Streak: White; Cleavage: Perfect; Fracture: Conchoidal; Hardness: 3; Sp. gravity: 2.7.	Used for making cement, as a flux in the smelting of metallic ores, as a fertilizer, and for use as a building stone/construction material..
Magnesite $Mg(CO_3)$	Rarely magmatic, mainly hydrothermal metasomatic and metamorphic. Association: Calcite, dolomite.	Crystallography: Hexagonal 32/m; Crystal habit: Massive, granular, compact, earthy, fibrous; Colour: Colourless to white; Lustre: Vitreous to earthy; Transparency: Transparent to translucent; Streak: White; Cleavage: Perfect; Fracture: Conchoidal; Hardness: 4; Sp. gravity: 3.0.	Used as a slag former in steel-making furnaces, in conjunction with lime; used as a catalyst and filler in the production of synthetic rubber and in the preparation of magnesium chemicals and fertilizers. Use in refractory materials; used as a binder in flooring material; used in fire assay, for cupellation.
Siderite $Fe(CO_3)$	Hydrothermal, sedimentary. Association: Clay, shale, coal and quartz, carbonatite, pegmatite.	Crystallography: Hexagonal 32/m; Crystal habit: Rhombohedral crystals, colloform, fibrous, oolitic, earthy; Colour: Yellow brown to greyish brown; Lustre: Subadamantine; Transparency: Transparent to translucent; Streak: White; Cleavage: Perfect; Fracture: Uneven to conchoidal; Hardness: 4-4.5; Sp. gravity: 3.9.	Ore of Fe and in pigments.

(*Contd.*)

Table 7.7 (*Contd.*)

	Occurrence and association	*Properties*	*Applications*
Rhodochrosite $Mn(CO_3)$	Rarely pegmatitic, hydrothermal, sedimentary and metamorphic. Association: Rhodonite, garnet, manganese oxides.	Crystallography: Hexagonal 32/m; Crystal habit: Massive, granular, botryoidal; Colour: Shade of pink, brown or yellow; Lustre: Vitreous to pearly; Transparency: Transparent to subtranslucent; Streak: White; Cleavage: Perfect; Fracture: Conchoidal; Hardness: 3.5-4; Sp. gravity: 3.3-3.7.	Ore of manganese, ornamental use.
Smithsonite $Zn(CO_3)$	Supergene. Association: Hemimorphite, cerussite, malachite, azurite or anglesite.	Crystallography: Hexagonal 32/m; Crystal habit: Colloform, earthy; Colour: Variable; Lustre: Subadamantine; Transparency: Translucent; Streak: White; Cleavage: Good; Fracture: Uneven to subconchoidal; Hardness: 4-4.5; Sp. gravity: 4.4.	Used as an ornamental stone; used as an astringent and excipient in shampoo, as a fireproofing filler for rubber and plastics, as a feed additive, as a pigment, in cosmetics and lotions, and in the manufacturing of porcelain, pottery, and rubber.
Dolomite group			
Dolomite $CaMg(CO_3)_2$	Magmatic in pegmatite, hydrothermal, metasomatic, sedimentary and metamorphic. Association: Siderite, magnesite, calcite.	Crystallography: Hexagonal 3; Crystal habit: Rhombohedral crystals; Colour: Colourless to white; Lustre: Pearly; Transparency: Translucent; Streak: White; Cleavage: Perfect; Fracture: Subconchoidal; Hardness: 3.5-4; Sp. gravity: 2.9.	Chiefly used as refractory, ramming, and fettling material in steel melting shop, and as fluxing material in blast furnace operation in secondary steel and ferromanganese manufacture. To a lesser extent it is used in the glass industry especially in sheet-glass manufacture; also used in the manufacture of mineral wool.

(*Contd.*)

(*Contd.*)

Ankerite $CaFe(CO_3)_2$	Hydrothermal, sedimentary and metamorphic. Association: Siderite and other minerals.	Crystallography: Hexagonal 3; Crystal habit: Rhombohedral crystals, granular; Colour: White, yellowish, brown; Lustre: Pearly; Transparency: Translucent; Streak: White; Cleavage: Perfect; Fracture: Conchoidal; Hardness: 3.5-4; Sp. gravity: 3.	
Aragonite group			
Aragonite $Ca(CO_3)$	Primary as late hydrothermal, secondary, sedimentary and metamorphic. Association: Limonite, siderite, calcite, malachite, azurite, smithsonite, lawsonite, glaucophane and jadeite.	Crystallography: Orthorhombic 2/m 2/m 2/m; Crystal habit: Columnar, aciculur crystals, colloform, fibrous, pisolitic; Colour: Colourless to white; Lustre: Vitreous to resineous; Transparency: Translucent; Streak: White; Cleavage: Imperfect; Fracture: Subconchoidal; Hardness: 3.5-4; Sp. gravity: 2.9.	Used in soil-treatment.
Witherite $Ba(CO_3)$	Hydrothermal. Association: Barite, galena.	Crystallography: Orthorhombic 2/m 2/m 2/m; Crystal habit: Globular, colloform, columnar or granular; Colour: Colourless to white; Lustre: Vitreous to resineous; Transparency: Transparent to translucent; Streak: White; Cleavage: Distinct and imperfect; Fracture: Uneven; Hardness: 3-3.5; Sp. gravity: 4.3.	Used to make barium compounds; used in case-hardening of steel and also in sugar refining.
Strontianite	Hydrothermal and sedimentary. Association: Barite, calcite, limestone, marl.	Crystallography: Orthorhombic 2/m 2/m 2/m; Crystal habit: Pseudohexagonal prismatic crystals; Colour: Colourless to grey; Lustre: Vitreous to resineous; Transparency: Transparent to translucent; Streak: White; Cleavage: Perfect; Fracture: Uneven to subconchoidal; Hardness: 3.5; Sp. gravity: 3.8.	Used as a secondary ore of strontium.

(*Contd.*)

Table 7.7 (*Contd.*)

	Occurrence and association	*Properties*	*Applications*
Cerussite $Pb(CO_3)$	Secondary. Association: Anglesite or iron oxides.	Crystallography: Orthorhombic 2/m 2/m 2/m; Crystal habit: Massive, granular to compact, earthy; Colour: Colourless to white; Lustre: Subadamantine; Transparency: Transparent to translucent; Streak: White; Cleavage: Distinct and fair; Fracture: Conchoidal; Hardness: 3-3.5; Sp. gravity: 6.6.	Used as an ore of lead.
Other carbonates			
Malachite $Cu_2(CO_3)(OH)_2$	Supergene. Association: Azurite, cuprite, limonite, calcite, chalcedony, chrysocolla.	Crystallography: Monoclinic 2/m; Crystal habit: Massive, encrusting, colloform; Colour: Bright green; Lustre: Subadamantine; Transparency: Translucent to opaque; Streak: Pale green; Cleavage: Perfect and distinct; Fracture: Subconchoidal to uneven; Hardness: 3.5-4; Sp. gravity: 4.	Used as an important ore of copper, as an ornamental stone, a pigment and for jewellery.
Azurite $Cu_3(CO_3)(OH)_2$	Secondary. Association: Copper oxides and sulphides.	Crystallography: Monoclinic 2/m; Crystal habit: Massive, encrusting, earthy; Colour: Light to dark azure blue; Lustre: Vitreous; Transparency: Transparent to opaque; Streak: Blue; Cleavage: Perfect; Fracture: Conchoidal; Hardness: 3.5-4; Sp. gravity: 3.8.	Used as Cu-ore, as a gemstone and rarely as pigment.

7.7 BORATES

Minerals of the borate class contain boron-oxygen groups that can link together, in a phenomenon known as polymerization, to form chains, sheets, and isolated multiple groups. The silicon-oxygen (SiO_4) tetrahedrons of the silicates polymerize in a manner similar to the $(BO_3)^{3-}$ triangular groups of the borates. A single oxygen atom is shared between two boron cations (B^{3+}), thereby linking the BO_3 groups into extended units such as double triangles, triple rings, sheets, and chains. The oxygen atom is able to accommodate two boron atoms because the small boron cation has a bond strength to each oxygen that is exactly one-half the bond energy of the oxygen ion.

Anhydrous borate group	
Ludwigite	$Mg_2Fe^{3+}BO_5$
Hydrous borate group	
Borax	$Na_2(B_4O_7).10H_2O$
Colemanite	$Ca_2(B_6O_{11}).5H_2O$

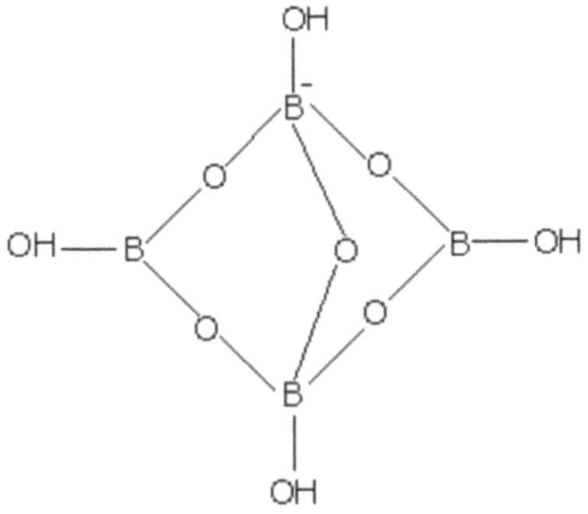

Fig. 7.14: Structure of hydrous borate.

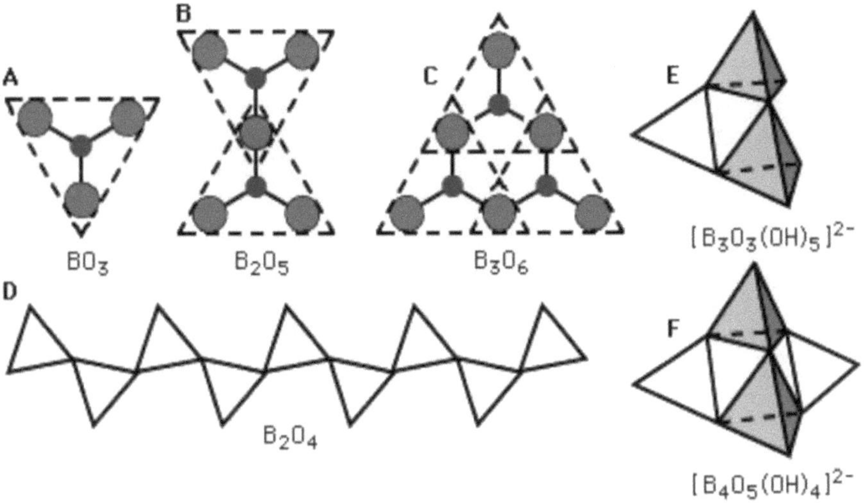

Fig. 7.15: Various possible linkages of (A) BO_3 triangles to form (B,C) multiple groups and (D) chains in borates. Complex (E) triangle and (F) quadrangle groups are also shown. The group depicted in (F) occurs in borax.

Table 7.8: Mineralogical description of borate minerals

	Occurrence and association	*Properties*	*Applications*
Anhydrous borate group			
Ludwigite $Mg_2Fe^{3+}BO_5$	Metamorphic in skarns and dolomitic marbles. Association: Magnetite and other borates.	Crystallography: Orthorhombic 2/m 2/m 2/m; Crystal habit: Prismatic, fibrous, massive, granular; Colour: Dark green, black-green, black; Lustre: Silky to dull; Transparency: Opaque; Streak: Blue-green; Cleavage: Perfect; Fracture: Uneven; Hardness: 5; Sp. gravity: 3.9.	Used for scientific and mineralogical study.
Hydrous borate group			
Borax $Na_2(B_4O_7).10H_2O$	Sedimentary. Association: Halite and other borates.	Crystallography: Monoclinic 2/m; Crystal habit: Granular aggregates; Colour: Colourless to white; Lustre: Vitreous to resineous; Transparency: Translucent to opaque; Streak: White; Cleavage: Perfect; Fracture: Conchoidal; Hardness: 2-2.5; Sp. gravity: 1.7.	Used for bufferin, co-complexing, flux, food additive, vaccine etc.
Colemanite $Ca_2(B_6O_{11}).5H_2O$	Sedimentary. Association: Other borates.	Crystallography: Monoclinic 2/m; Crystal habit: Massive, granular, compact; Colour: Colourless to white; Lustre: Vitreous to adamantine; Transparency: Transparent to translucent; Streak: White; Cleavage: Distinct and perfect; Fracture: Uneven to subconchoidal; Hardness: 4-4.5; Sp. gravity: 2.4.	Principal source of boron.

7.8 CHROMATES

Crocoite $PbCrO_4$

Table 7.9: Mineralogical description of chromate mineral

	Occurrence and association	*Properties*	*Applications*
Crocoite $PbCrO_4$	Secondary. Association: Cerussite, pyromorphic and wulfenite.	Crystallography: Monoclinic 2/m; Crystal habit: Massive, granular; Colour: Hyacinth red, orange-red, orange; Lustre: Vitreous to adamantine; Transparency: Transparent to translucent; Streak: Orange-yellow; Cleavage: Imperfect; Fracture: Absent; Hardness: 2.5-3; Sp. gravity: 6.1.	Only as mineral specimens and as a minor ore of chromium.

7.9 PHOSPHATES

Although this mineral class is large, most of its members are quite rare. Of the phosphates, only **apatite** [$Ca_5(PO_4)_3(F, Cl, OH)$], the most important and abundant, can be considered as truly common. The members of this group are characterized by tetrahedral anionic $(PO_4)^{3-}$ complexes, which are analogous to the $(SO_4)^{2-}$ groups of the sulfates. The phosphorus ion, with a valence of positive five, is only slightly larger than the sulfur ion, which carries a positive six charge.

Monazite	$(Ca,La,Nd,Th)(PO_4)$
Amblygonite	$LiAl(PO_4)F$
Montebrasite	$LiAl(PO_4)(OH)$
Apatite	$Ca_5(PO_4)_3(F,Cl,OH)$
Turquoise	$CuAl_6(PO_4)_4(OH)_8 \cdot 4H_2O$

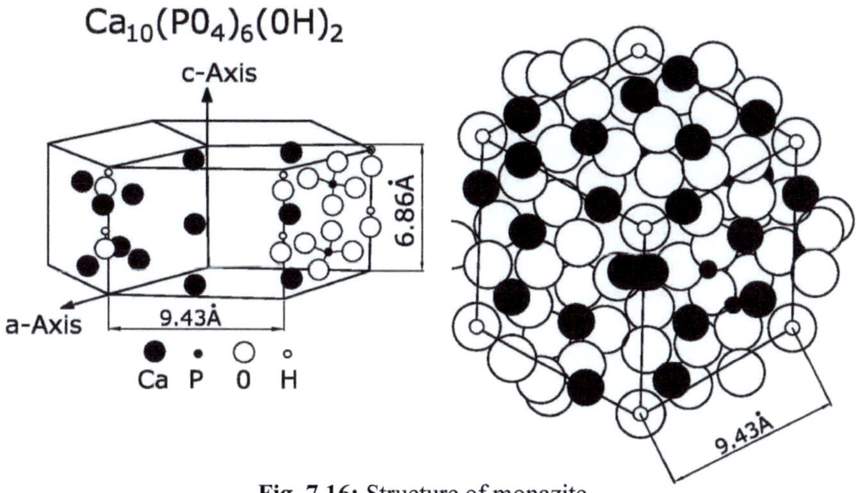

Fig. 7.16: Structure of monazite.

Table 7.10: Mineralogical description of phosphate minerals

	Occurrence and association	*Properties*	*Applications*
Monazite (Ca,La,Nd, Th)(PO_4)	Magmatic, metamorphic, commonly placers. Association: Apatite, xenoime, zircon.	Crystallography: monoclinic 2/m; Crystal habit: Crystal masses and rounded grains; Colour: Colourless or white; Lustre: Vitreous to dull; Transparency: Transparent to translucent; Streak: White; Cleavage: Imperfect and distinct; Fracture: Conchoidal to uneven; Hardness: 5.5-6; Sp. gravity: 3.1.	Used in high performance magnets, as pigment in ceramics, in robot motors, x-ray screens, fibre optics, energy efficient lanthanum lamps and colour television tubes.
Amblygonite LiAl(PO_4)F	Magmatic and hydrothermal. Association: Other phosphates.	Crystallography: Triclinic 1; Crystal habit: Short prismatic crystals, granular, massive; Colour: Yellow or red-brown, brown, green or white; Lustre: Vitreous to adamantine; Transparency: Translucent; Streak: White; Cleavage: Imperfect and distinct; Fracture: Conchoidal to uneven; Hardness: 5-5.5; Sp. gravity: 5.0-5.3.	Used as a source of lithium and phosphorous and also as a gemstone.
Montebrasite LiAl(PO_4) (OH)	Magmatic and hydrothermal. Association: Amblygonite.	Crystallography: Triclinic 1; Crystal habit: Columnar to compact massive; Colour: Yellow or white; Lustre: Vitreous to greasy; Transparency: Transparent to translucent; Streak: White; Cleavage: Perfect, good and distinct; Fracture: Subconchoidal to uneven; Hardness: 5.5-6; Sp. gravity: 2.9-3.2.	Used as a source of lithium and phosphorous.
Apatite $Ca_5(PO_4)_3$ (F,Cl,OH)	Pegmatitic, magmatic, hydrothermal and metamorphic. Association: Titanite, zircon, pyroxenes, amphiboles, spinel, vesuvianite and phlogopite.	Crystallography: Hexagonal 6/m; Crystal habit: Massive, granular or colloform; Colour: Shades of yellow-green, blue-green, violet-blue, blue or white; Lustre: Vitreous; Transparency: Transparent to opaque; Streak: White; Cleavage: Imperfect; Fracture: Conchoidal to uneven; Hardness: 5; Sp. gravity: 3.1-3.4.	It is the main source of phosphorous; used in the manufacture of phosphate fertilizers and is very important in the chemical and pharmaceutical industries. Apatite is also a popular collector mineral and some transparent specimens are faceted for collectors.

(Contd.)

(*Contd.*)

Turquoise $CuAl_6(PO_4)_4(OH)_8.4H_2O$	Secondary. Association: Kaolinite, limonite and chalcedony.	Crystallography: Triclinic 1; Crystal habit: Massive, compact or colloform; Colour: Sky blue to apple-green; Lustre: Waxy; Transparency: Translucent; Streak: White to pale green; Cleavage: Perfect and good; Fracture: Conchoidal to smooth; Hardness: 5-6; Sp. gravity: 2.6-2.8.	Used as a decorative stone and gemstone.

7.10 ARSENATES

Erythrite $Co_3(AsO_4)_2.8H_2O$

Table 7.11: Mineralogical description of arsenate mineral

	Occurrence and association	*Properties*	*Applications*
Erythrite $Co_3(AsO_4)_2.8H_2O$	Secondary. Association: Cobaltite and skutterudite.	Crystallography: Monoclinic 2/m; Crystal habit: Prismatic, aciculur, reniform; Colour: crimson, pink, purple-red; Lustre: Adamantine; Transparency: Transparent to translucent; Streak: Pale purple; Cleavage: Perfect basal; Fracture: Sectile; Hardness: 1.5-2.5; Sp. gravity: 3.06.	Used as a mineral specimen, and as an indicator in prospecting for cobalt and nickel.

7.11 VANDATES

Vanadinite $Pb_5(VO_4)_3Cl$

Carnotite $K_2(UO_2)_2(VO_4)_2.3H_2O$

Table 7.12: Mineralogical description of arsenate mineral

	Occurrence and association	*Properties*	*Applications*
Vanadinite $Pb_5(VO_4)_3Cl$	Secondary. Association: Galena, cerussite or limonite.	Crystallography: Hexagonal 6/m; Crystal habit: Prismatic, rounded or globular; Colour: Ruby red, orange, brown, yellow; Lustre: Resinous; Transparency: Translucent; Streak: White; Cleavage: Absent; Fracture: Subconchoidal; Hardness: 3; Sp. gravity: 6.9.	Used as steel alloys, auto and railway equipment, as mineral specimens, as an ore of vanadium.
Carnotite $K_2(UO_2)_2(VO_4)_2.3H_2O$	Secondary. Association: Sandstone and conglomerate.	Crystallography: Monoclinic 2/m; Crystal habit: Fine powder to crumbly aggregates; Colour: Yellow, yellow-green; Lustre: Dull; Transparency: Translucent; Streak: Yellow; Cleavage: Perfect; Fracture: Uneven; Hardness: 1; Sp. gravity: 4.5.	Used as an important ore of uranium and vanadium and as mineral specimens.

7.12 TUNGSTATES AND MOLYBDATES

Ferberite	$Fe(WO_4)$
Scheelite	$Ca(WO_4)$
Wulfenite	$Pb(MoO_4)$

Table 7.13: Mineralogical description of tungstates and molybdates

	Occurrence and association	*Properties*	*Applications*
Ferberite $Fe(WO_4)$	Pegmatitic, hydrothermal, placer. Association: Pyrrhotite, pyrite, chalcopyrite, arsenopyrite, scheelite, native bismuth, quartz, siderite, barite, fluorite etc.	Crystallography: Monoclinic 2/m; Crystal habit: Granular; Colour: Black, brownish black or yellowish black; Lustre: Sub-metallic; Transparency: Transparent to opaque; Streak: Brown black to black; Cleavage: Perfect; Fracture: Uneven; Hardness: 4-4.5; Sp. gravity: 7.1.	Used as an ore of tungsten.
Scheelite $Ca(WO_4)$	Sedimentary and hydrothermal. Association: Garnet, diopside, tremolite, hornblende, epidote etc.	Crystallography: Tetragonal 4/m; Crystal habit: Massive, granular and columnar; Colour: Colourless to white; Lustre: Sub-metallic; Transparency: Subadamantine; Streak: White; Cleavage: Distinct; Fracture: Subconchoidal to uneven; Hardness: 4.5-5; Sp. gravity: 6.1.	Used as an important source of tungsten, rarely cut as gemstones and as mineral specimens.

(*Contd.*)

(*Contd.*)

Wulfenite $Pb(MoO_4)$	Secondary. Association: Pyromorphite, cerussite, limonite, calcite, galena etc.	Crystallography: Tetragonal 4/m; Crystal habit: Granular or compact; Colour: Orange to yellow; Lustre: Resineous to adamantine; Transparency: Transparent; Streak: White; Cleavage: Distinct; Fracture: Subconchoidal to uneven; Hardness: 3; Sp. gravity: 6.5-7.0.	Used as a minor ore of molybdenum and as mineral specimen.

7.13 SILICATES

Silicates form the most abundant and most fundamental group of minerals, found on earth. The principal unit of the silicate structures is the silicon-oxygen $(SiO_4)^{4-}$ tetrahedron. The central silicon (Si^{4+}) atom is bonded to four surrounding oxygen atoms of a regular tetrahedron. Each oxygen ion (O^{2-}) is capable of bonding with silicon cation of another tetrahedron. So by the process of polymerization, i.e. by sharing oxygen atoms, SiO_4 tetrahedrons are linked to one another. On the basis of the type of polymerization, silicates can be divided into various groups. It is the most essential rock-forming mineral group.

7.13.1 Nesosilicates

- In these silicate structures, the Si : O is 1: 4.
- Only ionic bonds are active here.
- Crystal habits are equidimensional with no predominant cleavage directions.
- Relatively high specific gravity and hardness due to dense packing of atoms in these structures.

Common nesosilicates

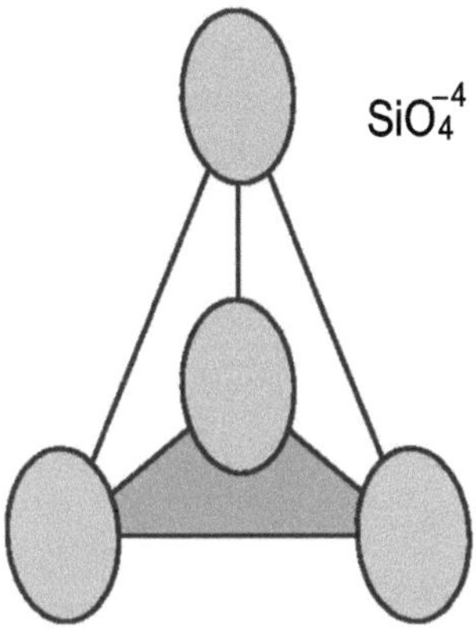

Fig. 7.17: Structure of nesosilicate.

Olivine group	Y_2SiO_4 Y
Forsterite	Mg
Fayalite	Fe
Tephroite	Mn
Garnet group	$X_3Y_2(SiO_4)$
Zircon	$ZrSiO_4$
Topaz	$Al_2(SiO_4)_4(F,OH)_2$
Chloitoid	$(Fe^{2+},Mg,Mn)_2(Al,Fe^{3+})Al_3O_2(SiO_4)2(OH)_2$

Table 7.14: Mineralogical description of nesosilicates

	Occurrence and association	*Properties*	*Applications*
Olivine group Y_2SiO_4 Y Forsterite = Mg Fayalite = Fe Tephroite = Mn	Magmatic in some ultrabasic rocks. Association: Diopside, magnetite, pyrope.	Crystallography: Orthorhombic 2/m 2/m 2/m; Crystal habit: Embedded grains or granular masses; Colour: Olive green, brownish-green; Lustre: Vitreous; Transparency: Transparent to translucent; Streak: White; Cleavage: Not distinct; Fracture: Conchoidal; Hardness: 6.5-7; Sp. gravity: 3.3.	Used in CO_2 sequestration. The aluminium foundry industry uses olivine sand to cast objects in aluminium. As gemstones, industrial uses as refractory sands and abrasives.
Garnet group $X_3Y_2(SiO_4)$	Magmatic, metamorphic and also hydrothermal. Association: Variable.	Crystallography: Isometric 4/m 32/m; Crystal habit: Well-formed crystals, also massive; Colour: red, brown, yellow, white, green; Lustre: Vitreous to sub-adamantine; Transparency: Transparent to translucent; Streak: White; Cleavage: Not distinct; Fracture: Subconchoidal to uneven; Hardness: 6.5-7.5; Sp. gravity: 3.5-4.3.	Garnet is ground to a variety of sizes to be used as an abrasive. Garnet sandpaper was the original application of this mineral. It is also used to make a number of similar products, including sanding belts, discs, and strips. Today, the vast majority of garnet is used as an abrasive blasting material, for water filtration, in a process called water jet cutting, and to make abrasive powders..

(*Contd.*)

(*Contd.*)

Zircon $ZrSiO_4$	Magmatic, metamorphic and rarely hydrothermal, also sedimentary and placer. Association: Found as an accessory mineral.	Crystallography: Tetragonal 4/m; Crystal habit: Well-formed crystals, tiny and embedded grains; Colour: Shades of brown, colourless, grey, grey or red; Lustre: Vitreous to subadamantine; Transparency: Transparent to opaque; Streak: White; Cleavage: Imperfect and poor; Fracture: Conchoidal to uneven; Hardness: 7.5; Sp. gravity: 4.7.	Used in ceramics, refractories, TV glass, chemicals and as gemstone.
Topaz $Al_2(SiO_4)_4$ $(F,OH)_2$	Magmatic, hydrothermal and in placers. Association: Lithium minerals, fluorite, cassiterite, tourmaline, apatite, beryl, quartz, mica, feldspar.	Crystallography: Orthorhombic 2/m 2/m 2/m; Crystal habit: Stubby prism with a square, 8-sided or diamond shaped; Colour: Colourless or light yellow; Lustre: Vitreous; Transparency: Transparent to translucent; Streak: White; Cleavage: Perfect; Fracture: Subconchoidal to uneven; Hardness: 8; Sp. gravity: 3.5.	The only significant commercial use of topaz is as a gemstone. It also serves as the Mohs Hardness Scale standard for a hardness of eight.
Chloritoid $(Fe^{2+},Mg,Mn)_2$ (Al,Fe^{3+}) $Al_3O_2(SiO_4)_2$ $(OH)_2$	Metamorphic and hydrothermal. Association: Muscovite, chlorite, biotite, garnet, staurolite and hematite.	Crystallography: monoclinic 2/m; Crystal habit: Coarsely foliated masses or thin scales; Colour: Dark green to black; Lustre: Pearly; Transparency: Translucent; Streak: Black; Cleavage: Perfect and fair; Fracture: Absent; Hardness: 6.5; Sp. gravity: 3.5-3.8.	

7.13.2 Sorosilicates

- In these silicate structures, the Si : O is 2 : 7.
- Two SiO_4 tetrahedrons are joined by sharing one apical oxygen.

Common sorosilicates

Melilite group	$X_2Y(Z_2O_7)$
Gehlenite	X= Ca, Y= Al, Z= Si, Al
Akermanite	X= Ca, Y= Mg, Z= Si

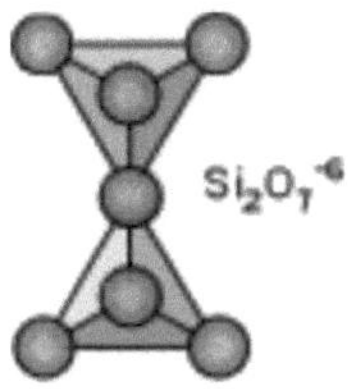

Fig. 7.18: Structure of sorosilicate.

Hemimorphite	$Zn_4(Si_2O_7)(OH)_2.H_2O$
Vesuvianite	$Ca_{10}(Mg, Fe)_2Al_4(Si_2O_7)_2(SiO_4)_5(OH, F)_4$
Epidote serirs	$X_2Y_3(SiO_4)(O, OH)$
Clinozoisite	X= Ca, Y= Al
Epidote	X= Ca, Y= Al, Fe
Piemontite	X= Ca, Y= Al, Mn, Fe
Allanite	X= Ce,Ca, Y= Al, Fe

Table 7.15: Mineralogical description of sorosilicates

	Occurrence and association	*Properties*	*Applications*
Melilite group $X_2Y(Z_2O_7)$ Gehlenite Akermanite	Magmatic, metamorphic. Association: Nepheline, leucite, augite and hornblende.	Crystallography: Tetragonal 42 m; Crystal habit: Tubular crystals and embedded grains; Colour: Colourless, white, yellow, grey or red; Lustre: Vitreous to resinous; Transparency: Transparent to opaque; Streak: White; Cleavage: Indistinct and poor; Fracture: Conchoidal to uneven; Hardness: 5; Sp. gravity: 3.0.	Used in microelectronic, photolithography, and medicine.
Hemimorphite $Zn_4(Si_2O_7)(OH)_2.H_2O$	Secondary. Association: Sphalerite, smithsonite and cerrusite.	Crystallography: Orthorhombic 2 mm; Crystal habit: Stalactitic or colloform, also massive or earthy; Colour: White or blue, green, light yellow to brown; Lustre: Vitreous; Transparency: Transparent to translucent; Streak: White; Cleavage: Perfect; Fracture: Uneven; Hardness: 4.5-5; Sp. gravity: 3.5.	Used as minor ore of zinc and specimens.
Vesuvianite $(Ca,Na)_{19}(Al,Mg,Fe)_{13}(SiO_4)_{10}(Si_2O_7)_4(OH,F,O)_{10}$	Metamorphic, hydrothermal, rarely magmatic. Association: Wollastonite, grossular, diopside.	Crystallography: Tetragonal 4/m; 2/m 2/m; Crystal habit: Massive, granular or fibrous; Colour: Brown to green, or red; Lustre: Vitreous to resinous; Transparency: Transparent to opaque; Streak: White; Cleavage: Very poor; Fracture: Subconchoidal to uneven; Hardness: 6.5; Sp. gravity: 3.4.	Used as a gemstone.
Epidote series $X_2Y_3(SiO_4)(O, OH)$ Clinozoisite Epidote Piemontite Allanite	Metamorphic, hydrothermal, rarely magmatic. Association: Iron oxides, quartz, feldspar, actinolite, chlorite, hornblende.	Crystallography: Monoclinic 2/m; Crystal habit: Striated columnar, massive, granular or fibrous; Colour: Brown to green, black, red; Lustre: Vitreous to resinous; Transparency: Transparent to opaque; Streak: Black; Cleavage: Perfect basal; Hardness: 6-7; Sp. gravity: 3-4.6.	Used as semi-precious stone and as mineral specimen.

7.13.3 Cyclosilicates

- In these silicate structures, the Si : O is 1:3.
- Silicon-oxygen tetrahedrons are linked into rings.

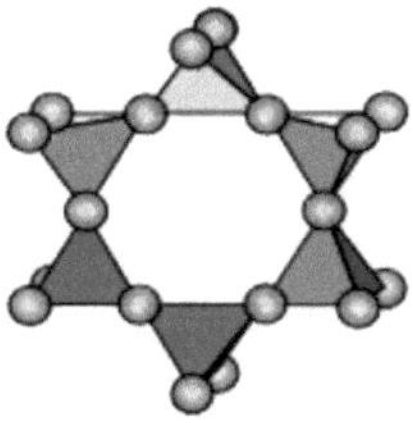

Fig. 7.19: Structure of cyclosilicate.

Common cyclosilicates

Tourmaline group	$WX_3Y_6(BO_3)_3(Si_6O_{18})(OH,F)_4$
Schorl	W= Na, X= Fe^{2+}, Y= Al
Dravite	W= Na, X= Mg, Y= Al
Elbaite	W= Na, X= Li,Al, Y= Al
Beryl	$Be_3Al_2(Si_6O_{18})$

Table 7.16: Mineralogical description of cyclosilicates

	Occurrence and association	*Properties*	*Applications*
Tourmaline group			
$WX_3Y_6(BO_3)_3$ $(Si_6O_{18})(OH,F)_4$ Schorl Dravite Elbaite	Pegmatitic and metamorphic. Association: K-feldspar, quartz, micas, beryl, apatite, fluorite, topaz and cassiterite.	Crystallography: Hexagonal 3 m; Crystal habit: Parallel or radiating, massive and compact; Colour: Black, blue, green, or red; Lustre: Vitreous to resinous; Transparency: Transparent to opaque; Streak: Uncoloured; Cleavage: Absent; Fracture: Subconchoidal to uneven; Hardness: 7-7.5; Sp. gravity: 3.0-3.2.	Used as gemstone.
Beryl $Be_3Al_2(Si_6O_{18})$	Magmatic, metamorphic, hydrothermal. Association: Feldspar, mica, tourmaline, spodumene and garnet.	Crystallography: Hexagonal 6/m 2/m 2/m; Crystal habit: Distinct crystals, columnar aggregates, massive; Colour: Blue, green, green, blue, yellow, white; Lustre: Vitreous to resinous; Transparency: Transparent to translucent; Streak: Uncoloured; Cleavage: Absent; Fracture: Conchoidal to uneven; Hardness: 7.5-8; Sp. gravity: 2.6-2.9.	Main source of beryllium; used as gemstone; alloys of beryllium are used in telecommunications, computers, automobiles, oil and gas drilling equipment, and aerospace industries.

7.13.4 Inosilicate (Single-chain)

- In these silicate structures, the Si : O is 1: 3.
- Formed by sharing two oxygen atoms from each tetrahedron.

Common inosilicates

Enstatite series	
Enstatite	$Mg_2(Si_2O_6)$
Ferrosilite	$Fe_2(Si_2O_6)$
Diopside series	
Diopside	$CaMg(Si_2O_6)$
Hedenbergite	$CaFe(Si_2O_6)$
Augite	$XY(Z_2O_6)$
Spodumene	$LiAl(Si_2O_6)$
Rhodonite	$Mn(SiO_3)$
Chrysocolla	$(Cu,Al)(SiO_3).nH_2O$
Wollastonite	$Ca(SiO_3)$

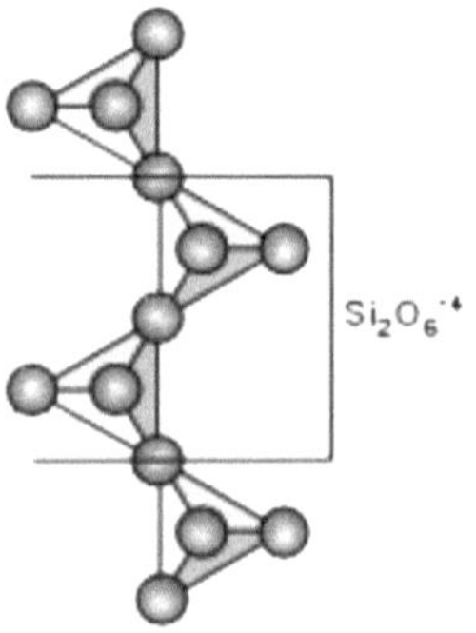

Fig. 7.20: Structure of inosilicate.

Table 7.17: Mineralogical description of inosilicates (single-chain)

	Occurrence and association	*Properties*	*Applications*
Enstatitie series			
Enstatite $Mg_2(Si_2O_6)$ Ferrosilite $Fe_2(Si_2O_6)$	Magmatic and metamorphic. Association: Pyrope and olivine.	Crystallography: Orthorhombic 2/m 2/m 2/m; Crystal habit: Fibrous, massive, lamellar; Colour: Greyish, yellowish, or greenish white; Lustre: Vitreous to pearly; Transparency: Translucent; Streak: White; Cleavage: Good; Fracture: Uneven; Hardness: 5-6; Sp. gravity: 3.2-4.0.	Used as minor gemstone, decorative purposes.
Diopside series			
Diopside $CaMg(Si_2O_6)$ Hedenbergite $CaFe(Si_2O_6)$	Magmatic metamorphic and hydrothermal. Association: Grossuler, epidote, plagioclase and magnetite.	Crystallography: Monoclinic 2/m; Crystal habit: Granular, lamellar or columnar; Colour: White to pale green; Lustre: Vitreous; Transparency: Transparent to translucent; Streak: White; Cleavage: Imperfect; Hardness: 5.5-6.5; Sp. gravity: 3.2-3.6.	Diopside creates fine collector specimens, and examples from famous localities are highly priced. Diopside is faceted into gems for collectors, although the emerald-green, chromium-rich variety is slowly seeping into the real gem market. Some polished specimens display asterism.

(Contd.)

(*Contd.*)

Augite $XY(Z_2O_6)$	Magmatic and rarely metamorphic. Association: Calcic to intermediate plagioclase.	Crystallography: Monoclinic 2/m; Crystal habit: Poorly formed stubby crystals; Colour: Black to dark green; Lustre: Vitreous; Transparency: Translucent; Streak: Grey-green; Cleavage: Distinct; Hardness: 5-6; Sp. gravity: 3.2-3.4.	Used for mineralogical and scientific study.
Spodumene $LiAl(Si_2O_6)$	Magmatic and hydrothermal. Association: Tourmaline, beryl, garnet, lepidolite, alkali feldspar, mica and quartz.	Crystallography: Monoclinic 2/m; Crystal habit: Cleavable masses and crystals; Colour: White to tinted green, grey, yellow; Lustre: Vitreous; Transparency: Translucent; Streak: White; Cleavage: Perfect; Hardness: 6.5-7; Sp. gravity: 3.0-3.2.	Important ore of lithium; some varieties used as gemstones; glass ceramics, heat resistant, container, pharmaceutical, television, lighting glass, fibre glass. Uses in ceramics frits, glazes, sanitary ware, porcelain, enamels and also in refractories, cements, metallurgy.
Rhodonite $Mn(SiO_3)$	Hydrothermal and metamorphic. Association: Pyrolusite, tephroite, rhodochrosite, zincite, willemite.	Crystallography: Triclinic 1; Crystal habit: Massive, compact discrete grains; Colour: Pink or red; Lustre: Vitreous to pearly; Transparency: Translucent to transparent; Streak: White; Cleavage: Perfect and distinct; Fracture: Conchoidal to uneven; Hardness: 5.5-6.5; Sp. gravity: 3.4-3.7.	Used as minor ore of gemstone and as a minor ore of manganese.
Chrysocolla $(Cu,Al)(SiO_3).nH_2O$	Secondary. Association: Malachite, cuprite, native copper.	Crystallography: Possibly amorphous; Habit: Encrustrations, earthy, colloform; Colour: Green to green-blue; Lustre: Vitreous; Transparency: Translucent to opaque; Streak: White; Cleavage: Absent; Fracture: Conchoidal; Hardness: 2-4; Sp. gravity: 2-3.	Used as a minor ore of copper and an ornamental stone.

(*Contd.*)

Table 7.17 (*Contd.*)

	Occurrence and association	*Properties*	*Applications*
Wollastonite $Ca(SiO_3)$	Metamorphic and rarely magmatic. Association: Grossularite, diopside, tremolite, Ca-plagioclase, vesuvianite, epidote and calcite.	Crystallography: Triclinic 1; Crystal habit: Cleavable mass to fibrous aggregates, compact; Colour: Pink or red; Lustre: Vitreous to silky; Transparency: Translucent; Streak: White; Cleavage: Perfect and good; Fracture: Uneven; Hardness: 4.5-5; Sp. gravity: 2.8.	Used in the ceramics industry where it is used as a principal ingredient in the ceramic bodies required for the manufacture of floor and wall tiles. As a filller in paint and paper and many other products. In metallurgy, it is used as a welding rod coating, to control the viscosity of the slag and alloying agents. A suitable mixture of finely pulverized wollastonitc with china-clay improves the properties of the clay to be used for paper-coating. It is found to increase the strength of Portland cement when mixed as an additive.

7.13.5 Inosilicates (Double-chain)

- In these silicate structures, the Si : O is 4 : 11.
- Formed when two adjacent single chains are joined by sharing an additional oxygen atom from alternate tetrahedrons.

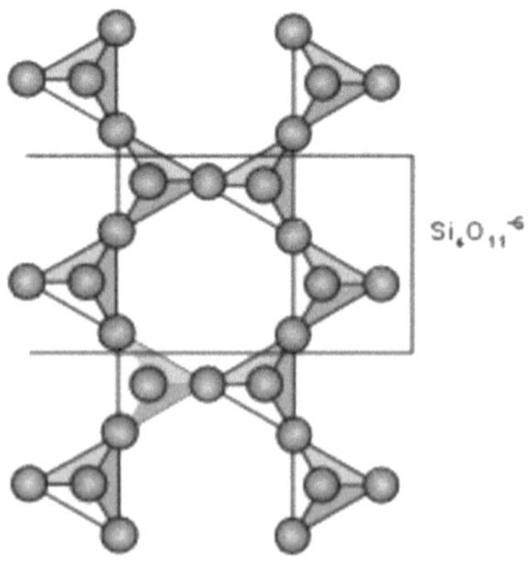

Fig. 7.21: Structure of inosilicate.

Common inosilicates

Anthophyllite	$X_2Y_5(Z_8O_{22})(OH)_2$
Cummingtonite	$(Mg,Fe)_7(Si_8O_{22})(OH)_2$
Actinolite	$X_2Y_5(Si_8O_{22})(OH)_2$
Hornblende	$Ca_2(Mg,Fe)_4Al(AlSi_7O_{22})(OH)_2$

Table 7.18: Mineralogical description of inosilicates (double-chain)

	Occurrence and association	*Properties*	*Applications*
Anthophyllite $X_2Y_5(Z_8O_{22})(OH)_2$	Metamorphic. Association: Actinolite, tremolite, cordierite.	Crystallography: Orthorhombic 2/m 2/m 2/m; Crystal habit: Fibrous, bladed, columnar; Colour: Shades of brown to green; Lustre: Vitreous; Transparency: Translucent; Streak: Uncoloured; Cleavage: Perfect; Fracture: Subconchoidal; Hardness: 5.5-6; Sp. gravity: 2.8-3.2.	Have some rare asbestos uses.
Cummingtonite $(Mg,Fe)_7(Si_8O_{22})(OH)_2$	Metamorphic. Association: Anthophyllite, hornblende, plagioclase, cordierite and garnet.	Crystallography: Monoclinic 2/m; Crystal habit: Fibrous, bladed, columnar crystals; Colour: Dark green or brown; Lustre: Silky; Transparency: Translucent; Streak: White; Cleavage: Perfect; Hardness: 5-6; Sp. gravity: 3.2-3.6.	Have some rare asbestos uses.
Actinolite $X_2Y_5(Si_8O_{22})(OH)_2$	Metamorphic. Association: Garnet, diopside, pyrite and calcite.	Crystallography: Monoclinic 2/m; Crystal habit: Prismatic, often flattened or columnar; Colour: White or grey to dark green; Lustre: Vitreous; Transparency: Translucent; Streak: White; Cleavage: Perfect; Hardness: 5-6; Sp. gravity: 3.0-3.3.	The finely fibrous variety of actinolite is used for industrial asbestos. Because it is not affected by fire and is a poor heat conductor, it is used in fire retardant devices and for heat protection. Some varieties are used as gemstones.
Hornblende $Ca_2(Mg,Fe)_4Al(AlSi_7O_{22})(OH)_2$	Magmatic. Association: Quartz, feldspar, pyroxene and chlorite.	Crystallography: Monoclinic 2/m; Crystal habit: Prismatic, bladed, columnar, fibrous; Colour: Black to dark green; Lustre: Vitreous; Transparency: Translucent; Streak: White; Cleavage: Perfect prismatic; Hardness: 5-6; Sp. gravity: 2.9-3.3.	Used in decoration, dimension stone.

7.13.6 Phyllosilicates

- In these silicate structures, the Si : O is 2 : 5.
- Have two-dimensional framework of infinite sheets of SiO_4 tetrahedrons.
- Three oxygen atoms are shared in each tetrahedron.
- Display properties like platy or flaky habit, prominent cleavage, low specific gravity, softness.
- Most commonly contains hydroxyl groups.

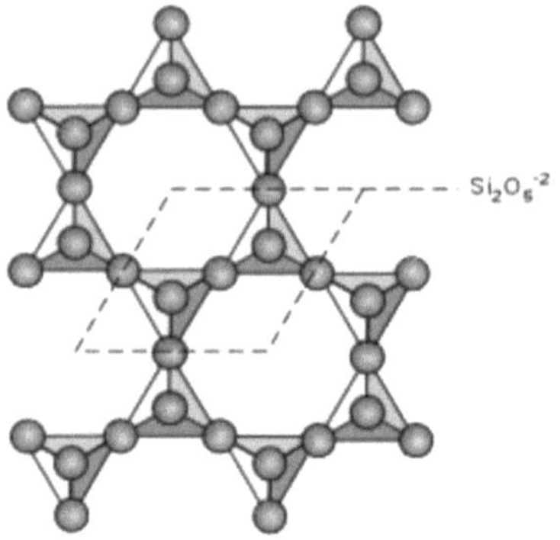

Fig. 7.22: Structure of phyllosilicate.

Common phyllosilicates

Clay mineral group	
Kaolinite group	$Al_2(Si_2O_5)(OH)_4$
Smectite group	
Talc group	
Pyrophyllite	$Al_2(Si_4O_{10})(OH)_2$
Talc	$Mg_3(Si_4O_{10})(OH)_2$
Chlorite group	$Y_6(Z_4O_{10})(OH)_8$ Y = Mg, Fe, Al, Ni, Mn, Cr; Z = Si & Al
Brittle mica group	
Margarite	$CaAl_2(Al_2Si_2O_{10})(OH)_2$
Elastic mica group	
Muscovite	$KAl_2(AlSi_3O_{10})(OH, F)_2$
Phlogopite	$KMg_3(AlSi_3O_{10})(OH, F)_2$
Biotite	$K(Mg,Fe)_3(AlSi_3O_{10})(OH, F)_2$
Lepidolite series	$KX_3[(Si,Al)_4O_{10}](OH, F)_2$ X= Li, Al, Fe, Mg, Mn
Miscellaneous	
Serpentine group	$(Mg,Fe)_3(Si_2O_5)(OH)_4$
Glauconite	$(K,Na)(Mg,Fe,Al)_2[(Al,Si)_4O_{10}](OH)_2$
Prehnite	$Ca_2Al(AlSi_3O_{10})(OH)_2$

Table 7.19: Mineralogical description of phyllosilicates

	Occurrence and association	*Properties*	*Applications*
Clay mineral group			
Kaolinite group $Al_2(Si_2O_5)(OH)_4$	Hydrothermal and secondary.	Crystallography: Triclinic 1; Crystal habit: Earthy masses, foliated, granular; Colour: White; Lustre: Pearly to dull and earthy; Transparency: Transparent to translucent; Streak: White; Cleavage: Perfect; Fracture: Absent; Hardness: 2-2.5; Sp. gravity: 2.6.	In the production of ceramics. As a filler for paint, rubber and plastics. In paper industry, to produce glossy paper as used in most magazines.
Smectite group	Secondary.	Crystallography: Monoclinic 2/m; Crystal habit: Earthy masses; Colour: White or grey; Lustre: Dull and earthy; Transparency: Transparent to translucent; Streak: White; Cleavage: Perfect; Fracture: Absent; Hardness: very soft; Sp. gravity: 2.0-2.7.	Used in foundry, oil-well drilling, wine and iron ore and feed palletizing industries; also in agricultural industries and for decolourization of oils.
Talc group			
Pyrophyllite $Al_2(Si_4O_{10})(OH)_2$	Metamorphic and hydrothermal. Association: Andalusite, kyanite, quartz, micas.	Crystallography: Monoclinic 2/m; Crystal habit: Foliated, massive, spherullitic or radiate; Colour: White or greenish or yellowish; Lustre: Dull and pearly; Transparency: Translucent; Streak: White; Cleavage: Perfect; Fracture: Hackly; Hardness: 1-2; Sp. gravity: 2.8.	Used as a machinable ceramic material used for electrical resistors, transducer cores, high vacuum gaskets and insulators in electron microscopes. Other uses for pyrophyllite are in ceramic formulations e.g. tile and refractory compositions, plastics and rubber as a filler, paint, insecticides.
Talc $Mg_3(Si_4O_{10})(OH)_2$	Metamorphic and hydrothermal. Association: Actinolite, dolomite, chlorite.	Crystallography: Monoclinic 2/m; Crystal habit: Compact or foliated massive; Colour: White or apple-green, brown; Lustre: Vitreous; Transparency: Translucent; Streak: White; Cleavage: Perfect; Fracture: Hackly; Hardness: 1-1.5; Sp. gravity: 2.7.	Talc acts as an ideal filler/extender for many industries like paper, ceramic, paints, plastics, cosmetics, detergents, pharmacy and animal feed.

(*Contd.*)

Table 7.19 (*Contd.*)

	Occurrence and association	*Properties*	*Applications*
Chlorite group			
Chlorite group $Y_6(Z_4O_{10})(OH)_8$ Y= Mg, Fe, Al, Ni, Mn, Cr; Z = Si & Al	Secondary and sedimentary. Association: Muscovite, biotite, talc, garnet, diopside, magnesite, magnetite and apatite.	Crystallography: Monoclinic 2/m; Crystal habit: Foliated mass, scaly aggregates; Colour: Dark green, yellow, white or rose; Lustre: Vitreous to pearly or dull; Transparency: Transparent to translucent; Streak: White; Cleavage: Perfect; Fracture: Absent; Hardness: 2-2.5; Sp. gravity: 2.7.	Used as a filler and as a constituent of clay.
Brittle mica group			
Margarite $CaAl_2(Al_2Si_2O_{10})(OH)_2$	Metamorphic. Association: Corundum, diaspore, tourmaline, staurolite.	Crystallography: Monoclinic 2/m; Crystal habit: Foliated massive; Colour: White, pink, brown; Lustre: Vitreous to pearly; Transparency: Translucent; Streak: White; Cleavage: Perfect; Fracture: Absent; Hardness: 3.5-5; Sp. gravity: 3.0.	
Elastic mica group			
Muscovite $KAl_2(AlSi_3O_{10})(OH, F)_2$	Metamorphic, magmatic and hydrothermal. Association: Quartz, K feldspar, albite, biotite.	Crystallography: Monoclinic 2/m; Crystal habit: Tabular, foliated mass, plumose, stellate or globular; Colour: Colourless, often tinted; Lustre: Vitreous to silky; Transparency: Transparent to translucent; Streak: White; Cleavage: Perfect; Fracture: ragged; Hardness: 2.5-3; Sp. gravity: 2.75-3.0.	It has been used as windows on high-temperature furnaces and ovens. It is an insulator and was used in the past to make circuit boards.
Phlogopite $KMg_3(AlSi_3O_{10})(OH, F)_2$	Metamorphic, magmatic. Association: Dolomite, diopside, anthophyllite.	Crystallography: Monoclinic 2/m; Crystal habit: Disseminated scales and plates; Colour: Reddish brown or yellowish brown; Lustre: Pearly to submetallic; Transparency: Transparent to translucent; Streak: White; Cleavage: Perfect basal; Fracture: Ragged; Hardness: 2.5-3; Sp. gravity: 2.8.	Used as heat and electrical insulator for industrial purposes.

(*Contd.*)

(*Contd.*)

Biotite $K(Mg,Fe)_3$ $(AlSi_3O_{10})$ $(OH, F)_2$	Metamorphic, magmatic. Association: Quartz, feldspar and muscovite.	Crystallography: Monoclinic 2/m; Crystal habit: Disseminated or foliated mass; Colour: Black, greenish black or brownish black; Lustre: Vitreous; Transparency: Transparent to opaque; Streak: White; Cleavage: Perfect basal; Fracture: Ragged; Hardness: 2.5-3; Sp. gravity: 2.7-3.1.	Used as a surface treatment in decorative concrete, plaster and other construction materials. It is also used in the potassium-argon method of dating igneous rocks.
Lepidolite series $KX_3[(Si,Al)_4$ $O_{10}](OH, F)_2$ X = Li, Al, Fe, Mg, Mn	Pegmatitic; Association: tourmaline, spodumene, muscovite, cassiterite, albite, quartz and topaz.	Crystallography: monoclinic 2/m; Crystal habit: coarse to fine-grained scaly or granular aggregates; Colour: rose-red or lialic red; Lustre: pearly; Transparency: translucent; Streak: white; Cleavage: perfect basal;Fracture: ragged; Hardness: 2.5-4; Sp. gravity: 2.9.	Used as ore of lithium, ornamental stone and heat insulator for industrial purposes.
Miscellaneous			
Serpentine group $(Mg,Fe)_3$ (Si_2O_5) $(OH)_4$	Secondary. Association: Magnesite, chromite, spinel and garnet.	Crystallography: Monoclinic 2/m; Crystal habit: Massive, platy or fibrous; Colour: Shades of green, yellow-green and yellow; Lustre: Greasy to waxy; Transparency: Translucent; Streak: White; Cleavage: Perfect; Fracture: Conchoidal to splintery; Hardness: 2.5-3.5; Sp. gravity: 2.2-2.4.	Used as thermal conductivity, industrial mineral, ornamental stone.
Glauconite $(K,Na)(Mg,Fe,$ $Al_2)[(Al,Si)_4$ $O_{10}](OH)_2$	Hydrothermal, volcanic exhalative. Association: Calcite, quartz.	Crystallography: Monoclinic 2/m; Crystal habit: Small pellets or scaly grains; Colour: Dull green; Cleavage: Perfect; Hardness: 2; Sp. gravity: 2.3-2.8.	Used as fertilizer, soil amendment.
Prehnite $Ca_2Al(AlSi_3$ $O_{10})(OH)_2$	Hydrothermal. Association: Calcite, albite, zeolite, epidote.	Crystallography: Orthorhombic 2 mm; Crystal habit: Barrel-shaped, colloform or stalactitic; Colour: Light green to yellow; Lustre: Vitreous; Transparency: Translucent; Streak: Uncoloured; Cleavage: Distinct and poor; Fracture: Uneven; Hardness: 6-6.5; Sp. gravity: 2.9.	Used as ornamental material, semiprecious gem.

7.13.7 Tectosilicates

- In these silicate structures, the Si : O is 1: 2.
- All oxygen atoms of SiO_4 tetrahedrons are shared with nearby tetrahedrons
- Except zeolites, all the members of this class are anhydrous.

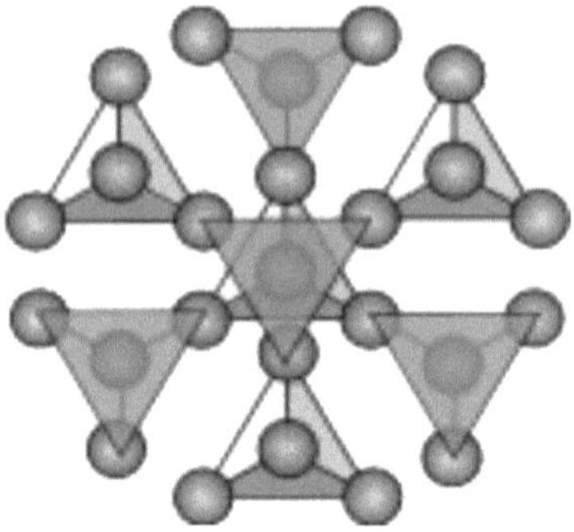

Fig. 7.23: Structure of tectosilicate.

Common tectosilicates

Silica group	
α-Quartz	SiO_2
Opal	$SiO_2 . nH_2O$
Feldspar group	
Orthoclase series	$KAlSi_3O_8$
Microcline series	$KAlSi_3O_8$
Plagioclase series	
Albite	$NaAlSi_3O_8$
Anorthite	$CaAl_2Si_2O_8$
Feldspathoid group	
Leucite	$KAlSi_2O_6$
Nepheline	$NaAlSiO_4$
Sodalite	$Na_4(AlSiO_4)_3Cl$
Zeolite group	
Stilbite	$NaCa_2(Al_5Si_{13}O_{36}). 14H_2O$
Chabazite	$Ca(Al_2Si_4O_{12}). 6H_2O$
Natrolite	$Na_2(Al_2Si_3O_{10}). 2H_2O$
Miscellaneous	
Scapolite group	
Cordierite	$Mg_2(Al_4Si_5O_{18})$
Analcime	$Na(AlSi_2O_6).H_2O$

Table 7.20: Mineralogical description of tectosilicates

	Occurrence and association	*Properties*	*Applications*
Silica group			
α-Quartz SiO_2	Magmatic. Association: Feldspar, mica.	Crystallography: Hexagonal 32; Crystal habit: Massive and as crystals; Colour: Colourless, white; Lustre: Vitreous; Transparency: Transparent to translucent; Streak: Uncoloured; Cleavage: Absent; Fracture: Irregular to conchoidal; Hardness: 7; Sp. gravity: 2.65.	Used in glass making, abrasives, foundry sand, in petroleum industry, as filler. Electronics grade quartz crystals used in filters, frequency controls, timers, electronic circuits that become important components in cell phones, watches, clocks, games, television receivers, computers, navigational instruments and other products. Optical-grade crystals used as lenses and windows in lasers and other specialized devices.
Opal $SiO_2 . nH_2O$	Magmatic. Association: Feldspar, mica.	Crystallography: Amorphous; Habit: Encrustrations, stalactitic; Colour: Colourless, white; Lustre: Vitreous; Transparency: Transparent to translucent; Streak: Uncoloured; Cleavage: Absent; Fracture: Conchoidal; Hardness: 5-6; Sp. gravity: 1.9-2.2.	Used as a gemstone and ornamental stone.
Feldspar group			
Orthoclase series $KAlSi_3O_8$	Magmatic and hydrothermal. Association: Quartz and muscovite.	Crystallography: Monoclinic 2/m; Crystal habit: Massive or as crystals; Colour: Colourless, white or flash-red in colour; Lustre: Vitreous; Transparency: Transparent to translucent; Streak: White; Cleavage: Perfect; Fracture: Conchoidal to uneven; Hardness: 6-6.5; Sp. gravity: 2.6.	Used in the manufacture of glass and ceramics and also as gemstone.

(*Contd.*)

Table 7.20 (*Contd.*)

	Occurrence and association	*Properties*	*Applications*
Microcline series $KAlSi_3O_8$	Magmatic, metamorphic and hydrothermal. Association: Quartz and muscovite.	Crystallography: Triclinic 1; Crystal habit: Massive or as crystals; Colour: Colourless, white or yellowish, reddish or green; Lustre: Vitreous; Transparency: Transparent to translucent; Streak: White; Cleavage: Perfect; Fracture: Conchoidal to uneven; Hardness: 6; Sp. gravity: 2.54-2.57.	Used as ornamental stone, in the manufacture of glass, enamel and porcelain products and as mineral specimens.
Plagioclase series Albite $NaAlSi_3O_8$ Anorthite $CaAl_2Si_2O_8$	Magmatic, metamorphic and hydrothermal. Association: Pyroxene, olivine, hornblende.	Crystallography: Triclinic 1; Crystal habit: Cleavable mass to irregular embedded grains; Colour: White, grey, reddish; Lustre: Vitreous; Transparency: Transparent to translucent; Streak: White; Cleavage: Perfect and distinct; Fracture: Conchoidal to uneven; Hardness: 6-6.5; Sp. gravity: 2.5 (albite) and 2.75 (anorthite).	Used as decorative stone, moon stone, ceramics and buildings.
Feldspathoid group			
Leucite $KAlSi_2O_6$	Magmatic. Association: Nepheline and sanidine.	Crystallography: Tetragonal 4/m; Crystal habit: Well-formed trapezohedral crystals; Colour: White or grey; Lustre: Vitreous to dull; Transparency: Translucent to opaque; Streak: Uncoloured; Cleavage: Poor; Fracture: Conchoidal; Hardness: 5.5-6; Sp. gravity: 2.6.	Used as mineral specimen and as a minor source of potassium and aluminum.
Nepheline $NaAlSiO_4$	Magmatic and rarely metamorphic. Association: Feldspar, biotite, sodalite, corundum and zircon.	Crystallography: Hexagonal 6; Crystal habit: Massive, compact or embedded grains; Colour: White, colourless, yellowish or grey, red; Lustre: Vitreous; Transparency: Transparent to opaque; Streak: Uncoloured; Cleavage: Distinct and poor; Fracture: Subconchoidal; Hardness: 5-6; Sp. gravity: 2.6.	Used as industrial mineral, glass industry.

(*Contd.*)

(*Contd.*)

Sodalite $Na_4(AlSiO_4)_3Cl$	Magmatic, and hydrothermal. Association: Nepheline, zircon and titanite.	Crystallography: Isometric $\bar{4}3m$; Crystal habit: Compact, cleavable, nodular; Colour: Blue, also white; Lustre: Subvitreous on faces and greasy on cleavages; Transparency: Transparent to opaque; Streak: Colourless; Cleavage: Distinct; Fracture: Conchoidal to uneven; Hardness: 5.5-6; Sp. gravity: 2.1-2.4.	Used as dimension stone, decorative purposes, ornamental material.
Zeolite group			
Stilbite $NaCa_2(Al_5Si_{13}O_{36}).14H_2O$	Hydrothermal, sedimentary and hot springs. Association: Calcite and other zeolites.	Crystallography: Monoclinic 2/m; Crystal habit: In sheaf-like aggregates; Colour: White, yellowish, brownish or reddish; Lustre: Vitreous; Transparency: Transparent to translucent; Streak: White; Cleavage: Perfect and poor; Fracture: Uneven; Hardness: 3.5-4; Sp. gravity: 2.2.	Used as mineral specimens and chemical filter. Can be used in oil spill clean-ups and in other industrial processes; can be used in water softening due to the capability of stilbite to attract calcium and magnesium.
Chabazite $Ca(Al_2Si_4O_{12}).6H_2O$	Hydrothermal and pegmatitic. Association: Calcite and other zeolites.	Crystallography: Hexagonal 32/m; Crystal habit: Crystal or crystal crusts; Colour: White to flash-red; Lustre: Vitreous; Transparency: Transparent to translucent; Streak: Colourless; Cleavage: Distinct; Fracture: Uneven; Hardness: 4-5; Sp. gravity: 2.1.	Used as mineral specimen and chemical filter.
Natrolite $Na_2(Al_2Si_3O_{10}).2H_2O$	Hydrothermal and pegmatitic. Association: Calcite and other zeolites.	Crystallography: Orthorhombic 2 mm; Crystal habit: Radiating, massive and granular; Colour: Colourless or white, greenish, yellowish or reddish; Lustre: Vitreous to silky or pearly; Transparency: Transparent to translucent; Streak: Uncoloured; Cleavage: Perfect; Fracture: Uneven; Hardness: 5-5.5; Sp. gravity: 2.2-2.4.	Used in softening of water.

(*Contd.*)

Table 7.20 (*Contd.*)

	Occurrence and association	*Properties*	*Applications*
Miscellaneous			
Scapolite group	Metamorphic. Association: Pyroxenes, amphiboles, garnet, apatite, titanite, zircon and biotite.	Crystallography: Tetragonal 4/m; Crystal habit: Massive and granular; Colour: White, grey or greenish, yellowish or reddish; Lustre: Vitreous to pearly; Transparency: Transparent to translucent; Streak: White; Cleavage: Distinct; Fracture: Subconchoidal; Hardness: 5-6.5; Sp. gravity: 2.5-2.8.	Used as gem stones and as mineral specimens.
Cordierite $Mg_2(Al_4Si_5O_{18})$	Metamorphic and rarely magmatic. Association: Anthophyllite, garnet, mica, quartz, andalusite, sillimanite, staurolite and spinel.	Crystallography: Orthorhombic 2/m 2/m 2/m; Crystal habit: Massive, compact; Colour: Shades of blue to shades of green; Lustre: Vitreous to greasy; Transparency: Transparent to translucent; Streak: White; Cleavage: Poor; Fracture: Subconchoidal; Hardness: 7-7.5; Sp. gravity: 2.6.	Used as gems, ceramics.
Analcime $Na(AlSi_2O_6).H_2O$	Magmatic, hydrothermal and rarely sedimentary. Association: Calcite, zeolite.	Crystallography: Isometric 4/m; Crystal habit: Granular, compact; Colour: Colourless to white, also grey, green, yellowish or reddish; Lustre: Vitreous; Transparency: Transparent to nearly opaque; Streak: White; Cleavage: Absent; Fracture: Subconchoidal; Hardness: 5-5.5; Sp. gravity: 2.3.	Used as building material. Use of analcime-rich tuffs as dimension stone or as pozzolanic addition to cement and concrete.

Concluding Remarks

Descriptive mineralogy summarizes results of studies performed on mineral substances. This chapter attempts to record the identification, classification, and categorization of minerals, their properties, and their uses, which will be beneficial to the readers.

Think for a while

1. With what intrusions is platinum genetically associated? What are its characteristic properties?
2. Characterize the origin and properties of diamond. What features are indicative of native sulphur and mention its uses.
3. What are the practical significance of sulfide minerals? What isomorphous impurities are possible in sphalerite? What is the origin and the diagonistic features of molybdenite?
4. Olivin is characteristic of what rock? What minerals result from its decomposition?
5. What is the origin of muscovite and gluconite? How to distinguish between phlogopite and biotite?
6. How to classify plagioclase? Mention the origin and practical significance of feldspars. What are the distinguishing features of orthoclase and microcline.
7. What is the difference of phosphorites and apatite? Mention the origin of apatite.

FURTHER READING

Deer, W. A., R.A. Howie and J. Zussman. An Introduction to the Rock-Forming Minerals. 2nd edition. Wiley, New York. [Standard in the field, used by mineralogists and petrologists who investigate rocks in thin section.] 1992.

Dana, E.S. and W.E. Ford. A text-book of Mineralogy with an extended treatise on Crystallography and Physical Mineralogy. John Wiley & Sons, Inc. London: Chapman & Hall Ltd. 1922.

Frye, K. (editor). The Encyclopedia of Mineralogy. Volume IVB of the Encyclopedia of Earth Sciences. Hutchinson Ross Publ. Co. Stroudsburg, PA: [Good overview of some interesting mineralogical topics, accompanied by short articles on those topics. In EPSc library, QE355.R6.] 1981.

Blackburn, W.H. and W.H. Dennen. Principles of Mineralogy. 2nd edition. Wm. C. Brown Publishers, Dubuque. 1994.

PART II

Mineral Transformations and Their Effects

CHAPTER

8

ENERGETICS, THERMODYNAMICS AND STABILITY OF MINERALS

Energy partition during mineral formation and subsequent recrystallization is controlled by thermodynamics. The minimum energy configuration gives the stability of a phase. To understand a system in terms of thermodynamic laws, energy distribution, entropy change is explained in this chapter.

8.1 INTRODUCTION

The formation and change of minerals and the rocks take place over a wide range of conditions with temperature, pressure and chemical potentials. The principles of thermodynamics developed in chemistry to quantify chemical transformations can be directly applied to these reactions. In the most general sense, thermodynamics is the study of energy—its transformations and its relationship to the properties of matter. In its engineering applications, thermodynamics has two major objectives. One of these is to describe the properties of matter when it exists in what is called an equilibrium state, a condition in which its properties show no tendency to change. The other objective is to describe processes in which the properties of matter undergo changes and to relate these changes to the energy transfers in the form of heat and work which accompany them.

These objectives are closely related. Thermodynamics is unique among scientific disciplines in that no other branch of science deals with subjects which are as commonplace or as familiar. Concepts such as "heat", "work", "energy" and "properties" are all terms in everyone's basic vocabulary. Although the stability relationships between various phases can be worked out using the experimental method, thermodynamics gives us a qualitative means of calculating the stabilities of various compounds or combinations of compounds (mineral assemblages). Thermodynamics not only allows us to predict what minerals will form at different conditions (*forward modelling*), but also allows us to use mineral assemblages and mineral compositions to determine the conditions at which a rock formed (*thermobarometry*) in geological environments.

Concept of Energy

The definition of energy from the viewpoint of thermodynamics would be "the capacity to induce a change in that which inherently resists change". This capacity encompasses a combination of an effort, expended in overcoming resistance to a particular type of change, with the change it produces. The combination is called energy.

The effort involved is measured quantitatively by what is defined as a "driving force" in thermodynamics. A driving force is a property which both causes and also controls the direction of change in another property. The quantitative value of this change is called a "displacement". The product of a driving force and its associated displacement always represents a quantity of energy, but in thermodynamics this quantity has meaning only in relation to a specifically defined system.

Relative to a particular system there are generally two ways of locating a driving force and the displacement it produces. In one way both the driving force and the displacement are properties of the system and are located entirely within it, so that the energy calculated from their product represents a change in the internal energy of the system. Similarly, both the driving force and its displacement could be located entirely within the surroundings so that the calculated energy is then a change in the total energy of the surroundings.

In another way, however, the displacement occurs within the system but the driving force producing it is a property of the surroundings and is applied externally at the system boundary. By definition, the boundary of a system is a region of zero thickness containing no matter at all so that the energy calculated in this way is not a property of matter either in the system or in its surroundings but represents a quantity of energy in transition between the two. In any quantitative application of thermodynamics it is always important to make a careful distinction between energy changes within a system or within its surroundings and energy in transition between them.

8.2 ENERGETICS

Energetics is the scientific study of energy under transformation. Because energy flows at all scales, from the quantum level, to the biosphere and cosmos, energetics is, therefore, a very broad discipline, encompassing for example thermodynamics, chemistry, biological energetics, biochemistry and ecological energetics.

The rearrangement of atoms occurring in a chemical reaction is virtually always accompanied by the liberation or absorption of heat. If the purpose of the reaction is to serve as a source of heat, such as in the combustion of a fuel, then these heat effects are of direct and obvious interest. It can be observed, however, that a study of the energetics of chemical reactions in general can lead us to a deeper understanding of chemical equilibrium and the basis of chemical change itself.

Energy Units

Energy is measured in terms of its ability to perform work or to transfer heat. Mechanical work is done when a force f displaces an object by a distance d: $W = f \times d$. The basic unit of energy is the *joule*. One joule is the amount of work done when a force of 1 newton acts over a distance of 1 m; thus 1 J = 1 N-m. The newton is the amount of force required to accelerate a 1-kg mass by 1 m/sec^2, so the basic dimensions of the joule are kg m^2 s^{-2}. The other two units in wide use, the *calorie* and the *BTU* (British thermal unit), are defined in terms of the heating effect on water.

Heat Capacity

Heat capacity of a substance is a measure of how sensitively its temperature is affected by a change in heat content; the greater the heat capacity, the less effect a given flow of heat q will have on the temperature. Temperature is the measure of the average kinetic energy due to translational motion of the molecules. If vibrational or rotational motions are also active, these will also accept thermal energy and reduce the amount that goes into translational motions. Because the temperature depends only on the latter, the effect of the other kinds of motion will be to reduce the dependence of the internal energy on the temperature, thus raising the heat capacity of a substance.

Vibrational and rotational motions are not active in case of monoatomic species like noble gas elements, so they have lowest heat capacities.

Metallic solids are a rather special case. In metals, the atoms oscillate about their equilibrium positions in a rather uniform way which is essentially the same for all metals, so they should all have about the same heat capacity. That this is indeed the case is embodied in the *Law of Dulong and Petit*. In the 19th century these workers discovered that the molar heat capacities of all the metallic elements they studied were around 25 J mol^{-1} K^{-1}, which is close to what classical physics predicts for crystalline metals. This observation played an important role in characterizing new elements, for it provided a means of estimating their molar masses by a simple heat capacity measurement.

8.3 THERMODYNAMICS

Thermodynamics is the study of energy and its transformations. Classical thermodynamics is based upon the equilibrium state. It is based upon the macroscopic (i.e., little underlying knowledge of the crystal structure is required) measure of the *intensive* and *extensive* properties of phases in the system. Intensive properties do not depend on the mass or number of particles in the system. This property is not additive in the sense that they do not require a specific quantity of the substance. Intensive properties can be specified at a particular point in the system. Included are:

- Temperature
- Density
- Pressure
- Solubility
- Heat capacity
- Viscosity
- Surface tension
- Melting/boiling point
- Colour
- Resistivity

Extensive properties are additive by virtue of the fact that their values constitute a property of the whole system and they depend on the mass or number of particles in the system. Examples are:

- Volume
- Mass
- Enthalpy (heat)
- Energy (calories, joules)

Using empirically derived parameters that describe the chemical and physical state of matter, thermodynamics predicts the energy changes for any given transformation. In essence, it tells us the most stable set of phases that should be present, given certain pressure (P), temperature (T), and chemical conditions (X).

Thermodynamics predicts what components of mineral assemblages should occur in a given environment assuming they are in chemical equilibrium. Therefore, principles of thermodynamics have great importance to understand several geological processes.

Phase

The phase is part of the system that is spatially uniform. The term phase can synonymously be used with mineral (if it is homogeneous at the atomic scale). A phase can be considered a solid, liquid, or gas with each having its own stability region or field in terms of chemical, pressure, and temperature conditions. An example is H_2O which may exist in three phases, e.g. ice, water and steam.

Phases are physically distinct, mechanically separable, and homogeneous and can be described by minimum number of independent chemical species known as components, e.g., quartz, SiO_2 or kyanite, Al_2SiO_5 are two phases whereas a quartz-kyanite rock is a $SiO_2 - Al_2O_3$ system.

Components are the smallest number of chemical entities to define the composition of all phases in a system.

System is a quantity of material defined by weights or numbers of molecules contained within a set of boundaries (i.e., imagine a container around the system, but the container is not part of the system). We generally classify systems into three conditions.

- *Isolated systems:* This is an ideal situation where there is absolutely no transfer of energy or matter across the boundaries of the system.
- *Closed systems:* In this case, there are possibilities for energy transfer but not matter. The matter can change in composition due to chemical reaction. We sometime assume this in certain geologic environments.
- *Open systems:* Exchange of both energy and matter. This is most often the rule in geologic environments.

Gibbs Phase Rule

Gibbs phase rule relates the number of phases and components with the various number of degrees of freedom

$$p + f = c + 2$$

where p = number of phases, f = the variance or number of degrees of freedom in the system and c = number of components.

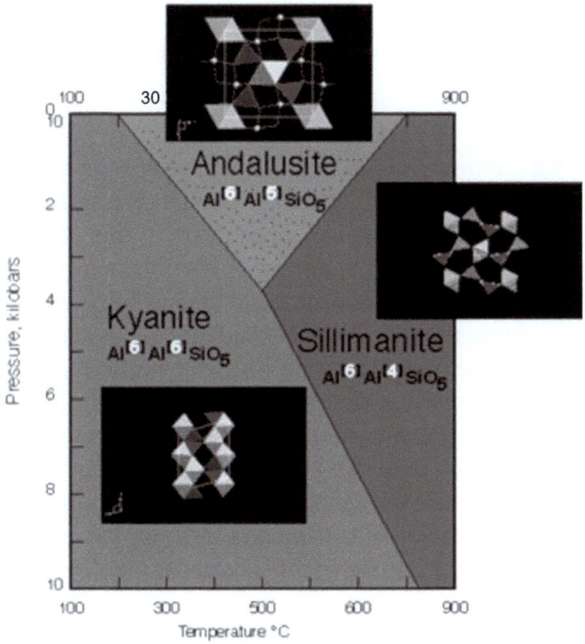

Fig. 8.1: Example with the kyanite-sillimanite-andalusite series minerals each having identical chemistry.

In the case of the kyanite-sillimanite-andalusite series there is a triple point or invariant point in T and P space. It is the unique set of T and P conditions under which all three phases can coexist.

This can be expressed by rearranging the Gibbs phase rule:

$$f = c - p + 2$$

In this case $c = 1$ (the component is Al_2SiO_5) and $p = 3$ (there are three phases at this point).

Therefore:

$$f = 1 - 3 + 2 = 0 \text{ (zero degrees of freedom)}$$

Along the curve or univariant line there is not one unique set of P and T conditions under which two phases can coexist.

In this case $c = 1$ (the component is Al_2SiO_5) and $p = 2$ (there are two phases at this point).

Therefore:

$$f = 1 - 2 + 2 = 1 \text{ (one degree of freedom)}$$

In the divariant or phase regions only one phase can coexist. No unique set of P and T conditions can be defined by the presence of one mineral phase.

In this case, $c = 1$ (the component is Al_2SiO_5) and $p = 1$ (there is one phase at this point).

Therefore:

$$f = 1 - 1 + 2 = 2 \text{ (two degrees of freedom)}$$

Entropy

Entropy is such a thermodynamic function whose change when multiplied by temperature gives a quantity of thermal energy. In classical thermodynamics there is no need to give any physical description of this property in terms of molecular behaviour. A change in entropy is defined simply as a quantity of thermal energy divided by the temperature driving force which propels it so that it always produces the thermal energy identically when multiplied by the temperature. Because temperature is an intensive property and this product is energy, the entropy must be an extensive property. Furthermore, thermal energy is a part of the total internal energy within a system so that the entropy change computed this way is a change in a property of the system.

In order to define the thermochemical properties of a process, it is first necessary to write a thermochemical equation defining the actual change taking place, both in terms of the formulas of the substances involved and their physical states (temperature, pressure or state of matter).

To take a very simple example, here is the complete thermochemical equation for the vapourization of water at its normal boiling point:

$$H_2O\ (1{,}383\ K,\ 1\ atm) \longrightarrow H_2O\ (1{,}383\ K,\ 1\ atm)\ \Delta H = 40.8\ kg\ mol^{-1}$$

The quantity 40.8 is known as the *enthalpy of vapourization* ("heat of vaporization") of liquid water.

Enthalpy

The enthalpy change for a chemical reaction is the difference

$$\Delta H = H_{\text{products}} - H_{\text{reactants}}$$

If the reaction in question represents the formation of one mole of the compound from its elements in their standard states, as in

$$H_f{}^\circ = \Sigma H_f{}^\circ{}_{\text{products}} - \Sigma H_f{}^\circ{}_{\text{reactants}} = -286 \text{ kJ} - 0 = -268 \text{ kJ mol}^{-1}$$

which defines the *standard enthalpy of formation* of water at 298 K.

The standard enthalpy of formation of a compound is defined as the heat associated with the formation of one mole of the compound from its elements in their standard states.

In general, the *standard enthalpy change for a reaction* is given by the expression

$$\Delta\, \Sigma H_f^\circ{}_{\text{products}} - \Sigma H_f^\circ{}_{\text{reactants}}$$

in which the ΣH_f° terms indicate the sums of the standard enthalpies of formations of all products and reactants.

A number of elements, of which sulfur and carbon are common examples, can exist in more than one solid crystalline form. The standard heat of formation of a compound is always taken in reference to the forms of the elements that are most stable at 25°C and 1 atm pressure. In the case of carbon, this is the graphite, rather than the diamond form.

$$\text{C(graphite)} + O_2(g) \longrightarrow CO_2(g)\ \Delta H^\circ = -393.5 \text{ kJ mol}^{-1}$$

$$\text{C(diamond)} + O_2(g) \longrightarrow CO_2(g)\ \Delta H^\circ = -395.8 \text{ kJ mol}^{-1}$$

Laws of Thermodynamics

Here we will discuss some basic concepts and principles of thermodynamics in relation to mineralogy. There are three main laws of thermodynamics.

The *First Law of Thermodynamics* is simply the law of conservation of energy and mass. The ready acceptability of this law is apparent from the fact that the concept of conservation in some form has existed from antiquity, long before any precise demonstration of it could be made. At present, when mass and energy can be shown to be mutually interchangeable, conservation of mass and conservation of energy should be combined into a single conservation law which, as far as we know, is universal. At the magnitude of energies involved in thermodynamic state changes the accompanying mass changes are negligible so that mass and energy are always considered to be conserved separately.

The First Law of Thermodynamics states that "the internal energy, *E*, of an isolated system is constant". In a closed system, there cannot be a loss or gain of mass, but there can be a change in energy, dE. This change in energy will be the difference between the heat, Q, gained or lost, and the work, W done by the system. So,

$$dE = dQ - dW \tag{1}$$

The work, W, is defined as force × distance. Since pressure, P, is defined as force/unit area, force = P × area, and thus $W = P$ × area × distance = $P \times V$, where V is volume. If the work is done at constant pressure, then $W = PdV$. Substitution of this relationship into (1) yields:

$$dE = dQ \times PdV \tag{2}$$

This is a restatement of the first law of thermodynamics.

The *Second Law of Thermodynamics* is a concept proposed by Rudolf Clausious in 1850. In its intuitive perception the Second Law is a sense of the uniqueness of the direction of the change which results from the action of a particular thermodynamic driving force i.e. it accommodates the irreversible processes also. Irreversibility reflects the changes in the value of entropy S and for all reversible processes $\Delta S = Q/T$ for all irreversible processes $\Delta S > Q/T$.

For example, the earth's gravitational potential is the driving force and causes water to flow from a tank on top of the hill to one at the bottom, but it alone will never cause the reverse to occur and hence is an irreversible process. This direction of water flow is always the same unless some work is being done, as for example with a pump, or unless a change in the properties of some region outside the two tanks, such as the change in the water level in the reservoirs. The earth's gravitational attraction at a given water level is a driving force because when the water levels are the same in each tank there is no further transfer of water and the rate of transfer increases with water level difference in elevation.

The Second Law states that there exists no physical process whose sole result is the absorption of heat from a reservoir and its conversion into an equivalent amount of work.

The Second Law in a way implies that the change in heat energy of the system is related to the amount of disorder in the system. Entropy is a measure of disorder, and so at constant temperature and pressure:

$$dQ = TdS$$

Thus, substituting into (2) we get:

$$dE = TdS - PdV \tag{3}$$

The *Gibbs Free Energy*, *G*, is defined as the energy in excess of the internal energy as follows:

$$G = E + PV - TS \tag{4}$$

Differentiating this we get:

$$dG = dE + VdP + PdV - TdS - SdT$$

Substituting (3) into this equation gives:

$$dG = TdS - PdV + VdP + PdV - SdT - TdS$$

or

$$dG = VdP - SdT \qquad (5)$$

The *Third Law* states that the limit of the entropy of a substance is zero as its temperature approaches zero, a concept necessary in making absolute entropy calculations and in establishing the relationship between entropy as obtained from the statistical behaviour of a multi-particle system, and the entropy of classical thermodynamics.

Hess's Law and Its Application

Two or more chemical equations can be combined algebraically to give a new equation. Even before the science of thermodynamics developed in the late nineteenth century, it was observed that the heats associated with chemical reactions can be combined in the same way to yield the heat of another reaction. For example, the standard enthalpy changes for the oxidation of graphite and diamond can be combined to obtain $\Delta H°$ for the transformation between these two forms of solid carbon, a reaction that cannot be studied experimentally

$$\text{C(graphite)} + \text{O}_2\text{(g)} \longrightarrow \text{CO}_2\text{(g)}\ \Delta H° = -393.51\ \text{kJ mol}^{-1}$$

$$\text{C(diamond)} + \text{O}_2\text{(g)} \longrightarrow \text{CO}_2\text{(g)}\ \Delta H° = -395.40\ \text{kJ mol}^{-1}$$

Subtraction of the second reaction from the first (i.e., writing the second equation in reverse and adding it to the first one) yields

$$\text{C(graphite)} \longrightarrow \text{C(diamond)}\ \Delta H° = 1.89\ \text{kJ mol}^{-1}$$

This principle, known as *Hess' law of independent heat summation,* is a direct consequence of the enthalpy being a state function. Hess's law is one of the most powerful tools of chemistry, for it allows the change in the enthalpy (and in other thermodynamic functions) of huge numbers of chemical reactions to be predicted from a relatively small base of experimental data.

Clausius-Clapeyron Equation

The *vapourization curves* of most liquids have similar shape. The vapour pressure steadily increase as the temperature increases. A good approach is to find a mathematical model for the pressure increase as a function of temperature. Experiments showed that the pressure P, enthalpy of vapourization, ΔH_{vap}, and temperature T are related,

$$P = A \exp(-\Delta H_{vap}/R\,T)$$

where R (= 8.3145 J mol^{-1} K^{-1}) and A are the gas constant and unknown constant. This is known as the Clausius-Clapeyron equation. If P_1 and P_2 are the pressures at two temperatures T_1 and T_2, the equation has the form:

$$\text{In}\left(\frac{P_1}{P_2}\right)=\frac{\Delta H_{\text{vap}}}{R}\left(\frac{1}{T_2}-\frac{1}{T_1}\right)$$

The Clausius-Clapeyron equation allows us to estimate the vapour pressure at another temperature, if the vapour pressure is known at some temperature, and if the enthalpy of vapouization is known.

Solid-Solid Reactions

A solid-solid reaction only involves the solid phases as both reactants and products, with no fluid phases showing up in the chemical reaction. Most solid-solid reactions appear as straight lines on Pressure-Temperature diagrams. The reason for this comes from the Clausius-Clapeyron equation.

$$dP/dT = \partial S/\partial V$$

where ∂S is the change in entropy of the reaction i.e., $\partial S_{\text{products}} - \partial S_{\text{reactants}}$ and ∂V is the change in volume of the reaction i.e., $\partial V_{\text{products}} - \partial V_{\text{reactants}}$.

In general, both the entropy, S, and the volume, V of any phase varies with temperature and pressure. As temperature increases, both S and V tend to increase (things become more disorganized at high temperature, increasing the entropy and molecules vibrate more at high temperature, increasing the volume). Similarly, both S and V tend to decrease with increasing pressure (less room to vibrate means better organization and lower volume). In addition, the change in volume and entropy at any given temperature and pressure tends to be small. The net result of this is that for solid-solid reactions the effects of increasing temperature tend to be offset by the effects of increasing pressure, and thus dP/dT remains more or less constant. A curve whose slope is constant is a straight line.

Let's use these principles to analyze some solid-solid reactions, such as those in the Al_2SiO_5 phase diagram. Note that for the solid-solid reaction andalusite ↔ kyanite, dP/dT is positive. This is what we expect, because the product kyanite occurs on the low T side of the reaction boundary and should have a lower entropy, making ∂S negative. Increasing the pressure causes a decrease in volume, so kyanite should have lower volume than andalusite and thus ∂V is also negative. With both ∂S and ∂V negative, the slope of the boundary curve dP/dT is positive.

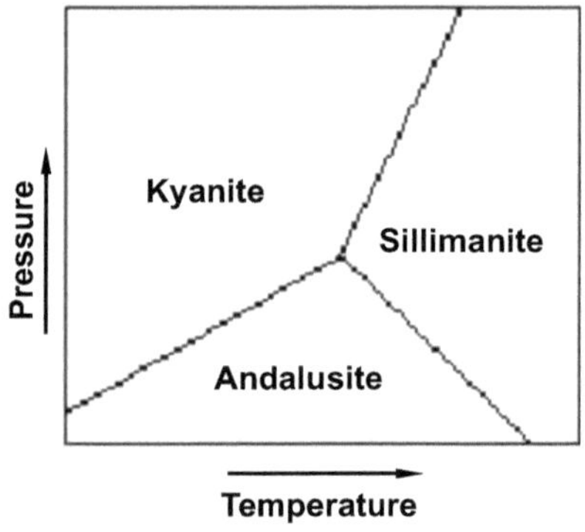

Fig. 8.2: Stability fields of Al_2SiO_5 minerals.

For the reaction kyanite ↔ sillimanite, the product sillimanite occurs on the high T side of the boundary, and thus $S_{Kyanite} < S_{Sillimanite}$, so ∂S is positive. Since kyanite occurs on the high P side of the boundary curve, $V_{kyanite} < V_{sillimanite}$, so ∂V is also positive. Thus, dP/dT is also positive.

But, note that the reaction boundary for andalusite ↔ sillimanite has a negative slope on the diagram. The product of the reaction, sillimanite, has a smaller volume than the reactant andalusite. So, ∂V is negative. But, sillimanite occurs on the high temperature side of the reaction and thus has higher entropy than andalusite. Thus, since the reactant, andalusite, has lower entropy than the product, sillimanite, ∂S is positive, making dP/dT negative.

Reaction Boundaries

Another useful relationship is:

$$G = H - TS$$

where G is the Gibbs Free Energy, H is the enthalpy, T is the absolute temperature in Kelvin, and S is the entropy.

For a chemical reaction, we can rewrite this as:

$$\partial G = \partial H - T\partial S$$

where again: ∂G = the change in free energy of the reaction i.e. $\partial G_{products} - \partial G_{reactants}$, ∂H = the change in enthalpy of the reaction i.e. $\partial H_{products} - \partial H_{reactants}$ and ∂S is the change in entropy of the reaction i.e. $\partial S_{products} - \partial S_{reactants}$.

In general ∂G, ∂H, ∂S, and ∂V are dependent of pressure and temperature, but at any given T & P:

If $\partial G < 0$ (negative) the chemical reaction will be spontaneous and run to the right,

If $\partial G = 0$ the reactants are in equilibrium with products, and if $\partial G > 0$ (positive) the reaction will run from right to left.

As stated above, G, H, and S depend on temperature and pressure. But, because G depends on H and S, it is usually more convenient to consider the temperature dependence of H and S, so that if we know H and S at any given temperature, we can calculate G.

$$\left(\frac{\partial H}{\partial T}\right)_p = C_p$$

where C_p is the heat capacity at constant pressure. The heat capacity is the amount of heat necessary to raise the temperature of the substance by 1° K.

Tables of thermodynamic data are usually compiled at some known reference temperature and pressure, most commonly at a temperature of 298 K, and pressure of 1 bar (= 0.1 MPa ~ 1 atm). Thus, we if we need to know H at some temperature, T, other than 298 K, we can use the above equation to determine H at the new temperature:

$$H_T = H_{298} + C_p (T - 298)$$

For a reaction, the above equation can be rewritten as:

$$\Delta H_T = \Delta H_{298} + \Delta C_p (T - 298)$$

Now let us take an example. We are assuming that $C_p = 0$, that V is not a function of pressure, that sillimanite is the Al_2SiO_5 mineral produced in the reaction, that muscovite, *k*-spar, quartz and sillimanite are pure phases (not solid solutions), and that H_2O behaves as an ideal gas and is the only gas present in the fluid phase.

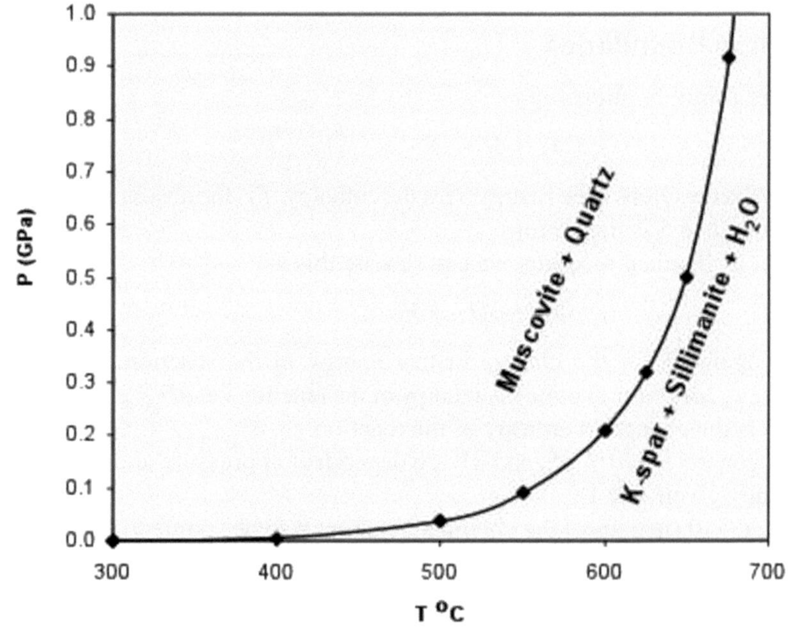

Fig. 8.3: Reaction boundary depicting decarbonation reaction in metamorphic minerals.

Note how the reaction boundary is curved and how it asymptotically approaches the temperature axis at low temperature. This is typical of dehydration and decarbonation reactions.

The assumptions about the lack of solid solution in the solids is pretty good for these solids to a first approximation (with the possible exception of *k*-spar), but in the earth, it is likely that the fluid phase is not pure H_2O, and is more likely a mixture of H_2O and CO_2.

All phases, whether mineralogical or not, have an associated Gibbs free energy of formation value abbreviated ΔG_f. The ΔG_f value describes the amount of energy that is released or consumed when a phase is created from other phases.

We can calculate the Gibbs free energy of any reaction (ΔG_{rxn}) by summing the energies of the right-hand side (products) of the reaction and subtracting the energies of the left hand side (reactants). If the calculation reveals that $\Delta G_{rxn} < 0$, the reaction proceeds to the right. If the $\Delta G_{rxn} > 0$, the reaction proceeds to the left.

Consider the reaction:

albite = jadeite + quartz

Under normal earth surface conditions, the Gibbs energy of this reaction is greater than zero. Consequently, albite (Na-feldspar) is stable and the assemblage jadeite (Na-pyroxene) + quartz is unstable. This is the reason that we shouldn't expect Na-pyroxene to be formed at normal surface condition and jadeite is obtained in rocks formed at substantially high pressure (e.g. eclogite).

The ΔG_f of a mineral varies with changes in pressure (*P*), temperature (*T*) and mineral composition (*X*). Consequently, the ΔG_{rxn} for any reaction will vary with *P*, *T* and *X*, being positive in some portions of *P-T-X* space and negative in others. The result is that we can plot reactions on phase diagrams.

Along any reaction line, such as the one separating the albite and jadeite + quartz fields as shown in Fig. 8.4, $\Delta G_{rxn} = 0$.

Doing thermodynamic calculations requires reliable thermodynamic data. Additionally, although the calculations can be done by hand or with a calculator, they are complicated and time consuming.

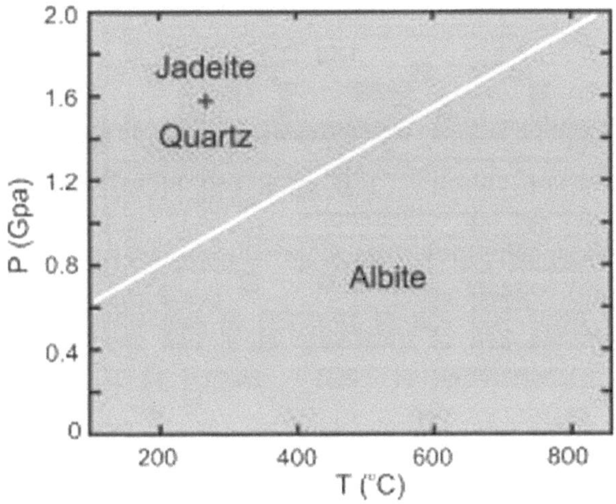

Fig. 8.4: P-T diagram showing relationship between albite, jadeite and quartz.

8.4 KINETICS

Kinetics is the area of chemistry concerned with reaction rates. The rate can be expressed as:

Rate = change in substance/time for change to occur (usually in M/s)

There are several factors that determine the rate of a specific reaction and those are expressed in the "collision theory" that states that for molecules to react, they must:

1. collide
2. have the right energy and
3. have the right geometry.

To increase the rate, you must make the above more likely to occur. This is possible by changing other factors such as:

- Increasing the surface area (of solids); this allows for more collisions and gives more molecules the right geometry.
- Increasing the temperature; this gives more molecules the right energy (also called the activation energy, E_a).

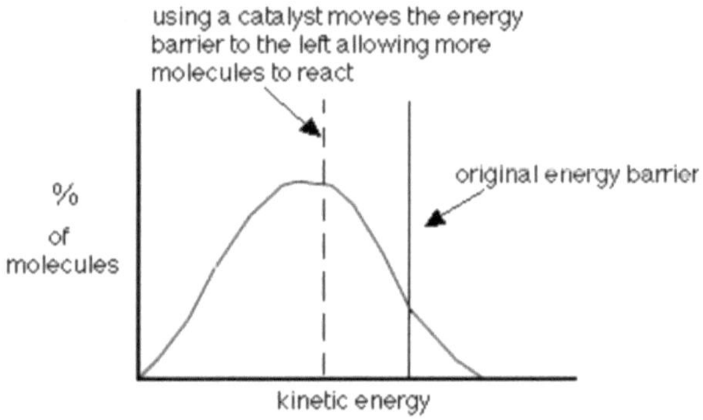

Fig. 8.5a: Relation between the molecule concentration and kinetic energy.

- Increasing the concentration (of gases and solutions); this allows for more collisions and more correct geometry.
- Using a catalyst; helps molecules achieve the correct geometry by providing a different way to react.

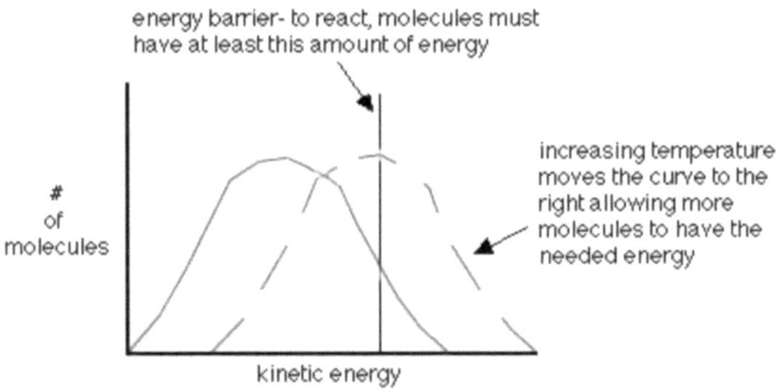

Fig. 8.5b: Relation between the molecule concentration and kinetic energy.

It should be mentioned here that kinetics is important for chemical reactions. There are many minerals which do not crystallize in their stability field and are not in chemical equilibrium. The reaction rate unexpectedly falls on cooling down and minerals, stable at high temperature, are preserved at room temperature. Some examples have been tabulated.

Table 8.1: High-temperature minerals preserved in low-temperature environment

High temperature minerals	*Low-temperture environment*
Cristobalite, tridymite	Sea water
Sanidine	Sediments
Aragonite	Sea-shells

Most minerals collected in the field or analyzed in the laboratory are not in chemical equilibrium. This property is thus used by the geologists to depict the conditions present during rock formations.

Phase Diagrams: The Results of Thermodynamic Calculations

Phase diagrams are graphical representations of the equilibrium relationships between minerals (or other phases). These relationships are governed by the laws of thermodynamics. Standard phase diagrams show how phases or phase assemblages change as a function of temperature, pressure, phase composition, or combinations of these variables.

Some phase diagrams (those for one-component diagrams) depict relationships involving multiple phases having the same composition (for example, the relationships between the vapour, liquid and solid forms of H_2O). Other diagrams (such as the one shown on the right) depict the relationships between a number of compounds having different compositions. Still others show how compositions of phases change under different conditions.

Standard phase diagrams are graphical representations of the equilibrium relationships between minerals (or other phases). These relationships are governed by the laws of thermodynamics. Standard phase diagrams show how phases or phase assemblages change as a function of temperature, pressure, phase composition, or combinations of these variables.

Condensed System

Condensed systems are those systems where the vapour phase is neglected. These condensed systems are very important in geology where vapour phase is not so important and liquid-solid equilibrium is the main point of focus. In case of those condensed systems (where there is no vapour phase), the Gibb's phase rule is modified to $F = C - P + 1$.

Examples: Diopside-anorthite system, orthoclase-albite system and albite-anorthite system (in all these systems, we use the modified phase rule $F = C - P + 1$ to calculate degrees of freedom or F at different parts of the system).

Pseudosections are a type of phase diagram that shows the fields of stability of different equilibrium mineral assemblages for a single bulk-rock composition. Standard phase diagrams may include many reactions but, depending on its composition, a particular rock may only experience a few (or none) of them. Pseudosections only include those reactions that affect a particular composition. Fields on a pseudosection are labelled (specifying the equilibrium mineral assemblage) with the reaction lines unlabelled (although the specific reaction can be deduced).

Some examples: Most minerals are not very compressible; over the pressure range of the Earth's crust, the molar volume of a mineral is nearly constant. If the reaction involves only mineral (solid) phases, then to a good approximation ΔV is constant. With $P_0 = 1$ bar, we have

$$\Delta G = \Delta H_{1\text{bar}} - T\Delta S + (P - 1)\Delta V$$

where ΔV is the molar volume change of the reaction. If two substances A and B are in *chemical equilibrium*, then the difference in free energy between them will be zero.

$$\Delta G = \Delta H_{1\text{bar}} - T\Delta S + (P - 1)\Delta V = 0$$

We can use this to calculate the boundary in P, T space between two phases. Along this boundary, both phases will coexist.

Calcite-Aragonite Transformation

Two phases of $CaCO_3$, calcite and aragonite, are found in a wide range of geologic environments. Aragonite (density 2.95) is denser than calcite (density 2.81) and should thus be a higher pressure form of $CaCO_3$ than calcite. Natural aragonite on heating converts to calcite, the conversion being very rapid at temperatures above 400° C.

This would indicate that calcite is a higher temperature form than aragonite, a conclusion substantiated by heat capacity measurements which show that calcite has a higher entropy than aragonite.

Consider the equilibrium between calcite and its polymorph, aragonite (Fig. 8.6). The reaction relating the two is simply

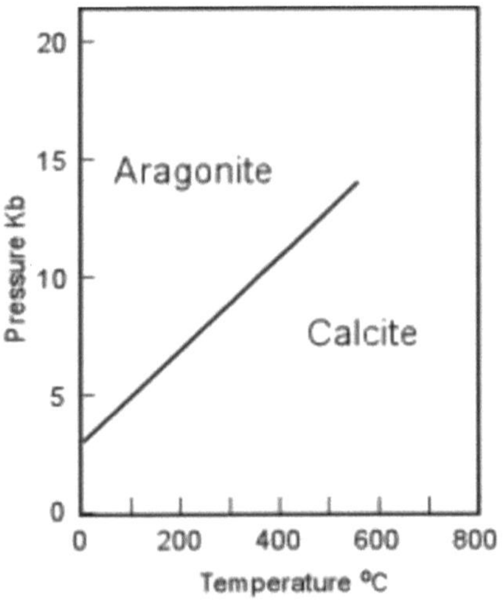

Fig. 8.6: P-T field of $CaCO_3$ polymorphs.

$$CaCO_3\ (cc) = CaCO_3\ (ar)$$

Species	ΔH_f *1bar* (*kJ/mol*)	S_{298K} (*kJ/mol-K*)	*V* (*cm³/mole*)
Calcite, $CaCO_3$	–1208.38	0.09181	36.93
Aragonite, $CaCO_3$	–1208.43	0.08899	34.15

The Gibb's free energy of the reaction $\Delta G = G_{aragonite} - G_{calcite}$

If $G_{aragonite} < G_{calcite}$, ΔG is negative and aragonite is stable. If $G_{aragonite} > G_{calcite}$, ΔG is positive and calcite is stable. Calcite and aragonite can coexist only when $\Delta G = 0$; the *P-T* conditions at which this is true are defined by the phase boundary.

Stability of Minerals

The assemblage of muscovite is so common in igneous and metamorphic rocks that its stability is an important consideration in the construction of any petrogenetic grid. Several recent workers have reported experimental data for the reaction:

Muscovite + quartz = *k*-feldspar + aluminum silicate + steam

Standard hydrothermal experiments using synthetic crystalline starting materials were conducted in order to locate the reaction:

Muscovite + quartz = sanidine + sillimanite + water

The indicated equilibrium temperatures at water pressures equal to total pressure are:

595 + 15°C at 1 kbar,

640 – 10°C at 2 kbar, and

662 ± 6°C at 3 kbar.

One self-consistent curve passing through 662°C, 638°C and 603°C at 3, 2, and 1 kbar respectively is also consistent with the observed calorimetric entropy of formation of muscovite at 298.15 K. The textures of the reaction products suggest that the reaction may proceed through nucleation of sanidine and aluminum silicate in the muscovite grains.

Concluding Remarks

From the point of view of thermodynamics and reaction kinetics, various aspects of mechanism e.g. formation, recrystallisation, development of planer defects etc. play important role in understanding the mineral structures and their impact on geological context.

Think for a while

1. Why are solid-solid reaction boundaries on pressure vs. temperature diagrams straight lines and dehydration or decarbonation reaction boundaries curved lines?
2. In order for all phases in a chemical reaction to be at equilibrium, what must be the necessary conditions required to be fulfilled?
3. The concentration of quartz and feldspar in a rock does not alter prior or after metamorphism. Explain with reasons.
4. The vapour pressure of water is 1.0 atm at 373 K, and the enthalpy of vaporization is 40.7 kJ mol^{-1}. Estimate the vapour pressure at temperature 363 and 383 K respectively.
5. What is implication of a condensed system in geology? How the modification of Phase Rule occurs in such system?

FURTHER READING

Anderson, G. M. and D.A. Crerar. Thermodynamics in geochemistry: The equilibrium model. Oxford University Press, New York. 1993. QE515.A6 1993.

Cemic, L. Thermodynamics in Mineral Sciences: An Introduction. Springer. 2005.

Fraser, D. Thermodynamics in Geology. NATO Science Series. C. Kluwer Academic Publishers. 1977.

Fletcher, Philip. Chemical thermodynamics for earth scientists. Longman Scientific & Technical, Harlow, Essex, England, J. Wiley, New York. 1993.

Greenwood, H.J. (ed.). Short Course in Application of Thermodynamics to Petrology and Ore Deposits. Mineralogical Association of Canada. 1977.

Kern, R. and A. Weisbrod. Thermodynamics for Geologists. Freeman, Cooper and Co. 1967.

Krauskopf, Konrad Bates. Introduction to geochemistry. 3rd ed. McGraw-Hill, New York. 1995. QE515.K7 1979.

Nordstrom, D.K. Geochemical Thermodynamics. Blackburn Press. 2006.

Spear, F.S. Metamorphic Phase Equilibria and Pressure-Temperature-Time Paths. Monograph of the Mineralogical Society of America. 1994.

Wood, B.J. Elementary Thermodynamics for Geologists. Oxford University Press. 1977.

Wood, B.J. and D.G. Fraser. Elementary Thermodynamics for Geologists. Oxford University Press. 1976.

CHAPTER

9

ORIGIN OF MINERALS AND THEIR TRANSFORMATIONS IN NATURE UNDER VARIOUS ENVIRONMENTAL CONDITIONS

Mineral formation process goes through many changes depending on the environmental condition (P-T) and elemental variation, available. On the basis of different P-T regime and mode of formation, mineral genesis has been grouped under igneous, sedimentary and metamorphic processes. Various aspects of these processes are explained here.

Minerals are naturally occurring, inorganic, homogeneous substances, having definite atomic arrangement and fixed chemical composition. Minerals are the product of complex natural processes, which will be discussed in this chapter.

9.1 GEOCHEMICAL CLASSIFICATION OF THE ELEMENTS

Table 9.1: Geochemical classification of elements

Elements	*Description*	*Examples*
Lithopile	Those elements which form ionic bonds generally have filled outer electron shells. They typically bond to oxygen in silicates and oxides.	Li, Na, K, Rb, Cs, Be, Mg, Ca, Sr, Ba, B, Al, Sc, Y, C, Si, Ti, Zr, Hf, Th, P, V, Nb, Ta, O, Cr, W, U, H, F, Cl, Br, I, Fe, Mn, Ga, Ge, Sn
Chalcophile	The elements that bond to S, Se, Te, Sb and As. These bonds are predominantly covalent in character.	Cu, Ag, ZN, Cd, Hg, Pb, As, Sb, Bi, S, Se, Te, Ni, Co, Fe, Mo, Re, Mn, Ga, Ge, Sn

(Contd.)

Table 9.1 (*Contd.*)

Siderophile	The metals near iron in the periodic table (i.e. the transition metals) that exhibit metallic bonding.	Pt, Ir, Os, Ru, Rh, Pd, Au, Fe
Atmophile	Elements and gases that remained in the gaseous state by the time earth condensed	Inert gases, N, O

9.2 PRINCIPAL PROCESSES OF MINERAL GENESIS

Minerals are produced by three major categories of processes: igneous, sedimentary and metamorphic.

Rock Cycle

The rock cycle consists of a series of constant processes through which earth materials change from one form to another over time.

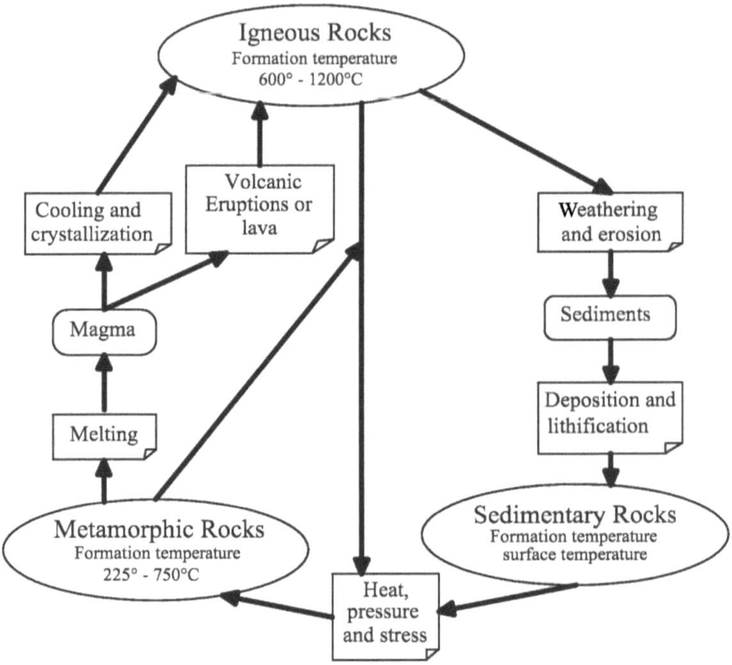

Fig. 9.1: Rock cycle.

9.3 IGNEOUS PROCESSES

Igneous rocks crystallize from magma, either underneath the earth crust or above it. The different rock-types are formed, based on cooling history of the magma, its composition, temperature, pressure and presence of volatiles. The

Goldschmidt classification in the Periodic Table

	1	2	3	4	5	6	7	8	9	10	11	12	13	14	15	16	17	18
1	1 H																	2 He
2	3 Li	4 Be											5 B	6 C	7 N	8 O	9 F	10 Ne
3	11 Na	12 Mg											13 Al	14 Si	15 P	16 S	17 Cl	18 Ar
4	19 K	20 Ca	21 Sc	22 Ti	23 V	24 Cr	25 Mn	26 Fe	27 Co	28 Ni	29 Cu	30 Zn	31 Ga	32 Ge	33 As	34 Se	35 Br	36 Kr
5	37 Rb	38 Sr	39 Y	40 Zr	41 Nb	42 Mo	(43) Tc	44 Ru	45 Rh	46 Pd	47 Ag	48 Cd	49 In	50 Sn	51 Sb	52 Te	53 I	54 Xe
6	55 Cs	56 Ba	57-71 Lan	72 Hf	73 Ta	74 W	75 Re	76 Os	77 Ir	78 Pt	79 Au	80 Hg	81 Tl	82 Pb	83 Bi	84 Po	85 At	86 Rn
7	87 Fr	88 Ra	89-103 Act	(104) Rf	(105) Db	(106) Sg	(107) Bh	(108) Hs	(109) Mt	(110) Ds	(111) Rg	(112) Uub	(113) Uut	(114) Uuq	(115) Uup	(116) Uuh	(117) Uus	(118) Uuo

Lanthanides	57 La	58 Ce	59 Pr	60 Nd	(61) Pm	62 Sm	63 Eu	64 Gd	65 Tb	66 Dy	67 Ho	68 Er	69 Tm	70 Yb	71 Lu
Actinides	89 Ac	90 Th	91 Pa	92 U	(93) Np	(94) Pu	(95) Am	(96) Cm	(97) Bk	(98) Cf	(99) Es	(100) Fm	(101) Md	(102) No	(103) Lr

formation of various rock-types has been depicted in Fig. 9.2. This graphic model describes the difference between nine common igneous rocks based on texture of mineral grains, temperature of crystallization, relative amounts of typical rock forming elements, and relative proportions of silica and some common minerals.

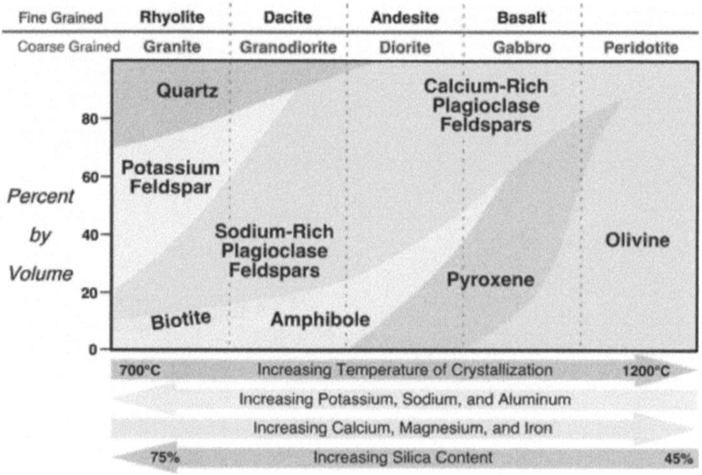

Fig. 9.2: The classification of igneous rocks.

The geothermal gradient which mainly regulates the formation of igneous rocks is controlled by pressure, water and minerals present.

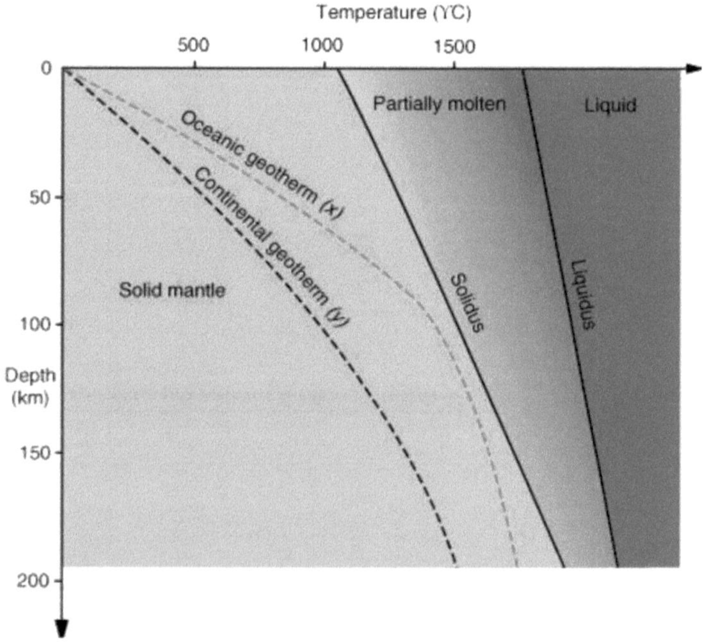

Fig. 9.3: An inter-relationship between geothermal gradient, depth and melting.

The pattern of crystallization of minerals in igneous rocks is well exemplified by Bowen's Reaction Principle. It explains the formation of igneous minerals by the process of fractional crystallization from the melt. Based on this principle, a continuous series and a discontinuous series of minerals can be formed within a fixed temperature range (Fig. 9.4).

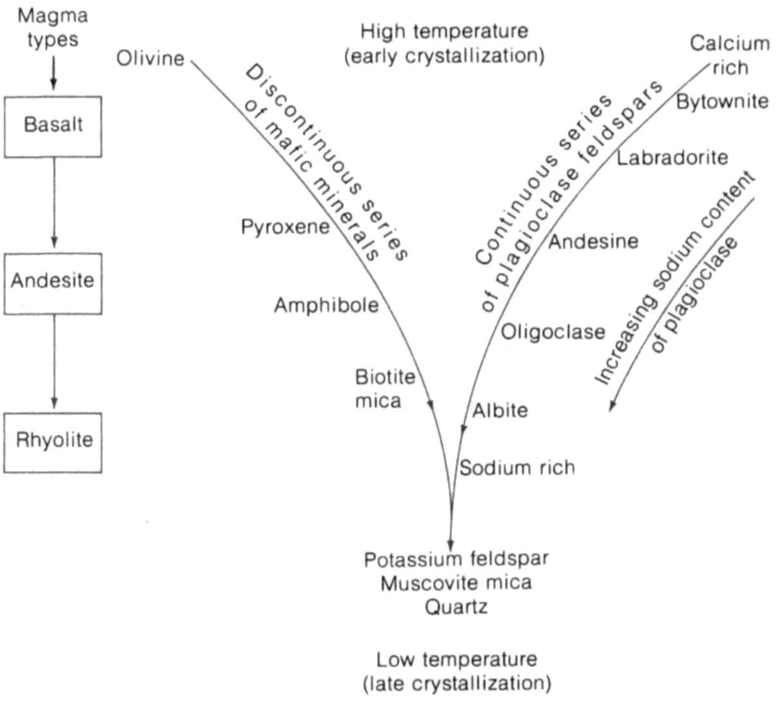

Fig. 9.4: Bowen's reaction series.

Processes of Crystallization from Magma

The processes are mainly of two types—differentiation and assimilation. *Differentiation* is defined as the process in which an originally homogeneous magma splits up into contrasted parts. Differentiation involves the processes listed in Table 9.2.

Assimilation is defined as the process of incorporation of foreign materials to the magma. There is also mixing of two liquids during this process. Under the action of heat and magmatic fluids, volatiles, foreign minerals, trace elements and wall-rock particles mix up with the ascending magma. The heat that the melt gains by excluding the quick-freezing refractories, which is an exothermic process, is sufficient to compensate for the heat loss in the endothermic reactions required for the assimilation of wall-rock components.

Table 9.2: Different processes of differentiation

Processes	*Description*
Fractional crystallization	A stagewise separation process, controlled by liquid-solid phase transition and enables multicomponent mixtures to be split into fractions of narrow chemical composition range.
Gravity separation	Affinity of the heavy minerals to sink to the bottom, while the lighter minerals rise up in the magma chamber is the gravity-controlled process. Also affected and controlled by the shape, size and specific gravity of individual crystals and the viscosity of the magma. Olivine is one such example.
Filter pressing	Formation of a framework of crystals with residual liquid in the interstices. Then extra pressure squeezes out the liquid to regions of least pressure. Thus the solid crystals separate out from liquid magma.
Liquid immiscibility	With lowering of temperature, a homogeneous magma separates out into two immiscible fractions, controlled by the difference in specific gravity.
Gaseous transfer	Volatiles serve as excellent solvents; gas bubbles also attach themselves to the growing crystals. These substances enable selective transfer of material from lower to higher level.

Table 9.3: Common igneous minerals

Essential minerals	*Accessory minerals*	*Common rock-type*
Quartz, feldspar (plagioclase, microcline, sanidine, orthoclase etc.), feldspathoid (nepheline, leucite, kalsilite etc.), olivine, pyroxene (clinopyroxene and orthopyroxene), amphibole (hornblende, actinolite, tremolite etc.), mica (biotite, muscovite, phlogopite etc.)	Magnetite, ilmenite, apatite, zircon, titanite, pyrite, pyrrhotite, allanite, tourmaline, sodalite, fluorite	Granite, rhyolite granodiorite, diorite, dacite, gabbro, basalt, andesite, syenite, trachyte, monzonite, nepheline syenite, phonolite, pyroxenite, peridotite, dunite, limburgite, ijolite, leucitite, nephelinite

Olivine Group

Olivine is a primary igneous mineral of nesosilicate group, crystallizing at high temperature and pressure from mafic (compositionally rich in magnesium and iron) magmas in the upper mantle. The olivines consist of a complete solid solution between Mg_2SiO_4 (forsterite, Fo) and Fe_2SiO_4 (fayalite, Fa). The phase diagram for the common end members of the olivine solid solution series shows that pure forsterite melts at 1890°C and pure fayalite melts at 1205°C. Thus, the olivines are sometimes seen to be zoned from Mg-rich cores to more Fe-rich rims, although such zoning is usually limited to 5 to 10% difference between the cores and the rims.

Pure forsterite is limited to metamorphosed Mg-rich limestones and dolomitic metamorphic rocks. Fo_{90-95} is found in ultrabasic igneous rocks, particularly dunites (>90% by volume olivine), and peridotites (Olivine + Cpx + Opx). Fo_{60-90} is found in basic igneous rocks like basalts and gabbros, and sometimes in andesites, where it occurs with plagioclase and pyroxene. Fa_{100-40} is found in Fe-rich siliceous igneous rocks like rhyolites and granites. Mg-rich olivines rarely occur in quartz bearing rocks and quartz rarely occurs with Mg-rich olivine because the reaction shown below runs to the right for most pressures and temperatures.

$$\underset{\text{Fo}}{Mg_2SiO_4} + \underset{\text{Qtz}}{SiO_2} \longleftrightarrow \underset{\text{En}}{2MgSiO_3}$$

Thus the crystallization of olivine results in a Si-enriched and Mg- and Fe-depleted melt, which thereby produces pyroxene, with completely different structure.

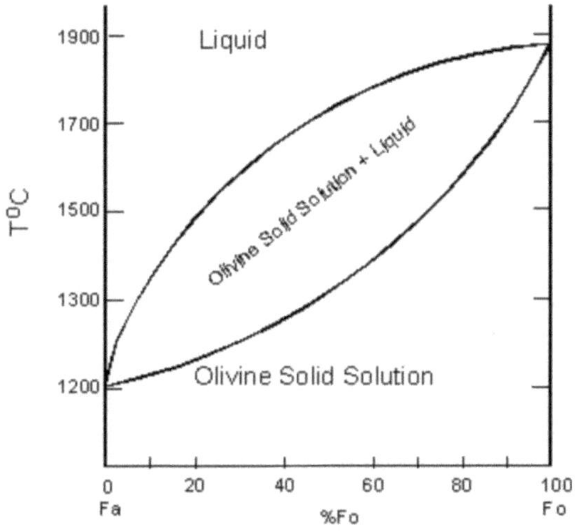

Fig. 9.5: Phase diagram of olivine group.

Pyroxene Group

Pyroxene crystallizes after olivine at a temperature range of 1100°-1250° C. They represent the inosilicate (single-chain) group. This group consists of many different elements with the typical end-members of:

Wollastonite ($CaSiO_3$),
Ferrosilite ($FeSiO_3$),
Enstatite ($MgSiO_3$),
Diopside ($CaMgSi_2O_6$) and
Hedenbergite ($CaFeSi_2O_6$).

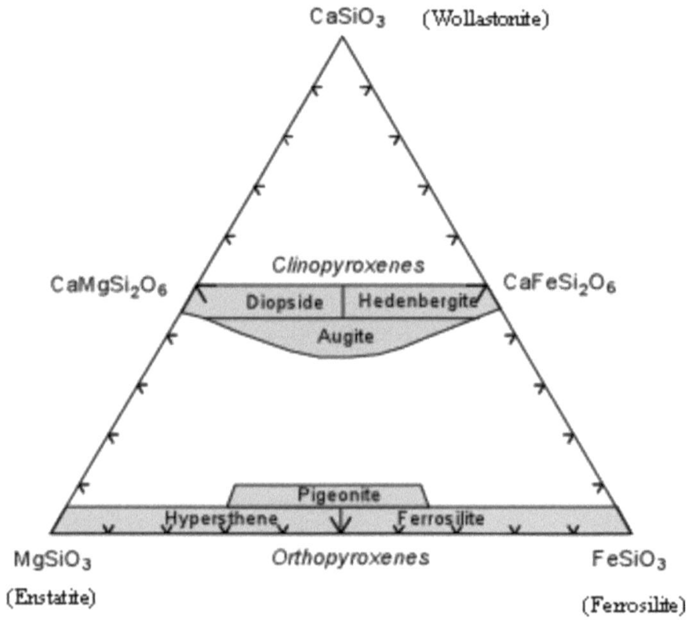

Fig. 9.6: Different end-members of pyroxene group.

The pyroxene quadrilateral forms part of a larger ternary system ($CaSiO_3$-$MgSiO_3$-$FeSiO_3$) that includes another single chain silicate wollastonite ($CaSiO_3$) in the diagram. The Ca-rich clinopyroxenes are separated from the orthopyroxenes, and from the Ca-poor clinopyroxene pigeonite by a solvus. The solid solutions between end members ferrosilite and enstatite are called orthopyroxenes as they exhibit orthorhombic crystal structure, while the solid solutions between end members diopside and hedenbergite are called clinopyroxenes, due to their monoclinic crystal structures. The Mg-rich end members crystallize at higher temperatures than the Fe-rich end members.

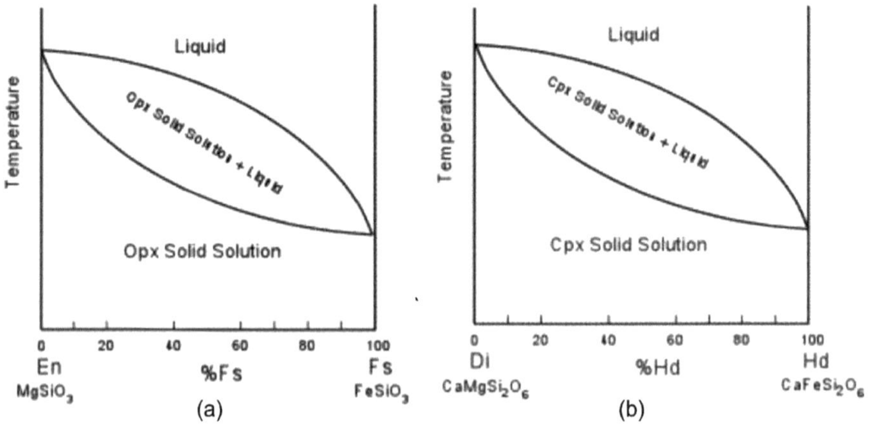

Fig. 9.7: Phase diagrams: (a) Orthopyroxene group and (b) Clinopyroxene group.

Solid immiscibility is present between the Diopside-Hedenbergite series and the Orthopyroxene series.

The solvus with pigeonite closes with increasing Fe-content at any given temperature. The position of the solvus is temperature dependent, with the solvus gap expanding with decreasing temperature. Pigeonite is only stable at higher temperature and it inverts to orthopyroxene if cooled slowly to lower temperatures at less than 1100-950°C. Thus, pigeonite is only found in volcanic and shallow intrusive igneous rocks, or as exsolution lamellae in a host augite or opx (more commonly in augite).

Subcalcic augite and pigeonite are stable only at high temperature. Along the $FeMg_{-1}$ axis the "quadrilateral" pyroxene solid-solutions have been divided into a host of mineral names. The common orthopyroxenes are orthoferrosilite, eulite, ferrohypersthene, hypersthene, bronzite and enstatite. The common Ca-rich clinopyroxenes are hedenbergite, ferroaugite, ferrosalite, augite, salite and diopside. In addition to the $MgCa_{-1}$ V_{-1} and $FeMg_{-1}$ exchanges connecting diopside to the other "quadrilateral" pyroxenes, $NaAlCa_{-1}Si_{-1}$ forms another important exchange in the pyroxene group, that connects diopside with omphacite and jadeite. Thus, much of the compositional variation in the pyroxene group can be represented in the $MgCa_{-1}$-$FeMg_{-1}$-$NaAlCa_{-1}Si_{-1}$ space shown in Fig. 9.8.

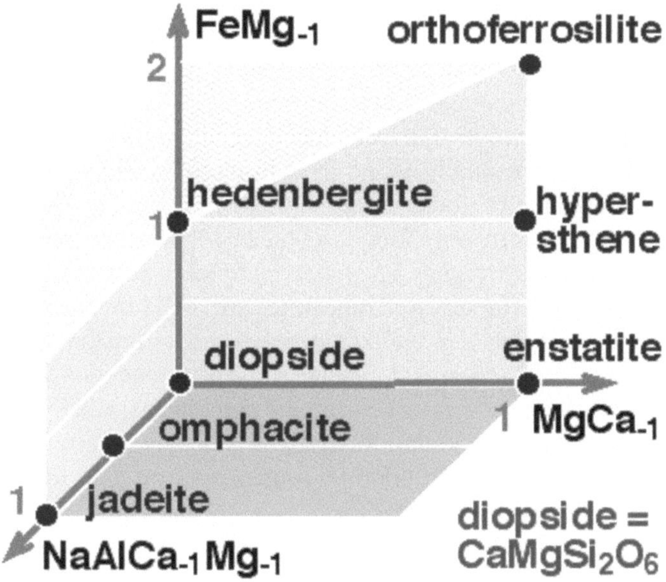

Fig. 9.8: Compositional variation in the pyroxene group represented in the $MgCa_{-1}$-$FeMg_{-1}$-$NaAlCa_{-1}Si_{-1}$ space.

Sodic-pyroxene Classification

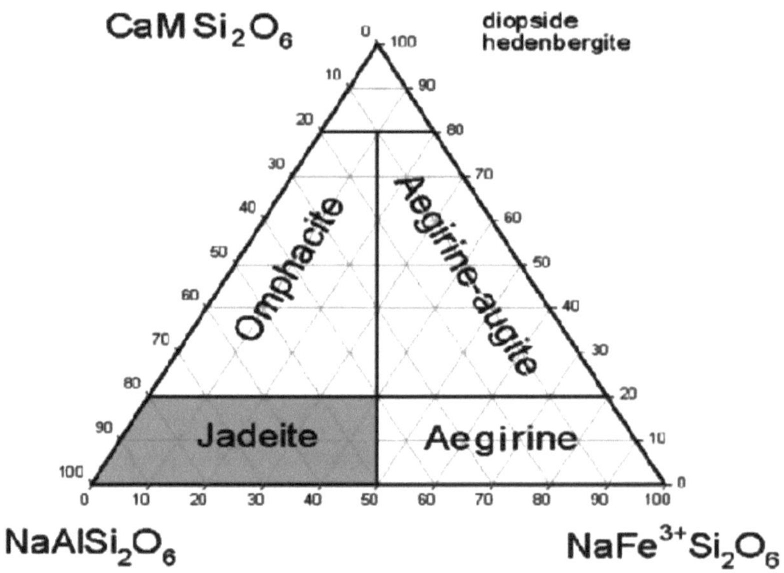

Fig. 9.9: Classification of sodic pyroxenes.

The quadrilateral pyroxenes form an important mineral in mafic and ultramafic igneous rocks, and in high temperature "granulite facies" metamorphic rocks.

Augite is commonly found in both plutonic and volcanic igneous rocks, as well as high grade meta-igneous rocks like gneisses and granulites.

Hypersthene is commonly found in both plutonic and volcanic igneous rocks and in meta-igneous rocks as well.

Pigeonite is generally only found in volcanic igneous rocks, although, as mentioned above, it can occur as exsolution lamellae in augites of more slowly cooled igneous rocks. But inverted pigeonites are found in layered ultramafic complexes and, very rarely, in high temperature meta-ironstones.

Aegerine (acmite)-Aegerine Augite are sodic pyroxenes and thus are found in alkalic igneous rocks associated with sodic amphiboles, alkali feldspars and nepheline. The mineral is common in alkali granites, quartz syenites and nepheline syenites (all alkalic plutonic rocks), and are also found in sodic volcanic rocks like peralkaline rhyolites.

Jadeite is a sodium aluminum pyroxene that is characterized by its presence in metamorphic rocks formed at relatively high pressure. It can form by a reaction of albite to produce:

$$\underset{\text{Albite}}{NaAlSi_3O_8} + \underset{\text{Jadeite}}{NaAlSi_2O_6} + \underset{\text{Quartz}}{SiO_2}$$

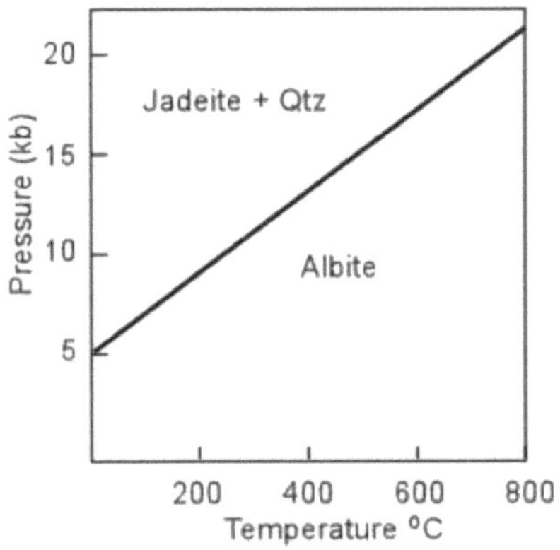

Fig. 9.10: P-T field of jadeite.

Diposide occurs in amphibolte facies calc-silicates and marbles.

Amphibole Group

Amphibole crystallizes after pyroxene during the crystallization history of a magma at around 900° to 1100° C. Amphiboles are considered as essential minerals of igneous rocks.

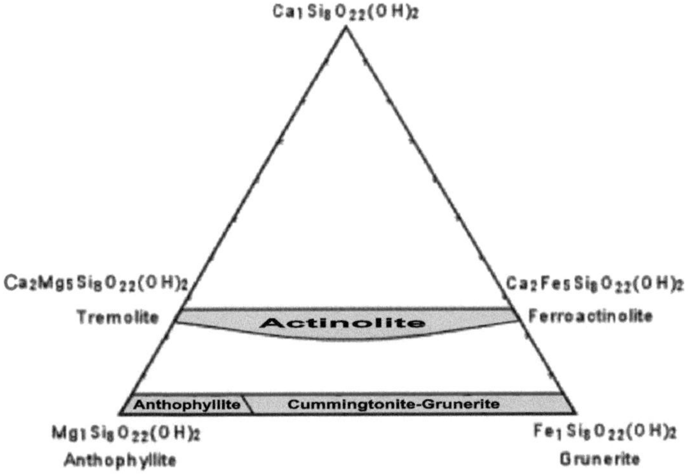

Fig. 9.11: Compositional diagram of amphibole group.

Tremolite occurs almost exclusively in low grade metamorphic rocks, particularly those with a high Ca concentration, such as meta-dolomites, meta-ultrabasic rocks.

Actinolite also occurs almost exclusively in low grade metamorphic rocks, particularly in meta-basalts and meta-gabbros where it is commonly associated with chlorite.

Hornblende is a common mineral in both igneous and metamorphic rocks. In igneous rocks it is found in andesites, dacites and rhyolites, as well as in gabbros, diorites and granites. In metamorphic rocks it is a common constituent of meta-basalts that have been metamorphosed to intermediate grades of regional metamorphism (amphibolites). It is also found in some ultrabasic rocks.

Basaltic Hornblende (also called Oxy-hornblende) is a dark brown to reddish brown variety of hornblende that results from oxidation during crystallization of basalts, andesites, dacites, and rhyolites. It usually has a dark reaction rim that consists of opaque oxide.

Anthophyllite does not occur in igneous rocks, but is a constituent of metamorphic rocks. It is the only orthorhombic.

Cummingtonite-Grunerite is more common in metamorphosed igneous rocks where members of the series occur with hornblende. It has been found in siliceous volcanic rocks as well.

Glaucophane-Riebeckite: Glaucophane is a common mineral in blueschist facies metamorphic rocks that result from low temperature, high pressure metamorphism along ancient subduction zones. Riebeckite is found in alkali granites, syenites and peralkaline rhyolites.

Arfvedsonite occurs most commonly in peralkaline volcanic rocks and alkaline plutonic igneous rocks, where it typically occurs with the sodic pyroxene aegerine.

Mica Group

Mica group of minerals is an important member of sheet silicates. The general formula for mica is $K(Al, Mg, Fe)_{2-3}(AlSi_3O_{10})(OH)_2$. The important minerals are muscovite, biotite, annite, paragonite and phlogopite. The formation temperature is from 700° to 850° C with primary formation of biotite, followed by muscovite.

Muscovite, $KAl_3Si_3O_{10}(OH)_2$, and Paragonite, $NaAl_3Si_3O_{10}(OH)_2$, are two potential end members of the solid solution series involving K and Na. But, there is a large miscibility gap between the two end members with muscovite being between 65% and 100% of K-rich end member, and paragonite showing compositions between about 80% and 100% of the Na-rich end member.

Biotite is a solid solution between the end members phlogopite $KMg_3AlSi_3O_{10}(OH)_2$ and annite $KFe_3AlSi_3O_{10}(OH)_2$, although pure annite does not occur in nature. In addition, small amounts of Na, Rb, Cs and Ba may substitute for K, and like in other minerals, F can substitute for OH and increase the stability of biotite to higher temperatures and pressures.

Muscovite is common constituent of Al-rich medium grade metamorphic rocks where it is found in Al-rich schists and contributes to the schistose foliation found in these rocks. Muscovite is also found in siliceous, Al-rich plutonic igneous rocks (muscovite granites), but has not been found as a constituent of volcanic rocks. In these rocks it is commonly found in association with alkali feldspar, quartz, and sometimes biotite, garnet, andalusite, sillimanite, or kyanite.

Nearly pure phlogopite is found in hydrous ultrabasic rocks like kimberlite, and is also found in metamorphosed dolomites. Biotite, with more Fe-rich compositions, is common in dacitic, rhyolitic and trachytic volcanic rocks, granitic plutonic rocks, and a wide variety of metamorphic rocks.

SiO_2 Group

SiO_2 is one of the most significant mineral of igneous rocks. It generally occurs in several polymorphic forms, among which quartz is the most stable mineral under Earth's surface conditions. Other polymorphs which remain in metastable condition on the Earth's crust, change into low quartz, if given enough time. Figure 9.12 shows different pressure-temperature regimes for different polymorphs. Like low quartz it is stable for a very wide P-T range. Stishovite and coesite are stable at very high pressures. Tridymite and cristobalite form at a temperature less than 900° C.

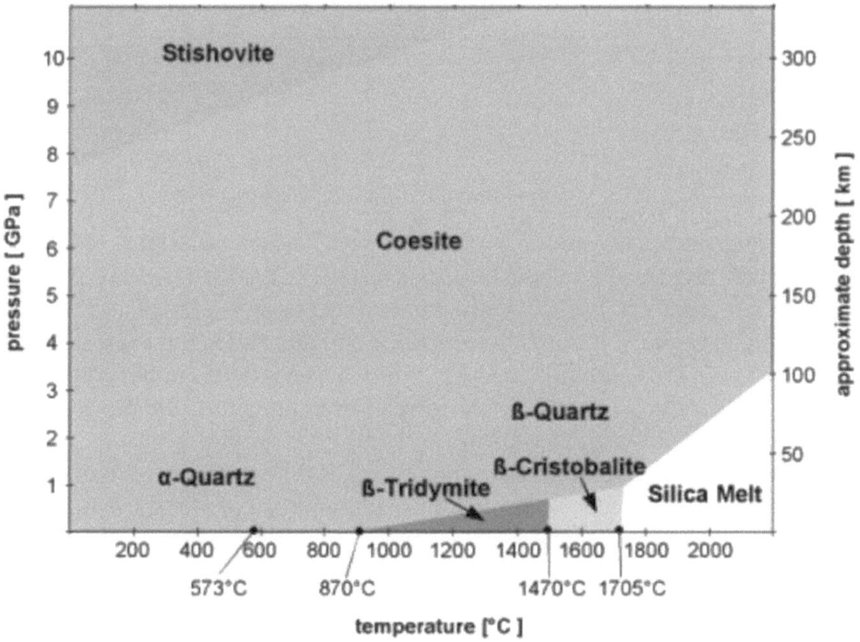

Fig. 9.12: Polymorphs of SiO_2.

Feldspar Group

Feldspars are important rock-forming minerals and are stable over a wide range of pressure-temperature conditions. The general formula for feldspar is (Ca, Na, K) (Al, Si)$_4$O$_8$ and they represent framework silicates.

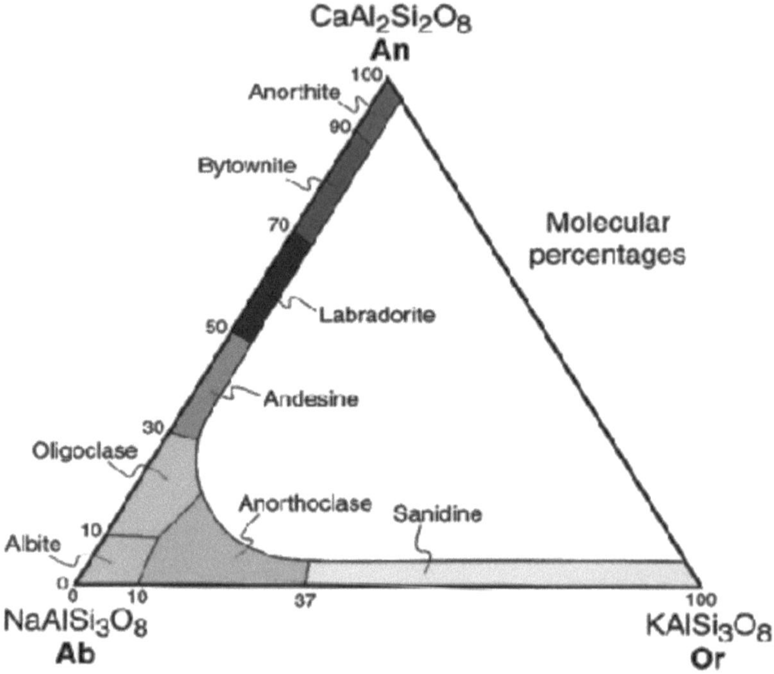

Fig. 9.13: Compositional diagram of feldspar group.

The feldspars are commonly represented by ternary solutions, with three end members namely albite ($NaAlSi_3O_8$), anorthite ($CaAl_2Si_2O_8$) and orthoclase ($KAlSi_3O_8$). These can further be divided into two series—plagioclase feldspar series which is solid solution between albite and anorthite; and alkali feldspar series with solutions $NaAlSi_3O_8$ and $KAlSi_3O_8$. It has been explained through the phase diagram (Fig. 9.13). The formation temperature for the feldspars varies between 550° C and 1100°C.

Feldspar also exists in polymorphic forms. Albite at higher temperature is known as monalbite. The low-temperature polymorphs of albite are all known by the same name of albite. K-feldspar exists in three stable polymorphic forms—sanidine, orthoclase and microcline at high, medium and low temperature respectively. The definite temperature boundary among these polymorphs is not known, the transformation from one form to another generally happens due to cooling and exemplifies order-disorder polymorphism.

Intermediate feldspar between K-feldspar and albite, called anorthoclase, is stable only at high temperature. On lowering of temperature, two separate feldspar form due to the presence of miscibility gap (a composition range where no mineral is stable under particular range of pressure-temperature conditions) between albite and orthoclase. The process of separating single feldspar into two compositions is called exsolution.

The phase diagram for the plagioclase series is given in Fig. 9.14 which shows that the anorthite component has a higher melting temperature than the albite component. Thus, on crystallization, higher temperatures will favour more An-rich plagioclase which will react with the liquid to produce more Ab-rich plagioclase on cooling.

Plagioclase occurs in basalts, andesites, dacites, rhyolites, gabbros, diorites, granodiorites and granites. In most of these igneous rocks, it always shows the characteristic albite twinning. Plagioclase also occurs in a wide variety of metamorphic rocks, where it is usually not twinned. In such rocks where the plagioclase is not twinned, it is difficult to distinguish from the alkali feldspars. Plagioclase can be a component of clastic sedimentary rocks, although it is less stable near the Earth's surface than alkali feldspar and quartz, and usually breaks down to clay minerals during weathering.

As an alkali feldspar cools from high temperature to lower temperature, the crystal structure changes from that of sanidine, through orthoclase to microcline. These transformations are order-disorder transformations, and thus require large amounts of time. Furthermore, if the feldspar is allowed to cool very slowly, then exsolution will occur, and the solid solution will separate into a Na-rich phase and a K-rich phase. Thus, sanidine is found in rocks that were cooled very rapidly from high temperature, i.e. volcanic rocks. Orthoclase and microcline will be found in plutonic igneous rocks (cooled slowly at depth in the earth) and in metamorphic rocks. In addition, in the plutonic rock types if the cooling takes place slowly enough, then perthitic exsolution lamellae may also form. Anorthoclase is a Na-rich feldspar with approximately equal amounts of the anorthite (Ca) and orthoclase (K) components. Generally anorthoclase occurs in Na-rich volcanic rocks.

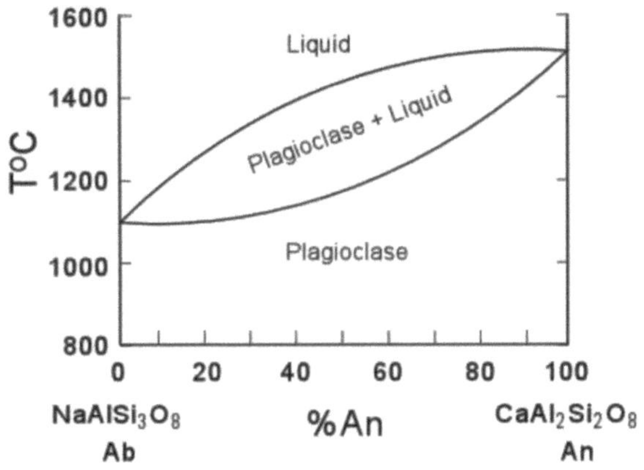

Fig. 9.14: Phase diagram of plagioclase feldspar.

Feldspathoid Group

The main members of this group are analcime ($NaAlSi_2O_6 . H_2O$), leucite ($KAlSi_2O_6$) and nepheline ($NaAlSiO_4$). They represent tecto-silicates. These minerals, on reacting with SiO_2, form feldspar. So they are not found with quartz in any rocks. They are quite uncommon than other minerals and are more noticeable in silica-poor rocks. The temperature and pressure range for this group of minerals is 800°-1050° C and at approximately 500-5000 bars.

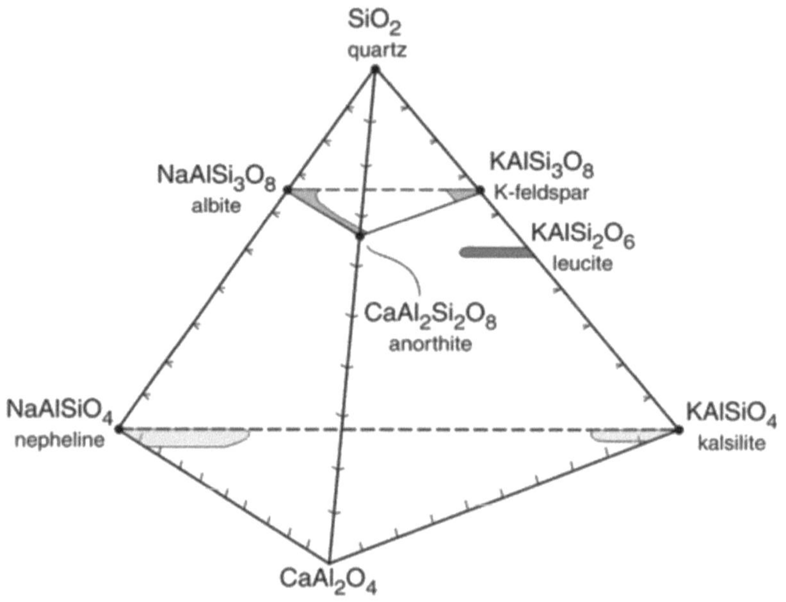

Fig. 9.15: Compositional diagram of feldspathoid group.

Nepheline occurs in both volcanic and plutonic alkaline igneous rocks. Leucite is found in alkalic volcanic rocks, and is rarely found in plutonic rocks. Sodalite occurs predominantly in alkali-rich plutonic igneous rocks, like syenites, but can also be found in volcanic rocks.

There are a number of gemstones that crystallize from igneous processes. Diamonds are found in kimberlite, a mixture of gas-charged, highly serpentinized and altered porphyritic peridotite, containing fragments of mantle and crustal rocks incorporated during the supersonic ascent to the Earth's surface (In Australia, diamonds are contained in lamproite, a potassium- and magnesium-rich mafic volcanic rock).

Zircon, monazite, tourmaline, beryl and topaz crystallize in water- and volatile-rich granitic pegmatites. Pegmatites are coarse-grained rocks, formed through the crystallization of the last residual melts that are rich in volatiles such as water and fluorine and the incompatible elements (boron, lithium, beryllium, niobium, tantalum, uranium, thorium and the rare earth elements) that do not fit in the crystal structures of minerals crystallized at higher temperatures. The separation and expansion of volatile gases in a near-surface magma, lava, or hot volcanic ash deposits during the last stages of crystallization may produce open void spaces. Beautiful well-formed crystals of topaz, red beryl, spessartine (manganese-rich garnet), and other minerals crystallize in miarolitic cavities in pegmatites or void spaces in rhyolitic volcanoclastic rocks from volatile-rich gases.

9.4 SEDIMENTARY PROCESSES

Sedimentary processes involve weathering, erosion, deposition, compaction and lithification. The main sedimentary minerals are calcite, chalcedony, dolomite and glauconite.

The above-mentioned sedimentary processes are important in placer deposits. A placer deposit typically is a water-sorted and water-deposited aggregate of sand or gravel that has concentrated gemstones, or other heavy minerals of value. Diamonds, rubies, sapphires and other hard precious gemstones and precious metals such as gold and platinum that were formed by igneous, metamorphic and hydrothermal processes are sometimes later concentrated by the action of river or sea-waters into placer deposits, after being eroded from the original host rock.

Another form of sedimentary minerals, that need to be mentioned here, are evaporites, which are mineral sediments, formed by the evaporation of surficial water.

Table 9.4: Sedimentary processes

Processes	*Description*
Weathering	Breaks rocks into finer processes by the action of wind, river, glacier.
Erosion	Combination of weathering and movement of the loose particles.
Deposition	Erosion and weathering products are thereby taken and deposited to other places by the action of wind, river, glacier etc.
Lithification	Alteration of sediments to rocks.
Compaction	Weight of the overlying sediments squeezes the loose particles together; water is completely removed.
Diagenesis	The textural and chemical changes that are undergone by the sedimentary rocks; new minerals formed by this process are termed as authigenic minerals. Examples are zeolites, clays, feldspar, pyrite and quartz.

Weathering

Weathering forces can be mechanical or chemical. Weathering rates follow a series roughly the inverse of the order of original crystallization for igneous minerals (Bowen's Reaction Series).

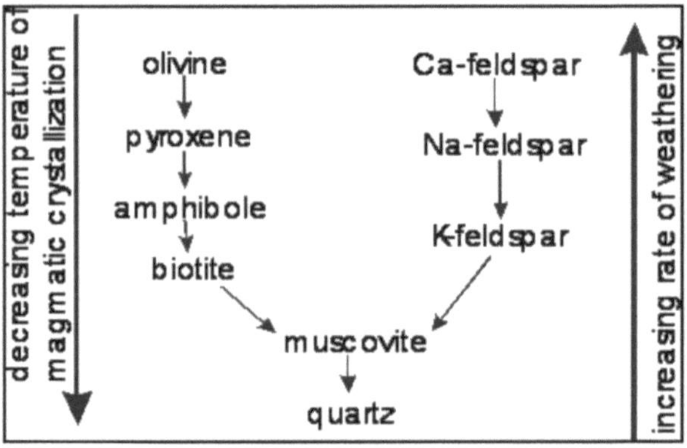

Fig. 9.16: Goldich's stability series.

Significance of Goldich's Stability Series

1. Goldich's series is in part a reflection of the crystal structures of the silicates and the molar ratio of silica to oxygen.
2. The higher the Si:O ratio, the more covalent the structure and the fewer other metal cations required for charge balance.
3. Quartz, the most resistant to weathering, has an Si:O ratio of 0.5, and olivine, the least resistant to weathering, has an Si:O ratio of 0.25.
4. Involves the release of Si to solution as H_4SiO_4 (silicic acid) and cations such as Ca^{2+}, Mg^{2+}, Na^+, and K^+.
5. Ferric iron (Fe^{3+}) and Al have very low solubilities, so when silicates containing these metals are weathered, secondary minerals (clays, oxides, hydroxides) form.
6. Thus weathering removes Si, alkalis and alkaline earths, but leaves behind Fe and Al.

The formation of sedimentary rocks is mainly regulated by the chemical weathering, rather than the mechanical, which only produces clastic materials, called detritus. Chemical weathering on the other hand, produces hydrolysate—the dissolved materials. The remaining materials, which remain undissolved, are called resistate. These resistate remain to become sediments, which are mainly quartz, clay, K-feldspar, garnet, zircon, rutile or magnetite.

Table 9.5: The major resistate minerals, resulting from weathering of common igneous minerals

Igneous minerals	*Resistate minerals*
Quartz	Quartz
Feldspar	Quartz, muscovite, clays
Olivine	Hematite, goethite, magnetite, quartz, muscovite, clays
Pyroxene	Hematite, goethite, magnetite, quartz, muscovite, clays
Amphibole	Hematite, goethite, magnetite, quartz, muscovite, clays

By the action of chemical weathering, clay minerals, like montmorillonite, illite and kaolinite are formed from feldspar. Such reactions are called hydrolysis reactions. The reactions of different minerals to weathering can be clearly depicted by Goldich weathering series. It shows that the high temperature, Si- and O-poor minerals are generally less resistant to weathering than those that crystallize at low temperature. Due to the presence of strong Si-O bonds, quartz, some feldspar and some non-silicates do not weather easily.

Transportation, Deposition and Lithification

By various geomorphic agents like flowing water, wind and gravity, sediments are transported and thereby deposited when the gravitational force is most dominant. Post deposition, the detrital sediments alter to sedimentary rocks by lithification. It is a process operated by compaction and cementation. Common cementing agents include the minerals quartz, calcite and hematite.

The step that follows lithification is diagenesis. It leads to the textural or chemical changes of the sedimentary rocks due to heating, compaction, or groundwater reaction. The new minerals, forming as a result of diagenesis, are called authigenic minerals. Examples of such minerals are zeolites, clays, feldspar, pyrite and quartz.

Sedimentary Minerals

The sedimentary minerals can be broadly categorized into groups depending on different sedimentary processes and environments.

Table 9.6: Classification of sedimentary minerals

Forms	*Sedimentary minerals*
Resistates	Quartz, accessory minerals
Hydrolysates	Clay minerals, Al-hydroxides, Glauconite, Chamosite
Oxidates	Hematite, Goethite, Pyrolusite, Psilomelane
Reduzates	Pyrite, Marcasite, Siderite, Sulfur
Precipitates	Calcite, Aragonite, Dolomite
Evaporites	Gypsum, Anhydrite, Halite, Sylvite, Carnallite, Kernite

Silicates

Silicates form an important part of sedimentary rocks, like that of the igneous rocks. The most essentials being quartz and feldspar.

Zeolites

Zeolites are inorganic, porous materials having a highly regular structure of pores and chambers that allows some molecules to pass through. So zeolites can be used as molecular sieves or as an ion exchange media. Zeolites are formed where volcanic rocks of specific chemical composition is immersed in water so as to leach away some of the components. It can be elaborated as, when volcanoes erupted, enormous amounts of ash-aluminosilicates of alkali and alkaline earths, some of the wind-borne ash settled to form thick ash beds. In some cases the ash fell into lakes and in others water percolated through the ash beds. Thereby as a result of chemical reaction between volcanic ash and salt water, zeolites are formed. The common zeolites include natrolite, laumonite, chabazite, clinoptilolite, heulandite, stilbite and sodalite.

Clay Minerals

Clay minerals constitute a bulk volume of the sedimentary minerals. Clay minerals are generally very fine grained and are composed of mixtures of finer grained clay minerals and clay-sized (0.04-0.15 micrometre) crystals of other minerals such as quartz, carbonate and metal oxides. Clay minerals are formed on or near the surface of the Earth. So clays crystallize mainly in the sedimentary environment and the source sediments control their composition. They are all sheet silicates and have high affinity for water. Clay minerals occur mainly in soil horizons, continental and marine sediments, geothermal fields, volcanic deposits and weathering rock formations. The presence of water is must for the formation of clay minerals. Clay minerals generally form at low temperature.

Clay minerals can be broadly classified into three main groups: kaolinites, smectites and illites.

Kaolinites

Kaolinite is the most common of this group and has the chemical formula $Al_2Si_2O_5(OH)_4$. Other members with similar structures are anauxite, dickite, and nacrite. Kaolinite is formed from K-feldspar by the process of chemical weathering through the following reaction:

$$\underset{\text{Orthoclase}}{4KAlSi_3O_8} + \underset{\text{Hydrogen ion}}{4H^+} + \underset{\text{Water}}{2H_2O} \longrightarrow \underset{\text{Potassium ion}}{4K^+} + \underset{\substack{\text{Kaolnute}\\ \text{(Clay material)}}}{Al_4Si_4O_{10}(OH)_8} + \underset{\text{Quartz}}{8SiO_2}$$

Thus kaolinite is formed by weathering or hydrothermal alteration of aluminosilicate minerals. Thus, rocks rich in feldspar commonly weather to kaolinite. In order to form ions like Na, K, Ca, Mg and Fe must first be leached away by the weathering or alteration process. This leaching is favoured by acidic conditions (low pH). Granitic rocks, because they are rich in feldspar, are a common source for kaolinite. Kaolinite is found in most weathering zones and soil profiles.

Smectites

The most important aspect of the smectite group is the ability to absorb H_2O molecules causing the volume of the minerals to increase when they come in contact with water. Thus, the smectites are expanding clays.

The most common smectite is montmorillonite, with a general chemical formula:

$$(\tfrac{1}{2}Ca,Na)(Al,Mg,Fe)_4(Si,Al)_8O_{20}(OH)_4 \cdot nH_2O$$

Other, less common, members of the smectite group include beidellite, hectorite, nontronite, sauconite and saponite.

Montmorillonite is prevalent in modern clay-rich sediments and sedimentary environments and illite is found in sedimentary rocks, that are more than 100 million years old. The more complexed montmorillonite are common in lower parts of the weathering profile, nearer the rock, where chemistry exerts a strong control on mineralogy. The formation of bentonite (beds containing smectite-group clay minerals, including montmorillonite) and Fuller's earth (a type of clay that has high capacity to absorb water) takes place due to diagenesis. Hydrothermal processes are also active somewhere. Bentonite beds generally form from altered volcanic ash, but other sources are also possible.

Illites

Illites are deficient in alkalis, with less Al substitution for Si. Thus, the general formula for the illites is:

$$K_yAl_4(Si_{8-y},Al_y)O_{20}(OH)_4$$

usually with $1 < y < 1.5$, but always with $y < 2$.

Because of possible charge imbalance, Ca and Mg can also sometimes substitute for K. The K, Ca or Mg interlayer cations prevent the entrance of H_2O into the structure. Thus, the illite clays are non-expanding clays.

Illite type clays are formed from weathering of K and Al-rich rocks under high pH conditions. Thus, they form by alteration of minerals like muscovite and feldspar. Illite clays are the main constituent of ancient mudrocks and shales.

Carbonate Minerals

There are about 50 known carbonate minerals, among which calcite and dolomite are most important. They are essential constituents of limestone and other sedimentary rocks.

Calcite is the principal constituent of limestone and its metamorphic equivalent i.e. marble. Deposits of fine grained calcite in powder form are referred to as chalk. It forms the cementing agent in many sandstones, and is one of the more common minerals precipitated by living organisms to form their skeletal structures. Calcite is also precipitated from groundwater where it forms veins, or in open cavities like caves and caverns can form the cave decorations—like stalactites and stalagmites, and encrustations. It is also precipitated from hot springs where it is called travertine. Calcite form from carbonate magmas. Calcite is also precipitated from hydrothermal fluids to form veins associated with sulfide bearing ores.

Most dolomites are secondary and are formed by reaction of calcite with Mg-rich solutions during diagenesis.

Siderite form complete solid solution series with magnesite, although the environment in which the two minerals occur usually determines that either Mg-rich magnesite or Fe-rich siderite will form, and one rarely sees intermediate end members. Magnesite ($MgCO_3$) is formed as an alteration product of mafic and ultramafic rocks.

Dolomite coexists with calcite, as can be seen in the magnesite-calcite phase diagram (Fig. 9.17). Two solvi exist at low temperatures. Thus, any high Mg-calcite- dolomite solid solutions that might exist at high temperatures would form nearly pure calcite and pure dolomite at surface temperatures, and similarly, any magnesite-dolomite solid solutions that might exist at high temperatures would form nearly pure magnesite and pure dolomite at low temperatures. Thus, magnesite and dolomite commonly occur together, as do calcite and dolomite.

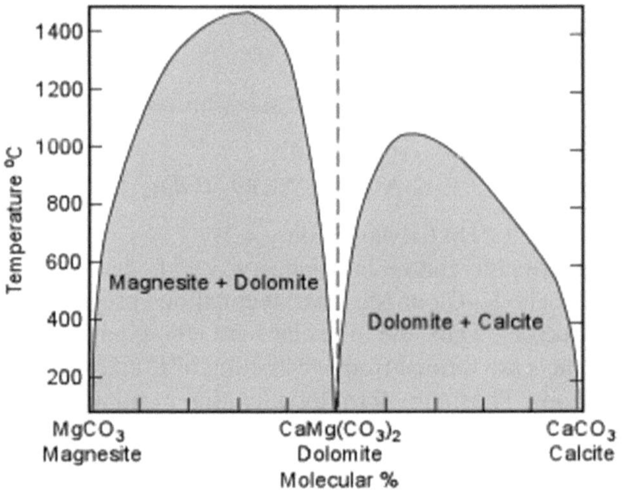

Fig. 9.17: T-regime of carbonate minerals.

Sulphate minerals: Sulphate minerals can be broadly divided into two groups—those contain no water (anhydrous) and those contain water (hydrous). Examples of anhydrous sulphates are anhydrite, barite, celestite and anglesite. Hydrous on the other hand includes gypsum and some rare species. Thick deposits of gypsum, anhydrite or both are associated with limestone, dolostone or halite beds. The sulphate minerals are generally found in vugs or fractures and also occur in soils. Some also occur as minor or major replacement minerals in ore deposits, like anglesite ($PbSO_4$) forms due to alteration or weathering as a result of replacement for galena (PbS).

Halides: The halides consist of minerals with one halogen element. The most common halides in sedimentary rocks are halite (NaCl), sylvite (KCl) and fluorite (CaF_2). Halite is typically found as rock salt in massive salt beds, in association with gypsum or anhydrite and occasionally with sulphur. Sylvite generally remains associated with halite.

Chert: Chert is fine-grained quartz. It is either massive or layered and is generally found as nodules or concretions in limestone. It can also form as a result of recrystallization of amorphous silica. Chert can form in parts of the deep sea where the tiny shells of siliceous organisms are concentrated or where underground fluids replace sediments with silica. Chert can be obtained in a variety of forms: jasper, with characteristic red colour, due to hematite inclusions; and flint, a darker form of chert, containing organic matter. Opal and chalcedony, both silica, are found to be associated with chert deposits.

Evaporites: Evaporite minerals are those that form by coming out of solution when seawater and the waters of large lakes evaporate. Rocks made of evaporite minerals are sedimentary rocks called evaporites. The evaporites include either borates (borax and ulexite) or sulfates (gypsum) or halides.

9.5 METAMORPHIC PROCESSES

Metamorphism occurs at temperatures and pressures higher than 200°C and 300 MPa. Rocks can be subjected to these higher temperatures and pressures as rocks become buried deeper in the Earth. Such burial usually takes place as a result of tectonic processes such as continental collisions or subduction.

The upper limit of metamorphism occurs at the pressure and temperature of wet partial melting of the rock in question. Once melting begins, the process changes to an igneous process rather than a metamorphic process. Metamorphic processes are mainly controlled by heat, pressure and time.

Metamorphic Grade

Metamorphic grade denotes the relative P-T condition, under which the metamorphic rocks form. The boundaries between the grades are chosen to correspond to the important discontinuous reactions (isograd) and are correlatable to the metamorphic facies.

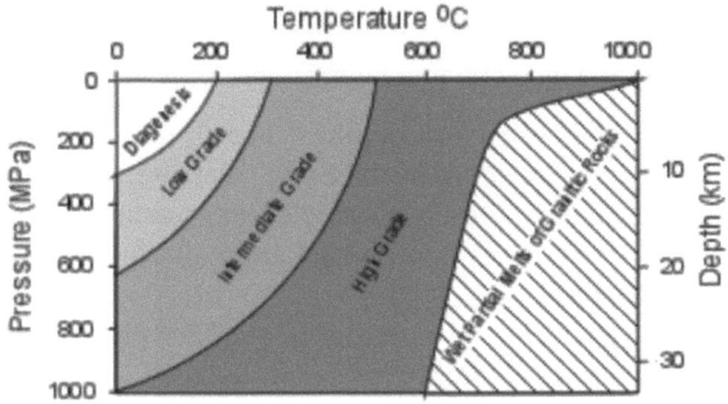

Fig. 9.18: Metamorphic grade.

Low-grade metamorphism takes place at temperatures between about 200 to 320°C, and relatively low pressure. Low grade metamorphic rocks are characterized by an abundance of hydrous minerals, minerals that contain water, H_2O, in their crystal structure.

Intermediate metamorphism takes place at temperatures between about 320 to 500°C, and relatively medium pressure.

High-grade metamorphism takes place at temperatures greater than 500°C and relatively high pressure. As grade of metamorphism increases, hydrous minerals become less hydrous, by losing H_2O and non-hydrous minerals become more common.

Table 9.7: Metamorphic minerals in rocks of different composition

Rock types	*Metamorphic minerals*
Pelitic	Quartz, kaolinite, pyrophyllite, chlorite, muscovite, kyanite, andalusite, biotite, chloritoid, garnet (almandine), staurolite, cordierite, K-feldspar, sillimanite, orthopyroxene
Mafic	Zeolites, prehnite, pumpellyite, plagioclase (Ca-rich), epidote, chlorite, actinolite, hornblende, garnet (almandine-pyrope), biotite, diopside, enstatite
Carbonate	Calcite, dolomite, quartz, biotite, tremolite, garnet (grossular), olivine (forsterite), diopside, periclase, wollastonite
Iron formation	Quartz, hematite, magnetite, minnesotaite, actinolite, grunerite, hedenbergite, fayalite
Ultramafic	Talc, brucite, serpentine, olivine (forsterite), anthophyllite, diopside, garnet (pyrope-almandine), enstatite
Granitic	Quartz, K-feldspar, biotite, hornblende, garnet (almandine), orthopyroxene

In Table 9.7, the minerals occur from low grade to high grade serially in each case.

Factors Controlling Metamorphism

Temperature: Increases with depth in the Earth along the geothermal gradient, with burial of rock and also due to igneous intrusions.

Pressure: Pressure increases with depth of burial, thus, both pressure and temperature will vary with depth in the Earth. Pressure is defined as a force acting equally from all directions. It is a type of stress, called hydrostatic stress, or uniform stress. If the stress is not equal from all directions, the stress is called a differential stress.

Differential stress during metamorphism can have profound effect on the textures of the metamorphic rocks.

1. Rounded grains can become flattened in the direction of maximum stress.

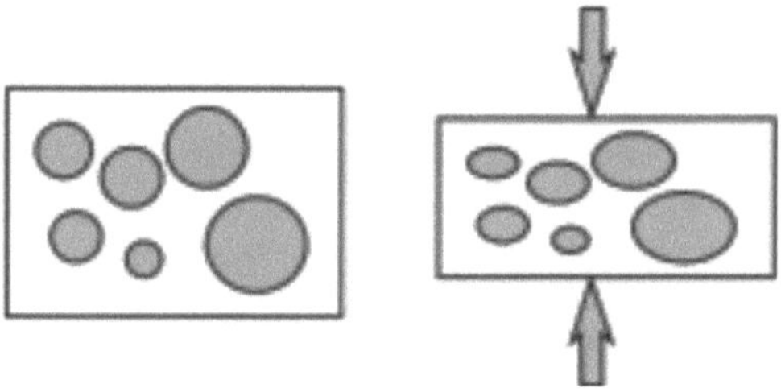

Fig. 9.19: Action of stress on grains.

2. Minerals that crystallize or grow in the differential stress field can have a preferred orientation. This is especially true of the sheet silicate minerals (the micas: biotite and muscovite, chlorite, talc and serpentine). These sheet silicates will grow with their sheets orientated perpendicular to the direction of maximum stress. Preferred orientation of sheet silicates causes rocks to break easily along approximately parallel sheets. Such a structure is called a foliation.

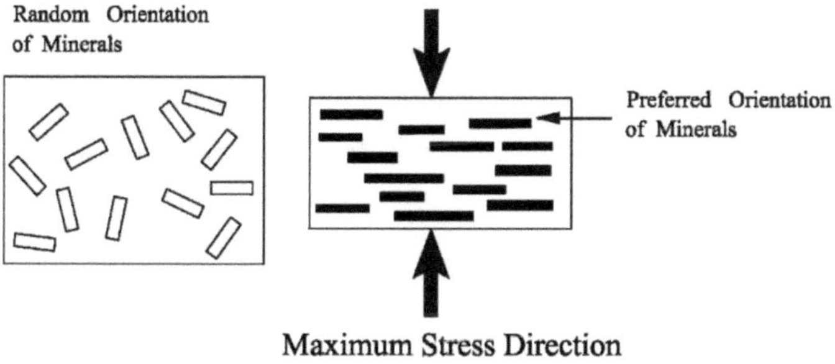

Fig. 9.20: Generation of foliated structure during metamorphism.

Time: The chemical reactions involved in metamorphism, along with recrystallization, and growth of new minerals are extremely slow processes. Laboratory experiments suggest that the longer the time available for metamorphism, the larger are the sizes of the mineral grains produced. Thus coarse grained metamorphic rocks involve long times of metamorphism. Experiments suggest that the time involved is millions of years.

Metamorphic Reactions

Metamorphic reactions depict the process of final formation of the metamorphic rocks. They can be of two types: Univariant reactions and Divariant reactions.

Univariant Reactions

Univariant reactions involve reaching a point in pressure temperature space where a reaction occurred resulting in a sudden change in mineral assemblage. This type of reaction can be plotted on a simple x-y diagram, explaining the formation of metamorphic rocks at specific temperature and pressure conditions. For example the formation of Al_2SiO_5 minerals can be shown with the reaction:

$$\underset{\text{Pyrophyllite}}{Al_2Si_4O_{10}(OH)_2} \longleftrightarrow \underset{\text{Ky or Andal}}{Al_2SiO_5} + \underset{\text{Qtz}}{3SiO_2} + \underset{\text{fluid}}{H_2O}$$

This reaction takes place when there is excess of SiO_2 and H_2O in the system. From this it can be inferred that if the mineral is andalusite, the rock was metamorphosed at a pressure less than about 2.5 kilobars.

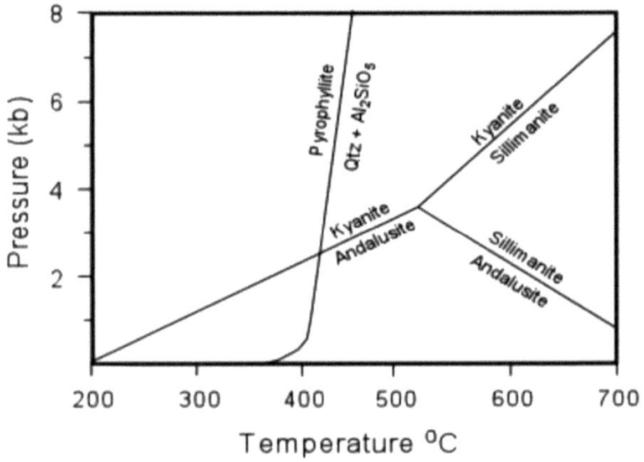

Fig. 9.21: P-T field of Al_2SiO_5 minerals.

If the mineral is kyanite, the pressure is greater than about 2.5 kilobars.

Another inference that could be drawn from this that a group of rocks are supposed to be buried along the geothermal gradient, shown in the diagram (Fig. 9.22). Rocks buried to a pressure less than about 4 kb and a temperature less than about 420 °C should have pyrophyllite so long as they have the right composition. Rocks buried to pressures between about 4 and 5 kb and temperatures between 420° and about 600 °C should have kyanite + quartz, and rocks buried to pressures along the geothermal gradient greater than about 5 kb and temperatures greater than about 600 °C should have sillimanite + quartz.

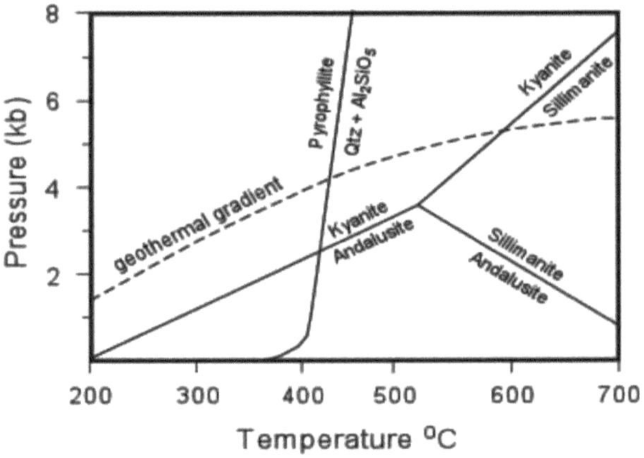

Fig. 9.22: Stability fields of Al_2SiO_5 minerals.

At metapellites, the stable mineral assemblage at low temperature is garnet + chlorite + biotite + muscovite + quartz. With increase in temperature, this assemblage becomes unstable and the new group of minerals are staurolite + biotite + quartz.

The reaction that occurred is the univariant reaction:

$$\text{Garnet} + \text{chlorite} + \text{muscovite} = \text{Staurolite} + \text{biotite} + \text{quartz} + H_2O$$

Now in some cases, on increasing the temperature, staurolite becomes unstable and it reacts to produce garnet, biotite and kyanite.

$$\text{Staurolite} + \text{quartz} + \text{muscovite} \longleftrightarrow \text{Garnet} + \text{biotite} + \text{kyanite} + H_2O$$

Here the disappearance of staurolite coincides with the appearance of kyanite in rocks. Thus it defines the kyanite isograd.

Divariant Reactions

The divariant reactions result in gradual change in composition of minerals and they occur over a wide range of temperature and pressure. The reaction that occurs with increasing temperature (at constant pressure) is:

$$\text{Chlorite} + \text{qtz} \longrightarrow \text{Garnet} + \text{Mg-richer chlorite} + H_2O$$

Metamorphic temperature-pressure estimation: Using combinations of reactions that have likely taken place during metamorphism, petrologists have been able over the years to determine the pressure and temperature of metamorphism in a variety of rocks, and in so doing have been able to place constraints on the fields of temperature and pressure for the various metamorphic facies. This pressure and temperature environment is referred to as metamorphic facies. Some of these reactions are shown in Fig. 9.23 with reaction boundaries superimposed over the facies diagram.

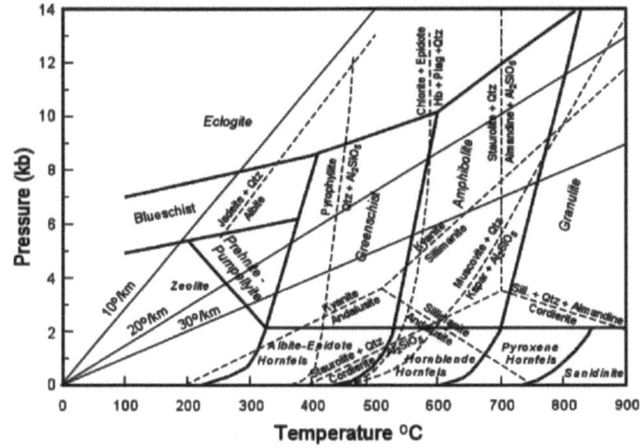

Fig. 9.23: Various geothermal gradients that would control the succession of facies encountered during prograde metamorphism if the rocks were pushed down into the Earth along one of these geothermal gradients.

- A low geothermal gradient of around 10°/km would cause prograde metamorphism to occur along a sequence of facies from zeolite to blueschist to eclogite. Such a progression is termed a facies series, and in general terms this would be called a high pressure facies series, as shown in Fig. 9.23. Such a facies series would be expected in areas near subduction zones where cool lithosphere is pushed to higher pressure.
- A geothermal gradient of around 30°/km, expected in areas undergoing an orogenic event, would produce a succession of facies from zeolite to prehnite pumpellyite to greenschist to amphibolite to granulite.

Along very high geothermal gradients, such as might be expected in the vicinity of intruding magmas the succession of facies would increase from the albite-epidote hornfels facies to the hornblende hornfels facies to pyroxene hornfels and sanidinite facies, the facies of contact metamorphism. This facies series is called the hornfels facies series or the contact facies series.

Metamorphism can be best explained under four headings: contact metamorphism, dynamic metamorphism, metasomatic metamorphism and regional metamorphism.

Table 9.8: Types of metamorphism

Type	*Description*
Contact metamorphism	Occured when magma intrudes into the surrounding solid "country rock", making contact and creating a metamorphosed zone called a "contact metamorphism aureole". Temperatures are highest at the boundary of contact, and decrease with distance from it.
Regional metamorphism	Typically marked by increasing temperature and directed pressure over a large area; occurs in a linear belt in the plate overriding the subducting plate due to increasing temperature and pressure caused by compression, thrusting, folding, and intrusion of magmas from greater depths.
Metasomatic metamorphism	Caused by the entry of foreign metamorphic fluids, thereby resulting in a change in rock composition.
Dynamic metamorphism	Caused by the sudden pressure, exerted by the faults or meteorite impacts, resulting in fracturing and granulation of rocks and occasional formation of high-temperature minerals like coesite or stishovite.

9.6 MINERAL PARAGENESIS

The time sequence of mineral deposition is known as the paragenesis of a deposit. The paragenesis, or chronological order of minerals, is determined by studies of mineral relationships, with the emphasis on microscopic textural features. Paragenesis can also be referred as the assemblage of minerals that crystallizes more or less at same time and in similar physical and chemical conditions. The study of paragenesis is particularly important in understanding the changing behaviour of an individual mineral under differential pressure, temperature and the chemical composition of the crystallization environment. A mineral can form from a continuous crystallization path, or may be interrupted by time-intervals, dissolution or precipitation of other minerals. Mineral or mineral crystals of same mineral, forming at separate interval of time are known as generations. For example, in porphyritic basalt, the phenocrysts of plagioclase feldspar represent 1st generation, while the fine-grained plagioclase of the groundmass is of 2nd generation. The mineral grains of different generations may differ in size, composition and properties.

The paragenesis of mineral formation in moving ore fluids produces changes in ore mineralogy along the course of deposition. Such changes are described as zoning, and are found in sedimentary deposits as well as in magmatic and metamorphic ores. In the ideal case of a radiating hydrothermal or pneumatolytic fluid, changes in chemistry, temperature and pressure along the fissure result in the deposition of different minerals in con-centric zones at increasing distances from the magmatic source. Syngenetic deposits, however, may be zoned parallel to a contemporaneous shore line or along a stream channel leading away from the source rock.

Some Examples of Mineral Paragenesis

Alkali feldspar: Alkali feldspars are essential constituents of alkali and acid igneous rocks, with predominance in syenites, granites, granodiorites and their volcanic equivalents. Alkali feldspars are also common in pegmatites and in many acid and intermediate gneisses. These feldspars are rarely found in basic rocks, like in some lamprophyres and in small amounts in teschenite, theralite, shonkinite and monchiquite. The percentage of potassium component is around 75% in volcanic rocks. The alkali feldspars exist in a wide P-T range in metamorphic rocks.

Plagioclase feldspar: Plagioclase feldspar, crystallizing as phenocrysts at higher temperature is prevalent in many igneous rocks of granitic (granodiorite, diorite etc.) and basaltic composition. These rocks may also contain groundmass of plagioclase, crystallizing during later part of crystallization history. The compositions also change from calcic to sodic in phenocrysts and groundmass respectively. Plagioclases are also main constituent of dolerites and hypabyssal rocks.

Plagioclase commonly occurs in metamorphic rocks. Its composition varies with the grade of the host rock. Albite is stable in chlorite and biotite zones of regional metamorphism. Anorthite does not appear prior to the garnet zone. Plagioclase of granulite facies ranges in composition between An_{30} to An_{35}. Plagioclase gets unstable under the PT regime of eclogite facies and gets transformed into omphacite pyroxene and garnet.

In sedimentary rocks, albite forms a common authigenic mineral.

Pyroxene: Mg-rich orthopyroxenes are prevalent in some ultrabasic rocks like pyroxenites, harzburgites, lherzolites and picrites. Fe-rich orthopyroxenes are found in differentiated dolerite sills and in some granites and norites.

Orthopyroxenes form an essential constituent in the rocks of charnockite series and a typical mineral of granulite facies. Clinopyroxenes are common constituents of igneous rocks like basalt, gabbro, peridotite etc. Representative members of diopside-hedenbergite series occur in metamorphic rocks.

Accessory minerals: Most of the common non-opaque accessory minerals (apatite, zircon, sphene) may be expected to crystallize late in igneous rocks since, other things being equal, constituents in small amount in a complex solution should crystallize late. This is confirmed by the association of these minerals with pegmatites, the presence of apatite and sphene in certain contact metamorphic deposits, and the lack of concentrations of these minerals in the case of rocks and ores produced by crystal settling at the base of complex differentiated sills and thick flows. Texture and distribution of these accessories in igneous rocks also demonstrate their late origin. The association of accessory minerals with biotite, hornblende, magnetite and, in some cases, quartz is demonstrated and attributed to their late magmatic crystallization. The more nearly equal distribution of accessory minerals among the essential constituents of paragneisses as compared with igneous rocks offers a criterion for determining the origin of gneisses. These conclusions have a bearing on the origin of late magmatic ore deposits.

Pegmatites minerals: Pegmatites mainly represent granitic composition and consist of minerals like microcline, oligoclase, subordinate quartz, and accessory apatite, muscovite, titanite, epidote, allanite and rarely thorite and zircon. Here zoisite occurs as an alteration product of microcline. At the contacts of the pegmatite with the limestone and with the zinc ore bodies, groups of reaction minerals and of pneumatolytic minerals have generally been developed.

Concluding Remarks

In this chapter, the various processes of mineral formation, in relation to their corresponding temperature and pressure has been discussed. A more scrutinized approach of mineralogy in respect to thermodynamics will be elaborated in Chapter 12.

Think for a while

1. How can you explain the presence of accessory minerals in igneous rocks?
2. If a rising magma is depleted in iron and magnesium, which mineral formation does it indicate?

3. The previous magma if again brought to previous P-T regime, what compositional difference will be observed?
4. Why do only the sulphates and halides constitute the common evaporite minerals?
5. How diagenesis and low-grade metamorphism can be differentiated?
6. Basalts show much varied range of metamorphism than granite rocks. Why?

FURTHER READING

Best, G. Myron Igneous and Metamorphic Petrology. Brigham Young University. Blackwell Publishing. 2003.

Blatt, H. Sedimentary Petrology, 2nd ed. W.H. Freeman and Co., San Francisco. 1992.

Ehlers, E.G. and H. Blatt. Petrology: Igneous, Sedimentary and Metamorphic. W.H. Freeman and Co., San Francisco. 1980.

Hatch, F.H., A.K. Wells and M.K. Wells. Petrology of the Igneous Rocks. Thomas Murby and Co., London. 1980.

Perkins, D. Mineralogy, 2nd ed. Prentice Hall of India. 2002.

Wenk, Hans-Rudolf and Andrei Bulakh. Minerals. Cambridge University Press. 2006.

Winkler, H.G.F. Petrogenesis of Metamorphic Rocks. Springer-Verlag, New York. 1976.

Yardley, B.W.D. An Introduction to Metamorphic Petrology. John Wiley and Sons, New York. 1989.

Internet Data Retrieved from:

1_6 The formation of metamorphic rocks - Practising science reading the rocks and ecology - OpenLearn - The Open University.htm

Pidwirny, M. (2006). "Characteristics of Igneous Rocks". Fundamentals of Physical Geography, 2nd Edition. http://www.physicalgeography.net/fundamentals/10e.html

http—earthsci_org-education-teacher-basicgeol-meta-metafacies_gif.htm

http—jaeger_earthsci_unimelb_edu_au-msandifo-Teaching-Minerals-pyroxeneQuad_gif.htm

http—www4_nau_edu-meteorite-Meteorite-Images-FeldsparOrdering_png.htm

http—www_tulane_edu-~sanelson-images-alibte-jadeiteqtz_gif.htm

CHAPTER

10

MINERAL DEPOSITS AND THEIR CHARACTERISTICS

This chapter will deal with the formation processes of mineral deposits, i.e. the minerals that can be extracted economically and profitably and are commonly designated as economic minerals. Minerals are important because they contain most of the elements we use. All the minerals found in nature however do not fulfill this criterion. There are a number of factors which categorize a mineral as an economic one. The most important of these is that an economic mineral is needed to be easily extractable and economically viable. For example iron can be found in a number of minerals, but it is hematite and magnetite, which are particularly important from economic point of view. Two categories of minerals fall under this group – ore minerals and industrial minerals. The economic feasibility of minerals also varies with their diversified zones of formation, which actually largely control their characteristics.

An ore is defined as a metalliferous, or an aggregate of metalliferous minerals, more or less mixed with gangue, which from the standpoint of the miner can be earned at a profit, or from the standpoint of the metallurgist can be extracted at a profit. The test of yielding a metal or metals at a profit seems to be the feasible one to employ. Ore minerals are chemical compounds formed naturally under favourable chemical conditions when their major constituents are available. Substitution of major constituents by trace metals may produce impurities in ore minerals. Both impurities and associated minerals in ore deposits can provide valuable metals as by-products, or create expensive problems for sustaining the environment.

Industrial minerals have been defined as any rock, mineral or other naturally occurring minerals of economic importance, excluding metallic ores, mineral fuels and gemstones (precious or semiprecious stones). These are designated as industrial, because either the mineral itself or its compounds/rocks has some industrial application. Asbestos, barite (minerals), granite, sandstone, limestone (rocks), bauxite, ilmenite, chromite (metallic ores), all have industrial aspects.

The distinction between the ore minerals and the industrial minerals is: the ore minerals comprise the metallic minerals while the industrial minerals include only the non-metallic ones. In some cases, a single mineral can be used both as an ore and industrial mineral. For example rutile (TiO_2) can be used as an ore of titanium and also in the pigment industry as an industrial mineral. Hence though their economic viability depends on their mode of formation, it is their individual characteristics and uses which distinguish between the two groups of economic minerals.

Here we are mainly concerned about the formation of different deposits and their characteristics.

The importance of metals to the industrialized world stems from the vast array of properties that they exhibit, both in their pure form and when used in combined form in alloys, chemical compounds and composite materials. The choice of a metal for a particular use depends on the suitability of its properties, its price and the availability of substitutes.

Three main processes, by which minerals are formed, are igneous, sedimentary and metamorphic. A separate process of mineral genesis is also worthy to mention, is the process of oxidation and supergene (secondary) enrichment. Further categorization of each of these processes will be discussed under separate headings.

10.1 IGNEOUS PROCESSES

A large number of mineral deposits are hosted by igneous rocks. Igneous processes take place in magmatic systems, and involve crystals, immiscible liquids (magma, sulphide melt, oxide melt, water), and gases (as formed when water boils). Chemical and physical processes associated with the separation of crystals or immiscible liquids from magma, give rise to magmatic fractionation, which can produce a range of igneous rock types and may concentrate metals to form ore deposits. Both mafic and felsic rocks are associated with mineral formation.

Mineral Deposit Formation

There are a variety of processes of mineral deposit formation which can be categorized in a number of sections.

Significance of Trace Elements in Ore-forming Process

Any element which is present in a rock at concentrations lower than 0.1%, is defined as a trace element. During partial melting of rocks, these elements separate themselves into two types—compatible (affinity for solid phase) and incompatible (affinity for liquid phase). Hence during cooling and solidification

Table 10.1: The formation of economic minerals from principal types

Magma type	*Formation*	*Typical deposits*
Basalt	Formed with or without felsic magma by partial melting of mantle material in high heat flow conditions.	Ore deposits contain concentration of siderophile and chalcophile elements like Ni, Co, Cr, V, Cu, Pt and Au.
Calc-alkaline	Crystallized from magma of intermediate composition with SiO_2 content between 53 to 63%. Can be formed by simple crystallization or by in situ fractionation.	Concentration of mainly trace elements.
Rhyolite	Felsic magma, crystallizing at depth, forming rocks of composition, varying from Na-rich tonalite to K-rich alkali granite, or erupted on surface to form dacitic or rhyolitic volcanic rocks may form by partial melting of crustal material.	Ore deposits contain concentration of lithophile elements like Li, Be, F, Sn, W, U and Th.
Alkaline	Depleted in SiO_2 and highly enriched in alkalies like Na, K and Ca. Alkali magmas can be of following types: nephelinite and kimberlite. The former in most cases is associated with carbonatites and the latter with diamonds.	Cu, Fe, P, Zr, Nb, REE, F, U, Th.

of magma, the former group are taken up by the crystals, while the latter get enriched in the residual melt. Thus the ore-forming minerals show a high concentration of the incompatible trace elements. It has also been seen that many of the ores, associated with igneous rocks are formed from elements like Cu, Ni, Cr, Ti, P, Sn, W, U etc., which were originally present as trace elements in magma and were subsequently enriched to ore grades during the course of crystallization.

Igneous rocks span a range of compositions—from peridotite to gabbro (basalt) to diorite (andesite) to granite (rhyolite)—the formation of which depends on the source of the magma and the extent to which magmatic fractionation has occurred. Different types of ore deposit are associated with different types of igneous rocks: magmatic segregation deposits are associated with peridotites and gabbros, pegmatite ore deposits with granites, and porphyry ore deposits with diorites, granodiorites and granites. The igneous processes which give rise to mineral deposits are grouped into three types:

- Magmatic
- Contact metasomatic
- Hydrothermal

Magmatic

Some magmatogenous deposits are formed either by simple crystallization and differentiation, during consolidation of magma. Liquid immiscibility also causes ore melts to form. In the magmatic stage itself minerals may form earlier or later and accordingly they are called early magmatic and late magmatic deposit. So in a whole three kinds of deposits are found as: Early magmatic, Late magmatic and Immiscible deposits.

1. Early magmatic: Early magmatic minerals have crystallized before the formation of the rock-forming constituents of the enclosing rocks. This type of concentration can be of three processes listed in Table 10.2.

Table 10.2: Types of processes of early magmatic deposits

Processes	*Description*	*Minerals formed*
Dissemination	Resulted from simple crystallization of a deep-seated magma.	Diamond-bearing kimberlite pipes and corundum.
Segregation (layered type)	Resulted from gravitational settling of the early-formed crystals during crystallization differentiation of the magma within the magma chamber.	Layered chromite deposits and platinum group of elements.
Injection	Resulted from the transport of the magmatic segregations from the original place into the surrounding rocks due to tectonic deformation.	Titanomagnetite, magnetite and some chromite deposits.

2. Late magmatic: This type of deposits are formed by crystallization of mobile residual magmas towards the end of the cooling process of magma. Thus the so formed ore minerals are later originated than the rock forming minerals of the enclosing rocks. These deposits are formed by two processes shown in Table 10.3.

Table 10.3: Types of processes of late magmatic deposits

Processes	*Description*	*Minerals formed*
Segregation	Resulted from the crystallization, due to accumulation in residual magma. Associated with mafic igneous rocks.	Iron and titanium ore bodies in mafic igneous rocks.
Injection	Resulted from the injection of the residual magma.	Most pegmatites and carbonatites.

Pegmatites and Carbonatites

Here the formation of pegmatites and carbonatites have been given special emphasis, due to their importance.

Pegmatites: Pegmatites formed from the injection of late magmatic and residual fluids of granitic composition, with high proportion of water, volatile components (phosphorous, fluorine, chlorine, sulphur, boron etc.), low melting point silicates (albite, mica) and a group of rare and exotic metals. Pegmatitic bodies generally remain in the form of dykes and lenticular bodies. The most common composition of pegmatite is quartz and feldspar (albite, microcline) and accessory mica and tourmaline. In some complex varieties, high concentrations of rare group of minerals are found.

Table 10.4: Metals and minerals obtained from pegmatites

Metals	*Minerals*
Beryllium	Beryl ($Be_3Al_2Si_6O_{18}$), bertrandite ($Be_4Si_2O_7(OH)_2$)
Lithium	Spodumene ($LiAlSi_2O_6$), amblygonite
Cesium	[$(Li,Na)Al(PO_4)(F,OH)$], petalite ($LiAlSi_4O_{10}$·), pollucite
Niobiumtantalum	[$(Cs,Na)_2Al_2Si_4O_{12}\cdot(H_2O)$]
	columbite-tantalite [$(Fe, Mn)(Ta, Nb)_2O_6$], fergusonite
	[$(Y,Re)NbO_4$], samarskite [$(Y,Fe^{3+},U)(Nb,Ta)_5O_4$]
Rare metals	Rubidium, gallium and scandium
Rare Earths	Monazite [$(Ce,La,Nd,Th)PO_4$], xenotime (YPO_4), allanite
	[$(Ce,Ca,Y)_2(Al,Fe^{3+})_3(SiO_4)_3(OH)$]
Tungsten	Wollframite [$(Fe,Mn)WO_4$], scheelite [$CaWO_4$]
Molybdenum	Molybdenite (MoS_2)
Uranium	Uraninite (UO_2)
Thorium	Thorianite (ThO_2), monazite
Industrial minerals	Sheet mica, ceramic and dental feldspar, quartz, fluorite
Semi-precious metals	Beryl, chrysoberyl ($BeAl_2O_4$) and topaz [$Al_2SiO_4(F,OH)_2$]

Important examples of pegmatitic deposits are Tanco (Manitoba, USA), Manono deposit (Zaire), Greenbushes deposit (Australia), Bancroft (Canada), Kobe (Japan) and Jharkhand (India).

Carbonatites: Carbonatites are late magmatic dykes formed from residual liquids generated due to differentiation of basaltic magma, derived from the mantle. Their general composition is calcitic and dolomitic carbonates and alkali, with a phosphorous content of about 2.35.

Table 10.5: Metals and minerals obtained from carbonatites

Metals	*Minerals*
Niobium	Pyrochlore
Titanium	Ilmenite, titanomagnetite
Rare Earths	Bastnesite
Zirconium	Zircon, baddeleyite
Copper	Chalcopyrite
Phosphorous	Apatite

Important examples of carbonatite deposits are Bayan Obo (China), Jacupiranga (Brazil), Mt. Pass (China), Kovdor (Russia) and Ambadongar (India).

3. Immiscible deposits: Immiscibility here means physical separation of a magma into two: one part is dominantly sulfur rich and the other is silicate. Immiscible melts formed irregular shaped segregations or may be injected as a dyke into a previously crystallized material. These deposits are formed by two processes as shown in Table 10.6.

Table 10.6: Types of immiscible deposits

Processes	*Description*	*Minerals formed*
Segregation	At the late stage of crystallization of mafic magmas, due to lowering of temperature, sulphides of some metals, like iron, nickel and copper separate out as immiscible liquid drops and remain as liquid till the silicates have crystallized out. Later on these immiscible liquid drops may accumulate at the bottom of the magma chamber and solidify there to form sulphide segregations.	Nickel, copper sulphides, Pt-group of minerals
Injection	When immiscible fraction is subjected to tectonic disturbance before consolidation, it reaches the region of lower pressure, where it consolidates to form injection deposits.	Sulphides

Hydrothermal Processes

Hydrothermal solutions: The hot aqueous solutions that emanate from magma in late stage of cooling and to which many mineral deposits owe their origin are called hydrothermal solutions.

Hydrothermal deposits: The hydrothermal solutions produce a number of ore minerals containing Pb, Zn, Cu, Ag, Mo, Hg etc. These fluids remain in solution within the magma under the unchanged pressure condition, but with the rise of magma to levels of lower pressure, hydrous fluids containing ore constituents separate out. These aqueous fluids, travelling away from the intrusives, give rise to different types of hydrothermal deposits with falling pressure-temperature, as the distance from the intrusive body progressively increases.

The hydrothermal deposits occur as cavity-filling deposits or replacement deposits. The former results from deposition of materials in the open space like fractures, fissures, shear zones, available in the country rocks; hence such deposits simulate the structure of banded or crustified appearance, symmetrical bandings and vein forms. The replacement type results from chemical interaction of the hydrothermal fluids with the country rocks and they are very irregular in shape and lack banded or crustified appearance. Since open cavities are rare in deeper parts of crust, hypothermal deposits are predominantly replacement

Table 10.7: Types of hydrothermal deposits formed on the basis of different parameters

Types	*Mode of occurrence*	*Pressure-temperature-depth*	*Ore minerals*	*Gangue minerals*	*Chief occurrences*
Epi-thermal	Mostly fissure-filled or cavity filled, showing comb structure, crustifications and symmetrical bendings. Replacement type rare.	$P = 100$ atm; $T = 50\text{-}200$ °C $D = <3000$ ft	Sulfantimonides, sulfarsenides, silver tellurides, stibnite, cinnabar	Quartz, chalcedony, adularia, calcite, dolomite, barite, fluorite	Hg deposits of Spain, Australia, Italy; Sb deposits of China, India
Meso-thermal	Mostly replacement types with some vugs, open cavity fillings.	$P = 100\text{-}300$ atm $T = 200\text{-}300$ °C; $D = 5000\text{-}10{,}000$ ft	Chalcopyrite, enargite, bornite, tetrahedrite, sphalerite, galena etc.	Quartz, pyrite, carbonates	Au deposits of California; Pb-Zn-Ag deposits of Zawar, Rajasthan
Hypo-thermal	Mostly replacement types and rarely cavity filling deposits.	$P = 300$ atm $T = 300\text{-}500$ °C $D = 10{,}000$ ft	Gold, wolframite, scheelite, pyrrhotite, pentlandite, pyrite, arsenopyrite, chalcopyrite, sphalerite, galena, stanite, cassiterite, uraninite etc.	Pyroxene, amphibole, tourmaline, garnet, sillimanite, kyanite, apatite, topaz, spinel etc.	Kolar gold mine of India; Cu deposits of Singhbhum, Khetri
Tele-thermal	Mainly cavity-filling types, but replacement types are also common.	$P = 40\text{-}240$ atm; $T = 50\text{-}150$ °C $D = 500\text{-}3000$ ft	Sphalerite, galena, chalcopyrite, marcasite, chalcocite	Calcite, dolomite, quartz, fluorite and barite	Pb-Zn deposits of Mississipi valley
Xeno-thermal	Mainly cavity-filling types.	$P = 80\text{-}700$ atm; $T = 300\text{-}500$ °C $D = 1000\text{-}4000$ ft	Cassiterite, wolframite, scheelite, magnetite, specularite (high temp) and silver sulfosalts (low temp)	Orthoclase, tourmaline, topaz, augite, diopside, phlogopite, chalcedony, apatite and alunite	Sn & Sn-Ag deposits of Bolivia

Source: Lindgrem, Bateman

types. On the other hand in epithermal deposits due to shallower depth there are much open spaces, so cavity filling types is common.

Porphyry deposits: Porphyry deposits are formed when metal rich watery fluids boil and are released explosively from a wet granodioritic magma being emplaced at a high level in the crust. The network of fractures produced, provides pathways for distribution of metal-rich fluids that are derived initially from the magma and subsequently, through convection of meteoric water, from surrounding rocks. The fractures also provide sites for the deposition of ore minerals, as fluids cool and react with wall rocks. Porphyry deposits provide much of the world's copper and molybdenum, as well as substantial quantities of tin, silver and gold as by-products. They are typically low grade but extremely large deposits.

Contact metasomatism: Metasomatism is a transformational process by which the chemical composition of a rock or rock portion is altered in a pervasive manner due to the introduction or removal of chemical components as a result of the interaction of the rock with aqueous fluids, released by a cooling magma body. During metasomatism the rocks remain in a solid state.

Difference between contact metamorphism and contact metasomatism: During the post-crystallization period, high temperature fluid outpourings escape from the magma chamber into the surrounding rocks and produce pronounced effects on them. The heat from these igneous body, when effective alone without any appreciable accession of materials from the igneous intrusive, gives rise to the phenomenon called *contact metamorphism*. This does not produce any mineral deposit, except in few rare cases of non-metallic deposits like kyanite, or sillimanite. On the other hand, when the effect of heat from these igneous emanations combines with accession of materials from igneous intrusives, this gives rise to *contact metasomatism*. This phenomenon gives rise to distinctive mineral deposits.

Processes and characteristics of contact metasomatic deposits

- Commonly associated with felsic rocks of intermediate composition, such as granodiorite, quartz-diorite, quartz-monazzite, monazite and occasionally syenite. Felsic rocks have a higher content of fluid than the felsic or mafic rocks.
- Rocks most susceptible to contact metasomatism and most favourable for formation of ore deposits are sedimentary rocks, preferably carbonate rocks. Impure calcareous rocks are mostly affected due to the presence of impurities like silica, alumina, iron which can enter easily into new combinations with CaO.
- Contact metasomatic rock produced at the contact of carbonate rock with the felsic intrusives is called skarn. It consists of several calc-silicate minerals like grossular garnet, andradite, pyroxene of diopside-hedenbergite series, tremolite-actinolite, epidote-clinozoisite-zoisite, wollastonite etc.

- Skarns host important metallic deposits like oxides (hematite, magnetite) and sulphides (sphalerite, galena, bornite, chalcopyrite, pyrite, pyrrhotite, molybdenite etc.), scheelite (radioactive mineral).
- In general, the contact metasomatic deposits are coarse-grained in texture, of comparatively small size, irregular in outline, having almost any shape and consist of several disconnected bodies.
- The contact metasomatic deposits constitute a distinctive assemblage of ore minerals and gangue minerals.

Table 10.8: Ore minerals hosted by contact metasomatic deposits

Deposit	*Chief minerals*	*Occurrences*
Iron	Magnetite, hematite	New Mexico, Hungary, Utah
Copper	Chalcopyrite, bornite	Arizona, Utah
Zinc	Sphalerite	New Mexico, Ontario, Japan
Lead	Galena	New Mexico, California
Gold	Arsenopyrite	British Columbia, Canada, Korea
Tin	Cassiterite, wolframite, scheelite	Finland, England, Alaska
Corundum	Corundum with magnetite and silicates	New York
Graphite	Graphites with contact silicates	New York, Ontario, Sri Lanka

Gangue minerals: Grosullar and andradite garnets, amphibole, diopside, wollastonite, epidote, zoisite, vesuvianite, anorthite, albite etc.

10.2 SEDIMENTARY PROCESSES

Processes at the Earth's surface, in addition to forming deposits of bulk resources, particularly industrial minerals, can also concentrate metals to form ore deposits. Minerals resistant to weathering are concentrated by physical transport and deposition to form placer deposits. Chemical processes concentrate metals through removal of soluble material to leave behind residual deposits, and by precipitation of dissolved material in a variety of different environments, to form bedded sedimentary deposits and encrusted deposits underwater. Chemical transport and precipitation of metals in surface and near-surface waters are controlled largely by changes in the pH and/or the oxidising conditions of the environment. These conditions are often influenced by the reducing properties of decaying organic matter. These surfacial processes give rise to different modes of ore deposits, depending on the process of formation, places of accumulation and their self-characterization. The deposits can be of following types:

1. Sedimentary deposit
2. Evaporate deposit
3. Residual deposit
4. Placer deposit

Sedimentary Deposits

Most sedimentary deposits are formed by the process of chemical precipitation, followed by diagenesis. Examples of such deposits are iron deposits [Banded Iron Formation (BIF) and phanerozoic ironstones], sedimentary manganese deposits and Pb, Zn and Cu deposits.

Iron Deposits

Banded iron formations: Banded iron formations are supposed to have formed during a very short interval of 2500—1100 Ma before. But present formations of older or younger ages are also known. They are characterised by fine layering of 0.5-3 cm thick and commonly exhibit laminations on a scale of millimetres or fraction of a millimetre. The composition of these layers is essentially silica (chert or crystallized silica), alternating with layers of iron minerals. The most common BIF has the composition of hematite and chert layers, with alumina content < 1%. James (1954) identified four important facies for BIF.

Table 10.9: Short description of BIF facies

Facies type	*Description*
Oxide	Principal iron phase may be hematite (more common) or magnetite; carbonate is calcite, dolomite or ankerite; chert is either cryptocrystalline or fine-grained quartz; contains 30-35% Fe.
Carbonate	Consist of chert and siderite in equal proportions; lacks oolitic or granular texture; accumulated as fine mud below the level of wave action.
Silicate	Iron silicates (greenalite, chamosite, glauconite, locally minnesotaite and pstilpnomelane) associated with magnetite, siderite and chert; iron is in ferrous state; contains 25-30% Fe, which cannot be exploited economically.
Sulphide	Consists of pyrite carbonaceous argillites, organic matter and carbon (7-8%); main sulphide is pyrite (around 37%); mined mainly for its sulphur content.

Phanerozoic ironstones: These ironstones are of low grade, so they cannot be mined economically. Moreover there is beneficiation problem due to their silicate mineralogy. These ironstones can be categorized into two types—clinton and minette.

Table 10.10: Types of phanerozoic ironstones

Types	*Description*
Clinton	Forms massive beds of oolitic hematite-chamosite-siderite rock. Fe content is about 40-50% and Al and P contents are higher than the BIF; absence of chert bands; form lenticular beds of 2-3 m thick and never >13 m; common in Cambrian and Devonian periods; formed in shallow water along continental margins, continental shelves or in shallow parts of miogeosynclines.
Minette	Consists of siderite and chamosite or another iron chlorite; Fe content is about 30%, lime is between 5-20% and silica above 20%; found in Mesozoic of Europe.

Sedimentary Manganese Deposits

The sedimentary manganese deposits form in similar environment of BIF and more often found to be closely related to BIF. Manganese in form of Mn^{3+} and Mn^{4+} stabilize as manganese oxides under relatively oxidising and alkaline conditions. Pyrolusite, MnO_2 is the dominant oxide phase at high Eh and over a range of pH. Manganese deposits are found in sediments, ranging in age from Palaeoproterozoic to recent.

Evaporate Deposits

Evaporates are chemical precipitates that form as a result of evaporation of a brine, derived from sea-water. Halides (halite [NaCl], sylvite [KCl]), borates and nitrates owe their sources to evaporates. The formation of evaporites can be explained by a simple process. As sea water or brine evaporates, the salinity of the residual solution increases and individual salts precipitate as their respective solubility limits are reached. Marine evaporites, formed from sea-water, tend to be dominated by the mineral assemblage of halite, gypsum/anhydrite, sylvite and alkali nitres.

Residual Deposits

In some cases, the action of weathering processes and groundwater action have concentrated materials, which are known as residual ore deposits. Intense chemical weathering, tropical climate and heavy rainfall favour the formation of such extensive deposits. Residual deposits can be high grade deposits of one metal only. Lateritic soils rich in insoluble oxides of iron and aluminium are a common form of residual deposit. Sometimes they contain ore deposits of nickel, cobalt, copper and gold, which are minable to quite low grades because they are easily accessible and poorly consolidated. The most important form of laterite is bauxite, the ore of aluminium, which is composed largely of hydrated aluminium oxides. The best sources of bauxite are rocks accessible to percolating rainwater and should be rich in Al and desirably low in Ti. Under warm, humid climatic conditions, silica produced during chemical weathering is more soluble than alumina in surface waters and is released out of the system leaving behind an Al-rich residual rock, i.e. bauxite. Broadly they are called lateritic deposits.

Table 10.11: The residual deposits of metals found on the earth crust

Metals	*Description*
Aluminium	Laterite, consisting of high Al_2O_3. Bauxite as the chief ore of aluminium. Al has a high affinity for oxygen. Due to weathering it accumulates in clay minerals or in purely aluminous varieties, like gibbsite, boehmite and diaspore.
Nickel	Residual nickel formed by the intense tropical weathering of rocks, rich in trace amounts of Ni, like peridotites, serpentinites.
Titanium	Found from anatase deposits in the weathered overburden of alkali rocks; mantles over titaniferous rocks are upto 200 m thick and have a basal phosphate-rich zone.

Placer Deposits

Placer deposits are described as the allochthonous sediments of economic interest. Due to differential weathering, the action of the moving water, natural gravity separation is achieved, whereupon the heavy minerals get concentrated. These minerals should possess high density, chemical resistance to weathering, and mechanical durability. Most of these deposits are of Tertiary to Recent age. They are generally ephemeral in nature, low-grade, but can be exploited, for they are loose and easily worked minerals. Mining usually takes the form of dredging, about the cheapest of all mining methods. At source the placer minerals could have occurred in an older deposit or as accessory minerals in the country rocks. The main minerals recovered are gold and cassiterite (SnO_2) from alluvial deposits and heavy oxide minerals such as rutile (TiO_2) and ilmenite ($FeTiO_3$), monazite from beach deposits. Some "fossil" placer deposits of Precambrian age are among the world's most important gold and uranium deposits. Recent, unconsolidated placer deposits are easy to mine by dredging or with high-pressure water jets.

10.3 METAMORPHIC PROCESSES

Metamorphic processes thermodynamically, but isochemically transform pre-existing mineral deposits (metamorphosed deposits) and form new ones (metamorphic deposits). The chief agencies involved are heat, pressure, time and various solutions. The substances operated upon are either earlier formed mineral deposits or rocks. Valuable non-metallic mineral deposits are formed from rocks chiefly by recrystallization and recombination of the rock-forming minerals.

Table 10.12: Types of placer deposits

Types	*Formation*	*Conditions of formation*	*Minerals*
Residual	Accumulation in situ due to weathering	Ground surface flat, slope present, creep will occur	Niobium, zircon, baddeleyite, magnetite, beryl, apatite
Eluviul	Concentration in a moving solid medium.	A rich source	Cassiterite
Alluvial Beach	Concentration near the source	Tidal zone of an unsheltered beach, wave-cut terraces	Casssiterite
Offshore		Continental shelf, within few kilometres of the coast; submergence of alluvial or beach placers	Casssiterite, diamond, gold, ilmenite, magnetite, monazite, rutile, xenotime and zircon
Aeolian	Concentration in a moving gaseous (air) medium	Presence of coastal dunes-foredunes, transgressive dunes and stationary dunes	Cassiterite, monazite Ilmenite, zircon, leucoxene and rutile

Metamorphosed Deposits

Due to metamorphism, early formed ore deposits undergo both textural and mineralogical changes. Ores undergo less mineral recombinations, while the textural changes are pronounced. Schistose or gneissic structures are induced, particularly with sectile minerals and flow structures are also common. The original texture and structure may be so obscure that it is difficult to determine to which class the original deposits belong. Such deposits are called metamorphosed deposits. For example, sedimentary low-grade manganese deposits due to metamorphism may be transformed to a high grade manganese ore with different structure and texture. Many Pb-Zn sulfide deposits belong to this type.

Metamorphic Deposits

By the process of recrystallization and recombination or both, several kinds of non-metallic ore deposits may form. Rarely water or CO_2 has been added. It is actually a phenomenon of regional metamorphism. No new materials are introduced unlike contact metasomatism. The chief deposits formed by this process are graphite, asbestos, talc, soapstone, andalusite, sillimanite, kyanite, garnet, Au, U-deposits etc.

10.4 SOME IMPORTANT ORE-FORMING PROCESSES

Process of Oxidation and Supergene (Secondary) Enrichment

When ore deposits become exposed due to erosion, they are weathered along with the enclosing rocks. The surface water oxidises many ore minerals and yield solvents, that dissolve other minerals. An ore deposit thus becomes oxidised and generally leached of many of its metals in solution. This leached zone thus becomes barren in ore minerals. The top surface of this zone is often marked by limonite capping formed by precipitation of insoluble oxides and hydroxides of iron during leaching reactions. This is called gossan.

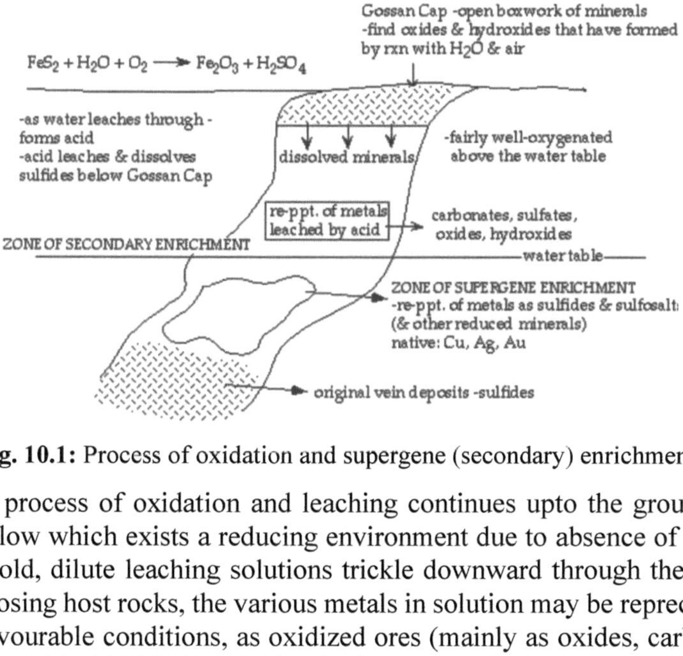

Fig. 10.1: Process of oxidation and supergene (secondary) enrichment.

The process of oxidation and leaching continues upto the groundwater level, below which exists a reducing environment due to absence of oxygen. As the cold, dilute leaching solutions trickle downward through the deposit and enclosing host rocks, the various metals in solution may be reprecipitated under favourable conditions, as oxidized ores (mainly as oxides, carbonates, silicates or chlorides) below the leached zone but above the water table (i.e. within the zone of oxidation itself). If the downward moving solutions penetrate below the water table, the metals in solution that escape precipitation in the oxidised zone above water table are now deposited as supergene sulphides by reacting with and replacing sulphides of metals in the primary, unoxidized sulphide ores, present below the level of water table. Thus an enriched zone of supergene sulphide ores develops just below the water table. This process of enrichment of a sulphide ore deposit by deposition of metals, removed from upper part through oxidation and solution into the lower unoxidized part is called the process of supergene enrichment.

Though it is applied more commonly to the enrichment of sulphide deposits, the process can also be observed in case of oxides and carbonate ores and rocks such as those of iron and manganese.

Process of Volcanic Exhalation

Some ore deposits often show spatial relationships to volcanic rocks. They are conformable with the host and are frequently banded, suggesting sedimentary processes. Principal constituent is pyrite with lesser chalcopyrite, sphalerite, galena, barite and Ag-Au. They are epigenetic to syngenetic (with respect to host rock) in origin.

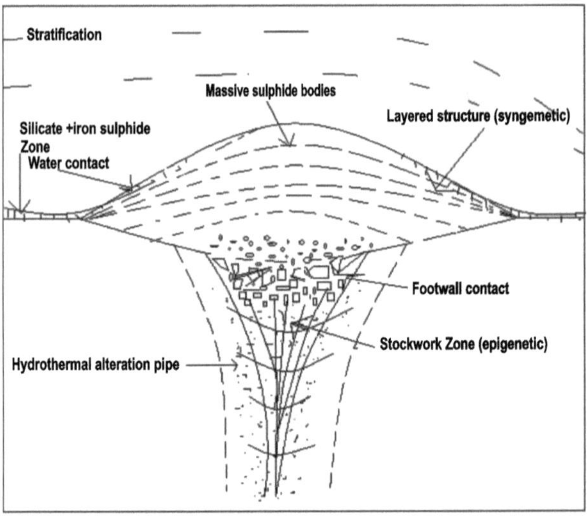

Fig. 10.2: A schematic diagram showing volcanic exhalative deposit.

They show a progression of types with three distinct end members:

1. *Cyprus type:* Associated with mafic volcanics and ophiolite sequences. Found in spreading centers and back arc basins. Consist predominantly of pyrite with lesser chalcopyrite. Typified by the Cyprus pyrite-Cu ores.
2. *Besshi type:* Associated with basaltic to dacitic volcanism. Thought to form during the initial stages of island arc formation. Many Besshi type deposits occur in Precambrian rocks and these may have been generated in entirely different tectonic settings. Pyrite is dominant, but chalcopyrite and sphalerite are also very common. These type of deposits are typified by many of the volcanogenic deposits of Japan (Sanbagava terrain, Shikoku) and Canada.
3. *Kuroko type:* Associated with dacitic to rhyolitic volcanics. Form during the waning stages of island arc volcanism. Pyrite occurs, but is not dominant. Usually galena or sphalerite are predominant with lesser chalcopyrite and tetrahedrite. Also significant silver is present in this type. Typified by the Kuroko deposits of Japan.

10.5 EXTRACTION OF METAL

A metal ore is a mineral or aggregate of minerals from which economically viable amounts of metal can be extracted, i.e. it has got to have enough of the metal, or one of its compounds, in it to be worth digging out. Ores are commonly oxides, or sulphides, sulfosalts and sulf-arsenides. In order to extract a metal, the ore or compound of the metal must undergo a process called reduction to free the metal i.e. the positive metal ion gains negative electrons to form the neutral metal atom, or the oxide loses oxygen, to form the free metallic atoms. The chemical that removes the oxygen from an oxide is called the reducing agent i.e. carbon, carbon monoxide, reduced sulfur or sometimes hydrogen.

Generally speaking the method of extraction depends on the metal's position in the reactivity series. The reactivity series of metals can be presented to include two non-metals, carbon and hydrogen, to help predict which method could be used to extract the metal. Metals above zinc and carbon in the reactivity series cannot usually be extracted with carbon or carbon monoxide. They are usually extracted by electrolysis of the purified molten ore or other suitable compound, e.g. aluminium from molten aluminium oxide or sodium from molten sodium chloride.

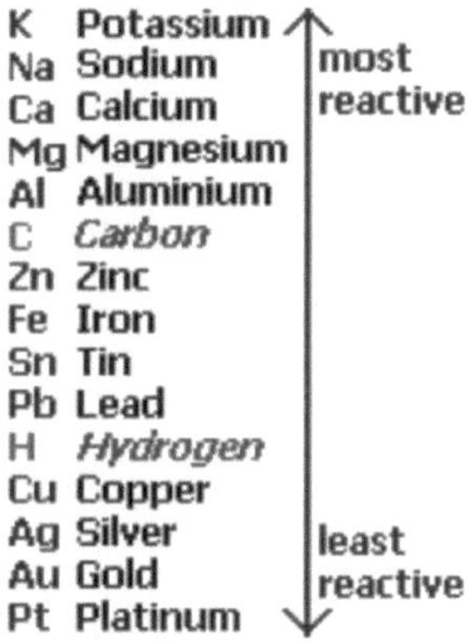

The ore or compound must be molten or dissolved in a solution in an electrolysis cell to allow free movement of ions (electrical current). Theory given in the appropriate sections.

Metals below carbon can be extracted by heating the oxide with carbon or carbon monoxide. The non-metallic elements like carbon will displace the less reactive metals in a smelter or blast furnace e.g. iron or/and metals lower in the series.

Metals below hydrogen will not displace hydrogen from acids. Their oxides are easily reduced to the metal by heating in a stream of hydrogen, though this is an extraction method rarely used in industry. In fact most metal oxides below carbon can be reduced when heated in hydrogen, even if the metal reacts with acid.

Some metals are so unreactive that they do not readily combine with oxygen in the air or any other element present in the Earth's crust, and so can be found as the metal itself. For example gold (and sometimes copper and silver) and no chemical separation or extraction is needed. In fact all the metals below hydrogen can be found as the 'free' or 'native' element.

Other methods are used in special cases using the displacement rule. A more reactive metal can be used to displace and extract a less reactive metal but these are costly processes since the more reactive metal also has to be produced in the first place.

10.6 COMMON METALS, THEIR EXTRACTION FROM ORES AND USES

The extraction processes of common metals from respective ores and their uses are listed in Table 10.13.

Table 10.13: Ores, extraction and uses of common metals

Metals	*Ores*	*Extraction*	*Uses*
Aluminium	Bauxite	Extracted by the process of reduction by electrolysis.	Alumina used in chemicals, abrasives and refractory products; Metallic Al used for making beverage cans, building construction, electrical applications and other products; Left-out slag used as low-cost raw materials for bricks.
Gold	Different alloys, particularly with Ag (± Hg, Cu)	By dissolving the ores in a very dilute solution of sodium cyanide and metal recovery is done by throwing scrap zinc in solution.	Used in ornaments, textiles, medicine, electronics and surgery.
Silver	Argentite, cerargyrite, polybasite, proustite, pyrargyrite	Extraction done by amalgamation process, wet process, smelting with lead or lead ores, smelting with copper ores.	Used in automatic devices, rockets and submarines, computers and nuclear installations. Used for making photographic, chemical and soldering materials.
Copper	Chalcopyrite, bornite, covellite, chalcocite	Extracted by the process of smelting, which includes concentration by froth-floatation, roasting in air, smelting, bessemerization.	Because of its high tensile strength, ductility, high thermal and electrical conductivity, it is used in electrical, and automobile industries.
Lead	Galena, cerrusite, anglesite	Extraction is done by smelting and roasting methods.	Used to store radioactive materials; used to make batteries for cars, trucks and other vehicles; used in electronics and communications, ammunition, television glass, construction and protective coatings; used for crystal glass production, weights and ballast and specialized chemicals.
Zinc	Sphalerite, smithsonite, zincite	Extracted by (i) roasting, involving preparation of zinc oxide and (ii) reduction with charcoal, involving release of zinc from its oxide.	Applied in thin layers on iron and steel to prevent rusting; used for making alloys like brass etc.; used for making paints, chemicals, agricultural applications, in the rubber industry, in TV screens, fluorescent lights and in dry cell batteries.
Manganese	Pyrolusite, rhodochrosite, psillomelane	Extracted by reduction of manganese dioxide with carbon.	Used in many alloys, as a depolarizer in dry cells, to decolourize glass and in the form of potassium permanganate, as an important reagent in analytical industry.

(*Contd.*)

(*Contd.*)

Iron	Hematite, magnetite, limonite, siderite	Extracted in blast furnace in the presence of coke and limestone, where the ore gets reduced to iron.	Wrought iron used for outdoor seating, decorative fences; alloyed steel used for construction of automobiles, hulls of large carrier and ships; cast iron used for making skillets, Dutch ovens, cake pans etc.
Tin	Cassiterite	Extracted by the process of crushing, water treatment, roasting, carbon reduction and liquation.	Used in electro-plating, in kitchen utensils, spray recipients and shaving foam, ink cans, electronic components, integrated circuits, clips, pins and many others; can also be used with ornamental purposes. As a pure metal can be used in storage tanks for pharmaceutical chemical solutions, in capacitors electrodes, fuse-wires, ammunitions, tinned iron sheets to protect virtuals, sweets or tobacco etc.
Nickel	Pentalandite, violarite, inillerite	Extracted by roasting and reduction processes.	Used as alloys, superalloys and as a coating on other metals to reduce corrosion; used for making coins, batteries (for mobile phones, power cordless tools), as catalyst, as a glass-colourant.
Chromium	Chromite	Extractcd by the thermite process, where chromic oxide is reduced by stochiometric amount of silicon and quick lime in an electric arc furnace.	Used in ferrochrome, in chrome vanadium steel (as tools), in nichrome, in stellite (for surgical instuments), as a plating metal, as its salts in the glass industry, in pottery and as a mordant in dyeing.
Cobalt	Smaltite, cobaltite and linaeite	Cobalt can be obtained from slag of copper mines by pyrometall-urgical and hydro-metallurgical process; can be extracted from waste water by liquid surfactant membrane process.	Used in making jet engines and gas turbine parts; alloyed with Cr and tungsten is used as high-speed cutting tools; used to make strong permanent magnets; used as pigments in pottery, glass enamels and paints; used as driers in paints, varnishes and printing inks.
Molybdenum	Molybdenite, wulfenite	By dissolving the ore in a sodium carbonate solution.	Alloyed with steel and used to make rifle barrels and filaments for light bulbs; used as catalysts, paint pigments, corrosion inhibitors, smoke and flame retardants, dry lubricant on space vehicles.

Concluding Remarks

A clear conception of mineral deposit formation processes helps to localize the deposits in a given environment. The mineralogical study of the deposit samples can be used to identify the economic element and the minerals that contain them. Mineralogical studies, particularly the textural analysis, help in evaluating the genesis and evolution of the ores as well as ore beneficiation during extractive metallurgy. A mathematical approach to quantify the deposit will help to decide its economic viability. Apart from these the rank, grade and tenor content of the concerned minerals can be estimated with the mineralogical analysis, using suitable analytical techniques.

Think for a while

1. Differentiate between the term mineral and ore mineral. From this elaborate the concept of ore.
2. The choice for ore minerals are mostly restricted to sulfide, oxide and native element groups. Why?
3. A large number of minerals have their origin related to hydrothermal processes. Why?
4. Minerals are specific of hydrothermal processes. Explain the cause for such specifications?
5. Number of ore minerals, resulting from metamorphic processes, is generally much less than the igneous and sedimentary minerals. Why?

FURTHER READING

Evans, A.M. An introduction to Ore Geology. Elsevier, New York. 1980.

Guilbert, J.M. and C.F. Park, Jr. The Geology of Ore Deposits. Friedman and Co., New York. 1986.

James, Harold Lloyd. Sedimentary facies of iron-formation. *Economic Geology*, **49(3),** 235-293. 1954.

Jensen, M.L. and A.M. Bateman. Economic Mineral Deposits, 3rd ed. John Wiley and Sons, New York. 1981.

Sherman, D.M. Hydrothermal Solutions and Ore Deposits: Physical Chemistry of Minerals and Aqueous Solutions. University of Bristol, 2005/2006.

CHAPTER

11

MARINE MINERALS IN DIFFERENT ENVIRONMENTS

At the new millennium, utilization of marine minerals is accelerating and knowledge of new types of marine mineral resources is expanding with significant present and potential scientific and economic benefits. The utilization of marine minerals is driven by growing societal and industrial needs, which may be met by turning to the sea for materials that are in short supply, strategically vulnerable, environmentally sensitive to recover on land, or can be recovered more economically from the seafloor. So a clear concept about the variety of minerals obtained from different marine regimes, their extraction, utility and the overall impact on environment has been concise in this chapter.

Ocean covers about 70.2% of the Earth's total surface area. It is home to a large variety of minerals, which are essential for supporting the existence of the living world in many aspects. With the rapid progress of industrialization, the land mineral resources are getting exhausted and the underwater mineral resources have become a serious matter of concern nowadays. Marine minerals are distributed in a vast array of environments, starting from the coastal belt up to the deeper part of the ocean. The chief tool for underwater mineral exploration is seismic and magnetic profiling, dredging and coring tools, depth sounding, laboratory analyses, and geological and bathymetric mapping. At the present, less than 5% of the coastal seafloor of the world has been scientifically surveyed. In the search of the deep sea for minerals, if all of the dredge samples taken to date were averaged over the deep sea floor, this would give three dredge hauls per million square kilometres.

11.1 MARINE ENVIRONMENT

Geochemistry of Sea Water

Ocean water is naturally alkaline and its bicarbonate buffering results in a normal pH of 8.0-8.3. Metal bicarbonates such as magnesium bicarbonate and calcium bicarbonate will only form in the natural aqueous, alkaline state of the ocean. These mineral combinations do not occur in solid or dry form. The complex mixture of minerals and liquid that make up sea water consists of 96.5 per cent water, 3.4 per cent salts, and smaller amounts of other substances including dissolved inorganic and organic materials, particulates, and a few atmospheric gases. It is the various salts that gives sea water its taste. Salt is a chemical compound of elemental atoms that act as molecules and ions when in solution. Eighty-four of the Earth's basic elements have been identified in sea water as either macro or trace mineral ions. The complex mixture of macro and trace elements found in sea water, result from the interaction of natural forces over millions of years.

Physiography of the Sea Floor

To understand the mineral potential of the marine environment, it is important to understand the physiography of the sea floor.

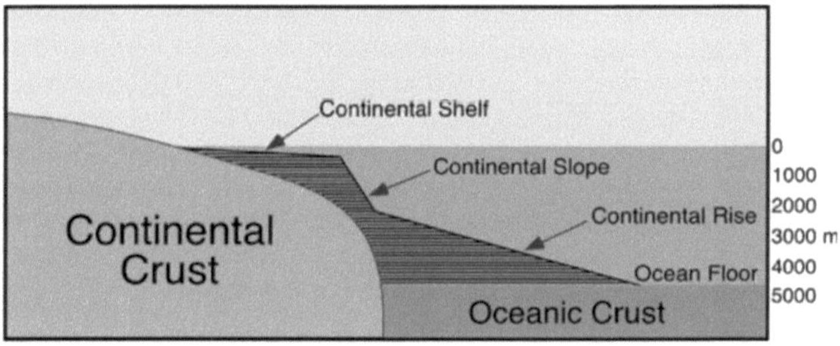

Fig. 11.1: Physiography of sea floor.

Some of the dominant topographic features associated with the ocean basins include:

Continental shelf is a shallow (average depth 130 metres) gently sloping part of the **continental crust** that borders the continents. The extent of this feature varies from tens of metres to a maximum width of about 1300 kilometres.

The **continental slope** extends from the continental shelf at an average depth of about 135 metres. The base of this steeply sloping (from 1 to 25°, average about 4°) topographic feature occurs at a depth of approximately 2000 metres, marking the edge of the continents. The width of the slope varies from 20 to 100 kilometres. Both the continental shelf and slope are considered structurally part of the continents, even though they are below the sea surface. The boundary between the continental slope and shelf is called the continental shelf break.

Submarine canyons are V-shaped canyons cut into the continental slope to a depth of up to 1200 metres. The submarine canyons are cut perpendicular to the running direction of the continental slope.

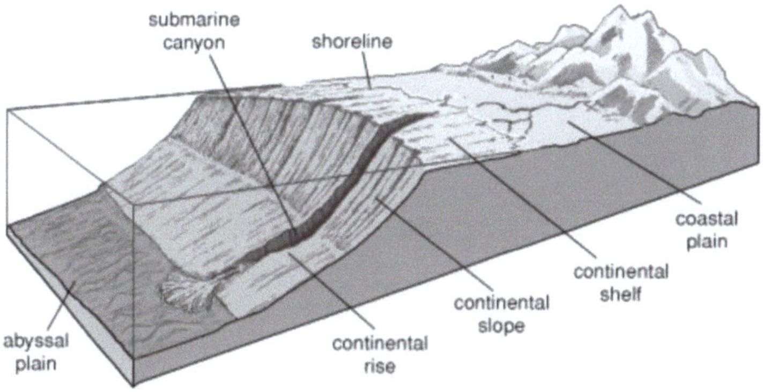

Fig. 11.2: Submarine canyon.

The **continental rise** is found at the base of the continental slope. The depth of the rise ranges from 2000 to 5000 metres deep. Its breadth is up 300 kilometres wide. This feature was created by the merging of accumulated deposits at the mouths of the many submarine canyons. Each canyon's thick fan-shaped sedimentary deposit is called an **abyssal fan**.

The ocean floor is found at the base of the continental rise in water 4000 to 6000 metres deep. The ocean floor accounts for nearly 30% of the Earth's surface. The composition of the ocean floor consists of a relatively thin layer (on average 5 kilometres thick) of **basaltic** rock with an average density of 3.0 grams per cubic centimetre (while in continents it is granite rocks with density 2.7 grams per cubic centimetre).

Mid-oceanic ridge is normally found rising above the ocean floor at the centre of the ocean basins. These features are involved in the generation of new oceanic crust from volcanic fissures produced by mantle up-welling. Some volcanic islands are part of the mid-ocean ridge system (Iceland). The mid-oceanic ridge constitutes 23% of the Earth's surface. In the centre of the mid-oceanic ridge is a rift valley, between 30 to 50 kilometres wide, that dissects 1000 to 3000 metres deep into the ridge system.

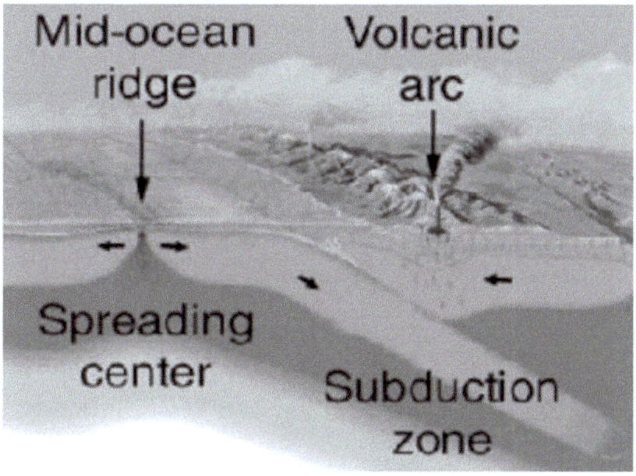

Fig. 11.3: Oceanic ridge.

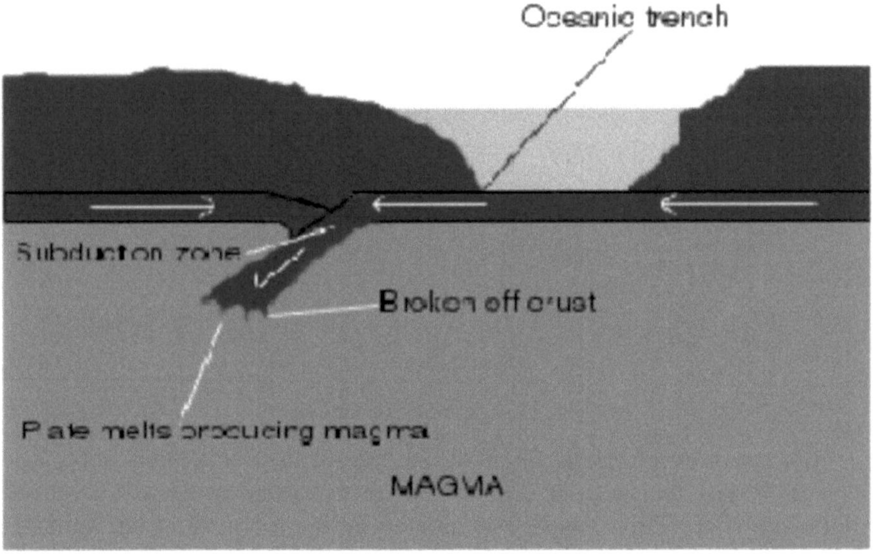

Fig. 11.4: Oceanic trench.

Ocean trenches are long, narrow, steep-sided depressions found on the ocean floor that contain the greatest depths in the ocean (11,000 metres—western Pacific). There are 26 oceanic trenches in the world: three in the Atlantic Ocean, one in the Indian Ocean, and 22 in the Pacific Ocean. Generally, the trenches mark the transition between continents and ocean basins, especially in the Pacific basin. Trenches are also the tectonic areas.

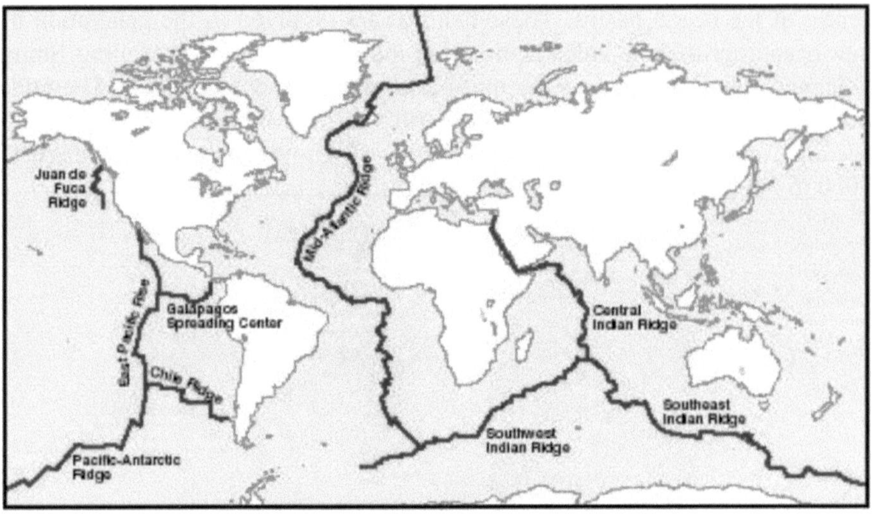

Fig. 11.5: Global distribution of mid-oceanic ridge.

Table 11.1: Association of potential mineral resources with types of plate boundaries

Type of plate boundary	*Potential mineral resources*	*References*
Divergent: Ocean ridges	Metalliferous sediments (copper, iron, manganese, lead, zinc, barium, cobalt, silver, gold; e.g., Atlantis II Deep of Red Sea)	Rona and Scott, 1993
	Stratiform manganese and iron oxides and hydroxides and iron silicates (e.g., sites of Mid-Atlantic Ridge and Galapagos Spreading Centre)	
	Hydrocarbons at early stage of rifting and opening of ocean basin (e.g., Red Sea; Atlantic continental margins; marginal seas of western Pacific region)	
	Polymetallic massive sulphides deposits (copper, iron, zinc, silver, gold; e.g., sites on East Pacific Rise, Galapagos Spreading Centre, Mid-Atlantic Ridge, Central Indian Ridge and marginal seas behind volcanic island chains of western Pacific)	
	Polymetallic stockwork sulphides (copper, iron, zinc, silver, gold; e.g., sites on Mid-Atlantic Ridge, Carlsberg Ridge, Galapagos Spreading Centre)	
	Other polymetallic sulphides in disseminated or segregated form (copper, nickel, platinum group metals)	
	Asbestos	
	Chromite	
Convergent: Offshore	Sections of oceanic crust containing types of mineral resources formed at divergent plate boundaries uplifted onto land (e.g., sites at volcanic island chains of western Pacific); Tin, uranium, porphyry copper and possible gold mineralization in granitic rocks	Mitchell and Garson, 1981; Sawkins, 1984
Convergent: Onshore	Porphyry deposits (large relatively lowgrade deposits of copper, iron, molybdenum, tin, zinc, silver, gold; e.g., deposits at sites in the Andes volcanic mountain chain)	Mitchell and Garson, 1981;
	Polymetallic massive sulphides (copper, iron, lead, zinc, silver, gold, barium; e.g., Kuroko deposits of Japan)	Sawkins, 1984
Offset: Offshore	Mineral deposits similar to those formed at divergent plate boundaries (ocean ridges) may occur at offshore transform plate boundaries; e.g., sites on Mid-Atlantic Ridge in Atlantic Ocean and Carlsberg Ridge in Indian Ocean	Bonatti, 1981

Lakes

Lakes are also important for they also serve as a storehouse of a number of minerals. Among the noteworthy minerals, placers and aggregates are significant, because they travel short distances during post-discharge period and the processes operating in the shallow areas promote their concentration. At the peripheral areas of the lakes, ferromanganese oxides precipitate due to the existing oxidizing conditions.

Sources of Materials to Oceans

The chemical composition of sea water is influenced by a wide variety of chemical transport mechanisms.

Rivers add dissolved and particulates to the oceanic margins, and wind-borne particulates are carried to mid-ocean regions thousands of kilometres from their continental source. Hydrothermal solutions that have circulated through crustal materials beneath the sea floor add both dissolved and particulate materials to the deep ocean.

Organisms in the upper ocean convert dissolved materials to solids, which eventually settle at greater depths. Volcanism supply elements to the sea in form of solid volcanic detritus, hydrothermal solutions or by the alteration of volcanic rocks (halmyrolysis). Particulates in transit to the sea floor, as well as materials both on and within the sea floor, undergo chemical exchange with the surrounding solutions.

Through these local and regional chemical input and removal mechanisms, each element in the oceans tends to exhibit spatial and temporal concentration variations. Physical mixing in the oceans (thermohaline and wind-driven circulation) tends to homogenise the chemical composition of sea water. The opposing influences of physical mixing and of biogeochemical input and removal mechanisms result in a substantial variety of chemical distributions in the oceans.

11.2 MAJOR TYPES OF UNDERWATER MINERALS

Major mineral assemblages are from placer, aggregates and clays. In marine mineral resources, phosphorites, evaporites, silica polymorphs, Mn-nodules, Fe-oxides, zeolites and other potentials are available. The underwater mineral exploration is first done using remote sensing techniques and after accessing the potentials, deposits are explored.

A brief survey of marine minerals may be placed in perspective, as follows:

Placer deposits: The profitable mining of tin off southeast Asia and of diamonds off southwest Africa is literally only scratching the surface in terms of the potential that exists for heavy metals and gemstones in sediments of continental margins.

Phosphorites: Extensive deposits on continental shelves in past and present areas of oceanic upwelling in low latitudes constitute an untapped resource most accessible to developing countries that can fertilize agricultural crops as land deposits dwindle.

Aggregates: Sand and gravel are the most widely utilized and accessible of marine mineral resources with growing needs for construction materials, shore protection, and beach replenishment.

Water: The need for a reliable and adequate source of fresh water for consumption and agriculture in light of depletion by population growth, industrialization and climate change is considered to be the most pressing environmental problem for humankind (Revenga et al., 2000). The ocean is by far the largest reservoir on Earth contingent on development of cost-effective desalination processes.

Metalliferous sediments: The Atlantis II zinc-silver-copper metalliferous sediment deposit of the Red Sea is the only seafloor hydrothermal deposit in which grade and tonnage are known anywhere near to mining standards on land. Legally its recovery rests with the bordering coastal states that share the intervening 200 mile zone; a pilot mining test has been proposed that should address the issue of refining the fine-grained material. Initiation of commercial mining is subject to market forces. The metallifeous sediments of the Atlantis II Deep represent a type of deposit that should be present at sites buried under sediments of Atlantic and other rifted passive continental margins.

Massive sulfides: The immediate significance of massive sulfides concentrated by high-temperature hydrothermal systems at plate boundaries is as analogs to guide economic geologists in the search for and mining of ancient VMS (Volcanogenic Massive Sulfide) deposits that formed at similar seafloor settings in the geologic past and have subsequently been uplifted onto land. A surge of discovery of these VMS deposits on land in P.R. China (Rona and Hou, 1999) and other places is attributable to insights gained from the actual submarine ore-forming systems. The mining of sea-floor massive sulfide deposits is likely to start at fore- and back-arc settings of volcanic island arcs within the 200 mile zone of coastal states of the western Pacific before those deposits on ocean ridges in the international area. Present technical limitations on drilling to adequately determine grade and tonnage will have to be overcome and favourable market conditions relative to land sources must evolve. While individual mineralized chimneys have been observed to grow quickly (days to years), sea-floor massive sulfide deposits as a whole are not renewable resources because they require thousands of years to accumulate prospective grade and tonnage. Furthermore, directly tapping the discharging metal-rich solutions would result in accumulating iron and little else.

Manganese nodules: The players are positioned with the International Seabed Authority in the richest international zone (nickel and copper content >2.5 weight percent) in the eastern equatorial Pacific. A strategic need and/or favourable market conditions relative to presently adequate land sources of these metals are conditions necessary to drive the considerable investment required to support production.

Cobalt-rich ferromanganese crusts: Cobalt is considered to be the potential driver, but the prospect of recovery and refining from a hard-rock substrate adds an extra challenge.

Marine clays: Marine clays are soils that were deposited by rivers flowing into oceans millions of years ago. Marine clays shrink during dry periods of the year and swell during wet periods. The pressures exerted by the swelling action can exceed the strength of the foundation on these clays.

Placers and Aggregates

Placers

A placer deposit is an accumulation of mineral grains concentrated by sedimentary processes. When pebbles, sands and silts are sorted by wave action or stream flow, minerals with higher specific gravity and resistance to weathering become concentrated, especially in beaches and drowned river mouths. Marine *placer mineral* deposits are found on the continental shelf from the beaches to the outer shelf. These transported deposits may have formed in a marine environment, but have generally been emplaced while the shelf was emergent under a lower sea level. The glacial sea level lows exposed most of the shelf to subaerial processes and stream placer deposits extended across the shelf.

The concentration of minerals depends on the source rock, weathering, transportation and trapping by subaerial processes and subsequent reworking and modification by marine processes. Locating potential marine placer resources is often less difficult than determining their exploitability, because varying littoral drift rates and directions, altering wave-energy distributions, and changing water levels contribute to regional and local variations in placer distributions.

Many mineral-bearing *offshore and beach placers* (diamond, gold, platinum, tin, chromite, iron sand, zircon, ilmenite, rutile and monazite) are now mined or have been mined in the past. The major placer materials are tin, titanium, gold, platinum and chromium. Diamonds are the only gemstone presently mined offshore.

Classification of Placers

Main five types of placers are as follows:

Residual placer: The long continued weathering of mineralized veins and surface rocks and minerals breaks down the constituents. The more resistant minerals will collect nearer to the outcrop while the softer and less resistant minerals are carried away by the forces of erosion.

Eluvial placer: The concentrations of valuable minerals along the hillside, downslope from the outcrop are called eluvial placers. These placers are also very important to the electronic prospector.

Stream placer: Stream placers consist of valuable minerals concentrated in sand and gravel sorted by the action of running water. Placers formed in recent time will occupy the active stream channel. The deposits will form where the water slows or around obstructions in the channel. If formed in ancient times, the placer deposits may occupy benches elevated well above the present day stream channel.

Eolian placer: Eolian placers are formed by wind action. Bajada placers can show surface enrichment due to the lighter materials being removed by wind and sheet flooding in desert regions. These areas offer very good potential for the modern nugget-shooter, as after ages of wind whipping away surface materials gold nuggets are left well within the range of metal detectors.

Beach placer: Gold and diamond occur in various locations along coast lines as the result of the action of shore currents and waves. Materials broken down from cliffs by waves or washed into the sea by streams are sorted and distributed by their size and specific gravity.

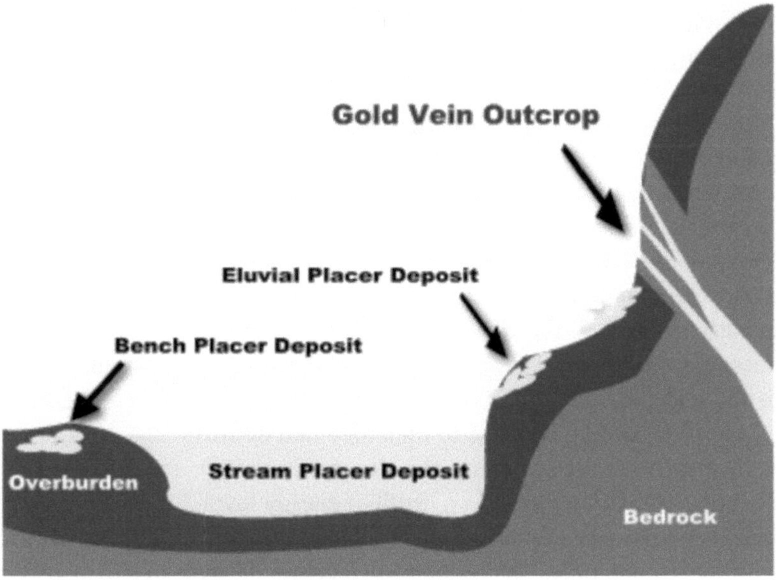

Fig. 11.6: Placer deposit formation.

Table 11.2: Classification of placer minerals

Type of placer minerals	*Sp. gravity*	*Examples*
Very heavy minerals	6.8-21	Gold, platinum, tin oxide
Heavy minerals	4.2-5.3	Ilmenite, rutile, zircon, monazite, magnetite
Gems	2.9-4.1	Ruby, sapphire, diamond

Emery and Noakes (1968)

Table 11.3: List of marine placer minerals

Minerals	*Sp.gravity*	*Hardness*
Cassiterite	6.8-7.1	6.0-7.0
Diamond	3.5	10
Fergusonite	5.4	5.5-6.5
Gold	19.3	2.5-3.0
Ilmenite	4.5-5.5	5.0-6.0
Magnetite	5.2	5.5-6.5
Monazite	4.6-5.4	5.0-5.5
Platinum	14-19	4.0-4.5
Rutile	4.25	6.0-6.5
Wolframite	7.1-7.5	4.0-4.5
Xenotime	4.4-5.1	4-5
Zircon	4.6-4.7	7.5

Controlling Factors for Placer Formation

Primary source: Primary source is igneous rocks. Light heavy minerals consist of resistant accessory minerals of plutonic rocks.

Weathering: Mechanical weathering is predominant, which leads to the liberation of the placer minerals from source rocks. Through chemical reaction also the less chemically resistant minerals in the rocks undergo selective dissolution resulting in the breakdown of the rock.

Transport: Running water is the most active means of transport. The clay minerals or similar unstable alteration products of primary rocks are selectively removed from the heavier and more resistant minerals by aqueous processes. The less resistant minerals are carried in suspension down streams and rivers to the sea and the heavy minerals are most commonly concentrated in traps within the river system because of their size and density.

Concentration: Placer minerals, due to their high specific gravity, are usually concentrated in traps. As a result of energy level difference, lighter and smaller minerals are separated from the placer minerals, which can be concentrated in sink holes, depressions on beaches, in areas of sluggish currents or in other sediment traps.

Marine Environment of Placer Mineral Deposition

Marine placer deposits occur on the continental shelf within five miles of the coast. Submarine placer deposits occur in the near-shelf region are of late Tertiary to Recent age.

During the low stands of sea level in the Plieistocene, alluvial, eluvial and autochthonous and beach placers were formed on the exposed shelf areas. Eluvial occurrences of marine placer deposits are found where weathered source rocks of heavy minerals outcrop on the shelf.

Aggregates

Aggregates are deposits of sands, gravels or shells, extensively used by the offshore industry. They occur on beaches, in rivers, lakes and in the offshore area. They have been concentrated by normal hydrodynamic processes. They represent an important group of offshore mineral deposits.

Origin: Aggregates > 0.25 mm get concentrated below the euphotic zone. Phytoplankton-containing aggregates is about half of the total aggregates in the upper 40-251 m layer, but decline to about 5% at 400 m possibly due to offshore transport of sinking material, input of resuspended aggregates at depth, and/or intense biological recycling of aggregates in the midwater. Have been deposited in 'Ice Age' rivers in valleys now submerged by the sea.

Composition of Marine Aggregates

Sand and gravel: Underwater gravel consist of fragments of stable rocks such as quartzite, flint and chert which can be transported for considerable distances without breaking down. Less durable rock fragments are occasionally present.

Sands also consist of stable minerals, of which the most common is the quartz. Mixed sand assemblages containing quartz, feldspar and other minerals occur in some areas where the source rock is a granite or some other polymineralic rock.

Exploration: Exploration tools for sand resources are seismic profiling and side scan sonar; followed by vibrocoring. Recovery can be by a variety of dredges. Sand and gravel mining is done with two systems: clamshell type buckets and sand pumps (a hydraulic system). The *hydraulic hopper dredge* seems to be a universal type suited for open marine waters but it causes a higher degree of turbidity and environmental problems with ensuing regulation of operation. Besides turbidity, a possible problem with sand extraction is a change in bottom bathymetry, which could cause a change in wave regime, in turn a problem of coastal erosion. In Japan, this is eliminated by restricting extraction to water depths greater than 20 metres.

Pure silica sands are dredged for the glass and chemical industry off Japan, northern New Zealand, and in the Baltic.

Shells

Shells are calcareoeus in nature, formed by the break up of marine shells and their concentration into shell banks etc. by hydrodynamic processes. These shells are used in the manufacture of lime and cement. For generation of shells, a source of shell debris is required. Due to their differential shapes and sizes they respond differently to wave and current action of the sea floor. This may result in their separation from detrital sediments and their concentration into deposits containing high concentrations of calcium carbonate.

Marine shell material has been used as an alternate to sand and gravel and has been used as a source of lime for cement. Because of Iceland's lack of sedimentary rock, the government has developed a state-owned cement-works and an agricultural lime establishment that use seashells as their raw material.

Brazil began producing lime from seashells in the 1950's. Several areas in the U.S. produce, or have produced seashells, including California, Texas, Louisiana, Mississippi, Alabama, Florida, Virginia and Maryland.

Use of Marine Aggregates
Manufacture of concrete; Production of asphalt and coated products; Production of masonary and paving blocks; Used for drainage and fill materials and other industrial applications; and Production of leisure and sport facilities.

Distinction between Terrestrial and Marine Aggregates
The main differences between the majority of terrestrial and marine aggregates are the presence of chloride (from sea salt) and shell in marine material.

Miscellaneous Authigenic Minerals

Authigenic minerals form directly from reactions between dissolved species in sea-water or fresh water, or from the alteration of pre-existing phases in subaqueous sediments. The precipitation actually results from the supersaturation of the element or compound required to form the mineral. Formation of these minerals depends on local geochemical conditions, including elemental abundances, water characteristics, proximity of hydrothermal sources and rate of sediment accumulation.

Common authigenic minerals are Zeolites, Barite, Feldspar, Garnet.

Zeolites

Zeolites are microporous crystalline solids with well-defined structures. Generally they contain silicon, aluminium and oxygen in their framework and cations, water and/or other molecules within their pores. Zeolites form typically in marine and laccustrine environments.

Structure of zeolite: A defining feature of zeolites is that their frameworks are made up of four connected networks of atoms. It represents a tetrahedra, with a silicon atom in the middle and oxygen atoms at the corners.

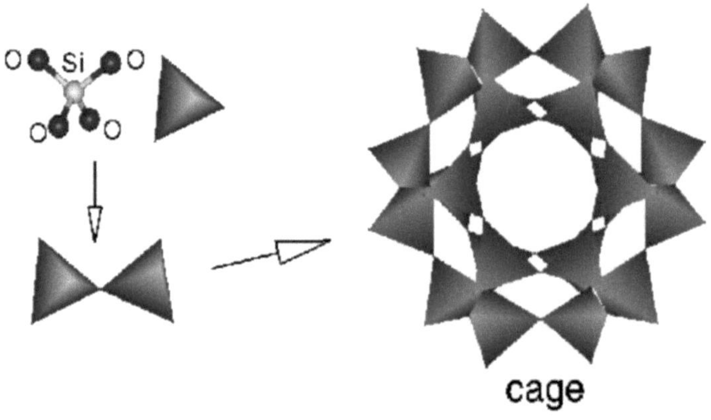

Fig. 11.7: Structure of zeolite.

Common members of zeolite group: Phillipsite ($KCa[Al_3Si_5O_{16}]. 6H_2O$); Clinoptilolite [$(Na,K)_4CaAl_6Si_{30}O_{72}. 24H_2O$]; Harmotome ($Ba[Al_2Si_6O_{16}]. 6H_2O$); and Analcime ($NaAlSi_2O_6. H_2O$).

Table 11.4: Types of zeolite minerals

Zeolite minerals	*Properties*	*Distribution*
Phillipsite ($KCa[Al_3Si_5O_{16}]. 6H_2O$)	Forms elongated prismatic crystals; generally colourless to yellow brown; may contain inclusions; Si/Al ratio varies between 2.3 and 2.8 with enrichment of alkali (K> Na and Ca).	Found mainly in marine sediments, where sedimentation rate is low and below carbon compensation depth.
Clinoptilolite ($(INa,K)_4CaAl_6Si_{30}O_{72}. 24H_2O]$)	Forms fine-grained euhedral crystals, occurs as radiolarian casts; Si/Al ratio varies between 4 and 5.25.	Found mainly in surface sediments of marginal areas of the oceans.
Harmotome ($Ba[Al_2Si_6O_{16}]. 6H_2O$)	Contain Ba as the principal cation and has similar structure to that of phillipsite.	Occurs in areas of low sedimentation rate, in association with phillipsite.
Analcime ($NaAlSi_2O_6. H_2O$)	Rare mineral, occurs as discrete euhedral crystals.	Found in deep-sea sediments with basaltic material.

Barite

Barite ($BaSO_4$) occurs in crystalline or microcrystalline phases or as replacement material in fecal pellets in deep-sea sediments. Barite concentrations average 1% in deep sea sediments, but can make up as much as 10% by weight of the carbonate-free fraction on the East Pacific Rise, where it is associated with hydrogenous iron oxide. Most (80%) of the elemental barite in the oceans enters through rivers, about 20% comes from hydrothermal vents. A major conduit of barium to ocean sediments is secretion by a group of deep-sea protozoans, the xenophyophores that produce barite crystals in large quantities. Elemental barite is found in biogenic sediments and has been attributed to production by these organisms or by concentration in organic matter following the death of the organism. Deep-sea sediments tend to be richer in barite than slope-depth deposits. Sediment pore waters in the deep sea are saturated with respect to barite; preservation potential is estimated at 30% in oxidized sediment and much lower in anoxic sediments.

In the Pacific, barite is found in radiolarian oozes beneath the equatorial upwelling zone. In the Atlantic, elevated barite concentrations are found on the mid-ocean ridges in areas of low sedimentation rates and where there is an abundance of ferromanganese or iron oxide from hydrothermal sources. Barite is mined extensively off the Alaskan coast.

Feldspar

Authigenic feldspar occurs in both marine limestones and dolostones. They are most commonly found to replace igneous calcic plagioclase or volcanic glass. These minerals exhibit highly ordered structure. Authigenic albite is far more abundant than authigenic microcline. Illite and smectite are commonly associated with them.

Garnet

Evidence of metasomatic andraditic hydrogrossular garnet in sediments during the Deep Sea Drilling Project has been reported.

Phosphorites

Phosphorites are essentially sedimentary deposits, consisting of phosphate minerals. Phosphorites are cryptocrystalline apatites and vary in composition. A general formula is $Ca_{10} (PO_4, CO_3)_6 F_{2-3}$, with increases in carbonate running parallel to increases in fluoride (or hydroxide).

Phosphorite deposits occur on several continental shelves—Atlantic waters of the U.S., Morocco, Gabon, Congo, Namibia, and South Africa and in the Pacific Basin in the US, Mexico, Ecuador, Peru, and New Zealand. They are usually found in less than 1000 m of water in relatively tropical regions where they are linked to zones of coastal upwelling, divergence and biological productivity. Most deposits are of Miocene age, reflecting good environmental conditions for phosphorite deposition at that time.

Submarine phosphorites generally occur in water depths of less than 1000 metres on continental shelves, offshore banks or marginal plateaus. Marine phosphorites are enriched in some trace elements compared with marine shales. Modem phosphorites typically occur in areas of high productivity—off Southern and Baja California, off Peru, and off South Africa. Concentrations within marine P-rich deposits as reported from off Califomia, western South America, South Africa, or on the Chatham Rise, east of New Zealand, are up to 80 kg/m^2 phosphorites with up to 25% P_2O_5. They occur over large areas and in water depths of less than 400 m.

Form: They occur as black or brown nodules from pellets up to head size, and as irregularly shaped cakes. Sometimes they may have a layered, conglomeratic or oolitic internal structure. Phosphorites may contain a number of impurities like clay minerals, detrital mineral grains, glauconite, silica, carbonaceous matter, calcite, dolomite, iron oxides and occasional authigenic pyrite and fluorite.

Origin: The algae in the surface waters extract the phosphorus from the water. Also, crustaceans and fish concentrate it further in their bodies and excrement. During decomposition of organic debris on the sea floor, much phosphate is released to the interstitial water (and also to seawater). Thus, interstitial waters below the sea floor may become saturated with the phosphate mineral apatite. Precipitation of apatite, replacement of preexisting carbonate minerals, and impregnation of sediment can then proceed.

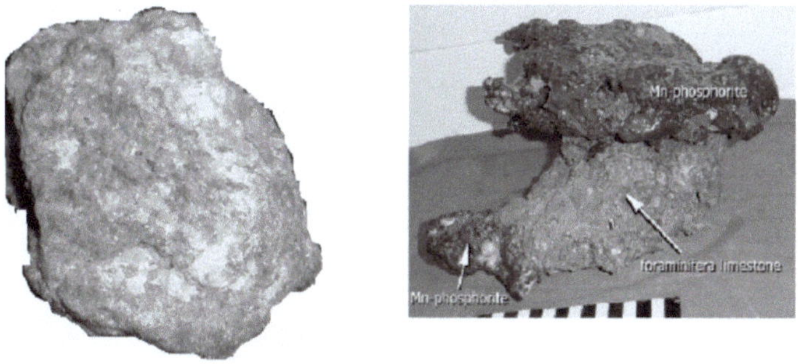

Fig. 11.8: Different types of phosphorite nodules.

Uses: Phosphorites are largely used for fertilizers, but also as a source of phosphorus in the chemical industry.

Manganese Deposits

Manganese oxide phases in marine environment are often metastable and grow in close association with other materials resulting in poor crystallinity which hinders their identification by conventional X-ray diffraction technique. Similarly numerous structural defects, disordered vacancies, domain intergrowth, extensive solution and cation exchange properties make X-ray diffraction further difficult. Recent methods used are scanning electron

Table 11.5: Various manganese minerals

	Manganese minerals	*Chemical formula*	*Crystallographic system*
Marine	Pyrolusite	MnO_2	Tetragonal
	Ramsdellite	MnO_2	Orthorhombic
Marine	δ-MnO_2	MnO_2	Hexagonal
	Hollandite	$(Ba,K)_{1\text{-}2}\ Mn_8O_{16}.xH_2O$	Tetragonal
Marine	Cryptomelane	$K_{1\text{-}2}\ Mn_8O_{16}.xH_2O$	Tetragonal or monoclinic
Marine	Romanchite (psilomelane)	$(Ba, K, Mn^{2+}, Co)_2\ Mn_5O_{10}.\ xH_2O$	Monoclinic or orthorhombic
Marine	Todorokite	$(Ca, Na, K)\ (Mg, Mn^{2+})\ Mn_5O_{12}.\ xH_2O$	Hexagonal
Marine	Buserite	$Na_4Mn_{14}O_{27}.\ 21H_2O$	Hexagonal
Marine	Lithiophorite	$(Al_4, Li_2)Mn^{4+}O_2(OH)_2$	Monoclinic
	Chalcophanite	$Zn_2Mn_6O_{14}.\ 6\ H_2O$	Triclinic
	Synthetic birnessite	$Na_4Mn_{14}O_{27}.\ 9H_2O$	Orthorhombic
	Synthetic birnessite	$Mn_7O_{13}.\ 5H_2O$	Hexagonal
Marine	Birnessite	$(Na, Ca, K)\ (Mg, Mn)\ Mn_6O_{14}.5H_2O$	Hexagonal
	Vernadite	$(Mn_4, Fe_3, Ca,Na)\ (O, OH)_2.n\ (H_2O)$	Hexagonal
	Rancieite	$(Ca, Mn)\ Mn_4O_9.3H_2O$	Hexagonal
	Groutite	α-MnOOH	Orthorhombic
	Manganite	γ-MnOOH	Monoclinic
	Hausmannite	Mn_3O_4	Tetragonal
	Pyrochroite	$Mn(OH)_2$	Hexagonal

microscopy, electron diffraction, high resolution transmission electron microscopy, infrared spectroscopy and extended X-ray absorption fine structure.

Manganese Nodules

The underwater accumulation of manganese oxides in the form of golf- to tennis ball-sized nodules lie on and in sediment on vast expanses of the abyssal plains that cover some 80 percent of the ocean basin at typical water depths of about 5 km.

Origin: The portion of the nodules that protrudes above the surface of the seafloor sediment is precipitated from metals dissolved in seawater (authigenic or hydrogenous origin), while the underside of the nodules accumulates from metals dissolved in the pore water of the sediment (diagenetic process) over millions of years. The metals come from two sources. As noted, manganese and other metals are dissolved by weathering of rocks on continents and transported into the ocean by rivers. At the same time hot springs at sites along submerged plate boundaries discharge an equivalent amount of manganese and variable amounts of other dissolved metals into the ocean. The result is a mixture of copper, nickel, manganese, cobalt and iron in the nodules that varies in different regions of the ocean related to proximity to sources and to concentration processes. Due to abundance of manganese and iron in marine environment, the nodules are sometimes referred to as ferromanganese nodules. There are three major presumptions about the sources of ore material supply to the ocean:

1. Main ore material supply to the ocean comes from terrigenous and volcanic, and also hydrothermal material.
2. Volcanic material is the major source of the supply of ore matter to the ocean.
3. Terrigenous is the major source of the supply of ore matter to the ocean.

These views have been debated and contested by several authors and it has been concluded that the accumulation rate of manganese should be studied in individual oceanic regions.

Table 11.6: Average concentration in manganese nodules from different oceans

Element	*Atlantic*	*Pacific*	*Indian*	*World Oceans*
Manganese (weight percent)	13.25	20.10	15.25	18.60
Iron	16.97	11.40	14.23	12.40
Nickel	0.32	0.76	0.43	0.6
Copper	0.13	0.54	0.25	0.45
Cobalt	0.27	0.27	0.21	0.27
Zinc	0.12	0.16	0.15	0.12
Lead	0.14	0.08	0.10	0.09
Iridium	9.32	6.64	3.48	–
Uranium	7.4	7.68	6.20	–
Palladium (parts per million)	5.11	72	8.76	–
Thorium	55.00	32.06	40.75	–
Gold (parts per billion)	14.82	3.27	3.59	–

Distribution: Nodule fields are concentrated between 9′30″S to 16′30″S and 72′30″E to 82′30″ E.

Classification: Depending on the major ore elements (manganese, iron and base metals) the nodules contain three major types distinguished:

1. *Hydrogenous:* Formed due to slow deposition of metals out of sea water and characterized by a high concentration of base metals and varying Mn/Fe ratios (from 0.5-5).
2. *Hydrothermal:* Rich in iron and depleted of other metals, and turns out to be having an extremely wide range of Mn/Fe ratios.
3. *Diagenetic:* Characterized by high Mn/Fe ratios and relatively low concentration of base metals.

Fig. 11.9: Different types of manganese nodules.

Mineralogy: Manganese nodules mainly consist of various minerals of manganese and iron, amorphous components and a variety of accessory silicate phases.

Table 11.7: Constituent mineral phases in manganese nodules

Minerals	*Examples*
Manganese oxide	Todorokite, birnessite and δ-MnO_2. Less common are pyrolusite, psilomelane, cryptomelane and lithiophorite.
Iron oxide	Goethite (α-FeOOH) and lepidocrocite (γ-FeOOH). Others are hematite (α-Fe_2O_3), maghemite (γ-Fe_2O_3) and akaganeite (β-FeOOH).
Accessory component	Rock fragments, volcanic glass, fossil test, quartz, feldspar, pyroxenes, amphiboles, clay minerals and zeolites.

Manganese encrustations

The manganese oxide accumulations in the form of blanket-like encrustations of different dimensions occur in various oceanic settings and at various depths. These are known as manganese encrustations. The metal supply for the formations, come from ambient seawater, from volcanic exhalations, or from both.

Table 11.8: Classification of manganese encrustrations

Division	*Sub-division*	*Properties*
Hydrogenetic	Shallow water occurrence	Form at water depth shallower than 2,500 m; have high contents of cobalt along with platinum, silver, gold, titanium and lead; occur above CCD; mineralogically it is δ-MnO_2, amorphous FeOOH. xH_2O.
	Deep water occurrence	Restricted to abyssal hills, seamounts and sediment hardgrounds at a depth range of 4,800-5,300 m, but abundant above 5,000 m; the crusts display botryoids, cusps and laminations; botryoids are rich in δ-MnO_2 and cobalt, while cusps and laminations are rich in todorokite and copper.
Hydrothermal	Formed by hydrothermal deposition and found mainly on or near actively spreading mid-oceanic ridges, volcanic arcs, off-axis seamounts and fracture zones; composition-wise it has high Mn/Fe ratio, very low trace metal contents, todorokite and birnessite; it serves as an indicator in the search for sea-floor sulphide deposits, which result from fractionation in the proximity of hydrothermal crusts.	

11.3 IMPACT OF MARINE MINERAL EXPLORATION ON ENVIRONMENT

Deterrents to marine mining have been aesthetic and environmental concerns, development of technology, cost-benefit factors, and the lack of clear government regulations. The removal of the sand can cause severe erosion and loss of marine habitat. Research has shown that dredging operations can cause impacts that range from detrimental to beneficial. The impacts are predictable and careful planning can minimize undesirable ecological effects.

Since we are moving coastal or sea bed materials in a mining operation, all of the environmental problems associated with dredging and filling apply. In the case of construction materials, large amounts of sediment are moved and the local environment may be altered. Mining of sand from beach dunes and from river mouths will affect beach sources and equilibrium. Indiscriminate removal of sand will result in beach erosion. Problems also exist in offshore bulk material mining. Both the direct removal, and the debris released to the water column can have an effect on local marine life. Large removals can change the bottom configuration enough to change wave and current patterns, possibly to an adverse effect on the coastal beaches.

Controversy has centered on the environmental impact of shell dredging. Control of allowable dredging sites has prevented kills of live oysters; nearly all sediments settle out within about 120 m of the dredge—actually less disturbance than a shrimp trawl. Dredging in many places has stopped (Texas and Florida, for example) because of: depleted shell supplies, competition from lower cost crushed limestones and other limestone sources, economic recession and increased costs for environmental protection.

Although metallic and precious minerals are less bulky themselves, large amounts of material must be removed to recover the minerals. This not only leads to the removal problems of bulk mineral mining, but adds return of the waste material as fill.

Concluding Remarks

Minerals from marine resources provide raw materials for the manufacturing and construction sectors, energy for industrial and domestic use, fertilizers for the agriculture sector, and importantly, fresh water for the very sustenance of life on Earth. The impact of marine mineral exploration has to be considered seriously for maintaining ecological and. environmental balance.

Think for a while

1. Which environments do you encompass within the term marine environment?
2. What are the primary controlling factors behind the accumulation of major mineral resources?
3. Placers also develop in terrestrial regime. How will you distinguish them from their marine counterparts?
4. Explain the term feldspar authigenesis in view of marine environment.
5. Distinguish between manganese nodules and encrustations.

FURTHER READING

Bonatti, E. Metal deposits in the oceanic lithosphere. *In:* Emiliani, C. (ed.), The Sea, Vol. 7. The Oceanic Lithosphere. Wiley, New York. 1981.

Broadus, J.M. Asian Pacific Marine Minerals and Industry Structure. *Marine Resource Economics*, **3(1)**, 1986.

Burns, R.G. (ed.) Marine Minerals. Reviews in Mineralogy. Vol. 6. Mineralogical Society of America. 1981.

Craig, James R., David J. Vaughtan and Brian J. Skinner. Resources of the Earth: Origin, Use, Environmental Impact, 3rd ed. Prentice Hall, Upper Saddle River. NJ, 2001.

Cronan, D.S. Underwater Minerals. Academic Press. 1980.

Ghosh, A.K and R. Mukhopadhyay. Mineral Wealth of the Ocean. Oxford and IBH Publishing Co., Calcutta. 1999.

Lahman, H.S. and J.B. Lassiter III. The Evolution and Utilization of Marine Mineral Resources. Books for Business. 2002.

Mitchell, A.H.G. and M.S. Garson. Mineral Deposits and Global Tectonic Settings. Academic Press, New York. 1981.

Rona, P.A. and S.D. Scott. (eds). A special issue on sea-floor hydrothermal mineralization: new perspectives. *Economic Geology*, **88(8)**, 1935-1975, 1993.

Sawkins, F.J. Metal Deposits in Relation to Plate Tectonics. Springer, New York. 1984.

Internet Data Retrieved from:

Marine Mineral Resources. International seabed authority. Pdf.

Marine Minerals Guidance Note 1 (Northern Ireland). Guidance on the Extraction by Dredging of Sand, Gravel and Other Minerals from the Northern Ireland Seabed. March 2007.

www.defra.gov.uk

Sharma, Rahul. Environmental impact analyses of mining of marine minerals. National Institute of Oceanography, Goa, India.

USGS Minerals Information: Mineral Commodity Summaries. U.S. Geological Survey. <http://minerals.usgs.gov/minerals/pubs/mcs/> .

USGS Minerals Information: Minerals Yearbook. U.S. Geological Survey. <http://minerals.usgs.gov/minerals/pubs/myb.html/>

CHAPTER

12

MINERALS AND MINERAL ASSOCIATIONS AS GEOTHERMOMETERS AND GEOBAROMETERS

Mineral assemblages and textures may provide information about the conditions at which a rock equilibrated. In metamorphic rocks, we use qualitative terms such as low-grade, medium-grade and high-grade (and even 'medium-low' or 'very high grade') to describe the approximate temperature conditions of metamorphism. These designations do not provide information about pressure, and are therefore not useful for describing subduction zone rocks. Other methods for characterizing metamorphic conditions include: (a) *Index minerals:* Characteristic minerals that provide an indication of the temperature (and, in some cases, pressure) conditions at which a rock is formed (e.g., kyanite in metamorphosed shale; magmatic epidote in plutons and volcanic rocks). Not all rocks have a suitable bulk composition to produce index minerals. (b) *Metamorphic facies:* Assemblages of minerals, each characteristic for a particular bulk composition and indicating the range of pressure-temperature conditions at which the rock equilibrated. For example, high-pressure and low-temperature conditions characterize the blueschist facies. Some igneous assemblages are also characteristic of crystallization within particular ranges of temperature/pressure conditions. This chapter deals with the determination of these methods.

12.1 INTRODUCTION

Geothermometers and geobarometers are mineral systems that may be used to estimate the absolute temperature and pressure that produced an equilibrium mineral assemblage in rocks of various litho-types, say, either igneous or metamorphic. Geothermometric and geobarometric methods are considered under two groupings: directly calibrated methods and internally consistent data set methods. Emphasis is placed on the importance of error propagation calculations. They allow the calculation of specific uncertainties on temperatures and pressures due to the quantifiable uncertainties involved in the calibration and application of methods. The possibility of systematic errors, due to the unquantifiable effects of the assumptions and approximations involved, must be evaluated using the criterion of geological possibilities. Thermobarometry or the determination of temperatures and pressures during the formation of minerals or mineral associations represents a practical application of geochemical thermodynamics to petrology.

12.2 GEOTHERMOBAROMETRY

Geothermobarometry is the quantitative determination of the temperature and pressure at which a metamorphic or igneous rock reached chemical equilibrium. The term "classical" thermobarometry refers to methods for calculating the P-T conditions of a specific chemical reaction.

Principle: Use measured distribution of elements in coexisting phases from experiments at known P-T to estimate P and T of natural samples.

Conceptual basis: The compositions of coexisting minerals in equilibrium are related by the *thermodynamic properties* of the minerals to the pressure and temperature conditions of equilibration. *Thermometers* are assemblages that form by reactions that are sensitive to temperature but not much to pressure. *Barometers* are assemblages that are sensitive to pressure but not much to temperatures. Not all rocks contain assemblages that make suitable thermometers or barometers.

To apply *a thermometer* or *barometer* to an equilibrium mineral assemblage, following parameters are needed to be known:

1. The compositions of all minerals that can have variable composition (e.g., solid solutions).
2. Thermodynamic data for the phases of interest.
3. A 'calibration' that allows you to relate mineral composition to pressure or temperature.

For igneous rocks, calculated pressures and temperatures likely represent the conditions at crystallization, particularly for rapidly cooled rocks. Metamorphic rocks have more complex thermal and pressure histories, but the most typical interpretation of thermobarometric results is that the calculated P and T represent the conditions at the thermal maximum (peak of metamorphism). It is important to recognize that the thermal maximum may not represent the pressure maximum. Furthermore, it is important to examine and determine the textures of a metamorphic rock if the mineral assemblages/ textures are consistent with the interpretation that highest grade of metamorphic conditions are preserved and have not been overprinted during later events.

Igneous and metamorphic rocks bear witness to processes that occurred under conditions of pressure and temperature much higher than those at the Earth's surface. Indirect techniques that use the compositions and assemblages of minerals in rocks to determine the temperatures and pressures at which they formed are called *geothermometry* and *geobarometry* respectively.

Although there are many different geothermometers and geobarometers, most rely on equilibrium chemical reaction(s) between two or more phases (e.g. minerals) in a rock. Because the extent of chemical equilibrium is determined by kinetic factors, such as solid-state diffusion, reactions become inhibited below a threshold (or 'closure') temperature. Rocks thus commonly preserve chemical or mineralogical evidence for the highest temperature that they have experienced. The most widely used geothermometers depend on exchange reactions between two components (e.g. Fe^{2+} and Mg^{2+}, or ^{16}O and ^{18}O) whose kinetics are sufficiently sluggish to preclude re-equilibration during cooling. Geobarometers typically employ net transfer reactions in which one or more phases is consumed or produced. The ability of thermobarometry to determine the magmatic or peak metamorphic temperature and pressure of a rock depends on the cooling history of the rock and the extent of subsequent re-equilibration. Ideally, a number of geothermometers and geobarometers are applied to a given rock in an attempt to counter these effects. Careful consideration must also be given to textural evidence for equilibrium between the phases of interest.

Thermobarometers require thermodynamic calibration of the equilibrium constant for a particular reaction or set of reactions. In general, geothermometers depend on reactions with large enthalpy (heat) changes and barometers on reactions with large volume changes, so as to maximize the sensitivity of the equilibrium constant to changes in temperature or pressure. In nature there are far fewer reactions suitable for geobarometry than geothermometry. Thermodynamic data used to calculate the equilibrium constant may be derived calorimetrically or empirically. In either case a reliable description of activity-composition relationships in the relevant minerals or phases are essential in determining the equilibrium constant, and hence pressure or temperature, from natural mineral compositions.

In practice, thermobarometry entails the precise determination of mineral compositions in a rock, typically using the electron microprobe. Common exchange geothermometers include:

- Fe^{2+}-Mg between garnet and biotite (useful for metapelitic schists);
- Ca-Mg exchange between clinopyroxene and orthopyroxene (basic and ultrabasic igneous rocks);
- Fe-Ti exchange between ilmenite and magnetite solid solutions (acid and intermediate volcanic rocks);
- ^{16}O-^{18}O isotopic exchange between magnetite and quartz (many igneous and metamorphic rocks);
- NaSi-CaAl exchange between plagioclase and amphibole (amphibolites and some igneous rocks); and
- Ca-Mg exchange between calcite and dolomite (metamorphosed limestones).

Geobarometers for use with metapelitic rocks use net transfer reactions in the assemblages plagioclase-garnet-Al_2SiO_5-quartz; or garnet-rutile-Al_2SiO_5-ilmenite-quartz. MgSi-Al_2 exchange between garnet and orthopyrozene can be used as a geobarometer in basic and ultrabasic rocks. Many continuous reactions offer potential for relative thermobarometry: for example, during growth of zoned crystals. Some discontinuous reactions can be used to estimate minimum or maximum pressure or temperature; for example, the stability of graphite and diamond, or the Al_2SiO_5 polymorphs. Related techniques use exchange reactions to constrain the partial pressures of oxygen and water during rock formation.

Sources of Error in Thermobarometry

Common sources of error in thermobarometric calculations include:

- The assumption that the minerals record equilibrium conditions may not be valid.
- The choice of mineral compositions to use in thermobarometric calculations is easy if all phases are homogeneous, but if minerals are zoned, the selection of compositions to use can be more uncertain.
- The stability of minerals may be affected by the presence of elements that are difficult to analyze with standard techniques. For example, electron microprobe analyses do not distinguish between Fe^{2+} and Fe^{3+}, but the difference is important. A thermobarometric calculation may be substantially in error if all Fe is assumed to be Fe^{2+} for minerals that contain substantial Fe^{3+}.
- Thermobarometric calculations rely on thermodynamic data and a 'calibration' that allows mineral composition to be related to temperature or pressure. The mineral compositions and crystal structures in the rock being analyzed should not be too different from the compositions and structures on which the calibrations are based.

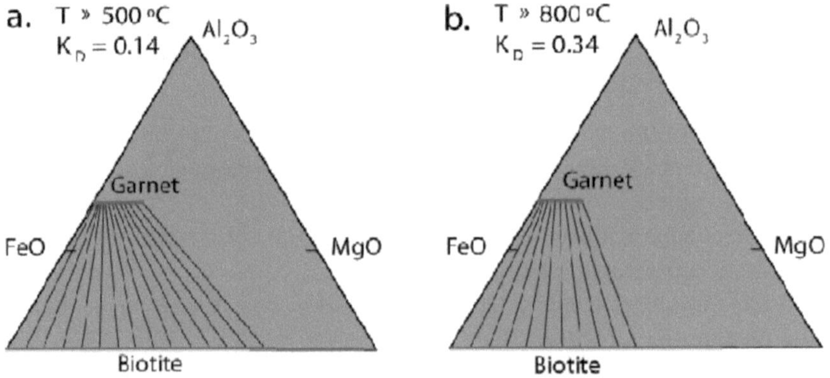

Fig. 12.1: Garnet-Biotite geothermobarometer: AFM projections to show relative distribution of Fe and Mg in garnet vs biotite at (a) 500° C and (b) 800° C.

- Temperature calculations require an estimation of pressure, and pressure calculations require an estimation of temperature. You therefore either need to know one variable, or use a thermometer or barometer that is very insensitive to the other variable. Hence, the best thermometers have nearly vertical slopes on a P-T diagram and the best barometers have nearly horizontal slopes.

12.3 GEOTHERMOMETERS AND GEOBAROMETERS

Geothermometry

Geothermometers are phase or reaction equilibria that depend strongly on temperature and negligibly on pressure (while in case of geobarometers there is significant pressure and almost minor temperature dependencies).

Geothermometric expression

The equation of a state for a solid-solid reaction at equilibrium P and T can be explicitly written as

$$\Delta G_r^{P,T} = \Delta H_S^{P,T} - T \cdot \Delta S_S^{P,T} + P \cdot \Delta V_S^{P,T} + RT\ln \frac{(a_C^{\gamma})^c\,(a_D^{\delta})^d}{(a_A^{\alpha})^a\,(a_B^{\beta})^b} \tag{1}$$

where $\Delta G_r^{P,T}$ = change in the free energy of the reaction; at equilibrium, it is equal to 0; ΔH_S = changes in enthalpy; ΔS_S = changes in entropy; and ΔV_S = changes in volume.

These changes are attributed to the solid phases at specific P and T. R is universal gas constant.

$$\frac{(a_C^{\gamma})^c\,(a_D^{\delta})^d}{\left(a_A^{\alpha}\right)^a\,\left(a_B^{\beta}\right)^b} = K \text{ (distribution coefficient)}$$

The equation (1) can be rearranged and written as

$$T(°C) = \frac{\Delta H_S + P\,\Delta V_S}{\Delta S_S\;R\ln K} - 273.15° \tag{2}$$

Equation (2) is the general form for any geothermometric expression where ΔS is reasonably large and $\Delta V \sim 0$.

Estimation of Geothermometry

Geochemical and isotopic geothermometers developed over the past two decades rely on the assumptions that the two species or compounds coexisted and have equilibrated within the geothermal reservoir that temperature is the main control on their ratio, and that re-equilibration has not occurred during ascent and discharge.

Methods used to determine the geologic thermometry can be broadly classified into two types.

Table 12.1: Estimation methods of geothermometry

Methods	*Types*	*Sub-types*
Non-isotopic	Direct measurement	*Hot springs:* Temperatures range from slightly above the mean annual temperature of the region in which they occur to the boiling point of water at the elevation of the outlets.
		Fumaroles: Temperatures near Vesuvius is about 560° C and up to 645° C in the valley of Ten Thousand Smokes (Alaska).
		Lava: Temperature of extrusion of andesitic and dacitic lavas is about 700 to 900° C and that of basaltic lava is 1200° C.
		Mines, boreholes and wells: Estimates of temperature at the centre of the Earth range from 1600 to 76,000° C.
	Indirect measurements	*Melting points:* The melting point of a mineral, corrected for the pressure under which it was formed, gives a maximum temperature of formation for the assemblage in which it grew because other substances lower the crystallizing temperature for a mineral.
		Transformation temperatures (inversions): Many minerals have two or more crystalline modifications which form, or exist in different temperature range.
		Dissociation and decomposition temperatures: Many minerals break up when they are heated. If one of the products is gas, the temperature of decomposition changes rapidly with pressure; the pressure being known, the mineral can be used as geologic thermometer.
		Solid solutions and exsolution pairs: The mineral assemblages exhibiting solid solution relations and exsolution on cooling can be used as a geothermometer. When minerals in an exsolution relationship occur in a rock or ore, it indicates that the temperature of formation was above their temperature of homogenization.
		Eutectics: The minerals exhibiting eutectic relationship are not widely used as geothermometer. Because it is difficult to be certain that a natural intergrowth of minerals was produced by eutectic crystallization.
		Mineral assemblages: Geothermometric indications may be based on syntheses, including hydrothermal experiments, known stability ranges of the individual minerals of assemblages.
		Inclusions: When minerals are present as inclusions and the associations are cooled at room temperature, stresses are developed in the mineral grains. By the determination of the conditions by which relative compression between host and inclusion is eliminated, conditions of formation can be estimated.
		Electrochemical methods: A cell is made using pairs of minerals that apparently formed in equilibrium and the temperature is varied until electromotive force is zero.

(*Contd.*)

(*Contd.*)

	Fossil assemblages: By determination of temperature of water in which certain types of organisms grow, it is possible to infer the temperature of the water at the time that strata containing fossils of same species or closely related species were laid down. *Properties dissipated by heating:* Properties like thermoluminescence, radiation colours and metamictization, which are exhibited by many minerals, are dissipated by heating. *Crystallography:* It has been postulated that crystals grown at relatively low temperature are simple in habit and those at higher temperature are comparatively more complex. *Fluid inclusions:* By heating plates of minerals containing fluid inclusions on a heating stage on a microscope and determining the temperature at which the solution fills the cavities, it is possible to estimate the temperature of formation if the pressure at formation was essentially the same as the vapour pressure of the solution. *Behaviour of organic material:* When organic compounds are heated in nature, characteristic changes take place that can be used to estimate the temperatures to which various kinds of organic matter have been exposed.
Isotopic	Chemical and isotopic geothermometers are equations or models based on temperature-dependent chemical reactions or isotope equilibrium fractionation relations from which equilibrium temperatures of these reactions can be calculated. Conventional chemical geothermometers based on solutes (Silica Geothermometers, Cation Geothermometers) and gases (Gas Geothermometers) as well as on the atoms in water molecules and dissolved substances (Isotope Geothermometers) in geothermal discharges have been used in the last three decades to predict the temperature at depth of geothermal systems (reservoir temperature), which is a key parameter in geothermal resources assessment. The major drawback of all the conventional geothermometry methods lies in their incapability in making a judgement on the equilibrium status of the studied systems. In this review emphasis is laid on two recent approaches in this field. The use of isotope geothermometers in the low-medium temperature range (<150° C) is discussed, with the oxygen geothermometer based or the sulphate-water pair as an example. The available calibrations of isotope fractionation between species of geochemical interest are reviewed. These calibrations are discussed in the light of simple theoretical considerations, and a new set of thermometric equation is derived for the most common mineral constituents of rocks. Examples of application are given for igneous and metamorphic rocks. The significance of 'isotopic' temperatures is discussed in terms of environmental and kinetic considerations. As an example of possible future developments we derive the relationship between observed isotope temperatures and the speed of ascent of a rock system during a post-orogenic isostatic readjustment.

Table 12.2: Typical age-dating methods for specific minerals

Isotopic methods	*Age equation*	*Minerals to be dated*
Rb-Sr method	$^{87}Sr = {}^{87}Sr_I + {}^{87}Rb\ (e^{\lambda t} - 1)$	Muscovite, biotite, K-feldspar, glauconite
Sm-Nd method	$^{143}Nd = {}^{143}Nd_I + {}^{147}Sm\ (e^{\lambda t} - 1)$	Garnet, clinopyroxene, omphacite
K-Ar method		Sanidine, orthoclase, plagioclase, biotite, hornblende, muscovite, pglogopite, glauconite
^{40}Ar-^{39}Ar method	$^{40}Ar^{*} = (\lambda_{EC}/\lambda_{total}) \cdot {}^{40}K\ (e^{\lambda total\ t} - 1)$	Biotite

Thermometric Methods

Cationic relationships: Studies of Na, K and Ca in aqueous systems suggest that cation concentrations are controlled by temperature-dependent equilibrium reactions with feldspars, mica and calcite. Several semi-empirical equations for the temperature have been determined on the basis of cation ratios (in ppm):

$$T\,°C = \frac{1217}{\log\,(Na/K) + 1.483} - 273 \text{ (Fournier, 1979)}$$

$$T\,°C = \frac{1647}{\log\,(Na/K) + \beta\left[\log(\sqrt{Ca/Na}) + 2.06\right] + 2.47} \quad 273$$

where $\beta = \frac{1}{3}$ for Na waters and $\beta = \frac{4}{3}$ for Ca waters (Fournier and Truesdell, 1973).

At temperatures less than about 200°C, the solubility of magnesium silicate increases and Mg plays a role in the controlling reactions, which must be accounted for (Fournier and Potter, 1979). In lower temperature or high salinity systems, other cations relationships including Na/Li and Mg are important, e.g.:

$$T\,°C = \frac{2220}{\log\,(\sqrt{Mg/Li}) + 5.47} - 273 \text{ (Kharaka and Mariner, 1987)}$$

The principal drawback to these chemical geothermometers is that given time or low water-rock ratios, re-equilibration can occur through exchange reactions at lower temperatures during ascent.

Silica solubility: The increased solubility of quartz and its polymorphs at elevated temperatures has been used extensively as an indicator of geothermal temperatures (Truesdell and Hulston, 1980; Fournier and Potter, 1982). In systems above about 180 to 190°C, equilibrium with quartz has been found to control the silica concentration, whereas at lower temperatures, chalcedony is the controlling phase (Arnasson, 1976). Temperature can be derived from the following relationships for equilibrium with these silica polymorphs from 0 to 250°C, where Si concentrations are in ppm (Fournier, 1981):

Quartz (no steam loss)

$$T^\circ \text{C} = 1309/(5.19 - \log \text{Si}) - 273$$

Quartz (max. steam loss)

$$T^\circ \text{C} = 1522/(5.75 - \log \text{Si}) - 273$$

Chalcedony

$$T^\circ \text{C} = 1032/(4.69 - \log \text{Si}) - 273$$

Amorphous silica

$$T^\circ \text{C} = 731/(4.52 - \log \text{Si}) - 273$$

^{18}O in sulphate-water exchange: Although several isotope exchange reactions can be used as indicators of subsurface temperatures, the exchange of ^{18}O between dissolved sulphate and water is often the most useful because equilibration is rapid at reservoir temperatures greater than about 200°C and pH less than 7, conditions that favour SO_4^{2-}–H_2O exchange. Below reservoir temperatures of 150 to 200°C, the abiotic exchange half-time increases exponentially from several years to some 10^6 years at 100°C and pH 7 (Chiba and Sakai, 1985). Only during bacterial sulphate reduction can ^{18}O exchange take place. Recall from Chapter 6 that during this process $\alpha^{18}O_{SO_4}$ increases asymptotically towards an equilibrium value with the water at the ambient temperature. If sulphate reduction in the discharge area is not important, then the reservoir temperature is often preserved. These two equations provide similar estimates:

$$10^3 \ln\alpha^{18}O_{SO_4\text{-}H_2O} = 3.251 \cdot 10^6\, T^{-2} - 5.1$$

(Mitzutani and Rafter, 1969)

$$10^3 \ln\alpha^{18}O_{SO_4\text{-}H_2O} = 2.88 \cdot 10^6\, T^{-2} - 4.1$$

(McKenzie and Truesdell, 1977)

^{34}S in SO_4^{2-}-H_2S exchange: In geothermal systems where both dissolved sulphate and sulphide are present, the large temperature-dependent fractionation of ^{34}S can be a useful geothermometer. While exchange is fast under acidic conditions, exchange at near-neutral pH can take hundreds of years (Truesdell and Hulston, 1980) and so ascending fluids can preserve their temperature record. The exchange follows the following reaction:

$$10^3 \ln\alpha_{SO_4\text{-}H_2S} = 5.07 \cdot 10^6\, T^{-2} + 6.33$$

(Robinson, 1973)

^{13}C in CO_2-CH_4 exchange: In groundwaters where methane coexists with dissolved inorganic carbon, ^{13}C will exchange through the reaction:

$$CO_2 + 4\,H_2 \longleftrightarrow CH_4 + 2H_2O$$

The temperature of equilibration can be calculated from the relationship of Ellis and Mahon (1977):

$$10^3\,\ln\alpha^{13}C_{CO_2\text{-}CH_4} = -\,8.949\,(10^6\,T^{-2})^2 + 181.264\,(10^6\,T^{-2}) - 90.888$$

or of Richet et al. (1977):

$$10^3\,\ln\alpha^{13}C_{CO_2\text{-}CH_4} = -\,0.62\cdot 10^9\,T^{-3} + 6.616\cdot 10^6\,T^{-2} + 6.04\cdot 10^3\,T^{-1} - 3.08$$

At high temperature, isotopic equilibrium is achieved quickly and is preserved in gases discharging from geothermal systems. However, re-equilibration often takes place as geothermal waters move into shallower zones at lower temperatures during ascent. In low temperature systems, isotopic equilibrium between CO_2 and CH_4 can be achieved during bacterial activity (Truesdell and Hulston, 1980).

H_2 in H_2-H_2O exchange: The H_2-H_2O pair is one of the few that give consistent agreement. Equilibration is faster for this exchange reaction than other geothermometers. Consequently, calculated temperatures are often closer to discharge temperatures than reservoir temperatures. Calculations can be made using a combination of the temperature equations for H_2 fractionation between H_2O vapour and water, and between H_2 and H_2O vapour.

Some widely-used geothermometers

Exchange thermometers

Garnet-biotite, Garnet-cordierite; Garnet-clinopyroxene; Garnet-hornblende; Garnet-orthopyroxene; Garnet-olivine; Biotite-tourmaline; Garnet-phengite; Garnet-ilmenite

Solvus thermometers

- *Two-pyroxene geothermometry:* It is based on the distribution of Ca and Mg between coexisting clino- and ortho-pyroxene.
- *Calcite-dolomite:* It is based on the distribution of Ca and Mg between coexisting calcite and dolomite.
- *Two-feldspar geothermometry:* It is based on the distribution of K and Na between coexisting alkali and plagioclase feldspar.
- *Muscovite-paragonite:* It is based on the distribution of K and Na between coexisting muscovite and paragonite.

Special cases

Aluminium in Quartz as a geothermometer: Quartz forming in aluminium-saturated environments has a trace aluminium content which varies linearly and directly with its temperature of crystallization. The aluminium content varies at a rate of approximately 1 ppm Al per 3.6 °C change in temperature.

Use of this geothermometer gives temperatures which are consistent with those expected by geologic reasoning or through the use of other geothermometers.

Clinopyroxene geothermometry of spinel lherzolites: Experiments in the system CaO—MgO—Al_2O_3—SO_2 have shown that the calcium tschermak's molecule ($CaAl_2SiO_6$) content of clinopyroxene in the spinel-lherzolite assemblage (forsterite + clinopyroxene + orthopyroxene + spinel) varies sympathetically with the $MgSiO_3$ content with changes in temperatures and pressure. Such variations are predominantly temperature-dependent, although both pyroxene end-members decrease slightly in amount with increase in pressure at constant temperature. Contrary to O'Hara (1967), the isopleths of $CaAl_2SiO_6$ and $MgSiO_3$ of clinopyroxene in temperature-pressure space are parallel, rendering these two pyroxene molecules indicators only of the temperature of equilibrium of spinel-lherzolites.

No information on the pressures of equilibrium can be obtained.

Using simple thermodynamic mixing models of the solid solution phases participating in the subsolidus exchange reactions, two temperature estimations can be made of natural spinel-lherzolites from the experimental data of the synthetic system. The first temperature estimation is based on the equilibrium exchange of $Mg_2Si_2O_6$ between clinopyroxene and orthopyroxene at 12 kbar (equilibrium constant K_1) and the second temperature estimation is based on the equilibrium exchange of $CaAl_2SiO_6$ between clinopyroxene, olivine, and spinel at 12 kbar (equilibrium constant K_2) (Herzberg, 1976).

These temperatures probably bear little, if any, relationship to those of a normal geothermal gradient.

Geobarometry

Measurement of the pressure at which a rock or suite of rocks reached thermodynamic equilibrium is termed as geobarometry.

Conditions for Geobarometry

Vapour-absent, solid-solid reactions involving breakdown of phases support for geobarometry. Large volume changes and small entropy changes characterize these reactions.

Geobarometric Expression

Continued from Eq (2)

$$P\,(\text{bar}) = \frac{-\,\Delta H_S + T\,[\Delta S_S\ \Delta R \ln K]}{\Delta V_S} \tag{3}$$

The Eq (3) is the general form for geobarometric expression where V is large as compared to S.

Some widely-used geobarometers

- Olivine-orthopyroxene-quartz
- Orthopyroxene-plagioclase-garnet-quartz
- Clinoopyroxene-plagioclase-garnet-quartz
- Olivine-plagioclase-garnet
- Clinoopyroxene-plagioclase-quartz
- Plagioclase-garnet-Al_2SiO_5-quartz
- Garnet-Al_2SiO_5-rutile-quartz
- Orthopyroxene-garnet

Special Cases

Quartz as geobarometer: As crystals grow from solution they may trap small amounts of the solvent and other solutes. It has long been known that minerals evolve gases on heating and it has been assumed that, except for alteration during weathering, the gases are incorporated from the growth medium at the time of crystallisation. If the amount and composition of the gases trapped in minerals are to be used to make detailed geological interpretations it is essential that the location of the gas within the crystal structure is established. In minerals like carbonates and hydrates, 'gases' form part of the structure. Sites in which gas can occur include primary and secondary fluid inclusions, point defects, lattice dislocations, grain boundaries, and structure holes. Gas can also be adsorbed on crystal surfaces. Each site has its own characteristic temperature at which gas loss occurs, so that on heating a mineral the gases are released at different temperatures. This provides a method for distinguishing between gases from different sites.

Olivine-orthopyroxene-spinel oxygen geobarometer: The common upper mantle assemblage olivine-orthopyroxene-spinel may be used to calculate the oxygen fugacity at which mantle-derived peridotites have equilibrated. The equilibrium has been calibrated using the large amount of existing data on the thermodynamic properties of each phase in this assemblage. A by-product of this procedure is a new calibration of the olivine-spinel Mg-Fe^{2+} exchange geothermometer. Application of the equilibrium to a variety of peridotite xenoliths indicates that the oxygen fugacity of the upper mantle lies between the quartz-fayalite-magnetite (QFM) and wüstite-magnetite (WM) oxygen buffers; the few apparent exceptions to this rule may be due to analytical error, particularly in the Fe^{3+} content of the spinel phase. In fact, the determination of Fe^{3+} in spinel is at present the limiting factor in the accurate application of the method: within this limitation, the presently available evidence suggests that the oxygen fugacity of the mantle may be laterally homogeneous over wide regions, but may also show small differences between these regions. The fluid species in the system C—H—O at such oxygen fugacities are predominantly CO_2 and/or H_2O, and not CH_4/H_2.

12.4 SELECTION OF GEOTHERMOMETER AND GEOBAROMETER

For geothermometric and geobarometric calculations, it is important to state that at low temperatures, mixing in multi-component solid solutions gets gradually deviated. Hence inaccuracies in the activity-composition relationships are propagated to P-T estimates from geothermometric and geobarometric formulations. The effect is clearly manifested at temperatures less than 500° C and pressures less than 4 kbar. At higher pressures and temperatures, like those of the lower crust and the upper mantle, thermobarometric relationships are complicated due to adverse effects of pressure on solid solutions and volumes of phases. As a result P-T estimations of such rocks are of lower order.

For any given mineral assemblage there may be several equilibria that can be used for geothermometry and geobarometry and each one of these equilibria may have several calculations. When the calibration calculation part is over (usually done by computer programs), it is commonly observed that for a single set of mineral compositions there is a range of temperatures calculated from the different thermobarometer calibrations. To overcome this difficulty, that is to determine the appropriate calibration, the convenient way is to examine the consequences of using different calibrations of a single thermometer or barometer. This procedure has been able to give satisfactory results in many cases.

12.5 GEOTHERMOMETRIC AND GEOBAROMETRIC PROBLEMS

Geothermometric and geobarometric applications to solve petrological problems can lead to various difficulties in such calculations. Such common problems are:

- P-T estimation from unreserved phase equilibrium experiments can give erroneous results. Such error can be seen in case of Al_2SiO_5 polymorphs. The transitionary P-T conditions during the change from andalusite to kayanite to sillimanite vary from one author to another. Thus the estimates from coexisting phases based on these various assumptions would lead to contradictory PT values.
- Over extrapolation of experimental conditions, in terms of temperature, pressure and chemical composition result in errors in PT estimates. It can be shown with the help of the experiment:

$$3CaAl_2Si_2O_8 = Ca_3Al_2Si_3O_{12} + 2Al_2SiO_5 + SiO_2$$

Although it is widely found assemblage, but the grossular content of these garnets is seldom more than five moles per cent. Since the geobarometric formulation of this assemblage is not constrained with respect to low concentration of grossular, pressure estimates based on this reaction should be checked against more dependable geobarometers, and random and systematic errors should be ascertained.

- Geothermometers and geobarometers which are quite sensitive to compositional diversity are likely to register scattered PT values around its mean, mainly because of: the effect of other components in the solid solution, analytical uncertainties and uncertain Fe^{2+}/Fe^{3+} ratios in mafic phases.
- Many granulite areas, such as Adirondacks, Inari complex and South Indian granulites, have been observed to have spatial gradients in P and T. In such cases PT estimates expressed as mean of PT values and its standard deviation is erroneous. But since most of the geothermometric formulations have uncertainties of + 60° C and the pressures obtained from geobarometers have uncertainties of +1.5 kbar, present-day geothermobarometers are unlikely to detect gradients in P and T of this order.
- Blocking is one of the serious problems in geothermobarometric formulation which is needed to be addressed separately. Prior to the exposure of the rocks on the surface, they have undergone complex processes. During its ascent a rock undergoes progressive changes in pressure and temperature, to which the phases respond by continuous readjustments in terms of its chemical composition. With decreasing temperature and pressure the energy requirement for element partitioning by cation-diffusion are seldom reached. As a result reequilibration among phases is blocked. Since the geo thermometers and geobarometers record PT values of the last equilibrating event, it can be easily assumed that these reactions will register pressures and temperatures depending upon the availability of energy budget. For example reactions based on intra-crystalline exchange will proceed to lower temperatures while those based on inter-crystalline exchange will be blocked at higher temperature due to its high energy requirement. Blocking effect is more pronounced in case of net-transfer reactions.
- From the preceding discussion it can be concluded that the sensitivities of different geothermometers and geobarometers are different to composition and P-T variations. Therefore it is quiet natural that all the P-T sensors will register unique value for the same rock. Hence comparison of PT conditions of formation of rocks in different areas should be based only on those thermometric and barometric formulations which have been found to yield realistic PT values. Moreover conditions should be restricted to PT values obtained from the same geothermometers and geobarometers.

Concluding Remarks

Methods of mineralogical geothermobarometry are widely employed in studies of igneous and metamorphic formations, providing the means of determination of evolution of physicochemical conditions of metamorphism in time and space. The P-T trends of this process reflect to a certain extent, the geothermal gradient line position of the region under study in the course of the formation development. That, when different-age objects are studied, makes it possible to follow their displacement in the geologic history.

Think for a while

1. Explain the use of a mineral as geothermobarometric indicator with suitable diagram.
2. With respect to geological field, elaborate the concept of geobarometry.
3. Explaining their behaviour and significance, cite some specific examples of mineral thermometers and mineral barometers.
4. In geothermometric estimation, explain the advantage of isotopic methods over non-isotopic methods.
5. Thermobarometric processes can be erroneous.What are the possible errors commonly encountered?

FURTHER READING

Althaus, E. Experimental evidence that reaction of kyanite to form sillimanite is at least bivariant. *American Journal of Science*, **267,** 273-277, 1969.

Boettcher, A.L. et al. Geothermometry-Geobarometry. *Amer. Mineral.*, **61(7 and 8),** 549-816. 1976.

Chatterjee, N.D. Applied Mineralogical Thermodynamics. Springer-Verlag, 1991.

Gupta, Aloke K. and Sisir K. Sen (eds). A Short Course on Elementary Thermodynamics for Earth Scientists. Department of Geology and Geophysics, University of Allahabad. 1986.

MacGregor, L.D. The effect of CaO, Cr_2O_3, Fe_2O and Al_2O_5 on the stability of spinel and garnet peridotites. *Earth Planet. Inter.*, **3,** 372-377. 1970.

Smith, D. Lherzolite inclusions from Green Knobs, McKinfey County, New Mexico (abstr.). *Geol. Soc. Am. Abstr.*, **6,** 1066. 1974.

Spear, F.S. Metamorphic Phase Equilibria and Pressure-Temperature-Time Paths. Monograph of the Mineralogical Society of America. 799 p. 1994.

Urey, H.C. The thermodynamic properties of isotopic substances. *J. Chem. Soc.*, 562-581, 1947.

Part III

Mineral Analysis, Industry and Environment

CHAPTER

13

COMMON ANALYTICAL TECHNIQUES IN MINERALOGICAL STUDIES

Analytical instruments now-a-days become an essential part of mineralogical studies. With increasing instrumental sophistications, the details of mineral interactions are opening up. An idea of these studies is briefed in the present chapter.

The study of complexity of mineral matter e.g. composition, effect of impurities, unmixing, ordering/disordering, micro inclusions, trace elements, rare isotopes, surface states and adsorbed molecules, ultrahigh temperature and pressure effects etc. involves various analytical techniques.

For any particular technique one should know: 1. Brief history. 2. Current uses. 3. How it works: (a) physical and chemical principles and (b) Instrumentation (modular approach), description of each major component and its operation. 4. What it does—analytical information—qualitative, quantitative,

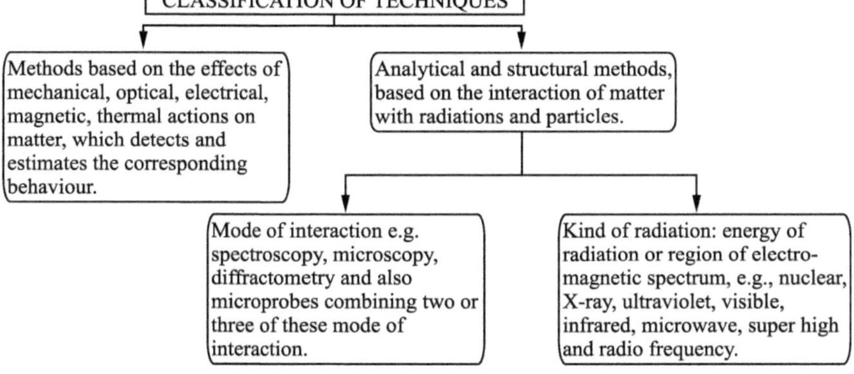

Fig. 13.1: Classification of techniques.

accuracy, precision and detection limits of the data. 5. Applications—general and specific cases. 6. Limitations. 7. Complementary and related techniques. 8. Sample: state, amount, preparation process and needed time. 9. Total analysis time. 10. Recent developments.

Table 13.1: Principles of methods of investigation

Radiation	*Interactions*		
	Microscopy	*Spectroscopy*	*Diffractometry*
Optical	Yes	Yes	No
Infrared microwave	Yes	Yes	No
X-ray electron	Yes	Yes	Yes
Nuclear	Yes	Yes	No
Superhigh frequency	No	Yes	No
Radio frequency	No	Yes	No
Thermal (heat)	Thermal analysis		

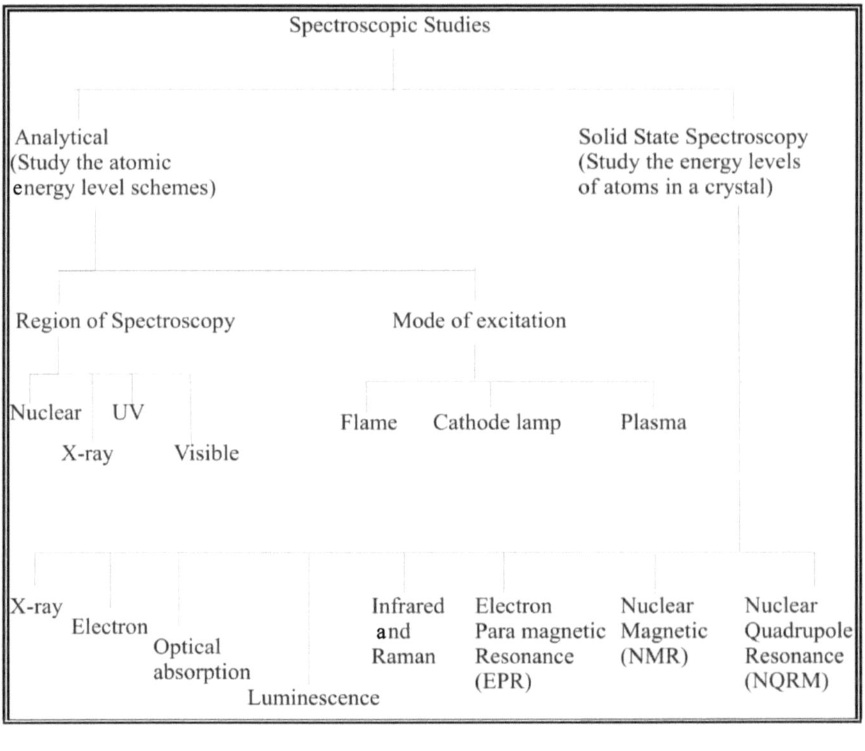

Fig. 13.2: Different types of spectroscopic studies.

Briefly, it can be summarised as TEM and SEM used for high resolution imaging study. Techniques suitable for chemical analysis are Microprobe, Atomic absorption, XRF and Mass Spectrometer. For crystallographic analysis, X-ray (and also Neutron) diffraction to be used are particularly important for clay studies. Various spectroscopic methods are there for investigating structural environments e. g. IR and Raman spectroscopy are used for molecular vibration studies, NMR to determine Al-Si distribution in minerals and glass, Mossbauer to estimate iron oxidation state or distribution over atomic sites.

In this chapter the following analytical techniques will be discussed:

1. Electron microscopes – SEM, TEM, EPMA
2. Mass spectrometry
3. Thermal analysis
4. X-ray analytical techniques – XRD, XRF
5. Atomic absorption
6. IR-spectroscopy

13.1 UNDERSTANDING THE PROCESS

General process of any analytical technique has the following important features:

- General applications
- Specific use
- Samples: (i) state, (ii) amount and (iii) preparation technique
- Analysis time
- Limitations
- Complimentary or related techniques.

For any particular technique, the additional required details are the following:

(A) Brief history
(B) Current use
(C) Procedure: (i) Physical and chemical principles; (ii) Instrumentation (modular approach); (iii) Description of each major component and (iv) Operation procedure.
(D) Methods—Analytical information: (i) Qualitative; (ii) Quantitative; (iii) Accuracy and precision; and (iv) Detection limits.
(E) Application domains: (i) General and (ii) Specific examples.

13.2 ELECTRON MICROSCOPE

In this microscope, electron beams are used replacing the visible light to increase the magnification and resolution of the system. Microscopes are instruments by means of which we see magnified image of very small objects which are not ordinarily visible to the eye with clear definition.

Simple Microscope

Small objects like small types, small prints etc. which are not clearly visible to the eye, may be seen distinctly with the help of a simple microscope or a magnifying glass.

Magnification, m = Angle subtended at the eye by the image/Angle subtended at the object placed at the position of the image.

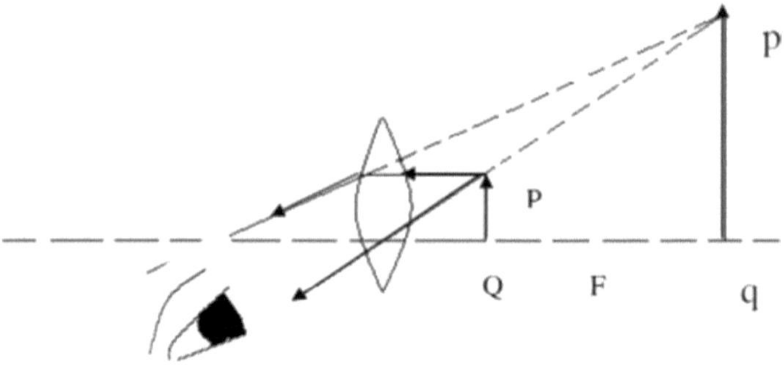

Fig. 13.3: Operation mechanism of simple microscope.

Compound Microscope

Resolving power: The smallest or limiting distance which can be seen separated or resolved by a microscope is given by

$$d = \lambda/2\mu \sin \theta$$

Thus by diminishing the wavelength λ of the incident light, increasing the refractive index μ of the object space and by increasing the inclination θ of the incident rays with the axis of the objective we can diminish the value d. That means the finer details of the object can be obtained by making d small. This explains the advantage of immersion objective (in which $\mu > 1$) over dry objective (in which $\mu = 1$). $\mu \sin \theta$ is called numerical aperture (N.A) of the objective.

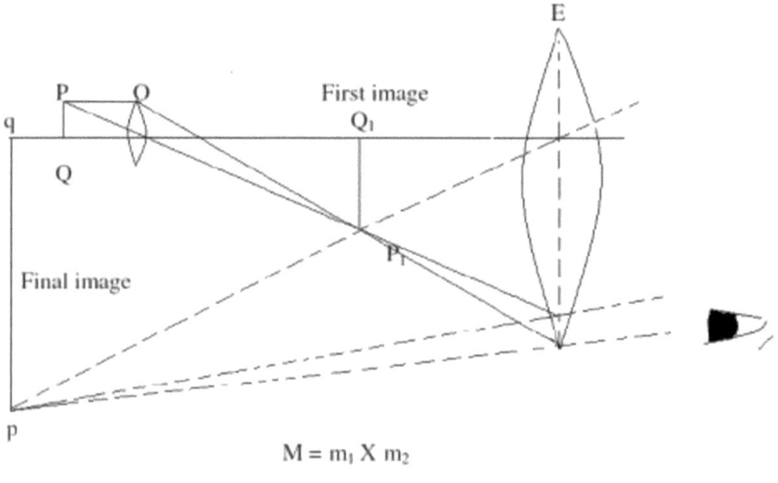

Fig. 13.4: Operation mechanism of compound microscope.

The maximum value of N.A for any lens system used in air is slightly less than unity; by using oil immersion objectives, for which μ is greater than 1, N. A can be made nearly equal to 1.5. Consequently the best we can do is to have

$$d = 1/3\ \lambda$$

Thus the actual value of *d* depends on the wavelength of the light used.

With optical microscopes, where the average wavelength of light used is 550 nm, a magnification of 500× to 1000× and a resolution of 0.2 can be achieved. With electric beam, with much shorter λ, the resolution is around 10Å and magnification is 1,000,000×.

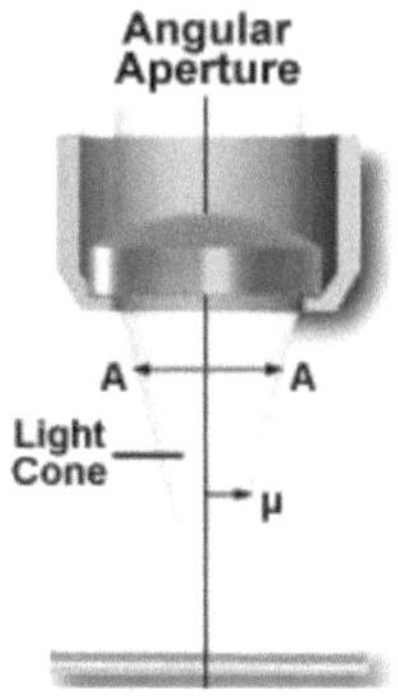

Fig. 13.5: Angular aperture.

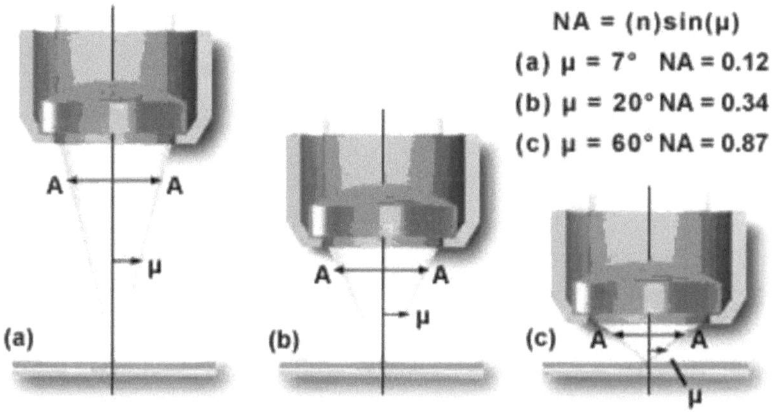

Fig. 13.6: Dependence of *d* values on the wavelength.

The various types of electron microscopes are:

- Transmission Electron Microscope (TEM) invented in 1931
- Scanning Electron Microscopy (SEM) invented in 1942
- Reflection Electron Microscope (REM)
- Scanning Transmission Microscope (STEM)
- Low Voltage Electron Microscope (LVEM)

The main observations made with these microscopes are:

1. *Topographic analyses*, i. e. surface features, its textures, relation between these features and material properties (hardness, reflectivity etc.).
2. *Morphological studies:* Shape and size of the particles in the object and relation between these structures and material properties, e.g. ductility, strength, reactivity etc.
3. *Composition determination:* Elements and compounds present in the sample and their relative abundances. These help to understand the relation between composition and material properties like melting point, reactivity, hardness etc.

4. *Crystallographic information:* It relates the properties like conductivity, electrical properties, strength etc. with the crystal structure of the compounds.

Working Principle

- An electron source (electron gun) produces electrons and is accelerated towards the sample using a positive electrical potential.
- A focussed and monochromatic beam is obtained using metal apertures and magnetic lenses.
- Another magnetic lens focusses this beam to the sample.

The irradiated sample starts interacting during this inelastic electron scattering process of the incident beam. Effects of these interactions are detected and transformed into an image.

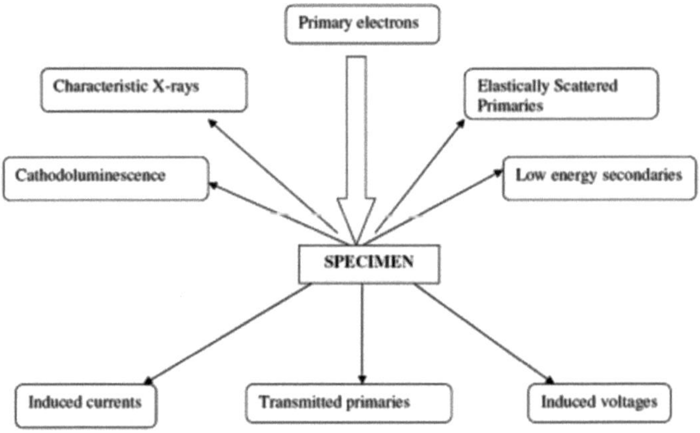

Fig. 13.7: Electron-specimen interaction in an electron microscope.

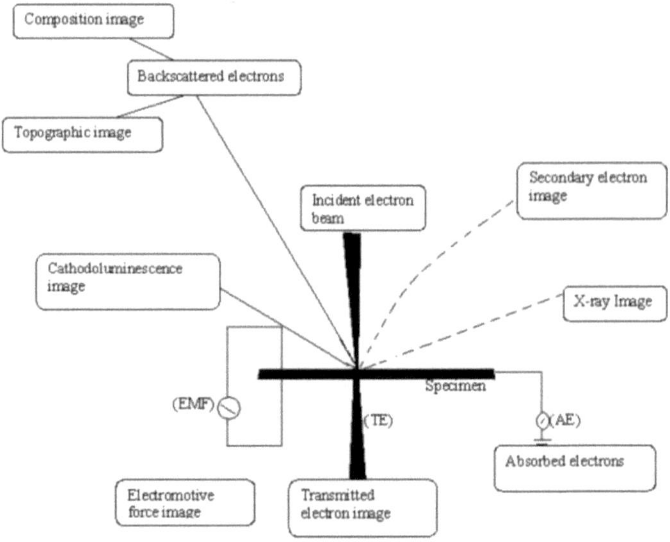

Fig. 13.8: Different images with signals in an EM.

Transmission Electron Microscope (TEM)

In the Transmission Electron Microscope (TEM), the electron beam passes through the sample specimen and the interactions are recorded for analysis and further study.

Working principle is same as that of an optical microscope but instead of light beam and optical lenses, electron beam and electromagnetic lenses are used. The basic operational principle is described in Fig. 3.9. Three essential parts of the system are:

1. An electron gun produces the electron beam and the condenser system focusses the beam on to the object. The source of electrons, the cathode is heated, V-shaped tungsten filament or in high performance instruments, a sharply pointed rod of lanthanum hexaboride. The filament is surrounded by a control grid, sometimes called Wehnelt cylinder, with a central aperture arranged on the axis of the column.
2. The image producing system, consisting of the objective lens, moveable specimen stage and intermediate and projector lenses which focus the electrons passing through the specimen to form a real, highly magnified image.
3. The image recording system, which converts the electron image into some form perceptible to the human eye. It usually consists of a fluorescent screen for viewing and focusing the image and a camera for permanent records. In addition, a vacuum system, consisting of pumps and their associated gauges and valves and power supplies are required.

Working Principle

Source of Illumination: A tungsten filament is heated to 2,700°C, in a vacuum to produce a beam of electrons (LM uses photons). Electrons behave in the same manner as light when in a vacuum. The entire microscope column is under vacuum otherwise the electrons would collide with air molecules and be absorbed.

Scanning Electron Microscopy (SEM)

This instrument uses a high energy focussed electron beam to scan the solid surface of the sample. The interaction between the electrons with the sample yields information about the chemical composition, external morphology, crystalline structure and orientation of the material making up the sample. Using conventional SEM (magnification 20 × 30,000) 1 cm to 5 microns width can be scanned with a special resolution of 50 to 100 nm). Selected point location analysis is also possible which is particularly useful to estimate the chemical composition, crystalline structure, crystal orientations qualitatively or semi-quantitatively.

SEM is very similar in design and functioning with EPMA.

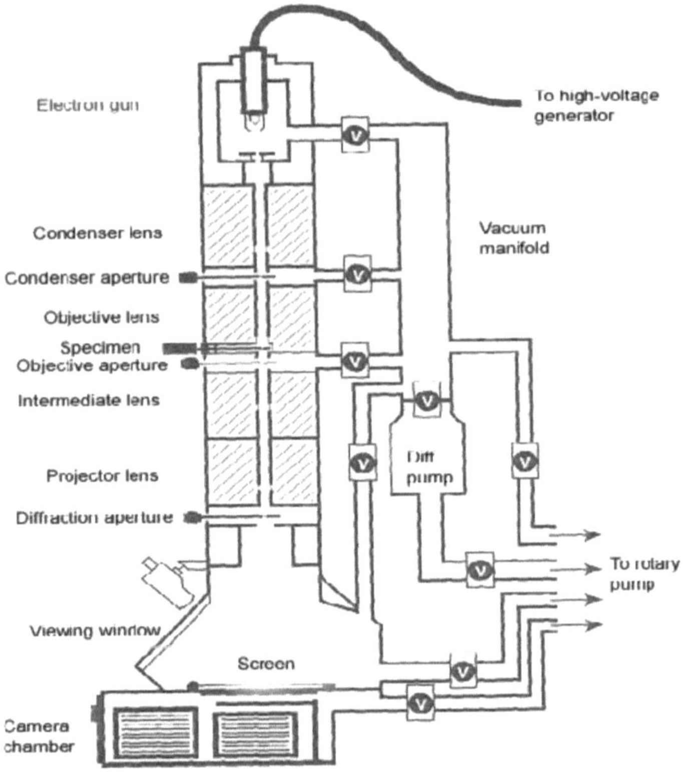

Fig. 13.9: Transmission Electron Microscope (TEM).

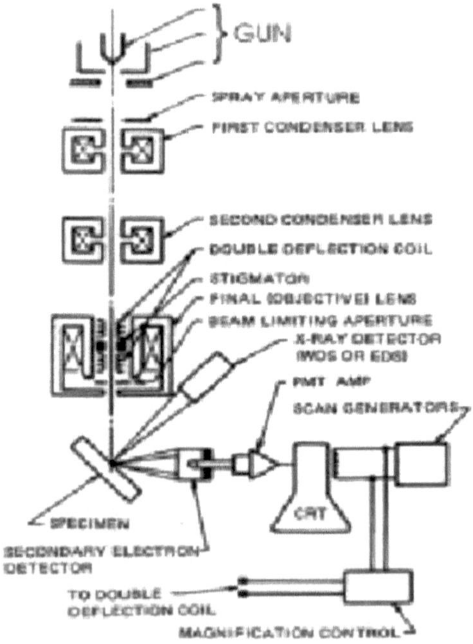

Fig. 13.10: Scanning Electron Microscopy (SEM).

Basic Principle

Accelerated electrons struck the solid sample got decelerated by transferring the energy. Consequently varieties of signals are produced e.g. secondary electrons producing SEM images, diffracted back-scattered electrons that gives information about the crystal structure and orientations of different minerals in the sample, photons e.g. characteristic X-rays which identify the elements present in the sample.

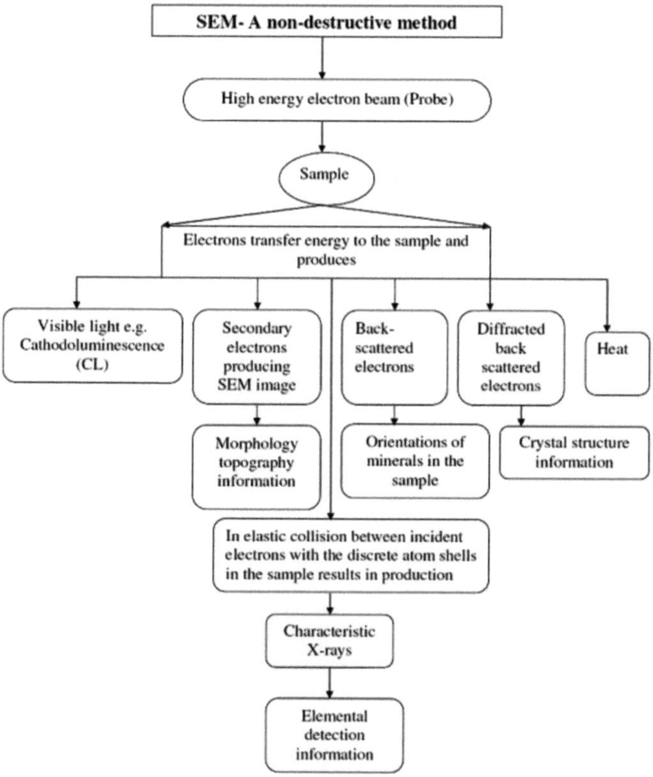

Fig. 13.11: Operational mechanism of SEM.

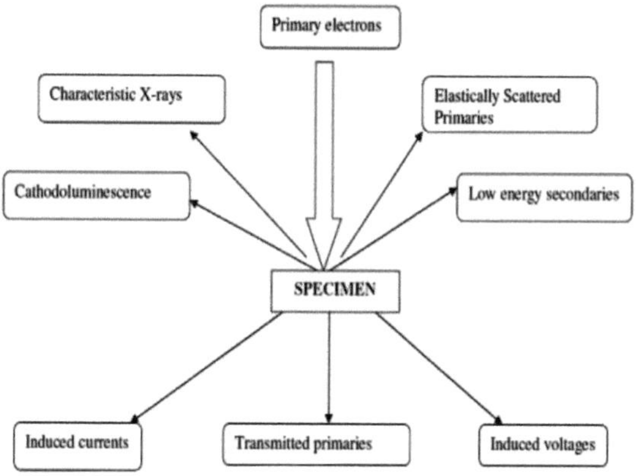

Fig. 13.12: Different branches related to SEM.

Applications

- Produces high resolution image of the shape of the object.
- Spatial variation in chemical composition e.g. elemental map or spectral chemical analysis using Energy Dispersive Spectrum (EDS).
- Differentiating phases based on mean atomic number (related to relative density) using BSE.
- Compositional maps showing the variation of different trace elements using CL.
- In a multiphase sample, backscattered electron (BSE) image distinguishes and discriminates the different phases.
- Very small features of objects (50 nm in size) can be analyzed using SEM.
- Qualitative chemical analysis and also crystallographic information can be determined using SEM.
- Microfabric and crystallographic orientation in materials can be studied in detail using diffracted backscattered electron images (EBSD).

Advantages

To characterize a solid material, in particular geological samples, SEM is one of the very useful tools. The instruments designed and available nowadays are user friendly, minimal sample preparation process and the available digital data are easily portable for further use.

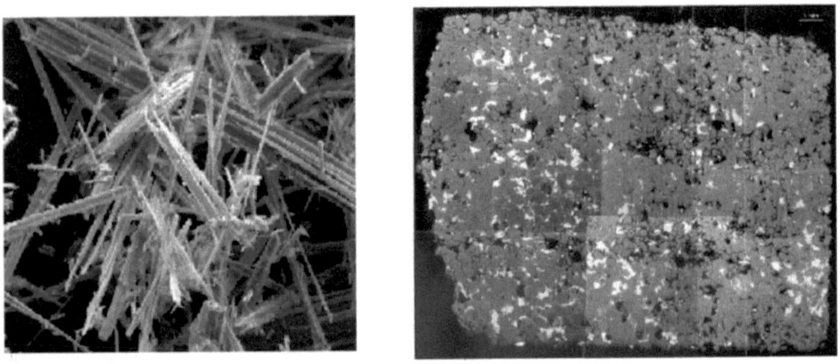

Figs. 13.13 and 13.14: The sample characteristics in typical SEM analysis with data.

Limitations

- Sample size is restricted by the microscope chamber size ~ 10 cm × 40 mm.
- Sample must be stable with a vacuum ~ 10^{-5} to 10^{-6} torr. Due to problem of outgasing at such a low pressure (rocks saturated with hydrocarbons), swelling clays, coal etc. are unsuitable for conventional SEM. With specialized SEM (environmental and low vacuum), these sample analyses are now possible.
- EDS detectors cannot detect elements with atomic number <11. Those solid state X-ray detectors are fast but relatively poor energy resolution and sensitivity is low for low abundant elements compared to wavelength.

Spectrum processing:
Peaks possibly omitted: 0.605, 0.930, 6.404, 15.645 keV
Processing option: All elements analyzed (Normalised)
Number of iterations = 1
Standard:
Au Au 1-Jun-1999 12:00 AM

Element	*App Conc.*	*Intensity Corrn.*	*Weight %*	*Weight % Sigma*	*Atomic %*
Au M	102.90	1.0000	100.00	0.00	100.00
Totals			100.00		

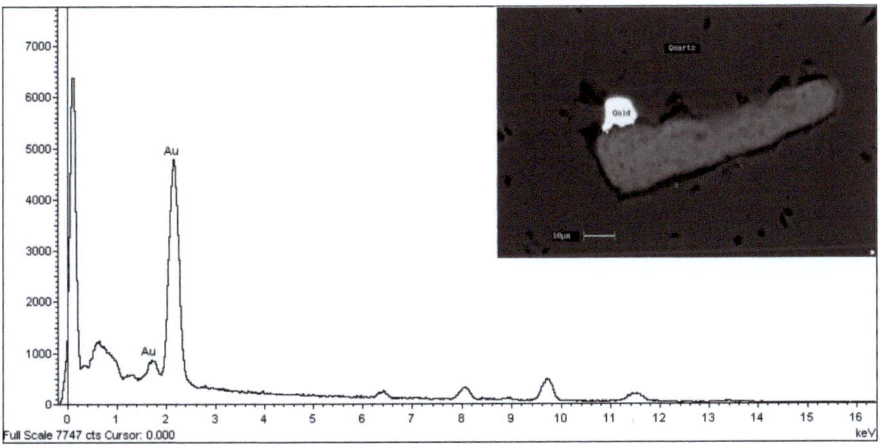

Fig. 13.15: Gold flake identified with EDX in quartz vein.

Spectrum processing:
Peak possibly omitted: 6.398 keV
Processing option: All elements analyzed (Normalised)
Number of iterations = 2
Standard:
Sb Sb 1-Jun-1999 12:00 AM
Pt Pt 1-Jun-1999 12:00 AM

Element	*App Conc.*	*Intensity Corrn.*	*Weight %*	*Weight % Sigma*	*Atomic %*
Sb L	51.03	0.8265	54.96	0.44	66.16
Pt M	42.29	0.8357	45.04	0.44	33.84
Totals			100.00		

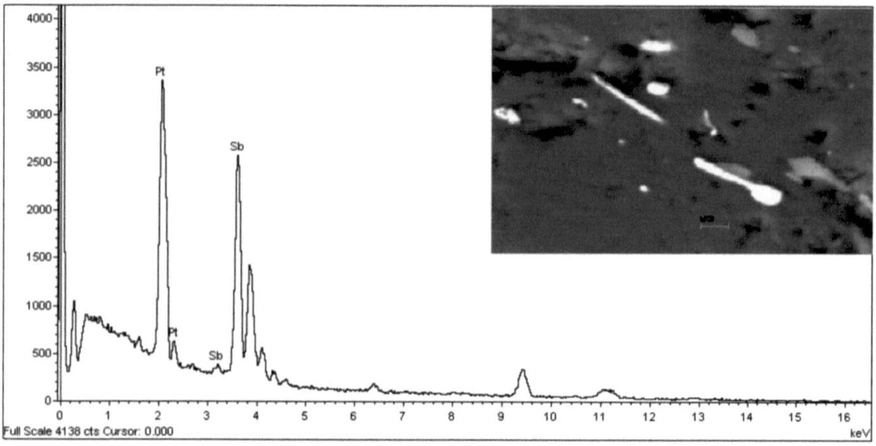

Fig. 13.16: SEM data of a specific sample in lab.

Spectrum processing:
Peak possibly omitted: 0.266 keV
Processing option: All elements analyzed (Normalised)
Number of iterations = 3
Standard:
S FeS_2 1-Jun-1999 12:00 AM
Fe Fe 1-Jun-1999 12:00 AM
Cu Cu 1-Jun-1999 12:00 AM

Element	*App Conc.*	*Intensity Corrn.*	*Weight %*	*Weight % Sigma*	*Atomic %*
S K	36.40	0.8949	38.83	0.35	54.11
Fe K	30.22	0.9726	29.67	0.34	23.74
Cu K	29.20	0.8850	31.50	0.42	22.15
Totals			100.00		

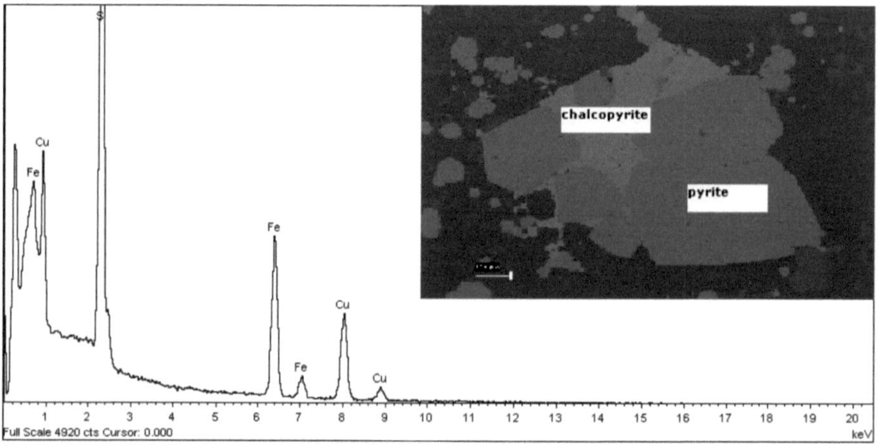

Fig. 13.17: SEM data of a specific sample in lab.

Electron Probe Micro Analysis (EPMA)

Electron Probe Micro Analyzer determines the chemical composition of a tiny material and also is able to image it with very high resolution compared to the optical one. This is a combination of SEM and XRF analytical systems with the feature of fine spot focussing (~ 3 μm) optical microscope imaging and precision sample positioning. This non-destructive method was developed by R. Casting during his Ph. D thesis in Paris in 1950. Commercial models started coming up from 1960. A schematic diagram is shown in Fig. 13.18.

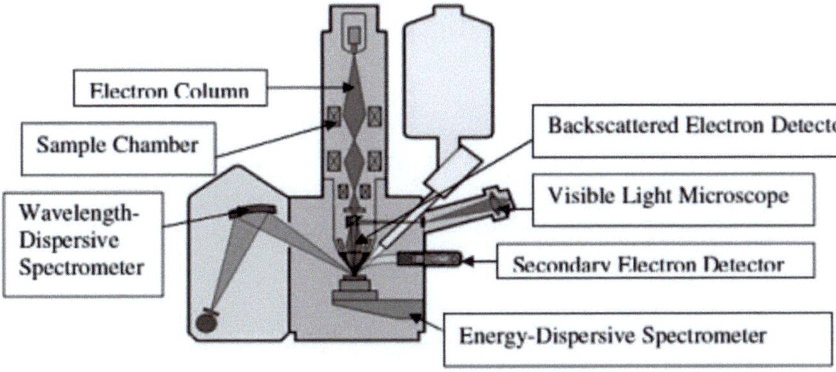

Fig. 13.18: Schematic diagram of Electron Probe Micro Analyzer.

The probing high energy electrons, produced from the electron gun consist of a tungsten filament cathode. The beam is focussed on the sample using the electromagnetic lenses located in the beam column, in a similar way around 5-30 KeV with the electron current of 10-200 nanoamps (about 100 times greater than SEM). The focussed beam radius on the sample is ~ 3 μm. The sample is taken in the form of a polished surface and must be vacuum compatible, size is in the range from 0. 5 mm × 0. 5 mm × 0. 5 mm to about 1 cm × 5 cm × 5 cm, bounded to the mounting block with epoxy. Non-conducting samples, a sputter coat of carbon develops the needed conduction property. Analysis locations are selected using a transmitted light optical microscope and the positioning can be made accurate within 1 μm (not possible in SEM).

Theory of Operation

When the high energy electrons fall on the sample, they ionize the inner shells of the elements in the sample. The decay of the ionized state results in the emission of characteristic X-rays (the basic principle of XRF analysis).

Using suitable analyzers (e.g. LiF crystal) and detectors (proportional counters), the energy spectrum of X-rays (EDS) identifies the elements present in the sample in the range of atomic numbers from $z = 9$ to $z = 92$ (fluorine to uranium). The qualitative detection is simple and fast with the detection limit about 300 ppm. For quantitative analysis, the intensities of the X-rays emitted from the sample is recorded and compared with standards. Applying the necessary conditions, using computer software, the results are obtained in wt% of atomic proportions.

To scan different locations in a sample, either the beam or the sample positions can be changed and the compositional variation may be observed and detected.

For light elements like C, N, O and F the detection limit and sensitivity is about 500-1000 ppm and for heavier ones, from Na to U, it is about 300 ppm. The spatial resolution for quantitative analysis is around 1-3 micron. Secondary and backscattered electron imaging depends on accelerating voltage and beam current ~100-200 nm. The film thickness can also be estimated if the density of the film is accurately known.

Fig. 13.19: Image showing exsolved clinopyroxene (CPX) and Cr-spinel (Cr-Sp) within host orthopyroxene (OPX) of lherzolite rock of ophiolites.

Applications

- Quantitative analysis of micrograins is possible. With a range of crystal spectrometers, this instrument gives better results of quantitative analysis than SEM (using Wavelength Dispersive System; WDS) at high sensitivity.
- Elemental scanning over an area of the sample can easily be done to view its distribution. Spot chemical in situ analysis can detect the compositional variations in zoning and textural studies. With an array of imaging detectors the image of the surface and internal compositional structure can be generated.
- Individual mineral phases in tiny amount can be studied and amalyzed, e. g. igneous and metamorphic minerals, sedimentary cement, volcanic glass, matrix of a meteorite etc.

- It analyzes synthetic minerals like optical wafers, thin films, microcircuits, semiconductors and super-conducting ceramics.

13.3 MASS SPECTROMETRY

The mass spectrometer is an instrument that will sort out charged gas molecules (ions) according to their masses. The name mass spectrometer/mass spectrograph is chosen by analogy with optical spectroscopy, because early instruments produced a photographic record resembling an optical line spectrum.

Instrumentation part consists of basically five distinct steps e.g. (1) ionization of the sample, (2) acceleration of the ions by electric field, (3) dispersion of the ions according to their mass to charge ratio, (4) detection of the ions to produce a corresponding electrical signal and (5) the study of ion trajectories in the analysis tube to be maintained under vacuum (10^{-4} Pa) which is of primary importance. It allows the determination of mass/charge ratio and eventually the nature of the ions.

This *Destructive Method* is extremely sensitive one.

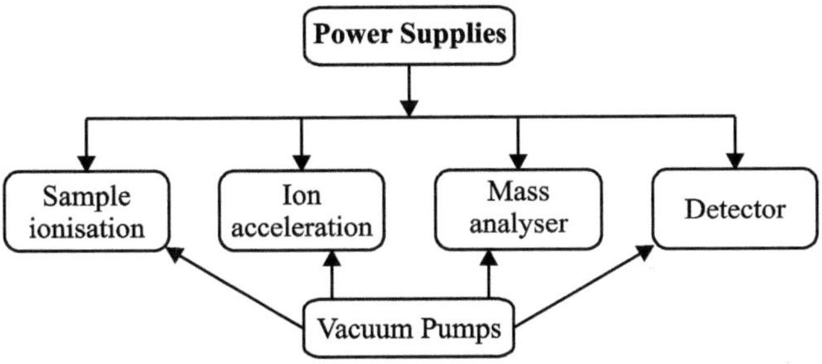

Fig. 13.20: Block diagram of a typical mass spectrometer.

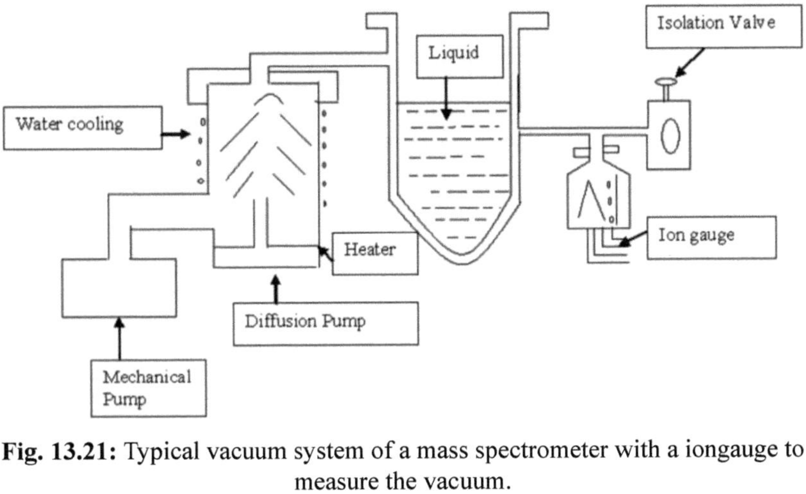

Fig. 13.21: Typical vacuum system of a mass spectrometer with a iongauge to measure the vacuum.

Ion Sources

Materials to be converted to gases at reduced pressure in the mass spectrometer was a problem till 1970. Later on the techniques for producing gaseous ions, solids and high boiling liquids have been introduced. Once the sample has been converted to a low pressure gas, ions can be formed by bombardment with a stream of electrons from a hot filament and is known as Electron Impact (EI) ionization. Ions can also be formed by a secondary process, originating in an ion molecule reaction in the gas phase, called Chemical Ionization (CI).

Gaseous ions can be obtained from solid samples by *Desorption*. The ionization energy is delivered to the surface of the sample by a beam of ions or neutral atoms. The impact causes a miniature explosion in which a small fraction of the surface molecules are ejected and ionized at the same time. Desorption can also be produced by a laser beam or by an application of an electrostatic field.

During mass spectral analysis, the sample undergoes the following process:

1. *Ionization:* The sample is vapourized and ionized in the ion source of the instrument by one of many possible ways (depending on the sample). At this stage molecules of the compound will lead to a statistical distribution of fragment ions.
2. *Acceleration:* Immediately after their formation ions are readily extracted from the ion source, focussed by a series of electronic lenses and accelerated to their kinetic energy.
3. *Separation:* The mass analyzer will filter ions according to their mass/charge ratio. Some instruments combine many mass analyzers in series.
4. *Detection:* After separation, ions terminate their path and strike the detector which measures electrical charge and amplifies the weak ionic current.
5. *Result:* Treatment of the resulting detector signals leads to display of the mass spectrum.

In this method, the final result gives mass/charge (m/q) ratio of the ions. To know m, q must be known $= ze$, where z is a small integer and e, the electronic charge. For singly charged ions ($z = 1$), increasing order of mass is the same as m/z. This is when the spectra are graduated in atomic units (u) or Daltons (Da).

Ion abundance can be recorded by two ways:

(a) The continuous spectrum of a selected mass or interval of masses; in this mode, ion signals appear as peaks of different widths depending on the instrument. Higher performance units can determine ion masses with a precision better than 10 ppm (= 10^{-5} Da).

(b) The fragmentation spectrum (bar spectrum) is produced by summing the intensities of ions at the nominal masses, closest to their exact mass and the intensities of ions are usually expressed as a percentage of the most intense peak, called the base peak.

Mass Analyzers

Passage of charged particles through a magnetic field results in a circular trajectory of radius given by $r = (mv)/zB$, where v = accelerating voltage and B = magnetic field. It can be shown geometrically that a homogeneous beam of ions diverging from a slit can be brought to a focus by a magnetic field in the shape; this is the basis of a magnetic sector mass spectrometer.

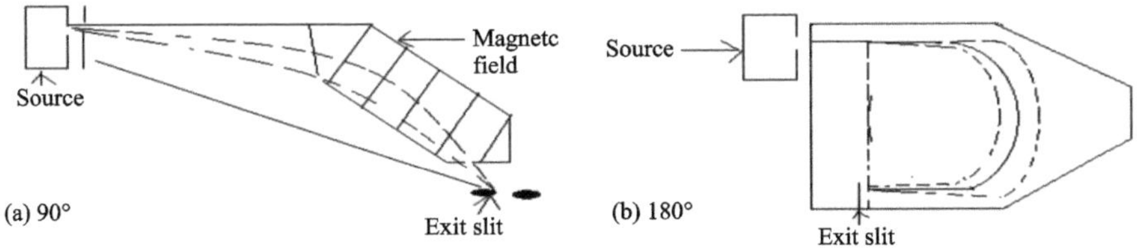

Fig. 13.22: (a) Sector at 90° and (b) Sector at 180°.

Resolution of the sector mass analyzer is limited by the spread of kinetic energies and also for the *not so sharp* boundaries of the magnetic field. Improving these two factors the resolution can be improved. The magnetic boundary effect can be eliminated through the use of a 180° sector where both the source and the detector are inside the magnetic field, but it has some other technical disadvantages.

In Fig. 13.23, the Bainbridge spectrograph with a magnetic deflection of 180° and a velocity filter is shown. This filter eliminates to have a heterogeneous mono-kinetic beam. A photographic recording of Ne spectrum is shown. The two series are due to stripping of 1 and 2 electrons from each isotope of Neon. m/z $^{20}Ne^{+}$, $^{21}Ne^{+}$, $^{22}Ne^{+}$, $^{20}Ne^{+}$, $^{20}Ne^{+P}$, etc. Here $m/z = RBe/v$ and $v = E/B$.

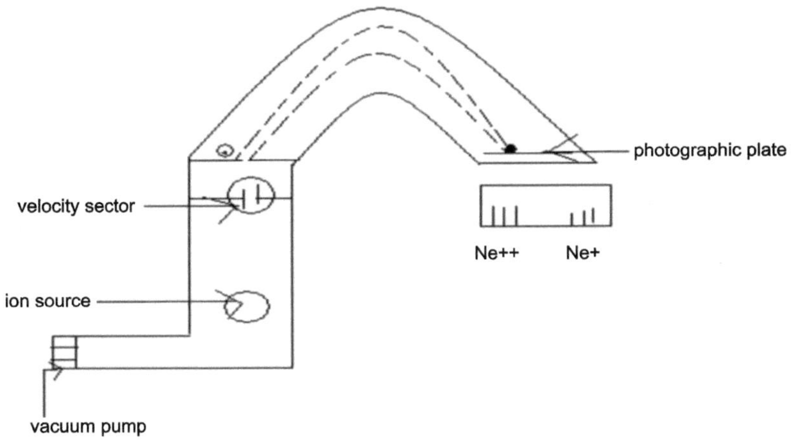

Fig. 13.23: Bainbridge spectrograph.

The process followed by a modern mass spectrometer is shown in Fig. 13.24.

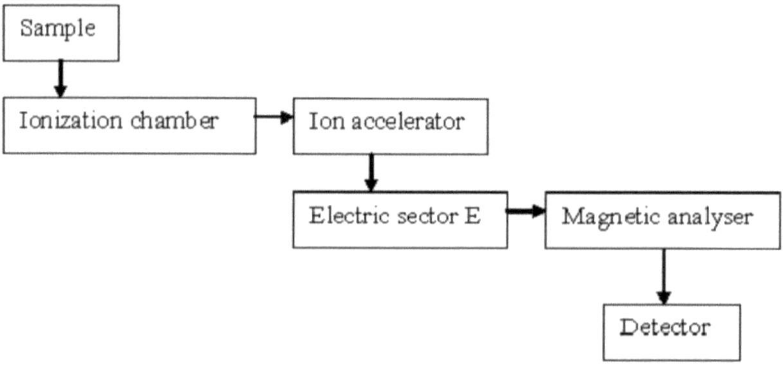

Fig. 13.24: Process followed by Modern Mass Spectrometer.

Double-focus instrument: The beam of ions can be made isoenergetics using an electric sector. The beam assumes a circular path in passing through the annular space between two concentric cylindrical electrodes. The radius r and the applied field E determine the energy of the particles that can pass as $1/2mv^2 = (ZEr/2)$

or $$(m/z) = (Er/v^2)$$

Fig. 13.25: A cylindrical electrostatic sector energy filter.

This shows that the mass number of ions passing through the sectors can be controlled by the applied potential E. This is also called the energy filter and can be used together with a magnetic sector to form a double focussing mass spectrometer, i.e. focusses the ions both in energy and mass.

Magnetic Analyzers

In II (Fig. 13.26) the radius of curvature in the magnetic field is in the same direction as that in the magnetic field. In III, the radius of curvature in the magnetic field is in the opposite direction of the electric field and the system is dispersive one.

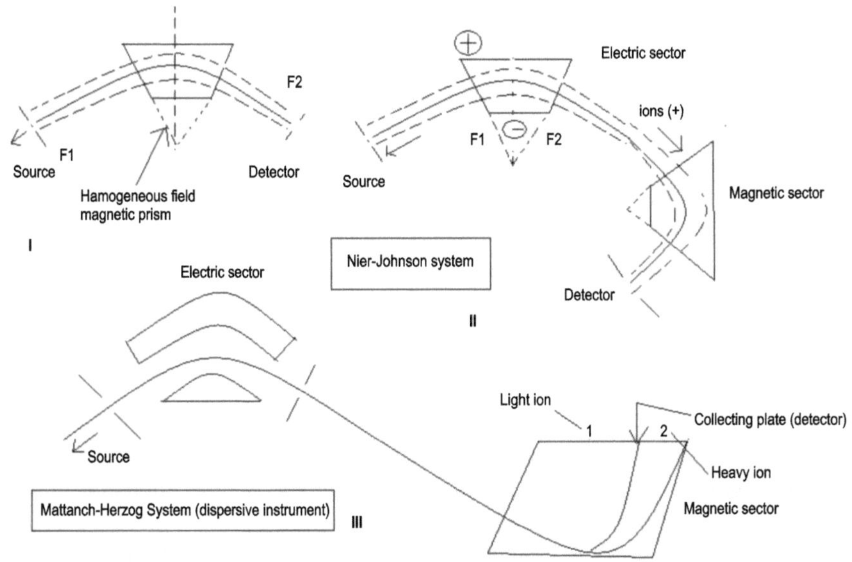

Fig. 13.26: I. Magnetic analyzer; II. Nier-Johnson system and III. Mattanch-Herzog system.

Geological Use of Mass Spectrometry

Estimating isotopic abundances (like $^{13}C/^{12}C$) for geological samples, its time of formation can be estimated. This is called "Geochronology" where the analytical data obtained from mass spectrometry will be used to unfold the formation history of a rock.

The heavy isotopes of oxygen and hydrogen give the source to hot spring water and ore-forming solutions; the isotopes of sulfur and carbon serve as indicators of the origin of salt dome sulfur deposits; and the isotopes of strontium serve as a possible guide to the history of granitic magma.

Isotopes means proton number Z is same but neutron number N is different. Chemical properties largely depend on the valence electron and so the isotopes have very similar chemical properties of the element. Some elements (e.g. F, Na, Co) have single natural isotopes, some have at least two and some have more, even eight or so (Sn, Te). Bombarding atomic nuclei with fast moving particles additional isotopes of all elements except hydrogen can be prepared.

Isotopes may be radioactive i. e. the atomic nuclei give out radiations and thereby change to nuclei of other elements. The radiations are (i) α-particles ($2p + 2n$); (ii) β-particles (e^+ or e^-) and (iii) γ-radiation i.e. highly energetic electromagnetic radiation. Some radioactive isotopes are continually being formed by reactions of stable nuclei with high energy particles in the atmosphere (^{14}C). The great majorities of naturally occurring isotopes is not radioactive and are called stable isotopes.

The accumulation of radiogenic isotopes by the decay of radioactive parent elements provided a powerful tool to estimate and establish the absolute ages of earth materials. Study of radiogenic isotopes of Sr and Pb yields information on the differentiation of planetary substance both in the early part of the history and later episodes. Distribution of stable isotopes has provided a very useful tool for measuring temperature and drawing inferences about sources of rocks and fluids (geothermometry).

The isotope ratios can yield a variety of geological information from absolute ages to temperature to sources of rocks and fluids and thus has become an interesting point of study. The isotope separation and abundance measurement is the aim of mass spectrometer analysis.

Radioactive and Radiogenic Isotopes

Because radioactive elements decay at a constant rate and is unaffected by temperature, pressure or chemical combinations found in geological environments, it can be made to serve as a clock for the measurement of geological time. Once a radioactive isotope is imprisoned in a structure of a growing crystal, its atoms decay to its daughter elements at a fixed rate. The ratio of daughter to parent thus steadily increases and measurement of the ratio gives a number from which the time elapsed since the crystal was formed can be calculated. This is based on the assumption that the mineral has not been altered since its formation. In such cases, ages are to be calculated from more than one parent-daughter pair. When alteration is suspected, e.g. metamorphism during the rock's history, the isotopic ratio, although not giving a precise age, often provides useful information about the time and nature of metamorphic event.

Consider the case of Rb-Sr dating system, consider Rb-Sr rich minerals (lepidolite, muscovite, biotite, K-feldspar) of Palaeozoic age or older. Rb forms no minerals of its own, but substitutes extensively for K in these minerals because of the close similarity of Rb^+ and K^+.

13.4 THERMAL ANALYSIS

In thermometric methods, some property of the system is measured as a function of temperature or where measurement of the heat evolved or absorbed by a reaction provides analytical information.

When, in a thermal reaction, heat is absorbed, it is called an endothermic one. Dehydration, decomposition, phase changes and crystalline inversions are generally *Endothermic*. When, in a thermal reaction, heat is evolved, it is called an *Exothermic* one. Crystallization, oxidation and some other chemical reactions are exothermic.

Other recently developed analytical method is Thermo-mechanical Analysis (TMA) which measures the temperature of deformation under a static load. Thermo-dilitometry determines the changes in the dimensions under a static load. The same instrument TMA is used for both measurements.

Table 13.2: Different types of thermometric methods

Sl No	*Name*	*Property measured*	*Apparatus*
1	Thermogravimetry (TG)	Change in weight	Thermobalance
2	Derivative Thermogravimetry (DTG	Change in weight	Thermobalance
3	Differential Thermal Analysis(DTA)	Heat evolved or absorbed	Differential thermometer
4	Differential Scanning Calorimetry (DSC)	Heat evolved or absorbed	Differential thermometer
5	Thermometric Titration	Change of temperature	Adiabatic calorimeter
6	Direct Injection Enthalpimetry	Heat evolved or absorbed	Adiabatic calorimeter

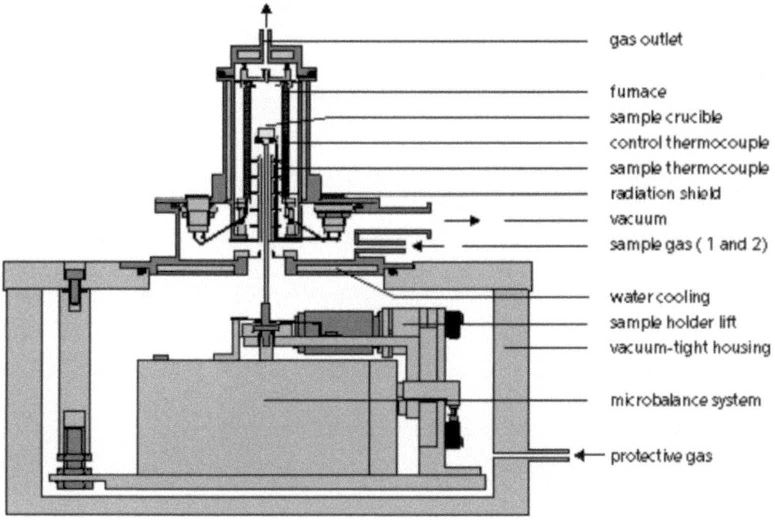

Fig. 13.27: Thermogravimeter (TG).

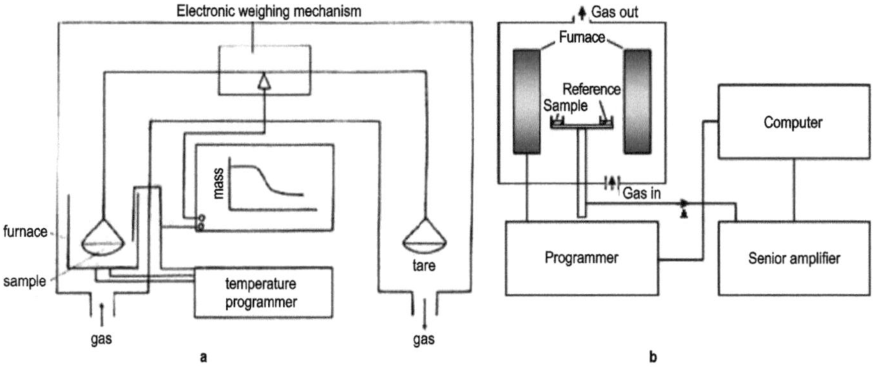

Fig. 13.28(a-b): Differential Thermal Analysis (DTA).

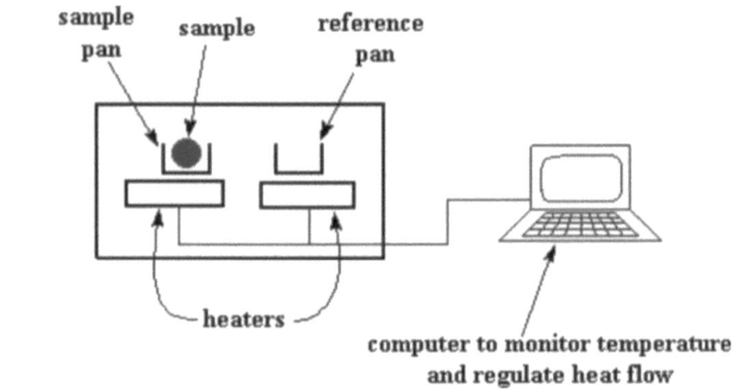

Fig. 13.29: Differential Scanning Calorimeter (DSC).

In Fig. 13.27, a sketch of the essentials of a thermal analyzer is shown. The crucibles A and B are for reference material (usually alumina Al_2O_3) and the sample to be analysed. They are placed over a set of thermo-couples, measuring their temperatures and two more thermo-couples to measure the furnace temperature. The whole system is placed in a furnace where the environment can be controlled using purge gas. The total system is kept in a thermally isolated chamber.

When some thermal reaction starts occurring in the sample, the difference of the heat content in A and B is measured and recorded. A typical thermogram is shown below for anglesite ($PbSO_4$) from 50°C to 1000°C. Differential thermal curve for anglesite containing about 5% of pyrite contamination is shown in Fig. 13.30. The thermal reactions recorded are: *A* = oxidation of pyrite

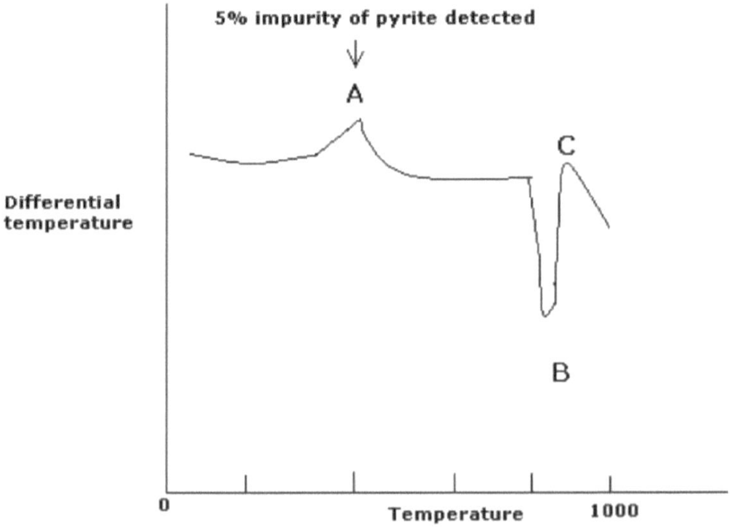

Fig. 13.30: Thermo-analytical curve for anglesite ($PbSO_4$) with small amount of Pyrite impurity.

exothermic) *B* = crystallographic inversion of anglesite (endothermic) from orthorhombic to monoclinic structure. Like many inversions, the change is rapid and hence the peak is sharp. Base line (where no reactions are taking place) is easily seen and can be extrapolated under the peaks. *C* = decomposition of anglesite, initially slow but becoming fast within a very short time.

Samples: Usually solid grain or powder can be studied. It allows for the study of materials as made for investigation of the effects of various processing conditions (e.g. for polymers). Sample amount is 1 mg to several grams depending on the test.

For non-homogeneous samples, size is an important factor in obtaining a representative sample. Sample preparation depends on the purpose of study. But for common geological samples, a powdered sample is to be taken. Effect of thermal history (preheating etc.) is to be taken into consideration.

Analysis time: for DSC scans, 20 to 40 min and for TG 30 min to several hrs. For TMA and dielectric measurements, usually of the order of hrs. and similar time is needed for kinetic studies.

Heating rate: Effects of slow heating rate—(i) little baseline drift, (ii) near equilibrium conditions, (iii) broad shallow peaks on ($\Delta T/t$) curves and (iv) short scanning time.

For normal work, heating rate from 8°C-12°C/minute, unless specified. 10°C/min is more or less a standard one. This gives peaks of a satisfactory size, overlap of neighbouring peaks is not excessive and the time per determination is good. The heating rate must be linear and this linearity is to be checked frequently.

Moreover, the temperature calibration of the furnace is also an important factor and may be calibrated using some standards e.g. using solidification of tin (on cooling at 231.83°C, melting of potassium dichromate at 397.5°C, $\alpha \leftrightarrow \beta$ transition of quartz at 573°C, melting of KCl at 770.8°C, melting of sodium sulphate at 884°C etc.

Reporting conditions of the experiment are:

1. Identification of all substances/sample, reference etc. by a name, empirical formula or equivalent compositional data.
2. Details of the sample source, pretreatment and chemical purity.
3. Heating/cooling rate applied during experiment, total range of temperature scan, temperature involving the phenomena of interest.
4. Identification of the sample atmosphere e.g. pressure, composition and purity. Whether it is a static atmosphere, self generated or dynamic or over the sample. In ambient condition, humidity to be mentioned, if not, the controlling details (rate of flow of purge gas, name of the gas etc.).
5. A statement of the methods used to identify the intermediate and final product.
6. Whenever possible, each thermal effect should be identified and supplementary supporting evidences, if available, to be quoted.

7. Sample wt. and dilution of the sample to be noted.
8. Time-temperature recording in the chart is now-a-days directly available from the attached computer in the instrument, where the convention of upward deflection as a +ve temperature differential and downward deflection as a –ve temperature differential w.r.t. reference is usually maintained.

Qualitative and Quantitative Estimation

Qualitative: From known tables of thermal reaction, the qualitative data can be estimated. Depending on some same related parameters, the peak shape may vary a little bit, thus affecting the peak temperature. It is usually avoided by choosing the "Characteristic temperature" for it is less subject to variation of peak temperature.

Quantitative estimation is done noting the peak area under consideration and the baseline plays an important role.

Simultaneous DTA-TGA diagram (Fig. 13.31) shows the decomposition of $CaC_2O_4 . H_2O$ in air and in CO_2. In DTA, the temperature difference between a sample and an inert reference material is monitored while both are subjected

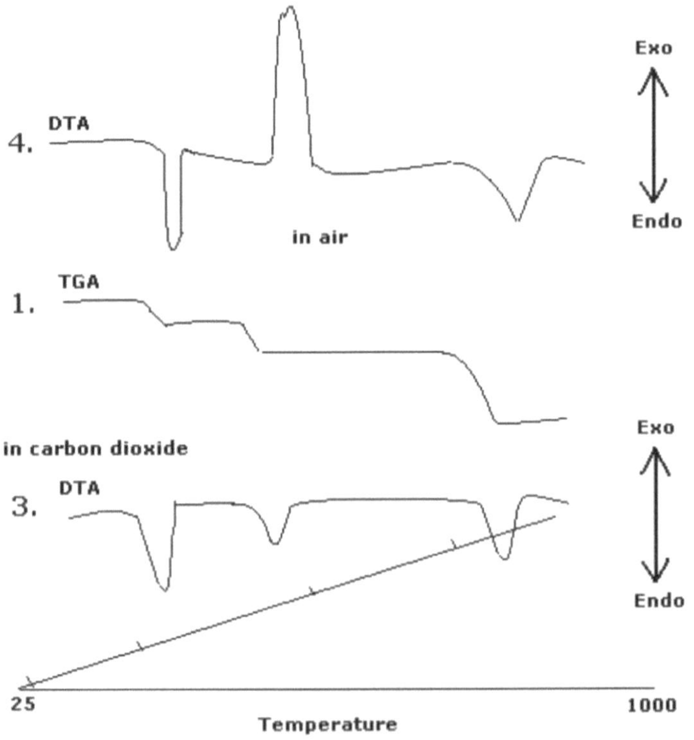

Fig.13.31: DTA-TGA diagram showing the decomposition of $CaC_2O_4{\cdot}H_2O$ in air and in CO_2.

to a linearly increasing environmental temperature. In two crucibles, the sample and reference material is kept and placed in the furnace with a temperature measuring system for each. The thermal stability of the reference material over the working temperature range is very important. The TG and DTA curves (Fig.13.31) show: TG curve records the weight loss and curve 3 is a DTA-one showing the decomposition of CaC_2O_4 in an atmosphere of CO_2. The three points of wt. loss correspond to three endothermic processes, as it needs energy to break the bonds in the successive elimination of H_2O, CO and CO_2. But, by contrast, the 2nd peak in curve 4 (in air) is sharply exo-one caused by the combustion of CO in air at the furnace temperature. DTA curve reveals the energy changes, regardless of change or consistency of atmosphere. This change or consistency of wt. is reflected in a DTG curve. In the thermograms of the $SrCO_3$, a pronounced endotherm in DTA (at 950°C) does not show in TGA or DTG. At this point the crystalline transition from rhombic to hexagonal modification of $SrCO_3$ is occurring without any weight change and hence no reflection in TGA or DTG. At ~1200°C, the decomposition of $SrCO_3$ is reflected in TGA and DTG.

Differential Scanning Calorimetry (DSC)

DTA gives good qualitative results showing the temperature and signs of transition, but it is difficult to obtain quantitative information about the sample or the heat of transition. This problem arises as the thermal conductivity and specific heat of the sample is unknown before and after transition. The heating rate and some other instrumental parameters affect the areas of endotherm/ exotherms.

To overcome this, DSC has been designed which is an isothermal calorimeter. Each sample holder (sample and reference both) is provided by own heat. Whenever some thermal reactions will start, depending on the type (endo/exo), the other one (either sample or reference) is compensated. It is heat content by supplying extra energy from its own heater. Thus, this extra power is exactly covering the area of the thermal reaction curve.

The decomposition of $CuSO_4.5H_2O$: Three endotherms correspond to the loss of 2, 2 and 1 molecules of water respectively. In the DTA curve the temperature curve is somehow perturbed by the thermal events occurring in the sample. No such effect is possible in DSC and the peaks are more regular in shape.

A scanning calorimeter provides a convenient means for the precise determination of impurity limits in highly purified compounds, through the observation of melting point depression. The average deviation of 14% (in old instruments) can be detected. But in recent ones, more accuracy is attained.

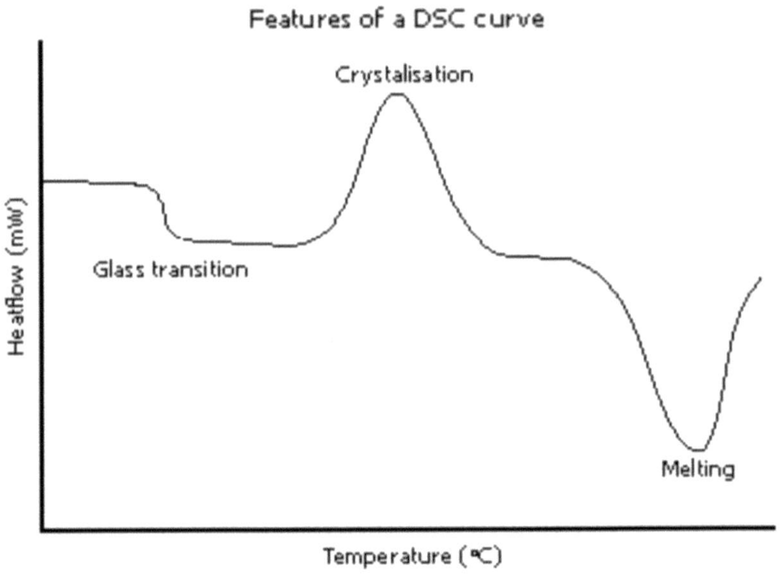

Fig. 13.32: DSC curves.

13.5 X-RAY ANALYTICAL METHODS

Table 13.3: Electromagnetic wave frequencies and wavelengths

Nature of waves	*Wavelength (in metre)*	*Frequency (in Hertz)*
Radio waves	1 to 1000	3×10^5 to 3×10^7
TV waves	1 to 10	3×10^7 to 3×10^8
Microwave, Radar	1×10^{-3} to 1	3×10^8 to3×10^{11}
Infrared waves	8×10^{-7} to 1×10^{-3}	3×10^{11} to 4×10^{14}
Visible light	4×10^{-7} to 7×10^{-7}	4×10^{14} to 7×10^{14}
Ultraviolet rays	1×10^{-8} to 4×10^{-7}	7×10^{14} to 3×10^{16}
X-rays	5×10^{-7} to 1×10^{-8}	3×10^{16} to 6×10^{19}
Gamma rays	1×10^{-13} to 5×10^{-12}	6×10^{19} to 3×10^{21}
Cosmic rays	$<1\times10^{-13}$	$>3\times10^{21}$

Emission of X-rays

X-rays, discovered by W.C. Roentgen in 1895, are generated by large energy transition of electrons within an atom from outer orbitals to core orbitals. Most X-rays are created by bombarding a metal target with energetic electrons. The electron beam typically ejects core electrons from the target metal, making the atoms unstable. The atoms relax from this position by dropping an outer valence electron to the core level. This large decrease in energy required for an electron to be able to drop to the core level requires the emission of the excess energy in the form of a photon. These photons have the energy of X-rays.

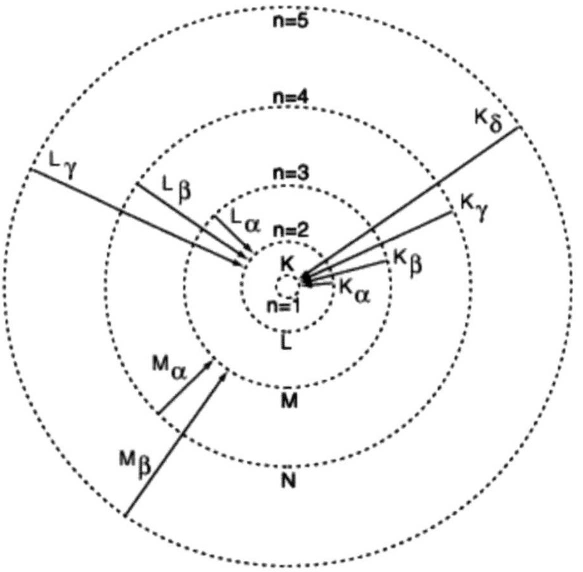

Fig. 13.33: X-ray radiations in an atom.

X-ray Transitions

Electron transitions to lower atomic levels in atoms have quantum energies which place them in the X-ray region of the electromagnetic spectrum. The X-ray emissions associated with these transitions are called characteristic X-rays. The labels on the illustration (Fig. 13.33) show the historical labelling of characteristic X-ray transitions.

Bremsstrahlung X-Rays

Accelerated charges give off electromagnetic radiation, and when the energy of the bombarding electrons is high enough, the radiation is in the X-ray region of the electromagnetic spectrum. It is characterized by a continuous distribution of radiation and becomes more intense and shifts toward higher frequencies (lower wavelength) when the energy of the bombarding electrons is increased.

The bombarding electrons can also eject electrons from the inner shells of the atoms of the metal target, and the quick filling of those vacancies by electrons dropping down from higher levels gives rise to sharply defined characteristic X-rays, which are over the bremsstrahlung background, as shown in the Fig. 13.34.

When vacancies are created, say for $n = 1$, i. e. K-shell of the atom, electrons drop down from above to fill the gap. From $n = 2$ to $n = 1$ transitions yields K-α and those from $n = 3>1$ transitions are called K-β X-rays. Transitions to the $n = 2$ or L-shell are known as L-X-rays. The continous distribution at the base is known as Bremsstrahlong radiation.

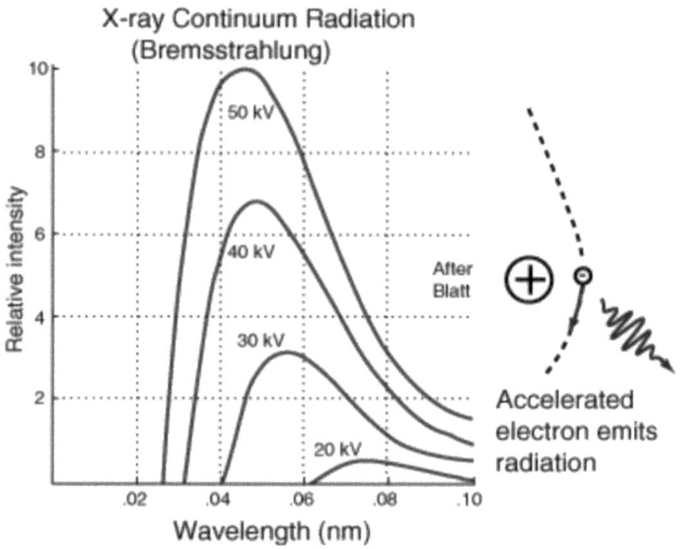

Fig. 13.34: Bremsstrahlung X-rays.

Characteristic X-Rays

The frequencies of the characteristic X-rays can be predicted from the Bohr model. Moseley measured the frequencies of the characteristic X-rays from a large fraction of the elements of the periodic table and produces a plot of them which is now called a "Moseley plot".

Characteristic X-rays are emitted from heavy elements when their electrons make transitions between the lower atomic energy levels. The characteristic X-rays emission which are shown as two sharp peaks in the illustration (Fig. 13.35) occur when vacancies are produced in the $n = 1$ or K-shell of the atom

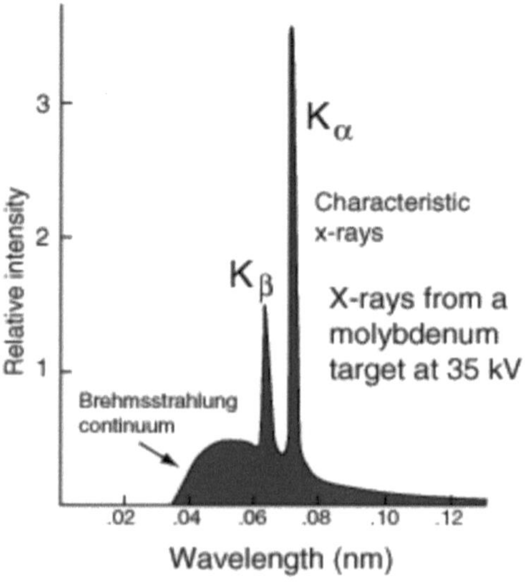

Fig. 13.35: Charcteristics of X-rays.

and electrons drop down from above to fill the gap. The X-rays produced by transitions from the $n = 2$ to $n = 1$ levels are called K-alpha X-rays, and those for the $n = 3 \geq 1$ transition are called K-beta X-rays.

Transitions to the $n = 2$ or L-shell are designated as L X-rays ($n = 3 \geq 2$ is L-alpha, $n = 4 \geq 2$ is L-beta, etc.). The continuous distribution of X-rays which forms the base for the two sharp peaks at left is called *"bremsstrahlung" radiation.*

X-ray production typically involves bombarding a metal target in an X-ray tube with high speed electrons which have been accelerated by tens to hundreds of kilovolts of potential. The bombarding electrons can eject electrons from the inner shells of the atoms of the metal target. Those vacancies will be quickly filled by electrons dropping down from higher levels, emitted X-rays with sharply defined frequencies associated with the difference between the atomic energy levels of the target atoms. As stated earlier, the frequencies of the characteristic X-rays can be predicted from the Bohr model.

Principal Properties

- The X-rays, like light, are electromagnetic radiations of extremely short wavelength (100-0.0001 A). They show reflection, refraction, interference, diffraction and also polarization under suitable conditions.
- They travel in straight lines like light, unaffected by electric and magnetic fields, indicating they are not charged particle beams, like cathode rays.
- They affect (blacken) photographic plate even when wrapped in black papers by several folds. The degree of blackening depends on the intensity of radiation. For this property, photography is a very suitable technique to detect/study X-rays.
- The X-rays are having high penetrating power. The shorter the wavelength, greater the penetration power. For low-density substances, the penetration is more compared to the high density ones. Usually passes through matter that is opaque to visible rays.
- Soft (long wavelength) X-rays are absorbed by materials to varying degrees. This property is used in the radiography of substances. The living cells can be destroyed by X-rays of suitable intensity.
- X-rays ionize gases through which it passes and the ionizing power depends on the intensity of the radiation. This property is used to measure the intensity of X-rays.
- X-rays cause fluorescence in many substances (e. g. ZnS, Ca salts etc.) and is used as a distinct property for identifying such materials.
- X-rays exhibit corpuscular nature. Compton Effect is the confirmation of this property.

X-ray analytical techniques use the characteristic X-rays. The techniques to be discussed here are:

X-ray fluorescence spectrometry which depends on characteristic secondary radiation emitted by materials when excited by a high-energy X-ray source.

X-ray crystallography which depends on the dual wave/particle nature of X-rays.

Major Uses of X-rays

1. *X-ray fluorescence spectrometry* is used primarily to determine amounts of particular elements in materials.
2. *X-ray crystallography* in which the pattern produced by the diffraction of X-rays through the closely spaced lattice of atoms in a crystal is recorded and then analyzed to reveal the nature of the crystalline structure of the material.
3. *X-ray microscopic analysis* which uses electromagnetic radiation in the soft X-ray band to produce images of very small objects.
4. *X-ray radiography* is used for creating images of light-opaque materials. It relies on the relationship between density of materials and absorption of X-rays. Applications include a variety of medical and industrial uses.
5. *X-ray astronomy*, which is an observational branch of astronomy, deals with the study of X-ray emission from celestial objects.

X-ray Fluorescence Spectrometry

General Uses

- Qualitative identification and quantitative determination of elemental composition of a variety of samples for solid and liquids. Minimal sample preparation, wide dynamic range and non-destructive methodology are the important aspects of XRF.
- Determination of sulfur in diesel fuel to meet environmental regulations.
- Process control of steel and cement production.
- Support of mineralogical and geological exploration and waste site field evaluation.
- Sorting of metal alloys.
- Forensic applications.
- Coating thickness and composition process monitoring for paper and metal industries.

Samples: State—Any solid or liquid sample can be analyzed. Hazardous samples need special sample cells(holders) to prevent instrument contamination.

Amount—Sample size ranges from micrograms to gram quantities depending on methodology used.

Sample preparation is determined by analytical needs such as level of accuracy and precision required. For simple analysis no preparation is necessary, may be analyzed as received, with rapid turn around for qualitative and semi-quantitative results. For minimal pretreatment, it is necessary to grind, homogenize and press into pellets for semi-quantitative and quantitative results. For maximum pretreatment, dry, fuse and cast into glass disks; this provides the highest accuracy and precision for quantitative results.

Analysis time: A qualitative spectrum on an energy dispersive XRF (EDXRF) instrument can be obtained within five minutes to provide gross elemental composition. Quantitative programme for high accuracy and precision on a sequential wave-length dispersive XRF (WD XRF) instrument can take upto 10 min for 15 elements, whereas a simultaneous instrument can do the same measurement within 5 min (the sample preparation time is extra).

Limitations

1. Elemental range is limited to Boron and up. Detection of light elements ($Z<11$) limited to solids and essentially surface composition.
2. Matrix interferences can prevent or limit detection of some elements.
3. Standards for quantitative analysis do not always match unknown matrix.

Accuracy: Depends on sample preparation and how well standards match the unknown matrix. In many situations, accuracy of less than 1% error can be achieved with proper care and attention to the determinations.

Sensitivity and detection limits: This varies with element and sample matrix. In general ppm (parts per million) detection limit can be achieved for normal matrices e. g. steels, soils, petroleum products, geological materials for transition row elements above. The lighter elements $Z<19$ have detection limits that are typically higher, ranging from tens of ppm to high hundreds of ppm for Mg and Na.

Synchrotron radiation, total reflection XRF and new methodologies for microsampling are pushing detection limits well below ppm level.

Complementary or related techniques are: (1) inductively coupled plasma (ICP), (2) neutron activation analysis (NAA) and (3) atomic absorption spectroscopy (AAS).

X-ray tube: X-rays are generated in an evacuated X-ray tube. Within the tube, a tungsten cathode filament releases electrons. They are accelerated under a voltage typically tens of kilovolt (30 kv to 50 kv), but the current is always kept low (20-40 mA). When the high velocity electrons strike the target metal with sufficient energy, they cause electrons in the target atoms to jump temporarily to higher energy level. As these electrons naturally return to lower levels, the energy difference between the two levels is given off as X-rays. X-ray tube generates a lot of heat and must be cooled constantly by cold-water flow.

The target metals may be Fe, Cu, Mo etc. Mostly Cu-target is used and the corresponding X-ray wave-length $\lambda_{Cu} = 1.54$.

X-ray fluorescence (XRF) spectrometry is an atomic spectroscopic method based on the detection of emitted X-radiation from excited atoms.

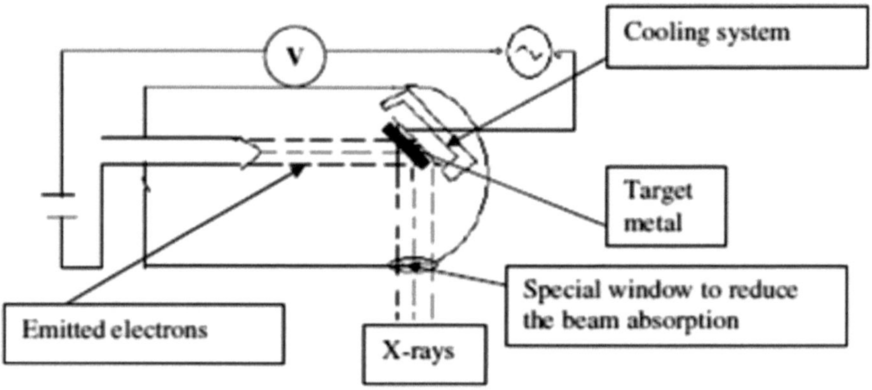

Fig. 13.36: Operational mechanism of X-ray tube.

In the above, X-ray fluorescence process and energy level diagram are shown in Fig. 13.36. An incident X-ray photon removes an inner shell electron. The vacancy is filled by an outer shell electron, which gives up an X-ray photon. The emitted X-ray photon is characteristic of the element. The energy or the wavelength of the photon is determined by the energy gap between the two energy levels of the electron filling the inner shell vacancy.

Thus, it is a two-step process:

1. The incident X-rays remove the inner shell electron of an atom. The resulting vacancy is filled by an outer shell electron.
2. The transition from the outer shell electron orbital to an inner shell electron orbital is accompanied by an emission of X-ray photon. This fluorescent photon is characteristic of the element and is equal to the difference in energy between two electron energy levels. Because the energy difference is always the same for given energy levels, the element can be identified by measuring the energy of the emitted photon. The intensity of the emitted photons determines the concentration of the element. Thus the measure of the photon energy provides the identification of the element and the intensity of the photon emission provides a measure of the amount of element.

This emission process is similar to other fluorescence techniques but restricted to X-ray region of the electromagnetic spectrum ranging from 0.1 to 120 keV or 11 to 0.1 nm. The typical analytical region is less than 50 keV.

To have monochromatic X-rays, filters or monochromator crystals be used. The absorption edge of one metal A when falls between the Kα and Kβ lines of metal B (target), metal A is used as a filter.

The photon energies detected are designated as K, L or M X-rays depending on the energy level being filled, e.g. a k-Shell vacancy filled by an L-level electron results in the emission of a Kα X-Ray. There are as many possible

Table 13.4: Filters for k-absorption of different targets

Target	$\lambda_{K\alpha}$	*Filter*	*Absorption of* K_β(%)	*Thickness*	
Mo	0.711	Zr	96.3	0.080	45-55 nv
Cu	1.542	Ni	97.9	0.015	20-40
Ni	1.659	Co	98.4	0.018	
Co	1.790	Fe	98.9	0.012	
Fe	1.937	Mn	98.7	0.011	35-40 kV
Cr	2.291	V	99.0	0.016	25 kV

X-ray lines as there are inner shell electrons. The most intense X-ray lines are of most analytical importance and they are the K-lines from Boron to Cerium. For the remainder of the periodic table L and some M lines are used for a clear spectra with a minimum interference from other lines.

It is used since 1922 and cover from trace level (ng/ml) to a major composition as high as 100%. If the standard and matrix are matched to unknown sample, the accuracy and precision are unique. So, it is widely used in various industrial and other scientific research programmes. If the detector part is attached with an energy dispersive XRF-spectrum, all elements from Na to U can be simultaneously detected. If the wave-length dispersive XRF (WDXRF) facility is available, quantitative estimation can be done easily.

The schematic diagrams of wavelength and energy dispersive XRF instruments are shown in Figs 13.37a and 13.37b.

Tube excitation source typically penetrates microns to millimetre within the sample. Particle induced X-ray emission (PIXE) uses protons or other heavy particles to induce X-ray emission. Radioisotopes are also used as excitation sources in XRF but their intensity is much less than tube excitation. These are good for portable instrumentation but with limited sensitivity (flux limitation).

Synchrotron radiation source (SR) includes intensity, polarization and collimation, but the accessibility of SR is very much restricted one.

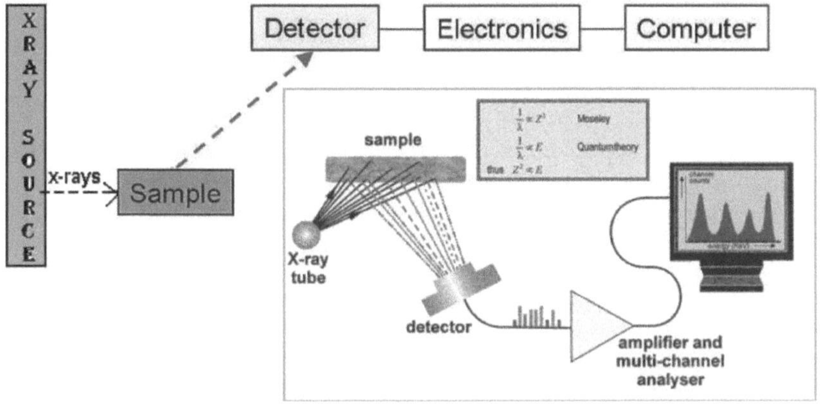

Fig. 13.37a: Wavelength Dispersive XRF-analyser (EDXRF).

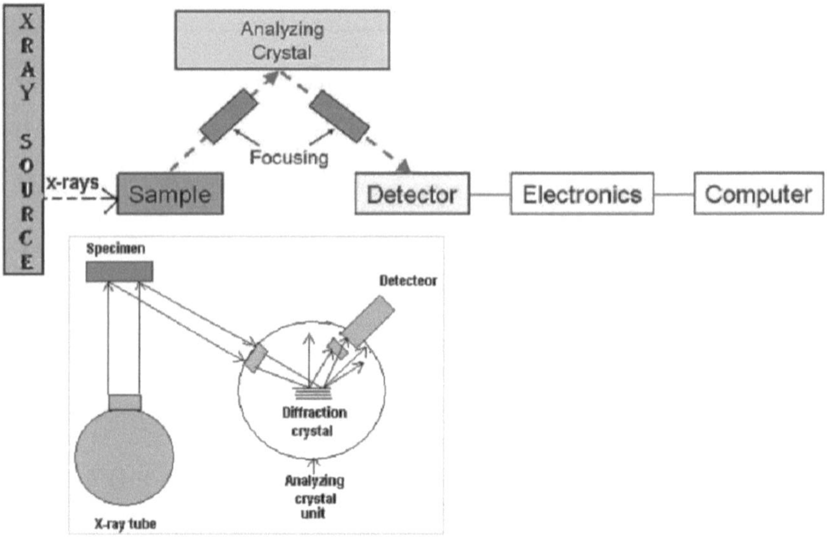

Fig. 13.37b: Wavelength dispersive XRF-analyzer (WDXRF).

Basic detectors used in XRF measurements are: 1. Gas ionization and 2. Scintillation (both used in WDXRF) and 3. Solid state semiconductors.

Gas ionization detector consists of two electrodes, the wire anode and a metal cylinder cathode. The chamber is filled with gas (90% argon + 10% methane). The X-ray photon enters through a window in the cylinder and ionizes the gas. The resulting ions and electrons are collected and the current is proportional to the intensity of the X-ray photon. These detectors are used for longer X-ray wave-lengths of the lighter elements ($Z < 27$). There are both seated and flow proportional counters. Flow proportional counters are used in most commercial instrumentation.

Scintillation detectors consists of thallium doped NaI crystals on the front of a photo multiplier tube. X-rays from the sample strike the NaI crystals, generate photons and are multiplied by the photomultiplier and then detected. The X-ray energy is proportional to the number of photons produced. Working range $25 < Z < 35$.

Solid state detector is a lithium drifted silicon (SiLi) wafer. X-ray strikes the detector and generates a series of pulses that correspond to X-ray energy. Here the pulse height is proportional to energy. By counting the pulses, the concentration of the element is estimated. The detector and preamplifier are cooled with liquid N_2 to minimize the current noise.

Quantitative Analysis

For quantitative analysis, standard samples i.e. samples of known compositions are of vital importance. Any variation of weight % of any element in the sample will be reflected in its intensity value. A calibration curve is to be drawn showing the variation of wt.% of any element (along x-axis) and corresponding intensity variation (along y-axis) with standard sample data. The linear part of the calibration is usually chosen for quantification. Due to the presence of many elements in the sample, the intensities are affected due to the presence of other elements (matrix effect, absorption etc.) and is accounted for by various correction factors.

This linear calibration curve is represented by

$$I = mc + b$$

where I = intensity for concentration c, b = background, m = slope of the curve.

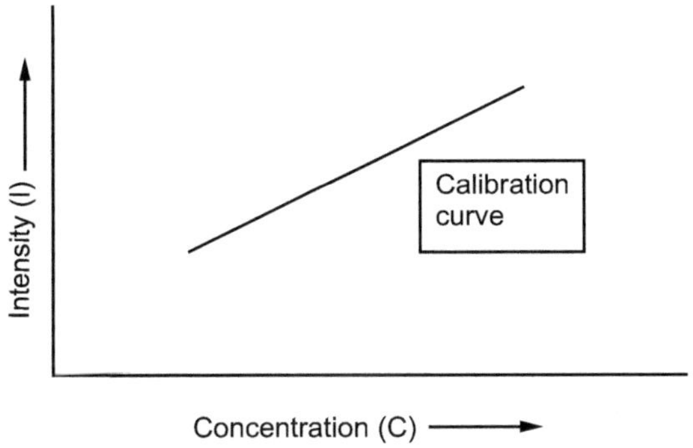

Fig. 13.38: Principle of Bragg's Law.

$$C_i = (A_i + B_i I_i) \quad (1 + \Sigma\alpha_{ij}\,\Delta C_j)$$
$$\text{(Term 1)} \qquad \text{(Term 2)}$$

where A = reciprocal of slope, B = background intensity (constant for element i). I_i = measured intensity of element I; + α_{ij} = absorption coefficient of element j on element i); ΔC_{ij} = average composition of sample. Second term (II) defines the matrix effects and ΔC_j may vary depending on the model used.

X-ray Diffraction

X-ray diffraction is a versatile analytical technique for examining *crystalline solids*, e.g. metals, ceramics, geological samples, electronic materials, multilayers, thin films, sheets, fibres or irregular shapes depending on the desired measurement.

The major components of X-ray study are: X-ray wave source (monochromatic or polychromatic), sample (powder, single crystal or a solid piece) and detector (photographic film or radiation counters).

The different methods of X-ray study are classified as below.

We will concentrate on powder diffractometry. The basic equation used is the Bragg's equation: $2d \sin\theta = n\lambda$, more specifically, where λ = wave-length of X-rays used, and $2d_{hkl} \sin\theta_{hkl} = n\lambda$.

Bragg angle θ_{hkl} = angle between incident beam and lattice plane hkl and d_{hkl} = distance between lattice planes.

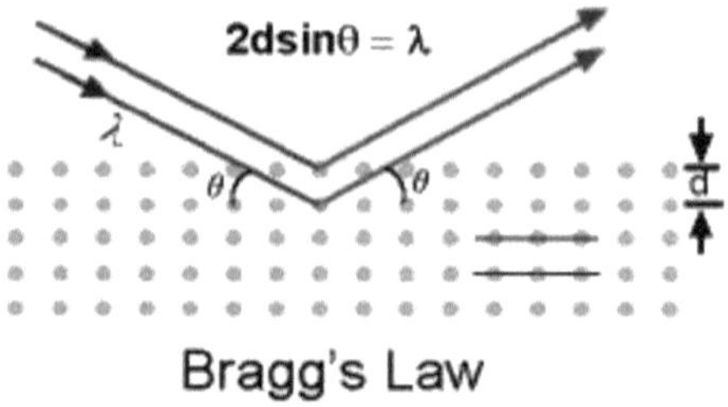

Fig. 13.39: Principle of Bragg's Law.

In general, the *d*-spacing is a function of the lattice parameters (*a*, *b*, *c*) and angles (α, β,) defining the unit cell and the Miller indices (h, k, l) denoting particular reflection. More precisely $2d_{hkl} \sin \theta_{hkl} = n\lambda = \lambda$ if $n = 1$

As such, it is the geometry of the crystal lattice that determines the position of the peaks in an X-ray diffraction pattern. In general, more symmetrical the material, the fewer peaks in its diffraction pattern. The diffracted intensities associated with those peaks are determined by the type and arrangement of atoms within the crystal lattice.

Examples of unit cells of minerals belonging to each of six crystal systems.

Lattice types can be: 1. Primitive, 2. Body centered (I), 3. Face centered (F), 4. Side centered (C), 5. Rhombohedral (R) etc.

Miller indices: Used to describe the different planes in the crystal with respect to its axes, and denoted by *h* (for axis *x* or *a*), *k* (for axis *y* or *b*) and l (for axis *y* or *c*). The corresponding planes are denoted by d_{hkl}.

Relation between the crystal system and d_{hkl} system:

(a) Cubic system: $1/d^2_{hkl} = 1/a^2(h^2 + k^2 + l^2)$ $[a = b = c]$
(b) Tetragonal: $1/d^2_{hkl} = (h^2 + k^2)/a^2 + (l^2/c^2)$ $[a = b \neq c]$
(c) Orthorhombic: $1/d^2_{hkl} = (h^2/a^2) + (k^2/b^2) + (l^2/c^2)$
(d) Hexagonal: $1/d^2_{hkl} = 4/3a^2(h^2 + hk + k^2) + (l^2/c^2)$

Rhombohedral:

$$\frac{1}{d^2_{hkl}} = \frac{1/a^2\,(h^2+k^2+l^2)\sin 2\alpha + 2(hk+kl+lh)(\cos 2\alpha + \cos \alpha)}{1 - 2\cos 3\alpha + 3\cos 2\alpha}$$

(e) Monoclinic: 1st setting:

$$\frac{1}{d^2_{hkl}} = \frac{(h^2/a^2) + (k^2/b^2) - (2\,hk\cos\gamma)/ab + l^2/c^2}{\sin^2\gamma}$$

2nd setting:

$$\frac{1}{d^2_{hkl}} = \frac{(h^2/a^2) + (l^2/c^2) - (2\,hl\cos\beta)/ac + k^2/b^2}{\sin^2\beta}$$

(f) Triclinic:

$$\frac{1}{d_{\text{hkl}}^2} = \frac{(h^2/a^2)/\sin^2\alpha + (k^2/b^2)/\sin^2\beta + (l^2/c^2)/\sin^2\gamma + 2hk/ab\,(\cos\alpha\cos\beta - \cos\gamma) + 2\,kl/bc\,(\cos\beta\cos\gamma - \cos\alpha) + 2\,lh/ca\,(\cos\alpha\cos\gamma - \cos\beta)}{1 - \cos^2\alpha - \cos^2\beta - \cos^2\gamma + 2\cos\alpha\cos\beta\cos\gamma}$$

α, β, γ, a, b and c are lattice constants and h, k, and l are miller indices.

General uses

1. Crystal structure determination
2. Phase identification
3. Quantitative phase analysis
4. Texture and stress analysis
5. High and low temperature and pressure studies

Common application

1. Identification of unknowns
2. Quality control: Qualitative and quantitative analysis
3. Characterization of polycrystalline thin films
4. Quantification of texture and orientation in metals, polymers and fibres (jute etc.)
5. Variable temperature studies to determine thermal expansion, stability and phase diagrams.

Sample

State: Crystalline solids as powders, single crystals, sheets, foils or fibres

Amount: Single crystal, uncracked, untwined crystal 0.1 to 0.5 mm on a side. Powder typically several hundred mg (or fractions of mg), grain size ~ 200 mesh

Preparation: Single crystal: mounting on a fibre or in a glass rod.

Powder: grinding to about 200 mesh size, to be filled in the slot in the sample holder, to be pressed (but not rubbed to create preferred orientation) so that the powder packing in the holder slot does not move or slide.

Analysis time: Depends on the type of experiment.

Limitations

1. Crystalline phases only
2. Peak overlap may hinder phase identification and quantitative analysis.
3. *Matrix effects:* Strongly diffracting materials (well crystalline), may obscure weakly diffracting one (poor crystallinity).
4. Preferred orientation affects diffracted intensity and is damaging for the quantitative analysis data.
5. Sample fluorescence will raise the background and also can affect the detector system.

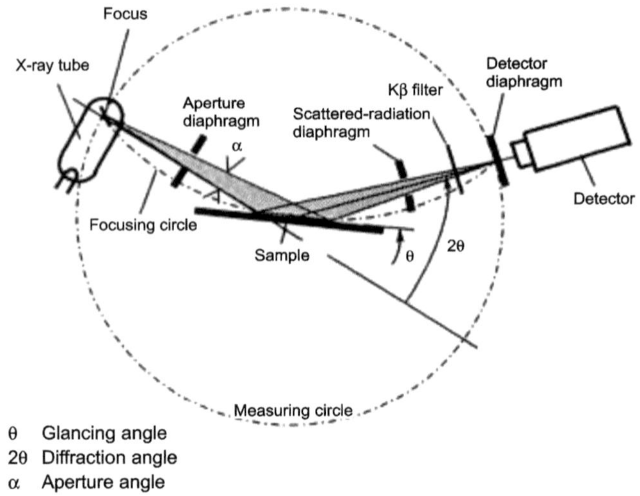

Fig. 13.40: Diffractometer beam path in θ/2θ mode.

13.6 ATOMIC ABSORPTION SPECTROSCOPY

Atomic absorption spectroscopic technique analyses and quantifies the elements present in a sample from major to trace level. The atomic absorption spectra was first observed in 1802 and in 1953, an Australian physicist, Sir Alan Walsh showed its utility as an analytical technique.

When an electromagnetic wave passes through an atomic medium, the atoms reaches their higher excited states by absorbing energy from the radiation passing through them. The absorption of energy is characterized by the atomic electronic configuration. Thus from the absorption spectra of the sample, elements present can be identified. The working process is shown in the block diagram (Fig. 13.41).

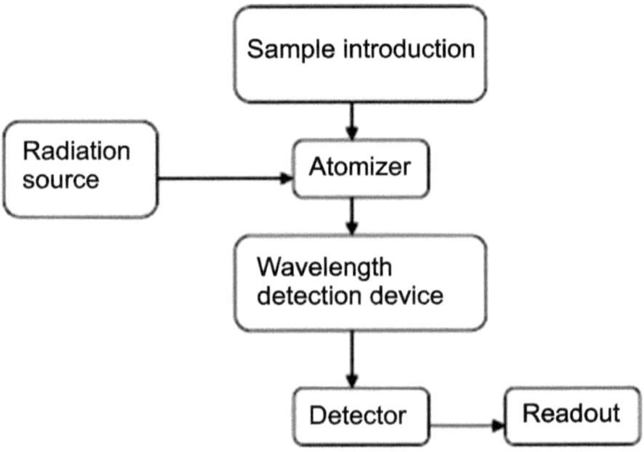

Fig. 13.41: Block diagram of AAS system.

The sample is vapourized by putting in a flame or graphite furnace and kept in a container (atomiser). The electromagnetic radiation (light) is then passed through the atoms to absorb energy from the incident UV or light source so as to reach their excited electronic states. The energy difference between the incident and transmitted light can be estimated, yielding quantitative information about the element present in the sample.

This quantification is guided by Beer-Lambert law, which relates the properties of light travelling through a material.

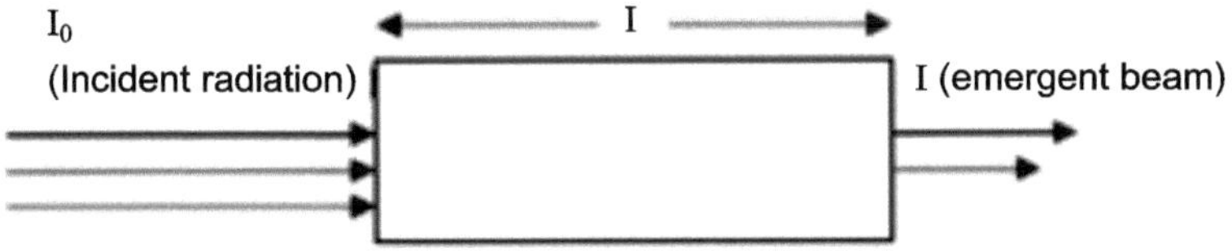

Fig. 13.42: Absorption of radiation while passing through a medium.

A beam of light with intensity I_0 start travelling in a medium of path length l and after transmission the beam intensity is I, then the transmissivity T can be written as

$$T = I/I_0$$

where I_0 = initial intensity and I = resultant intensity.

For liquid phases it can be shown as

$$T = I/I_0.\ 10^{-\alpha l} = I/I_0.\ 10^{-\varepsilon c l}$$

where l = path length of the radiation through the substance and α = absorption coefficient of the substance = $\varepsilon.c$ = (molar absorptivity) × (concentration of substance).

Or the transmissivity, in terms of absorbance, is defined as

$$A = -\log_{10}(I/I_0) \text{ i.e. } A = \varepsilon l c = \alpha\, l$$

and for gases

$$A' = -\ln(I/I_0)$$

i.e. $$A' = \sigma l N = \alpha' l$$

Therefore, knowing the path length (l) and molar absorptivity (absorption cross section), the concentration of the substance (number density of absorbers) can be estimated.

The basic instrumentation needs: (a) radiation source, (b) a nebulizer, (c) an atomizer, (d) a monochromator, (e) photomultiplier, (f) other instrumentation for recorder/detector. Each of the above parts has some specific purpose as given below:

(a) *Radiation source* is needed to provide the analytical line for the element of interest. This source must produce a steady and intense beam of the needed analytical line. Usually a hollow cathode lamp is used for this purpose. Ar-ions bombard to sputter the cathode metals of interest. Fraction of the sputtered atoms are excited to fluoresce.

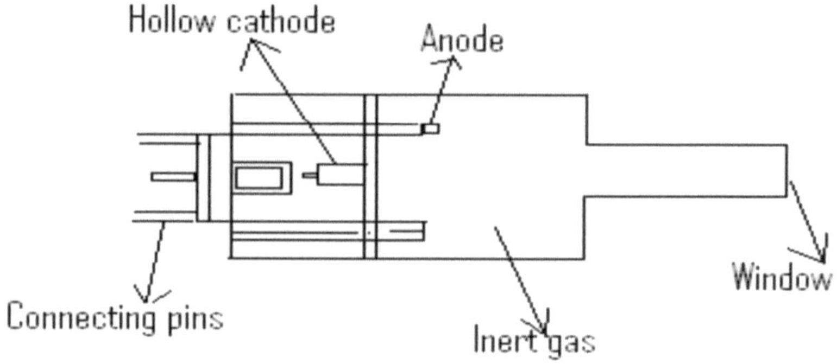

Fig. 13.43: Hollow cathode lamp.

A different lamp for different element restricts multielement detection in a single run. Hollow structure maximizes the redeposition on cathode but restricts light direction.

(b) *Nebulization* is conversion of a sample to a fine mist of finely divided droplets using a jet of compressed gas. The flow carries the sample into the atomization region.

Nebulizer creates a fine aerosol for introduction into the flame by sucking the sample (liq) at a controlled rate. The aerosol is then mixed thoroughly with fuel and oxidant for introducing it to the flame (or atomizer).

(c) The *atomizer* produces atoms (in gaseous state) of the element of interest. On absorption of ultraviolet or visible light, the free atoms undergo electronic transitions from ground state to higher excited electronic states. High temperature is necessary for atomization. The temperature stability and gas flow rate of fuel is very important. High temperature for atomization is necessary. The types of atomizers are flame atomizer (mostly used), electrothermal atomizer and graphite furnace.

Flame atomizer can analyze solutions only. The temperature range is 1700–2400° C. For refractory samples higher temperature is needed. A slot type burner is used to increase the path length, equivalent to increasing total absorbance. Sample is introduced from nebulization chamber as small droplets. The sensitivity varies with element and the burner position to be adjusted for optimum conditioning. Multielement detection is difficult.

A graphite furnace directly accepts very small quantities of sample. This system provides a reducing environment for easily oxidized elements. The furnace is electrically heated in several steps to dry the sample, ash (to remove the volatile hydroxides, sulfates, carbonates) and vapourize

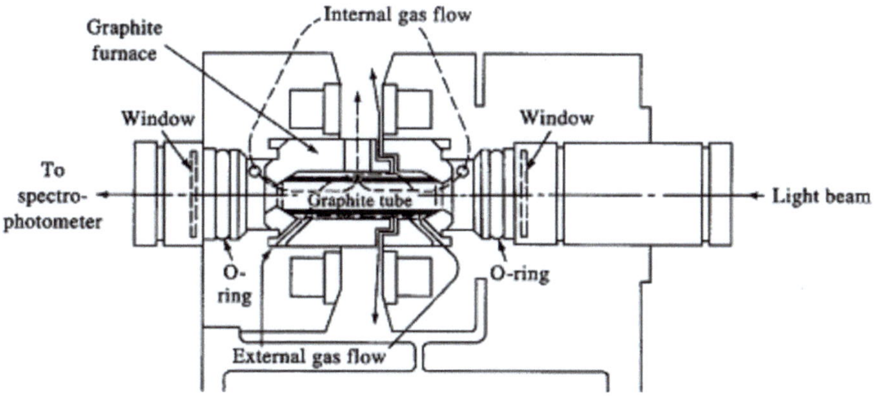

Fig. 13.44: Graphite atomizer.

the analyte atoms. External Ar-gas prevents tube damage and internal Ar-gas prevents tube damage and internal Ar-gas circulates the gaseous analyte.

Electrothermal atomizer energizes the entire sample (temperature range 2000–3000°C) in a very short time. A better sensitivity (10^{-10} to 10^{-13} g is achieved but a poorer reproducibility of 5-10%).

(d) *Monochromator* is used for UV and visible light. The main purpose of the monochromator is to isolate the absorption line from the background light due to interferences. Selected narrow spectral lines are allowed to pass through the monochromator and can interact with the detector (photomultiplier tube).

(e) *Photomultiplier* is used as the detector in AAS. The intensity of the photons of the lines of interest is estimated by the PMT.

(f) Other instrumentations like the computer oriented softwares to analyze the line and their intensity are nowadays a regular feature of any modern AAS.

Analysis Using AAS

Detection limit of this analysis is ppb level. Limitations are less suitable for weak absorbers (forbidden transitions) metals with low *I*. For quantitative analysis, a calibration curve is to be drawn with known concentration of the elements (mg/ml) with the absorbance of the element.

Unknown concentrations can be estimated from the calibration curve. Some correction factors should be considered for more precise result, e.g. background, unabsorbed radiation, stray light or disproportionate decomposition of molecules at high concentration etc.

Recent developments are:

(a) In electronics solid state detectors are used.

(b) High resolution continuum source AAS in both flame and graphite mode are available.

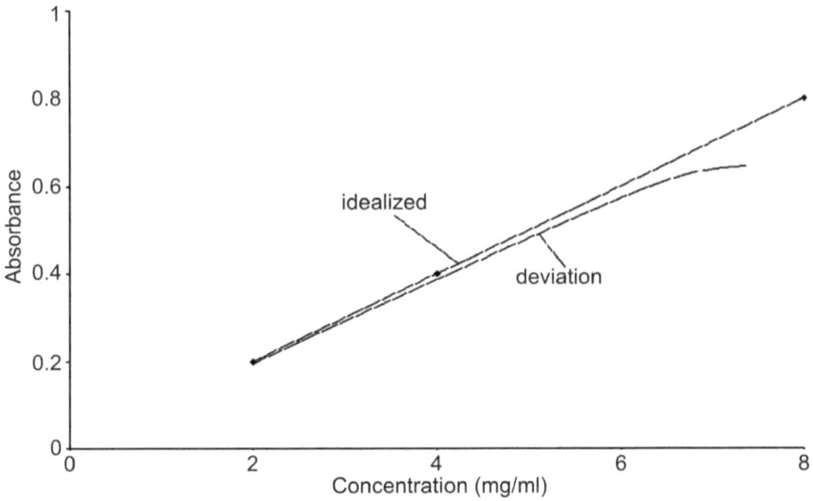

Fig. 13.45: The calibration curve.

(c) Simultaneous background correction can be measured.
(d) Extra lines of an analyte are available extending the working range.
(e) Instead of multiple hollow cathode lamps single xenon arc lamps are used covering the elements in the wavelength range 185-900 nm. This facility makes AAS as multielement analyzer with 10 elements per minute.

13.7 INFRA-RED SPECTROMETRY

Infrared spectroscopic analysis identifies the chemical functional group present in the sample and also the types of bonds in it. Atoms in molecules and crystals have their natural vibrational frequencies in the IR-range and they are quantized. When the molecules absorb infrared radiation, they vibrate more vigorously. The mode of vibrations is described as bending, stretching, rocking, twisting and scissoring.

Infrared wave ranges from visible to the radio waves. This is conventionally divided into the wave number range as

Near infrared: 13000–4000 cm^{-1} equivalent wave length is 0.78-2.5 μm
Medium infrared: 4000–200 cm^{-1} equivalent wave length is 2.5-50 μm
Far infrared: 200–10 cm^{-1} equivalent wave length is 50-1000 μm
Wave number and wavelength are related by the following equation:

$$\bar{\nu}\,(\text{in cm}^{-1}) = \frac{1}{\lambda\,(\text{in}\,\mu\text{m})} \times 10^4$$

The IR-wavelength range of minerals and inorganic compounds usually is from 2.5 to 50 μm. Ultra violate visible spectra covers generally from 0.2 to 0.8 μm and the near infrared region (NIR) covers the range 0.8 to 2.5 μm.

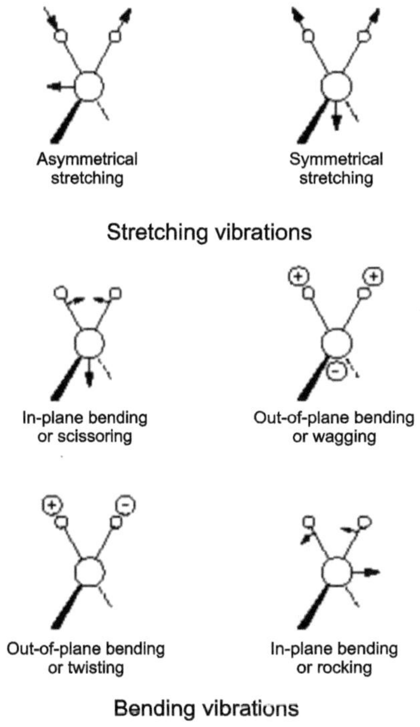

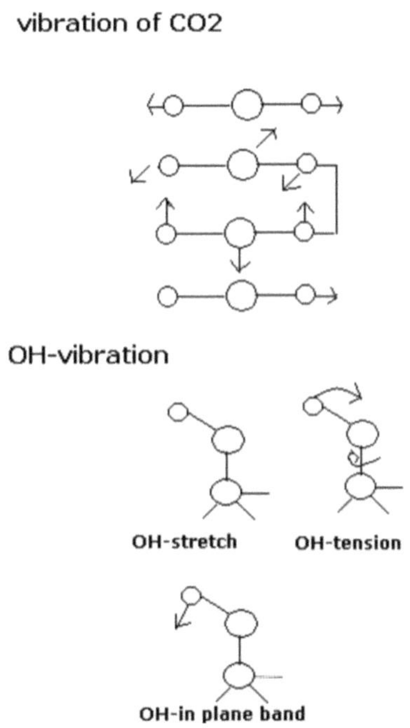

Fig. 13.46: Vibrational modes. **Fig. 13.47:** Vibrational modes in specific cases.

In this technique, molecules are exposed to infrared radiation and depending on the chemical functional group present (hence the chemical bonds) the characteristic frequency is absorbed. This yields the infrared absorption spectra, identifying the chemical functional group. For mineral samples, the IR-wavelength of interest is from 4000 to 250 cm^{-1}. Infrared spectra can be obtained from solid, liquid and gaseous phases. Sample preparation methods are accordingly different.

Block diagram of IR-Spectrometer arrangement is shown in Fig. 13.48.

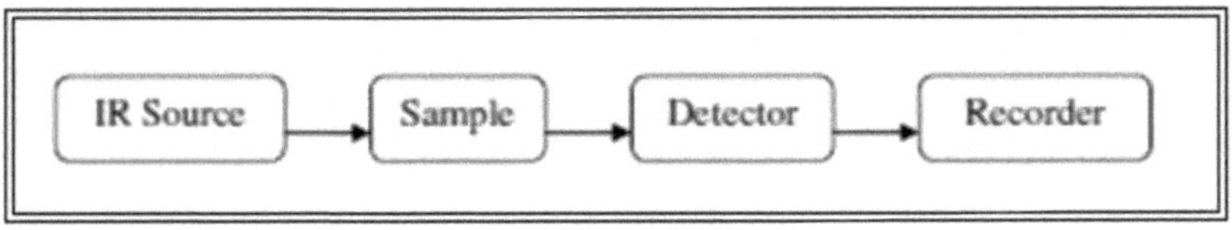

Fig. 13.48: Block diagram of IR analytical technique.

$$\text{The transmittance } T = \frac{\text{Powder transmitted by the sample } (I)}{\text{Powder incident on the sample } (I_0)}$$

And the absorbance A is related as

$$A = \log_{10}(1/T) = -\log_{10} T = -\log_{10} I/I_0$$

The transmittance range is from 0-100% whereas absorbance is from infinity to zero.

General Applications

- All types of organic and most of the inorganic compounds can be identified.
- Functional groups of compounds can be determined.
- Molecular composition of surfaces can be determined.
- Chromatographic effluents can be identified.
- Compounds in mixtures can be determined quantitatively.
- Molecular conformation (structural isomers) and stereochemistry (geometrical isomers) can be determined.
- Molecular orientation can be determined.
- Non-destructive method.

Specific Uses

- Fingerprinting: Compounds can be identified by matching spectrum of unknown compound with reference spectrum.
- Functional groups of unknown substances can be determined.
- Reaction components and kinetic studies of reactions can be identified.
- Molecular orientation in films can be determined.
- Molecular impurities or additives present in amounts of 1% and in some cases as low as 0.01% can be detected.
- Polymers, plastics and resins can be identified.

Limitations

- Minimal elemental information is given for most samples.
- Background solvent or solid matrix has to be transparent in the required spectral region.
- In the IR region molecules must be active.

Accuracy

In analysis of mixtures under optimum conditions, accuracy greater than 1% is obtained. In routine process it is ± 5%.

IR-Spectrometers

- *Sequential devices:* Information collected sequentially in time; single detector used to collect information as each spectral element is scanned at a time.
- *Spatial devices:* Multiple detectors used; a photographic recording is done as in spectrograph.
- *Multiplex devices:* Here the detector is a single viewer which receives the signals; signals are thereupon transmitted by a single channel and are capable of being decoded for identifying the individual spectral element.

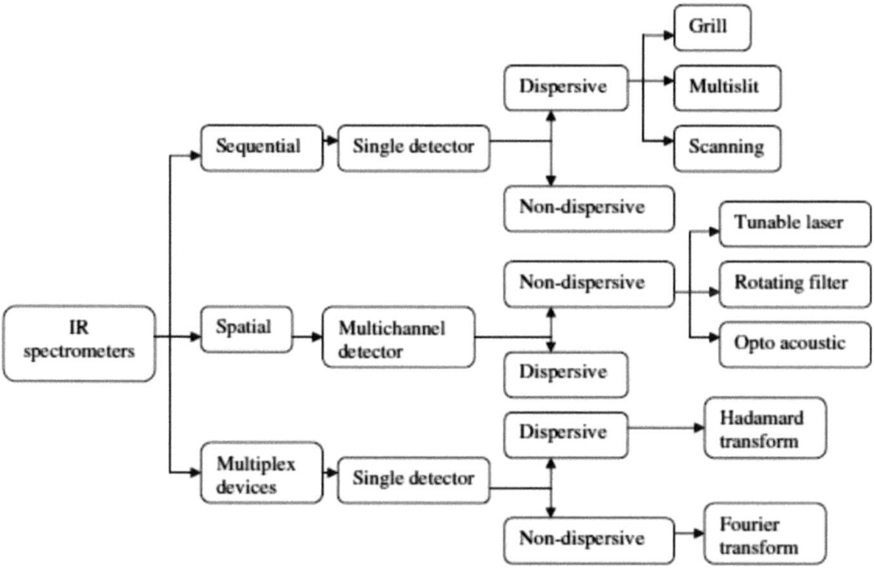

Fig. 13.49: Varieties of IR spectrometers.

A double beam IR-Spectrometer arrangement is shown in Fig. 13.50. In this arrangement, one of the splitted beam falls on the sample and the other on the reference. Commercially available standards are used as reference materials.

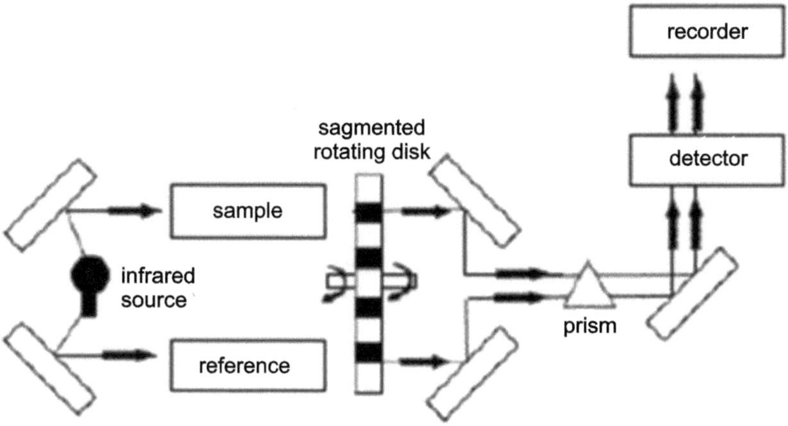

Fig. 13.50: A sketch of a double beam Infrared (IR) spectrometer.

Sample Preparation

The mineral powder is mixed homogeneously with KBr and pulverized and pressed into a clear disc. This disc is mounted in the sample position and examined directly.

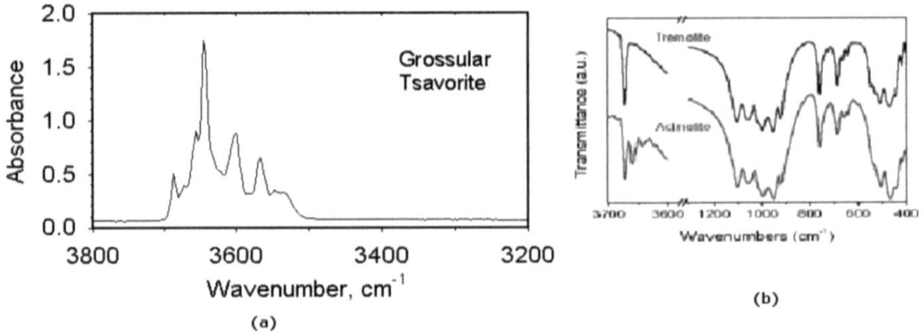

Fig. 13.51(a and b): Infra-red spectra of minerals.

Identification

Spectra of unknown samples are compared with bands of known samples for identification. For silicate minerals, strong absorption bands of vibration and rotation of atoms occur in the range of 9-12 μm. With the increase of Si:O ratio (isolated silica tetrahedron, like SiO_4 olivine) the strong absorption range is 0.25 and that in quartz (SiO_2) is 0.50. As the number of coordinated atomic bonds increases, possible vibrational or rotational modes will increase, increasing the wavelength range also.

Minerals containing hydroxyl group (like serpentine, goethite, lepidocrocite etc.) are characterized by an absorption band of the minerals at shorter wavelengths, greater than 2800 cm^{-1}. The absorption is due to stretching vibrations within the hydroxyl group.

The presence of H_2O in a mineral phase can be detected by two characteristic absorption bands in the 3600-3200 cm^{-1} and also in the 1650 cm^{-1} region. When lattice water is present, an additional band is observed at 600–300 cm^{-1} and for water coordinated with metal ions, another band is observed in the region 800-650 cm^{-1}. The fundamental vibrations of water molecule and its overtones combined to yield the additional characterized peaks. Band resolution can be improved at a lower temperature.

The characterized bands and the wavelength region represent the chemical group and its vibration type, yielding the fingerprint of the material under study.

Concluding Remarks

The instrumental analysis to be used judiciously with the necessary complimentary techniques. Knowing the types of interactions and their reasons proper choice of the analytical technique is possible for specific problrms in mineralogical studies.

Think for a while

1. Which analytical method distinguishes different polymorphs of a mineral and why?
2. Giving an application for each, explain the difference between TEM and SEM.
3. Mentioning various types of thermal analysis, explain which one is suitable for quantifying a mineral and how.
4. Chemical compositions of minerals can be estimated by various analytical methods. State the speciality of those methods with examples.
5. Explain the use of different radiometric dating method with examples.

FURTHER READING

Azaroff, Leonid. V. and M.J. Buerger. The Powder Method. McGraw Hill Book Company.

Charsley, E.L. Thermal analysis: Techniques and Applications. S.B. Warrington.

Ewing, G.W. Instrumental methods of Chemical analysis. McGraw Hill.

Faurie, G. Principles of isotope geology. John Wiley & Sons.

Makenzie, R.C. (ed.) Differential Thermal Analysis. Academic Press, New York. 1970-72.

Marfunin, A.S. (ed.) Advanced Mineralogy. Vol 2. Springer-Verlag, Berlin, Heidelberg. 1995.

Roussac, F. and A. Roussac. Chemical Analysis: Modern Instrumentation methods and techniques. John Wiley & Sons.

Sachinath Mitra. Fundamentals of Optical, Spectroscopic and X-ray Mineralogy. New Age International.

Todor, D.N. Thermal Analysis of Minerals. Abacus Press.

Internet Data Retrieved from:

University of Minnesota Electron Microprobe Lab—A comprehensive overall resource on EPMA lab instrumentation and sample preparation.

University of Massachusetts-Amherst Probe Lab—also a good site with particular emphasis on geochronological applications of EPMA.

Electron Microprobe—Dave Waters, Oxford University.

http://serc. carleton. edu/research_education/geochemsheets/techniques/EPMA. html

http://www. cameca. com/instruments-for-research/epma. aspx.

CHAPTER

14

PRECIOUS AND SEMIPRECIOUS STONES

Gems are found worldwide. When there is a sufficient quantity to be worked, it is termed a deposit. A deposit, as well as the location of discovery of a single find, is termed the gem occurrence. The gem cut defines the shape of a stone and determines whether it sparkles or shines. The cut of a gem is often the first consideration in choosing the proper setting for the stone. In addition to gemstone cutting, over the centuries countless other methods have been developed to improve upon the natural properties of gemstones. These techniques are known as *enhancements*. Synthetic gems have been manufactured since the 1960s as an attempt to substitute for naturally occurring gems. Advances in the methods of manufacture have made it possible to produce synthetic gems of equal or better appearance than naturally occurring gems. Thus these important aspects of gemstones—properties, cuts, enhancements, and synthetic gem preparation—are to be discussed in this chapter.

Gemstone is any of the various minerals, highly prized for beauty, durability and rarity. Most are cut from naturally occurring crystals. To qualify as a gem, the mineral should be hard and tough to resist scratching and have beautiful clarity and colour. There are different types of gemstones, found on earth's surface. On the basis of clarity, chemical composition, crystal structure, optical characteristics, specific gravity and durability, the quality of gemstones can be determined. A mineral can be designated as a gemstone with respect to certain qualities like beauty, large size and clarity. Gemstones can be differentiated into two distinct groups of precious and semi-precious stones. Diamonds, ruby, emeralds and sapphires can be designated as precious stones, while amethyst, garnet, opal, aquamarine, lapiz lazuli, jade, turquoise, onyx, agate and malachite are among the common semi-precious stones. Some gemstones are of *organic* origin, like *amber*, which is a fossil tree of resin. *Pearls* and *corals* are similar gemstones of organic origin. Out of so many minerals, identified so far, about

200 are considered as gemstones. In recent times, many synthetic gemstones like diamond, emerald, ruby and sapphire have been produced. In addition to their use in jewellery, gems are regarded by many civilizations as miraculous and endowed with mysterious power as well.

14.1 CLASSIFICATION OF GEMSTONES

Gemstones can be classified into two groups on the basis of its atomic arrangement and crystalline structures.

Table 14.1: Classification of gemstones

	Amorphous	*Crystalline*
Properties	No orderly internal atomic structure; no characteristic shape; products of rapid cooling; physical properties constant in directions.	Definite and regular internal atomic structure; geometrical external forms; directional properties; products of slow cooling.
Examples	Glass, amber, jade, opal	All the remaining gemstones.

The above classification (Table 14.1) is mainly based on mineralogy, but gemologists prefer to follow a different mode of classification for gemstones, which is furnished in Table 14.2.

Table 14.2: Conventional classification of gemstones after gemology

Mode of classification	*Divisional groups*	
Origin	Natural	Synthetic
Cutting style	Faceted	Cobachon
View of industry people	Diamond	Coloured stone
Historical view on value	Precious	Semi-precious
User	Collection item	Jewellery item
Treatment status		

14.2 PHYSICAL PROPERTIES

The mineral's composition and crystalline structure impart the various physical properties that characterize each specimen. Knowledge of the properties of gemstones is important for the gem cutter and setter, as well as to the consumer who can use that information to care for the gem. The common physical properties include colour, lustre, streak, specific gravity, hardness, cleavage, fracture, parting and inclusions.

Colour

The visual appeal and characteristics of a gemstone are determined by several factors including: brilliance (sparkle), colour, fire (light dispersion), and lustre (surface reflectiveness) of which colour is one of the diagnostic property. The colour of gemstones may be of idiochromatic, allochromatic or pseudochromatic in nature. The gemstones display wide variety of colours like metallic (hematite), colourless (diamond, opal, sapphire), white (diamond, pearl), blue (aquamarine, sapphire), red (ruby), green (emerald, zircon), yellow (chrysoberyl), orange (citrine, opal), brown (beryl), pink (spodumene, rose quartz), purple (amethyst, zircon [rare]), grey (diamond, sapphire), black (opal, pearl), and multicoloured (alexandrite, ametrine).

Table 14.3: Types of different gemstones identified with definite colour

Cause of Colour	*Gemstones*	*Colour*
Idiochromatic: "self-coloured" due to the presence of major chemical constituents.	Malachite [$Cu_2(CO_3)(OH)_2$]	Green
	Peridot [$(Mg, Fe)_2SiO_4$]	Green
	Rhodonite	Pink
	[$(Mn^{++}, Fe^{++}, Mg, Ca) SiO_3$]	
Allochromatic: Caused by chromophores from the "transition metal" trace impurities found within crystalline structures.	Emerald (Cr) [$Be_3Al_2Si_6O_{18}$]	Blue-green
	Emerald, Jade [$Ca_2(Mg, Fe)_5[(OH,$	Green
	$F)\vert Si_4O_{11}]_2.$], Tourmaline(Cr)	Red
	Alexandrite [$BeAl_2O_4$],	Green, greenish
	Ruby [Al_2O_3],	blue
	Topaz [$Al_2SiO_4(F, OH)_2$], Spinel	Green
	[$MgAl_2O_4$](Cr)	Green, yellow,
	Turquoise	brown
	[$CuAl_6(PO_4)_4(OH)_8 \cdot 4(H_2O)$],	Green, pink
	Malachite (Cu)	Pink
	Aquamarine [$Be_3Al_2Si_6O_{18}$],	Green
	Tourmaline (Fe)	Yellow
	Chrysoberyl [$BeAl_2O_4$],	Blue
	Citrine[SiO_2],	Green-red
	Jade [$NaAl(SiO_3)_2$] (Cu)	Green
	Tourmaline (Li)	
	Tourmaline (Mn)	
	Opal (Ni) [$SiO_2.n(H_2O)$]	
	Diamond (N)	
	Sapphire (Ti) [Al_2O_3]	
	Emerald, Alexandrite, Sapphire (V)	
	Tourmaline (V)	
Pseudo colouration by dispersion: caused by varying optical effects like crystal defects created by spectral dispersion and refraction.	Diamond [C]	Red
	Zircon [$ZrSiO_4$]	Blue, yellow
Pseudo colouration: caused by light scattering, generated by physical structure of the mineral.	Adularescence: Moonstone, Opal	Blue, Milky
	Asterism: Star ruby, Star sapphire	white
	Chatoyancy: Alexandrite	Yellow
	Opalescence: Agate, Opal	
	Play of colour: Opal	
	Fluorescence: Diamond	Blue
	Iridescence: Labradorite	
	Luster (pearlescence): Pearl, Talc, Gypsum	White

The wide range of colours that gemstones display are caused due to reasons listed in Table 14.4.

Table 14.4: Causes of colouration in gemstones

Causes of colouration	*Description*
By absorption or reflection	Caused due to absorptive or reflective properties of any trace elements or impurities like iron, chromium, copper, manganese, vanadium within the crystal. Can also be caused due to chromophores, crystallographic distortions or both.
Saturation colour	Caused due to the amount of light absorbed per some unit path length.
Extinction by light	Caused due to the varieties of chromophores, with different rate of absorption within the same mineral. It can also be ascribed as tone.
Chromophores and transitional metals impurities	Caused due to the colour of chromophores or major constituents within the chemical formula of the gemstones.

GARNET ($X_3^{2+} Y_2^{3+} Si_3 O_{12}$)
X stands for Ca, Fe2+, Mn, and Mg and Y stands for Al, Cr, and Fe3+

Star Ruby

Ruby: cut and uncut

(Contd.)

Fig. 14.1: Pictorial description of gem stones.

Fracture

Fracture describes any irregular breakage that does not occur along smooth cleavage planes that are inherent to the material's crystalline structure. All minerals, including those with perfect cleavage, exhibit the ability to fracture or shatter when stressed, but when strong cleavage is present the exact nature of the fracture's "joint" can be difficult to ascertain.

This type of fracture, also known as loss of cohesion, is defined as any separation or mechanical rupture of a material into two or more pieces, under the exertion of stress which is applied to a zone of structural discontinuity. Typically, a fracture has none of the geometric sharpness of cleavage. Some gemstones can fracture without the entire gem-body separating into two or more pieces. The inclusions these internal fractures create are referred to as feathers, which can form as stress fractures around included crystals that were undigested during crystal growth. Any internal fracture tends to reduce the strength of most gemstones, and they can also inhibit the transmission of light causing cloudiness.

The strain at which the fracture happens is typically determined by the purity of the material. In a ductile fracture, some of the energy from stress concentrations at the crack tips is dissipated by plastic deformation before the crack actually propagates into a total failure.

Specific Gravity

The specific gravity of a gemstone is the ratio of the weight of the material to the weight of the same volume of water at a temperature of 4 degrees Celsius. This property mainly depends on the elemental composition, atomic bonding, atomic packing and crystalline structure. There are several ways to directly measure the specific gravity. To arrive at a relative measure of specific gravity, heavy liquids are used. Gems are placed in liquids of a known specific gravity. If the gem floats, its specific gravity is less than that of the liquid; if it sinks, the gem is heavier than the liquid; and if the gem remains suspended, it is very close to the liquid's known specific gravity. Another useful specific gravity liquid is saturated salt solution (SG = 1.08) which is used to separate amber from most plastic imitations. Amber will float and the plastic imitations will sink.

There are drawbacks to these heavy liquids though. All of the heavy liquids used to determine specific gravity are poisonous and breathing the vapours is not advised. Also gems are susceptible to chemical attack, such as amber or hematite, could be damaged using this suspension method.

Table 14.5: Specific gravity of gemstones

Gemstones	*Sp. Gravity*	*Gemstones*	*Sp. Gravity*
Amber	1.08	Pyrope	3.78
Alexandrite	3.73	Peridot	3.40
Amethyst	2.66	Ruby, Sapphire	4.03
Diamond	3.52	Spinel	3.60
Emerald	2.74	Topaz	3.53
Hessonite	3.61	Tourmaline	3.10
Almandine	4.05	Turquoise	2.82
Opal	2.15	Zircon (lighter)	4.20
Kunzite	3.18	Zircon (heavier)	4.614

Lustre

Lustre refers to the manner and degree in which light is reflected from the surface of a material. It is an important property for some gemstones with characteristic lustre.

Table 14.6: Common lustre of specific gemstones

Gemstones	*Lustre*
Diamond	Adamantine
Garnet	Vitreous
Peridot, Hessonite	Oily
Turquoise	Dull, waxy
Malachite	Metallic

Transparency

Transparency refers the quantity of light transmitted through a material and it depends upon thickness of the material, presence/absence of inclusions and effects of tarnishing, alteration or weathering. For degree of transparency, see Fig. 14.2.

Very opaque minerals such as lapis lazuli, malachite or pyrite rarely occur in fully transparent form, but relatively opaque minerals such as Jade exhibit translucency. Minerals such as agate, carnelian and chalcedony all appear as translucent when the specimen is cut thin enough to transmit light.

The presence of inclusions can play a big part in the diaphaneity of a material. Varieties of corundum (ruby, sapphire) can be heavily included with rutile which can render them nearly opaque.

Fig. 14.2: Degree of transparency.

Tenacity and Durability

The tenacity and durability of gemstones are important points to be considered in the fields of use, marketability and value. These two properties mainly depend on physical properties like hardness, cleavage, parting, brittleness/toughness, inclusions, thermal shock, gem cut, and the physical environment (gem setting and location to be worn) of the gemstone.

Hardness

The hardness of different gemstones is generally determined on the basis of Mohs' scale of hardness. A measure of absolute hardness is also done nowadays with the help of an instrument called sclerometer.

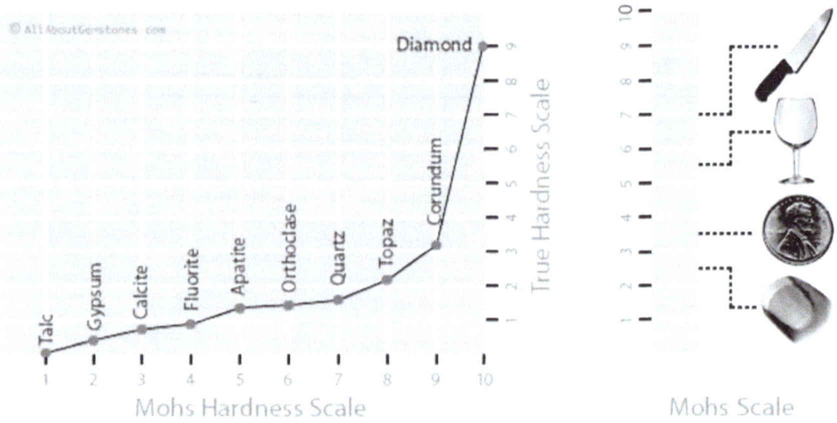

Fig. 14.3: Mohs' scale of hardness.

Mohs' scale of mineral hardness quantifies the scratch resistance of minerals by comparing the ability of a harder material to scratch a softer material.

Sclerometer: An instrument for determining absolute hardness, by means of a scratch with a diamond pyramid is called sclerometer. Here the diamond used, is standardized.

Table 14.7 shows a comparison between Mohs' hardness and absolute hardness of following minerals, since the Mohs scale is a ordinal or successive scale and therefore, does not measure or compare actual hardness. For instance, corundum is twice as hard as topaz, but diamond is almost four times as hard as corundum; yet there is only one step between each of these three minerals.

Table 14.7: Mohs' scale of hardness

Minerals	*Mohs' hardness*	*Absolute hardness*
Talc	1	1
Gypsum	2	2
Calcite	3	14
Fluorite	4	21
Apatite	5	48
Orthoclase	6	72
Quartz	7	100
Topaz	8	200
Corundum	14	400
Diamond	10	1500

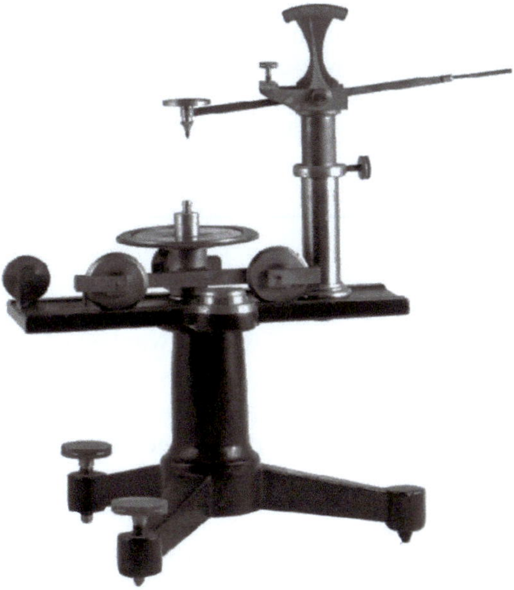

Fig. 14.4: Sclerometer.

The gemstones which can be naturally durable, range upto quartz, i.e. with hardness 7. The gemstones softer than quartz can be made durable only by neglecting touching their surfaces. Proper cleaning that minimizes contact with the stone can prolong the polish, but in time the touched areas of the stone will become dull. Continued contact with the stone, such as rubbing it during cleaning will lead to the erosion of the sharp faceted edges of the gem.

Cleavage: Cleavage refers to a weakness that is an intrinsic physical property of the chemical(s) that makes up the gemstone. Gems can have more than one cleavage plane, any one of which can lead to the instantaneous fracturing of the gem when enough energy is applied in the right location. The orientation and location of a gem's cleavage plane can also lead to the increased possibility of chipping along some of the edges of a faceted gemstone irrespective of its hardness.

Example: Tanzanite, a variety of zoisite, has distinct cleavage and hardness less than quartz. So it is less durable than many gemstones. While corundum (Ruby and sapphire) with hardness 14, high toughness and no cleavage, is a highly durable gemstone.

Table 14.8: Cleavage types found in gemstones

Cleavage	*Gemstones*
Perfect	Diamond, kunzite (spodumene), labradorite (feldspar group) malachite, moonstone (feldspar group), tanzanite (zoisite, epidote) topaz
Good	Alexandrite (chrysoberyl), chrysoberyl
Indistinct	Aquamarine (beryl), emerald (beryl), garnet group, morganite (beryl) peridot, spinel, tourmaline group, zircon
No cleavage	Agate (quartz), amethyst (quartz), chalcedony (microcrystalline quartz), citrine (quartz), jade, jasper (microcrystalline quartz) lapis lazuli, opal, ruby, sapphire, quartz

Parting

Parting is caused by the twinning of the crystal from which the gem was cut, or physical force applied to the crystal while it was being formed.Some gems are more prone to this type of weakness, but it is not universal to every sample of the chemical that makes up the gem, as is the case with cleavage.

Examples: Corundum-ruby (rhombohedral), hematite (two planes), tourmaline- elbaite (basal)

Brittleness/Toughness

The "brittleness" or "toughness" of a stone is a measure of how much impact energy can be applied to weak points such as sharp edges of facets and corners of a gemstone without breaking. All faceted gemstones are brittle to some degree, but some varieties of gems will chip excessively, even if given proper care. Although emerald is a "hard" gemstone with a hardness of 7.5 to 8, its toughness is rated as fair to poor.

Thermal Shock

Rapid changes in temperature can cause "thermal shock" in certain gemstones due to moisture content that is trapped within inclusions or fissures, as well as other structural and/or chemical features/defects that may be inherent in the gemstone. Any moisture that may be trapped within an inclusion can expand and contract due to extreme temperature shifts, causing fracturing. Two gemstones that are particularly affected by this type of thermal shock are opal and emerald. Opal can also be affected by low humidity which can cause fracturing. Gemstones that are susceptible to fracturing due to thermal shock should never be cleaned using a steam cleaner.

Inclusions

Inclusions refer to any imperfections like feathers, growth tubes, foreign crystals and other internal flaws that disturb the durability of the gemstones. Emerald can be sighted as an example which contains actinolite as inclusions that affect its durability as gemstones.

The type of cut that is used on a gemstone, and the cut's orientation within the rough stone, can make a gem less prone to breakage. All gemstones should be checked to see that their physical structure is sound before purchase. Employing proper faceting angles and a fine polish will keep gemstones bright and scintillating, even if its setting does not allow any light into the stone from the sides or back.

14.3 OPTICAL PROPERTIES

The way light will interact with gemstones depends upon the faceting of the gemstones.The "critical angle" at which a light beam intersects with a given series of facets is the angle at which total internal reflection is achieved. Light travelling through the crown and interior of the stone will eventually intersect the outer surface of the stone's pavilion facets from the inside. If the light intersects within the critical angle the facet will act like a mirror, and the light will exit, or return through the stone's crown. If the light intersects outside of the critical angle of incidence it will be internally reflected and lost.

Determination of critical angle is important for ascertaining how facets should be placed with respect to each other in order to control the path of light within the gemstones.

Optical Phenomenon 1—Refraction

When a light ray enters a gemstone at any angle except 140°, the ray bends with slowing. The degree of slowing is determined by the density of the gem and the degree of bending is specified both by density and the angle of its entry. The ratio of the speed of light in air to the speed of light in a gem is called the gem's *refractive index.* The colour of a gemstone, and the frequency (colour) of the light travelling through it, affect its refractive index. In natural gemstones it varies between 1.2 and 1.6 and is generally estimated by an instrument called refractometer.

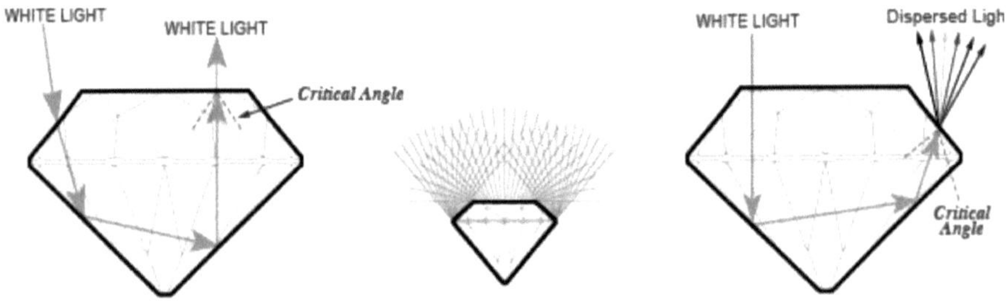

Fig. 14.5: Dispersion of light and critical angle.

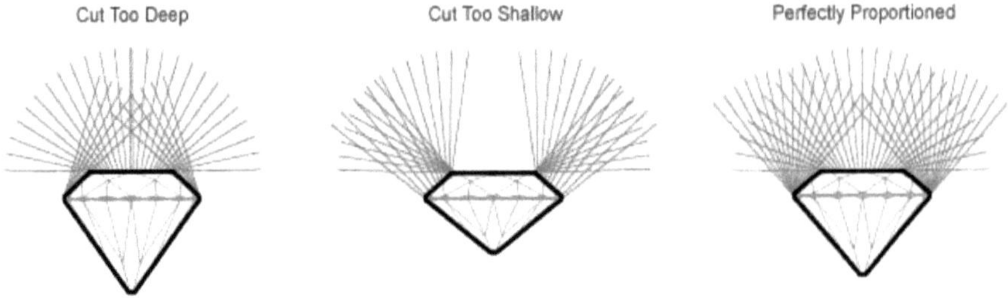

Fig. 14.6: Critical angle with respect to gem-cut.

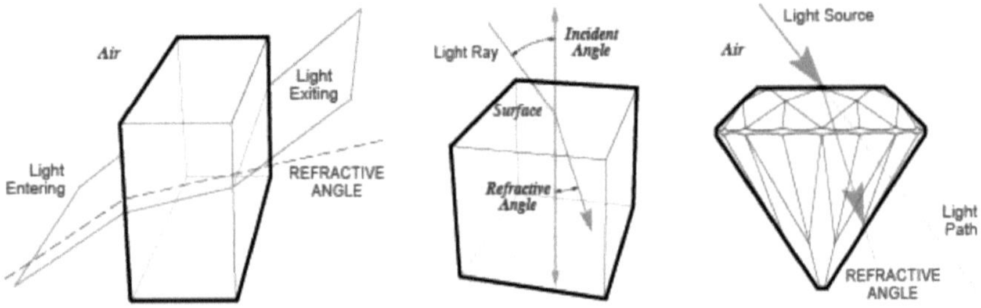

Fig. 14.7: Refraction in gemstones.

Table 14.8: Refractive index of some common gemstones

Gemstones	*R.I*	*Gemstones*	*R.I.*
Agate	1.544	Moonstone (Adularia)	1.518-1.526
Alexandrite	1.745	Morganite	1.577
Amber	1.5314 - 1.546	Onyx	1.486
Amethyst	1.532 - 1.554	Opal	1.450
Apatite	1.632 - 1.420	Pearl	1.530
Aquamarine	1.577	Peridot	1.654
Beryl	1.577 - 1.600	Quartz	1.544
Chalcedony	1.530	Rubelite	1.567
Chrysoberyl	1.745	Ruby	1.766
Citrine	1.532 - 1.554	Sapphire	1.766
Coral	1.486	Spinel	1.712
Corundum	1.766	Tanzanite	1.6142-1.700
Garnet Almandine	1.760	Tiger's Eye	1.544
Ivory	1.540	Topaz	1.620
Jasper	1.540	Tourmaline	1.624
Labradorite	1.560-1.570	Tourmaline (Paraiba)	1.610-1.650
Lapis Lazuli	1.610	Turquoise	1.610
Malachite	1.655	Zircon	1.800-1.1460
Moissanite	2.670		

Optical Phenomenon 2—Dispersion

Dispersion in gemstones is attributed to the differential refraction of various wavelengths of light as they travel through them. This behaviour is mainly dependant on the refractive index of the gems. The dispersion coefficient is determined by a stone's refractive index, quantifying its average optical dispersion and dispersive power.

Table 14.9: Dispersion of gemstones

Intensity of dispersion	*Gemstones with dispersion coefficient*
Slight	Fluorite (.007), quartz (.013)
Moderate	Tourmaline (.017), corundum (.018), spinel (.020)
Strong	Zircon (.038), diamond (.044), cubic zirconia (.066)
Very strong	Sphalerite (.156), synthetic rutile (.280)

Optical Phenomenon 3—Single and Double Refraction

The gemstones, which experience single refraction, are mainly amorphous in nature. Gemstones belonging to cubic system also undergo this phenomenon. Examples include opal, amber, diamond, garnet and spinel.

Certain anisotropic crystals exhibit a phenomenon known as double refraction, where the incident light is split into two separate rays, each with different refractive indices and velocities, and their refractive indices will typically vary between two extreme values. The difference of two extreme values of RIs, due to double refraction, gives the value of birefringence. It ranges from .003 to .287.

Optical Phenomenon 4—Pleochroism

Pleochroism is defined as an optical phenomenon in which light is focussed on stone splitting in two different components—ordinary ray and extraordinary ray polarized at perpendicular angle to each other. When different wavelength of light is travelling at different speed following different directions within a stone, it gets absorbed producing different colours when it leaves the stone. Pleochroic gemstones may have a different colour when looked at from different directions.

Factors Controlling Pleochroism

- Thickness of the gemstone
- Crystallographic structure
- Chemical properties of the stone

Table 14.10: Pleochroism in gemstones

Gemstones	*Pleochroic change in colour*
Andalusite	Yellow green to red brown
Zoisite	Pink to yellow
Kunzite	Pale pink to dark pink
Iolite	Blue to purple
Corundum	Violet to orange
Alexandrite	Red purple to orange to dark green
Peridot	Yellow green to green
Beryl	Violet to colourless
Elbanite	Pale colour to dark colour
Carletonite	Blue colour to colourless

Types of Pleochroism

Pleochroism is a general term referred to the processes of dichroism and trichorism.

Dichroism is found in uniaxial crystals with single optic axis. When ordinary light falls on uniaxial crystal any given wavelength of light is absorbed except along the optic axis displaying one colour in its direction and different colour at other angles. These dichronic stones show two colours or shades. Minerals in the hexagonal, tetragonal, and trigonal crystal system are dichroic. Examples are tourmaline, amethyst, rubies, emerald, citrine, apatite, zircon and sapphires.

Trichorism is found in biaxial crystals with two optic axes. These stones with two optic axes show sometimes three or more shades or colours. Minerals in the monoclinic, orthorhombic and triclinic crystal system are trichroic. Examples are tanzanite, peridot, andalusite and iolite.

Pleochroic halo: A spherical shell of colour, pleochroic halo is present around radioactive impurity in a stone. This ring or halo said to represent a site where crystal structure has got changed by the absorption of energy of radioactive alpha particles. As energy has got absorbed it produces an intense colour centre on the inclusion within a stone. These pleochroic halos are common in zircon, monazite, xenotime and apatite. The rings are distanced and identified according to the range of alpha particles within a stone. The beauty of a stone is enhanced when it displays different colours from different angles.

Pleochroism is commonly observed and measured by an instrument called dichroscope. One must view the gem being tested with the dichroscope from several different directions because:

1. Some directions will be optic axis directions, in which there would be no pleochroism shown, even if the gem were pleochroic. So, to avoid error, conclusion must be based by studying the stone on more than one direction.
2. Only two colours show at a time; so, although dichroism might be detected from a single directional view, it would take more than one to see all three colours in a trichroic stone.

Fig. 14.8: Dichroscope.

Optical Phenomena 5—Some Specific Behaviour of Gemstones

Fluorescence

Fluorescence is caused by impurities, so it is not an obvious phenomenon for every gemstone. When a gem absorbs either SW or LW UV, or both, and immediately emits visible light, the phenomenon is called fluorescence. In order to test for fluorescence, it is necessary to have a controlled source of SW and/or LW and a darkened viewing chamber. The specifics of the colour and intensity of fluorescence can sometimes be a useful diagnostic test in identifying gems. Fluorescence can be absent, in case of inert gem or present in weak, moderate or strong form. The light emitted by fluorescence can be the same colour or a different one from the colour of the gem itself and a gem can have the same or different reactions to LW and SW.

Table 14.11: Examples of gemstones, exhibiting fluorescence

Gemstones	*Fluorescent colours*
Diamond	Blue
Synthetic emerald	Red
Rubies (natural and synthetic)	Red

Labradorescence

It is unique for labradorite and is exhibited as a dramatic flash of blue and golden colour, which arises from light interference within microscopically thin layers of twinned crystals.

Iridescence

All sorts of rainbow effects, as well as schiller and labradorescence are referred as iridescence. It occurs due to self-interference of light in microscopically thin layers of material. Examples are pearl, fire agate, some obsidian as well as some synthetic gems.

Opalescence

The opalescence of opals, also known as adularescence, appears as a milky blue, pearly appearance. It is caused by scattering of light within the stone by thin microcrystalline layers. Usually it occurs in the blue spectrum. Examples are opal, moonstone, agate and milky quartz.

Play of Colour

The iridescent colouring in opal is called play of colour. It is controlled by the structural pattern of the gems. Due to the structural arrangement, the light interfere or diffract in gems and produce this phenomenon.

Table 14.12: The play of colour grouped by gemologists

Play of colour types	*Description*
Pinfire	A minute, close-set flash of colour
Mosaic	A large, regular, angular, close-set flash of colours
Flame	Sweeping reddish bands across the gems
Peacock	Blue and green colour effect

Chatoyancy

The reflection exhibits a single glowing band of light across the surface of a gemstone and helps us to see a silky glow, which is called chatoyancy. The displayed band of light looks like a slit of an eye. A finely marked eye line can be recognized, which runs right through the gem. In case of some good specimens, the eye seems to open and close when the gem is rotated.

Inclusions of foreign particles in gems are the main reason of reflection of light and create the phenomenon chatoyancy. The reflective inclusions of fibrous particles of foreign minerals in gems should be thin, needle-like parallel orientation. Sometimes occurrence of hollow tubes in parallel orientation is the reason of silky radiance of gems, which is known as chatoyancy. The inclusions of cat's eye gems are not well organized, scattered randomly. The silky radiance produce when the cat's eye band is perpendicular to the needle. To display the chatoyancy or cat's eye, the gems must be cut in a tall, domed shape, which is called as cabochon.

Chatoyance or cat's eye gems are transparent to opaque. The special optical phenomenon chatoyancy seems to glide magically across the surface of a gem, when it is moved. Gems that produce 'chatoyancy' allow the cosmic colour to pass through it and create a magical effect.

Examples include chrysoberyl (cymophane), crocidolite (tigers eye) and quartz. Many gems can exhibit an 'eye' including tourmaline, beryl, nephrite, jadeite etc.

Asterism

When light passes through these crystals, it exhibits a star-shaped figure, which is called "asterism". This figure is produced by reflected or transmitted light. It is a special type of chatoyancy as the cause is due to many small fibrous inclusions oriented at set angles to each other. Generally, these inclusions are the reason of the reflection of light and form a wavy star-like formation, which moves around when the mineral is rotated. All the inclusions in the gems are formed via exsolution. Exsolved inclusions form a pattern in the host mineral. The pattern may vary from mineral to mineral. These inclusions in the case of corundum are all parallel to the lateral axes of the crystal and at right angles to the vertical crystal axis. Star gems are translucent to opaque. The rays of star should cross at the top of these gems and be distinct.

Examples are commonly sapphire and ruby, garnet, spinel, diopside, rose quartz and crysoberyl.

14.4 ORIGIN OF GEMSTONES

Different gemstones with their respective possible processes of origin with description are listed in Table 14.13.

Table 14.13: Possible origin of gemstones

Processes	*Description*	*Example*
Formation from water at shallower level	Water at shallower level of the crust dissolves the minerals present. Then with change of physical conditions, the minerals precipitate from the solution.	Amethyst, opal, turquoise etc.
Hydrothermal deposits	Gems crystallize from solution of connate water and/or magmatic water, when it finds open spaces such as cracks etc.	Emerald, tourmaline
Pegmatites	Pegmatitic magma is rich in water, silica and unusual elements, which on cooling produce big crystals of gemstones.	Beryl, tourmaline
Magmatic deposits	Some gems crystallize in magmas or in gas bubbles in volcanic rocks.	Zircon, topaz, ruby
Metamorphic deposits	As a result of regional and contact metamorphism, attributed to different pressure-temperature range and presence of different impurities, gemstones can be formed.	Corundum, spinel, chrysoberyl, Garnet, cordierite Epidote, diopside, Corundum, spinel, lazurite, lapis lazuli
Mantle deposits	Gems crystallizing from upper mantle are brought to the surface through tectonic activity and volcanism. Deep mantle rocks like kimberlite under intense high temperature-pressure gives rise to diamonds.	Peridot Diamond
Alluvial deposits	On the earth crust, gems are released from the rocks by weathering. But gems, resistant to chemical weathering are deposited in stream beds and beach sands.	Diamond

14.5 IDENTIFICATION OF GEMSTONE

This is the most important aspect of gemology, which offers the main guidance to identifying a gemstone. Visual inspection, assisted by a 5 or 10 power hand lens is the easiest option available to examine a gemstone on the spot, and to the experienced this reveals a lot. In the usual gem trading, this is the only method that would be readily available. However, there are powerful scientific methods to identify gems. In these, various physical properties of crystalline materials are used, viz. crystal structure determination by X-ray diffraction, spectroscopic methods including light scattering, density, refractive index and birefringence—these have been already discussed earlier.

Table 14.14: Instruments and methods used for gem-identification

Instruments used	*Information obtained*
Standard Equipment	
Hand lens (or loupe)	10 × magnification of visual features
Binocular microscope	10 × –60 × magnification of visual features (often used with various illumination, immersion, polarization, photographic and other imaging techniques)
Refractometer	Refractive indices, optic character, birefringence
Spectroscope	Visible spectra (both prism and diffraction grating designs)
Polariscope	Optical character, birefringence, strain, twinning, pleochroism
Ultraviolet lamp	Fluorescence, UV transparency
Balance	Weight, specific gravity
Calcite dichroscope	Pleochroic colours
Colour Filters	Diagnostic colour appearance
Reflectometer	Surface luster (reflectivity)
Thermal conductometer	Thermal conductivity (thermal inertia)
Electrical conductometer	Electrical conductivity
Advanced Equipment	
Spectrophotometer	Ultraviolet, visible and near-infrared spectra (can also be used for colour measurement)
FTIR spectrometer	Infra-red spectra
Raman microspectrometer	Raman spectra of gems or inclusions
Luminoscope	Cathodoluminescence
X-ray fluorescence	Non-destructive chemical analysis of an area of sample
Electron microprobe	Non-destructive chemical analysis of a small area of a sample
Scanning electron microscope	High magnification images, element distribution mapping, non-destructive chemical analysis of a small area of a sample
Radiograph	X-ray transparency images
Diffractometer	X-ray diffraction patterns (or recorded photographically with a powder camera)

The detailed processes and techniques used for gem-identification and the methodology of the used instruments will be discussed in the next section.

14.6 GEM TESTING

The process of gemstone identification and testing involves two aspects:

(a) The correct identification of the species and variety to which a gemstone belongs.

(b) To express the measurements or distinctive characteristics by using gemological instruments.

Testing of gems using common gem testing instruments is done within a limited range. In some instances, further testing may be needed through the following methods:

(a) By applying some changes to the specimen, such as unsetting of a mounted stone, repolishing of facets or making a polished flat surface on the rough stone.
(b) By using advanced techniques. Such methods are used only when the common gem testing methods are not possible.

Although the diagnostic features of a gemstone are observed, the specimen should be named only when the identity can be proved in relation to its identified properties. It is always convenient to follow an ordered procedure, starting from observations with the unaided eye, and then with the use of instruments.

The common properties to be identified for gem testing are:

- Colour
- Fluorescence-phosphorescence
- Optical properties
- Specific gravity
- Inclusions
- Thermal conduction
- Water contact angle
- Quality of cut
- X-ray transparency
- Magnetism
- Hardness (for rough stones only)
- Non-destructive chemical

In case of some large rough stones, certain destructive tests such as chemical analysis are applied. But generally such tests are avoided.

The general instruments found in a well-equipped gem testing laboratory are:

- Stereo Binocular Microscope with digital imaging facility
- Stereo Binocular Microscope with fibre optic illuminator
- Refractometers
- Spectroscope
- Polariscope
- UV lamps
- Thermo diamond probe
- Digital weighing balance
- Other gem instruments include Chelsea Colour Filter and Dichroscope

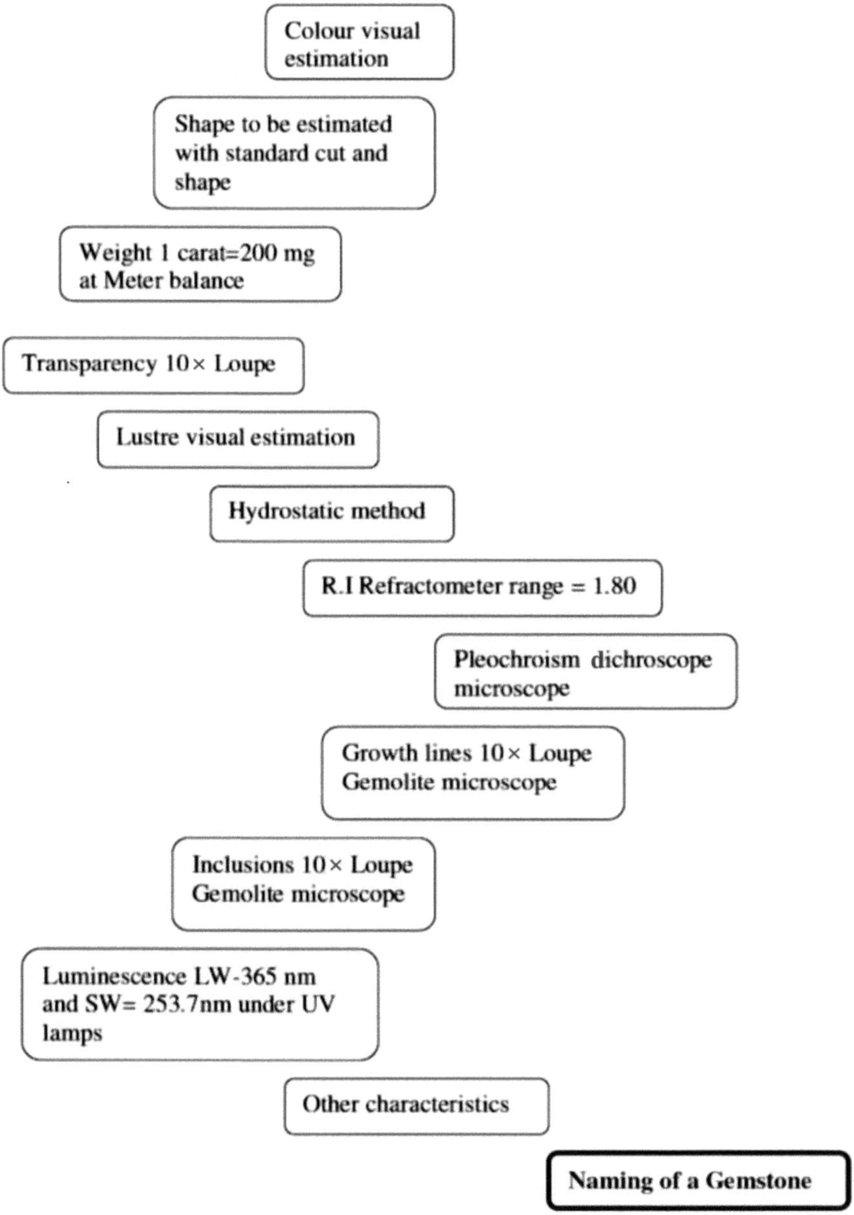

Fig. 14.9: Flow chart showing gem-testing and gem-testing instruments.

Loupe

Loupe is actually a sophisticated magnifying glass. It magnifies the gemstone in detail. The internal flaws, surface blemishes, cracks and any other cover ups are easily seen. It is a powerful convex lens that produces a magnified image of the gemstone. The standard loupe used is 10×, which magnifies the gem ten times its size. This standard is mandatory for testing gemstones in the USA.

Binocular Microscope

Binocular microscope is equipped to spread light from the bottom and from the sides. Its function is similar to the loupe. Presence of the lights and its superior optical strength provides excellent clarity. It usually has 10× magnification power though 200× can be got. Most common ones are 10× to 40×. There are stages of adjusting the iris diaphragm. It also provides coarse to fine focussing.

Refractometer

Refractometer measures the refractive index of the gemstone. This is needed in order to find out its composition and physical properties. It measures the extent to which the light that passes through the gemstone from the air is bent. It helps identify the gem by comparing the refractive index thus got with the established values. It assesses the purity of the gem by comparing it to the value of a natural gemstone. It also determines the amount of solute that has been added to the gemstone by comparing the refractive index thus obtained with the standard curve that is established. This is done by putting a drop of a liquid with a high RI, on a glass disc which is based on the tip of the refractometer. Natural or white light is passed through this. The RI is reflected in the magnified gauge.

Dichroscope

Dichroscope is used for the quick identification of gemstones. It can effortlessly separate gemstones based on dichroism, which is the property of a gemstone to show two contrasting colours when turned in two different directions, under a light. This is used to distinguish two physically similar looking gemstones. The dichroscope is used to view the dissimilar colour tones of the dichroic piece. It can easily differentiate a Ruby from a tourmaline, which is dichroic while a ruby is pleochroic. Some gemstones that are dichroic are sapphire, topaz and zircon, to name a few.

Polariscope

Polariscope is used to determine if a gemstone is natural or not. It can differentiate diamonds from moissanite. It also detects strains in diamonds. It can differentiate between gems that have double or single refraction i.e. birefringence. Double refraction is the breaking up of a ray of light into two rays, the ordinary ray and the extraordinary ray, when it passes through some materials like calcite also called Iceland spar. This effect can take place in gems that are anisotropic in structure i.e. having properties that differ according to the direction of measurement. The polariscope helps to track and see the path the rays of light are taking through the stone that is being tested. By measuring the optic characteristics of transparent to translucent gem material, the refractive index is found. Double refractive is only a property of solids. For transparent materials, the graph of index, unlike the wave length, is curved with a few general characteristics. A gem is held between two polarized plates and the light transmitted through it is studied and, based on the readings, the identification is made.

Some gemstones with double refraction are quartz, tourmaline, ruby, peridot, sapphire, zircon and rulite to name a few. Diamonds are isotropic, meaning single refraction. Topaz, mica and perovskite have triple refraction.

Ultra Violet Light source: Some gemstones start emitting a fluorescent glow after being exposed to short or long ultra violet radiation. Many gems can be easily recognized by the fluorescent colour and the locale. Some fluorescent minerals are agrellite, tugtupite, dolomite, fluorite and benotoite.

Colour Filter

Colour filter is also called a 'Chelsea filter' or 'Emerald filter'. This instrument helps in separating natural, synthetic and imitation gem material. It helps in establishing whether similar looking stones exhibiting the same colours are natural or not. It comes handy while dealing with some green, red or blue stones, like blue spinel and blue topaz. It detects dycs in gemstones also.

Spectroscope

Spectroscope is also used to separate natural gems from synthetic gemstones. This is used in specialized gem testing laboratories. The differences in the chemical composition are revealed by the absorption spectrum of the light transmitted through the gemstone that is being tested. The instrument breaks up the light that is being transmitted from a gemstone into its spectral colours. This helps in testing the various wavelengths that are being taken in i.e. being absorbed by the gemstone. This indicates the type of elements that are present in the stone, which are absorbing the wavelengths. It allows the gemologist to see which colour of light is being absorbed and thus identify the gemstone.

Specific Gravity Liquids

Specific gravity liquids: This is important for identifying a number of gemstones. A carat scale also called hydrostatic balance is used.

X-Rays

X-rays are sometimes conducted to separate natural pearls from cultured pearls.

Diamond Tester

Diamond tester is used for testing diamonds. The unit is switched on and the diamond that is to be tested is touched by a pointed tip. A light will come on to indicate if the stone is a genuine diamond or not. It uses thermal conductivity to differentiate diamond from its many stimulants. When the tip is touched to the diamond, the meter will show green if it is a real diamond; if not then it will remain in the red zone. This instrument can be used to test the minutest diamond, as small as 0.02 carats. It can also separate coloured stones from one another. By merely touching the thermo electric probe to the gemstone, the relative heat conductivity of the material is exhibited on the dial.

The use of a hand spectroscope is in determining the cause of colouration in most instances. Spectroscope is mainly used in identifying absorption bands resulting from chromophoric elements, differing from gem to gem. Apart from these certain advanced instruments such as ultraviolet-visible-infrared (UV-VIS-NIR), optical absorption spectroscope (OAS) and Electron Spin Resonance spectrometer (ESR) are also used.

Dichroscope (already mentioned) is used to determine the pleochroism of anisotropic gems. Filters such as Chelsea filter and crossed filters are employed in identifying stones like chromium-rich emeralds, cobalt-rich blue spinels and alexandrite, which look blue through these filters.

Fluorescent and phosphorescent colours are studied using ultraviolet lamps and sometimes under X-rays. Optical properties such as RI of the range up to 1.80 are determined using Rayner refractometer. Stones with RI > 1.80 are measured by relative reflectivity meters. Hydrostatic method comprising a sensitive mechanical balance is used to measure specific gravity.

Another important test is to study the gemstone under magnification, primarily under a loupe and followed by a microscope. This is used to detect inclusions which actually distinguish a natural stone from a synthetic one. Besides there are some other tests which are to be performed in order to clearly identify the gems.

14.7 UNIAXIAL GEMS AND BIAXIAL GEMS

Uniaxial Gems

- Crystal belonging to the tetragonal, hexagonal and trigonal system has only one optic axis, parallel to the crystallographic "c" axis direction.
- Tetragonal gems are zircon, idocrase, scapolite, cassiterite, scheelite, anatase, rutile and tugtupite.
- Hexagonal group comprises beryl, apatite, zincite, benitoite, painite, taffeite, covellite and nepheline.
- Trigonal division includes corundum, tourmaline, quartz, phenakite, dioptas, willemite, stichtite, calcite, rhodochrosite, haematite and silicon carbide.
- The optic sign is calculated on the refractometer by noting whether the ordinary ray or the extra-ordinary ray moves faster or slower.
- Uniaxial positive (U+): varying ray (extra-ordinary) has a higher R.I. than the fixed ray (ordinary).
- Uniaxial negative (U–): varying ray has a lower R.I. than the fixed ray.
- Both curves constant: At 0° polaroid angle, only the o-ray is seen
  (a) If low curve is seen = (+)
  (b) If high curve is seen = (–)

- Uniaxial gemstones are dichroic and two colours may be observed.
- Corundum shows habits like hexagonal prisms, rhombohedrons, hexagonal bipyramids and scalenohedron.
- Rubies are flattened parallel to basal pinacoids.
- Sapphires are elongated parallel to c-axis, displaying a shape of barrel.
- When quartz reflects reddish brown or mahogony colour under conoscopic illumination, it displays bull's eye.
- Cat's eye of apatite composition shows uniaxial interference figure when the dome of cobachon is kept sideways, perpendicular to the bright chatoyant line.
- In transparent star corundum, interference figure is straightway obtained when cobachon stone is placed with its dome pat upward.

Biaxial Gems

- Crystals belonging to orthorhombic, monoclinic and triclinic system have two directions of single refraction i.e. two optic axes.
- The angle between the optic axes varies from species to species. There is no simple relationship between optic axis directions and crystallographic axes.
- Common biaxial gemstones include chrysoberyl, alexandrite, cymophane, jade, jadeite, nephrite, andalusite, sillimanite, kyanite, feldspars, pyroxenes, amphiboles, topaz, olivine (peridot), cordierite (iolite), epidot, turquoise etc.
- The optic sign is calculated on the refractometer by noting whether alpha or gamma (in biaxial) moves beyond the half way mark of the total birefringence.
- If n_β is closer to n_α, the gem is (+).
- If n_β is closer to n_γ, the gem is (–).
- If n_β is halfway between n_α and n_γ, the gem is (±).
- If two possible betas exist, false beta will have a polaroid angle equal to 90°. True beta will have a polaroid angle unequal to 90°.
- Biaxial stones are trichroic and three colours may be seen.
- Most biaxial minerals possess relatively lower hardness, exceptions are topaz and chrysoberyl.
- Biaxial minerals like topaz, feldspar and pyriboles exhibit strong cleavage.
- Some biaxial minerals like sillimanite (fibrolite), chrysoberyl (cymophane) and feldspar (labradorite) exhibit chatoyancy and iridescence.

14.8 GEM ENHANCEMENT AND ITS CLASSIFICATIONS

Gem Enhancement

Enhancement is defined as any processing (other than fashioning) which improves the appearance or durability of a gem. Enhancement is done to make an unuseable piece of gem material useable, or an ugly one pretty, thereby increasing the value of that particular piece.

Enhancement Classifications

The designation for identifying enhancements on a gemological report is: "N" (not enhanced), "E" (normally enhanced), and "T" (nontraditional enhancements).

Table 14.15: Commonly used methods of enhancement

Enhancement techniques	*Description*	*Example*
Semi-Permanent or Temporary Gem Enhancements		
Bleaching (B)	Used to lighten and/or whiten gemstones	Pearl
Coating (C)	Surface enameling, inking, foiling, lacquering, sputtering, or vapour deposition	Druzy quartz
Dyeing (D)	Used to enhance the colour and uniformity of gemstones	Pearl, onyx, druzy quartz, coral, lapiz lazuli
Filling (F)	Filling surface cavities or fissures with colourless glass, plastic or resin	Opal, emerald
Impregnation (I)	Treating gems with oil, wax, or resin to improve clarity and appearance	Opal, emerald, peridot
Waxing (W)	Impregnation with colourless wax, paraffin or oil	Emerald, peridot, opal
Permanent Gem Enhancements		
Flux Healing (FH)	Heat enhancement used to heal fissures and fractures	Ruby
Fracture Filling (F)	Injecting plastics or glass into fractures	Ruby, sapphire
Heat Treatment (H)	Used to lighten, darken, or alter colour, to deliberately add stress fracture inclusions.	Amethyst, amber aquamarine, druzy quartz, citrine, rubellite, ruby, sapphire, tanzanite, topaz, tourmaline, zircon
Irradiation (R)	Used to add colour intensity to diamonds, gemstones, or pearls	Druzy quartz, diamond, morganite, pearl, topaz, tourmaline
Lattice Diffusion (U)	High-temperature heat treatment to produce colour and/or asterism	Sapphire
Lasering (L)	Laser and chemical treatment to remove inclusions	Diamond

Table 14.16: Changes induced by heat treatment

No	*Change due to heat-effect*	*Example*
1	Darken colour	Light blue sapphire to dark blue
2	Light colour	Dark pink tourmaline to light pink
3	Colour change	Amethyst to citrine
4	Removal of secondary colour	Removal of purple hue from ruby, green colour from aquamarine
5	Development of asterism/silk	In corundum
6	Removal of asterism/silk	In corundum
7	Structural change	Low zircon to high type zircon
8	Cracking	Irish quartz

Factors controlling such heat-induced changes

- Maximum temperature
- Time duration of maximum temperature
- Heating rate/change
- Chemical nature of atoms
- Nature of material contact in furnace with gemstone
- Quality of gemstone

Table 14.17: Possible changes in corundum by heat treatment

Atomic state	*Type of change*	*Temperature*	*Time*
Oxidising	Lightening of dark blue	1700°C	Prolonged hours (~20 hrs)
	Reduction of colour zoning/ yellow patch	1800°C	3 hrs
	Reduction of blue patch	1800°C	4 hrs
	Intensity yellow	1800°C	2 hrs
	Removal of silk	1650°C; 1000°C (at 40°C/min)	Cooling at 40°C/min
	Development of silk	1500°C	2 hrs
Reducing	Darkening of light blue	1770°C	6 hrs
	Development of pink	1800°C	3 hrs
	Development of blue	1600°C	2 hrs

Most of the 'temporary' enhancements or treatments are basic and self-explanatory, but most of the 'permanent' gem enhancements involve highly sophisticated techniques that can be difficult to detect. Most treatments leave some kind of telltale signature or fingerprint, but as treatment methods become increasingly subtle, they are more problematic. Some of the more involved treatments and enhancements are listed above. These treatments are permanent, but can have a significant impact on the value of a gem. Most gemstones, with the notable exception of garnet, have a particular treatment, or series of treatments that are commonly used to increase the marketability of the stone.

The general precautions that will protect almost any enhanced gem are:

- Avoidance of solvents,
- Ultrasonic and steam cleaning,
- Gentle wear,
- Protective settings,
- Avoidance of recutting, and
- Removal of gems from their settings prior to jewellery repair.

14.9 GEMSTONE CUTTING—PROCESS, TYPES AND TECHNIQUES

Coloured gemstones are generally cut to maximize the beauty of their colour. Cutting (also known as lapidary) is the process whereby a rough stone is turned into a gemstone. The process makes a gem assume a certain shape, bringing out its lustre and colour, enabling it to be set into jewellery.

A good cut enhances the gemstone's colour, diminishes its inclusions, and exhibits overall good symmetry and proportion. Good cutting also brings about its brilliance which is an unified result of lustre, clarity, fire and scintillation.

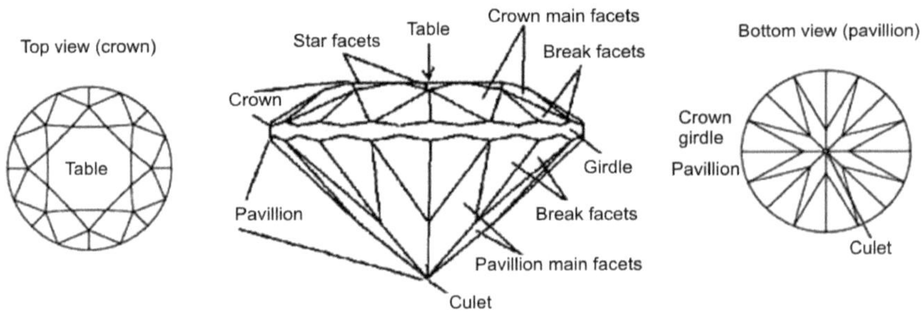

Fig. 14.10: Gemstone cutting.

Types of Gem-cutting

There are four basic styles of gem cutting: tumbling, cabbing, faceting and carving.

Gem-cutting Techniques

The techniques of cutting in faceting gemstones are:

Slicing: Also called cutting, slicing is one of the most crucial stages in the finishing of gemstones, as it will ultimately determine the size and beauty of the finished gem. Once the roughstone is selected, using a diamond-tipped circular steel saw, the gem slicer will determine how to cut, where to cut, and how many pieces to cut, in order to produce the highest quality.

Table 14.18: Types of gem-cutting

Types of cutting	*Description*
Tumbling	The rough material is put in a revolving barrel with abrasives. Progressively finer abrasives are used, until a polish is obtained.
Cabochon cut	Cabs are gems that are cut with a flat bottom and a curved or domed top. Ex: opal or turquoise.
Faceting	The geometrically arranged, flat surfaces on the surface of gemstones are called facets. The gem is faceted, by a faceter on a faceting machine. The purpose of faceting is to bring out the brilliance of a gem.
Carving	Is the most challenging of the lapidary arts and there are very few recognized experts in the field. Often cabochons are carved. If the design is cut into the top, it is called an intaglio, or a relief carving. If the design is carved on the back, it is a reverse intaglio.

Pre-forming: Once the roughstone has been carefully cut, pre-forming commences. This process requires tremendous experience and concentration. Pre-formers determine the most suitable shape for each gemstone. Apart from beauty, pre-formers also control the weight of the finished gem. Pre-forming is typically performed by using a vertical steel grinding wheel.

Shaping: The shaper uses a special type of heat activated resin to affix the pre-formed gemstone onto a metal rod, commonly called a "dob stick. " The shaper then delicately applies the gemstone to the shaping wheel to obtain a more accurate presentation of the facets and size. Due to the immense precision required by this process, the shaper is usually a very experienced pre-former. Shaping is completed using a hand-operated shaping wheel.

Polishing: The final step is known as polishing. Once gemstones have reached their ideal size and shape, they are taken to a steel (or steel and copper) horizontal polishing wheel where the polisher completes the faceting and gives them a final polish using fine diamond paste to reveal their hidden lustre, brilliance and fire.

Factors on which the cutting depends: The cut of a gemstone largely depends on the shape of the gem rough (i.e., the shape of the raw gem crystal as it comes from the earth). The oval cut is most frequently used, as it best balances beauty and carat weight retention. Factors to consider when choosing to facet a gem in another shape include design aesthetics, inclusions, carat weight loss and colour.

Some Common Cuts in Gemstones

Round brilliant cut: The round brilliant cut is also known as the round cut, American ideal cut or American standard cut. The standard number of facets of a round brilliant cut gemstone is 57. This cut is optically the most efficient. The round brilliant boasts one of the best recoveries for well shaped diamond and gemstone rough. The round brilliant cut is designed to provide maximum optics for brilliance and scintillation (play of light), making the gem sparkle and dance in the light. This cut was specially developed for diamonds but is today common for all gem types.

Oval cut: The standard number of facets of an oval cut gemstone is 614 and it has an elliptical shape when viewed from the top. For the oval cut, the ratio of the length to the width should be approximately 2:1, although this does vary slightly depending on the optical properties of different gem types. A well cut oval gemstone can be nearly as bright as a round brilliant cut. The oval cut is a particularly beautiful shape and if well proportioned gives great scintillation and fire.

Baguette cut: The baguette shaped gemstone is really only a special oblong shape and has approximately 20 facets. Most oblong cuts are "step" cut, which means that the facets on the pavilion have been cut in steps, parallel to the edges, in the manner of a pyramid with its top chopped off. The base and table are square with triangular facets. The baguette cut best suits gem types whose rough crystals occur in a similar shape (e.g., tourmaline).

Square cut: The standard number of facets of a square cut gemstone is 57 and is really only a special oblong shape where the sides are the same length. Most oblong cuts are "step" cut, which means that the facets on the pavilion have

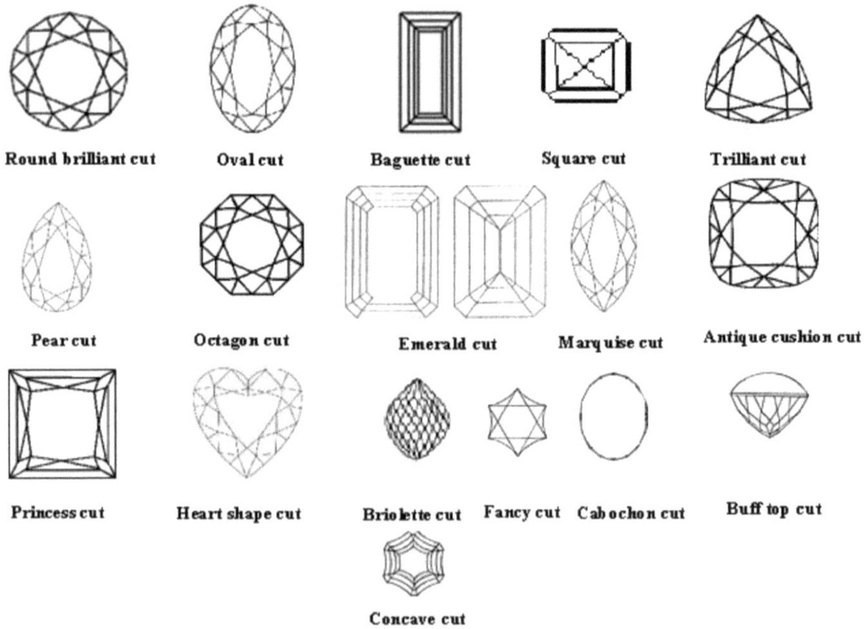

Fig. 14.11: Some common cuts in gemstones.

been cut in steps, parallel to the edges, in the manner of a pyramid with its top chopped off.

Trilliant cut: The standard number of facets of a trilliant cut gemstone is 43. Trilliant cut gemstones are based on a triangular shape. Usually with truncated corners and displaying a variety of facet designs, this cut creates a spectacular wedge of brilliant fire. The tips and culets of trilliants are pointed and thin. Trilliants work well with light-coloured gems—such as diamonds, aquamarines, beryls and white sapphires—where cutters try to maximize brilliance. Inversely, some cutters use trilliants to effectively lighten and brighten the appearance of darker gems such as tanzanite, spessartite garnet, rhodolite garnet and amethyst.

Pear cut: The standard number of facets of a pear cut gemstone is 71. A hybrid cut, combining the best of the oval and the marquise, it is shaped like a sparkling teardrop. A nice pear cut is generally one that is well cut with a polished girdle. Although it varies depending on the optical properties of each gem type, pear cuts should generally have a good depth such as 1.5:1 aspect ratio for a great look and a lively gem.

Octagon cut: The standard number of facets of an octagon cut gemstone is 53. This is another "step" cut but with the four corners mitered. The facets run in steps parallel to the gemstone circumference. This cut is differentiated from the emerald cut by steps on the pavilion that are not equidistant. With this cut, colour plays a very important role in the beauty of the gemstone as colour tends to show very dramatically in Octagon cut gemstones.

Emerald cut: The approximate number of facets of an emerald cut gemstone is 50. The emerald cut looks like a rectangle from the top, with truncated corners. These can be beautiful gemstones with stepped facets; the sheen tends to display large flashes of these stepped angles on the pavilion of the gem. This is another "step" cut and it has rows of facets that resemble a staircase and usually are four-sided or elongated. It is known as a step cut because its concentric, broad, flat planes that resemble stair steps. The emerald cut was developed specifically to reduce the amount of pressure exerted during cutting and to protect the gemstone from chipping. Today, modern cutting techniques make this less important and it is used for a wide variety of gem types.

Marquise cut: The standard number of facets of a marquise cut gemstone is 57. The marquise cut is also known as the "Navette" shape and looks like a long oval that has been stretched out to a point at each end like a football viewed straight down from the top. The general ratio of length to width should be 2:1. It is important that the marquise-cut gem not be too shallow or light will pass through the back of the gem diminishing its brilliance and colour. However, as with all coloured gems, this can vary from type to type.

Antique cushion cut: The approximate number of facets of an antique cushion cut gemstone is 64 and it is also known as "The Old Miner" or "Old European" cut. Primarily used for ruby and sapphires.

Princess cut: Technically known as the "square modified brilliant" cut, the princess cut is a square version of the round brilliant cut with 76 sparkling facets. Depth percentages of 70% to 78% are not uncommon in the princess cut and it is a "brilliant style" shape with sharp, uncut corners ("brilliant style" refers to the vertical direction of the crown and pavilion facets).

Heart shape cut: The standard number of facets of a heart shape cut gemstone is 514 and is a pear-shaped gemstone or diamond with a cleft at the top. Generally, a heart shape's length to width ratio is slightly over 1:1, approximately 1.1:1 in favour of length, but usually not over 1.2:1.

Briolette cut: The briolette cut is a drop or pear-shaped gemstone with shaped facets all the way around. There is no table, crown or pavilion. Considering the shape of the briolette, it is most difficult to cut and an experienced cutter can only cut and polish 5-10 briolette gemstones per day. Because of the specific number of cuts to show the facets, the briolette cut requires perfection from top to bottom. While the approximate number of facets of a briolette cut gemstone is 84, the more facets the drop has, the more brilliant it is.

Fancy cut: Fancy cuts include flowers, clover leaves, stars, triangles, kites and all manner of modern cuts.

Cabochon cut: A cabochon is a polished gemstone with a flat bottom (or slightly rounded bottom) and a convex or rounded domed top. The traditional cabochon is an oval but cabochons can also be fashioned into other shapes including triangles and rectangles. Cabochons, commonly known as cabs, are the oldest and most common form of gem cutting. Cabochons are used for making jewellery, often carved as intaglio (a gem carved in negative relief) or cameo (a gem carved in positive relief), and are also used in crystal healing. Today, the cabochon cut is applied to gems of limited transparency (e.g., turquoise, jade, agate, etc.) or as a result of predominant inclusions (e.g., relatively opaque sapphires, rubies or emeralds) or for gems where the cut's curved surface accentuates special characteristics (e.g., iridescence, chatoyancy or the cat's eye effect, asterism or the star effect, etc.).

Buff top cut: A cabochon variant for transparent gems, the buff top cut mixes a faceted cut with a non-faceted cut. This results in a gem with the typical domed top of a cabochon and a faceted pavilion, giving the illusion of depth as the eye is drawn into the centre of the gem. The cut shows good brilliance and has a crown that is less easily abraded than those of faceted gems.

Millennium cut: Possessing an incredible 1000 facets, the millennium cut is so named as it was created by Rogerio Graca as a unique and challenging symbol of the new millennium. Having 624 facets on the pavilion and 376 facets on the table, each facet has to be touched from one to four times during cutting and polishing.

Concave cut: The concave cut is a three dimensional conical shaped facet applied to the pavilion of the gem that creates depth as well as length and breadth. Instead of the facets being joined by an angle they are joined with a groove. This third dimension allows the gem to refract more light, thereby maximizing its brilliance. The concave cut also distributes light more evenly, giving the gem a homogeneous interior glow.

Mirror cut: The mirror cut is characterized by an extraordinarily large table and thick girdle consisting of as much as 140% of the width of the gem. This makes the gem highly refractive and literally gives it the properties of a mirror, hence the name.

Cutting of Diamonds

A diamond's cut grade is an objective measure of a diamond's light performance or more generally, its sparkle. When a diamond is cut with the proper proportions, light is returned out of the top of the diamond. If it is cut too shallow, light leaks out of the bottom; too deep and it escapes out of the faces.

During dispersion, a white ray splits into its seven composition colours. Splitting of rays takes place due to different R.I. of the rays, depending upon its wavelength λ. When a ray travels from a denser to rarer media (e.g. from glass to air), travelling path is not very long and the emerging surface is slanting then the splitting can be clearly observed (e.g. in prisms) and for longer paths (e.g. in glass blocks) the overlapping of these colours reunite to form the white light again. This particular phenomenon is used in gem-cutting to produce brilliance and rainbow.

Cut Grades

Our most brilliant cut, representing roughly the top 1% of diamond quality based on cut. The highest grades of polish and symmetry allow it to reflect even more light than the standard ideal cut.

Ideal cut: Represents roughly the top 3% of diamond quality based on cut. Reflects nearly all light that enters the diamond. An exquisite and rare cut.

Very good cut: Represents roughly the top 15% of diamond quality based on cut. Reflects nearly as much light as the ideal cut, but for a lower price.

Good cut: Represents roughly the top 25% of diamond quality based on cut. Reflects most light that enters. Much less expensive than a very good cut.

Fair cut: Represents roughly the top 35% of diamond quality based on cut. Still a quality diamond, but a fair cut will not be as brilliant as a good cut.

Poor cut: Diamonds that are generally so deep and narrow or shallow and wide that they lose most of the light out of the sides and bottom.

14.10 SYNTHETIC GEMSTONES AND THEIR PRODUCTION

A synthetic gem is created in a laboratory under controlled conditions. The gem is grown from a natural gemstone "seed". By controlling the environment in which the gem is grown, impurities and inclusions can be avoided which occur in natural gemstones. The created stones have perfect colour and clarity which is so highly desired. Most natural gemstones have imperfections due to uncontrolled growing conditions in nature. Synthetic stones will have the same chemical, physical, and optical properties of their natural counterparts without the imperfections. But they do not have the rarity of naturally coloured stones and they are less expensive than naturally mined stones.

The value of synthetic gems is two-fold: low production cost, and the ability to control characteristics. Synthetic gems contain the same mineral properties of naturally existing gems, including the trace elements that cause colour. They have the same brilliance and fire, and are often of better structure, since growth is controlled. The gems are cut to near-ideal proportion, and to the naked eye are essentially flawless.

Some Common Synthetic Production of Gemstones

Synthetic Corundum (Ruby)

The first synthetic ruby was made in late 1800s by A.V. Verneuil by flame fusion method. In this method the aluminium oxide is melted and is allowed to crystallize in a cone-shaped bowl; the colour is controlled by the addition of chemical additives. These synthetics can be differentiated from the original by the presence curve growth striations, and spherical gas bubbles or by Plato method.

Other methods of ruby production are flux-grown, Czochralski pulled and hydrothermal methods.

Synthetic Spinel

Verneuil method is used as an imitation for blue sapphire and aquamarine. Differentiated from the original by its R.I, spherical gas bubbles and identifying strain lines between crossed polars.

Synthetic Emerald

Produced by flux growth, hydrothermal and lechleitner overgrowth and sold under the names of Gilson and Chatham. Distinguished from the original by the presence of inclusions.

Other growth methods for synthetic gems include solution techniques for emeralds, hydrothermal method for aquamarine and others.

Synthetic Quartz

Quartz crystals are grown in a solution in autoclaves, where temperature and pressure are controlled to create the feed material called lascas in the hotter portion. Seed crystals are in the cooler portion upon which the lascas redeposits, forming synthetic quartz. This process can take 30 to 60 days, and is also used to grow amethyst, citrine, or rock crystal.

Turquoise, jade and lapis lazuli are grown via liquid-state reactions and phase transformations.

Of all the methods available, the Verneuil flame-fusion, the Czochralski pull method, and the skull melting processes are the ones most often used for gem materials.

Synthetic Diamond

Synthetic diamonds, grown from a metallic flux at high temperatures and pressures, have evoked concern in the jewellery trade ever since the first production of gem quality crystals in the early 1470s. However, due to the growth conditions involved, as well as the great expense and restricted availability of the growth equipment used, the actual number of gem-quality synthetic diamonds continues to be limited. Those that are encountered in the jewellery trade are mainly brownish-yellow crystals, weighing one carat or less (1 carat = 0.2 g), which can yield faceted pieces of 0.5 carat or smaller in size. In contrast to synthetic diamonds, colourless or near-colourless imitation materials are much more abundant.

Over the years, a number of natural gem minerals and synthetic materials have been used to imitate colourless diamond.

Cubic zirconium oxide (cubic zirconium or CZ) is the most widespread stimulant because of its low cost and similar appearance to a polished diamond. It (and most other stimulants) can be readily distinguished from diamond on the basis of a difference in heat conduction, which can be detected with a simple gemological test instrument.

Recently a new material, synthetic moissanite (silicon carbide), has been marketed for jewellery purposes. It has caused some serious identification problems among jewellery, because it cannot be distinguished from diamond by the heat conduction test mentioned earlier. However, synthetic moissanite displays optical features due to its anisotropic optical character (seen as an optical doubling when viewed with magnification) which allow it to be easily recognized by trained gemologists.

Distinction between natural and synthetic gemstones

- Viewing the refraction lines under a microscope,
- Viewing metallic inclusions through the microscope,
- Subjecting synthetic gems to shortwave ultraviolet light and
- Viewing patterns caused by seed crystals under the microscope.

14.11 USES OF GEMSTONES

Gemstones are very useful for humankind. These stones have been used since time immemorial. But with rapid advancement of civilizations and growth of various industries, the gemstones have surpassed the jewellery kingdom and have paved their ways through other uses as follows:

- Diamond (synthetic)-embedded cutters can slice through large pieces of concrete and steel.
- Diamond (synthetic) is also used for lathe machine cutting tools.
- Diamond, due to its high thermal conductivity, can be used to reduce the friction in many engineering parts.
- Diamond as a 'heat sink' helps in extending the life of the machinery since it avoids wear and tear due to friction and heat.
- 'Slices' of synthetic diamonds are also used for other industrial and surgical tools.
- Diamond is embedded in mechanical tools to enable the shaping of engine blocks and automotive components.
- Gemstones are recognized by modern science in the technological uses of crystals in watches, lasers and computers.
- Rubies are used in laser, due to their exceptional wave-transmitting abilities through entire range of electromagnetic spectrum.
- Due to high hardness, rubies are used to guide threaded needles in textile industries.
- Rubies are used as bearings to reduce friction in precious devices.
- Corundum is very hard and is used as an abrasive in the manufacture of sandpaper, polishing compounds and cutting tools.
- Gem varieties of corundum are sometimes used for pivot supporters in delicate scientific instruments, as jewel bearing in watches.
- Garnet is used in industry as coated sandpaper.
- Zircon also has industrial uses in the manufacture of ceramics, glass, metals and chemicals.

Concluding Remarks

Gemstones form an essential branch of mineral kingdom. This chapter gives a brief idea about their types, properties, origin, industrial synthesis, enhancement and the utilities. Recent development of the analytical techniques provides many methods of production of synthetic gems and enhancement of the natural ones. These analytical techniques also help to distinguish the artificial and natural ones.

Think for a while

1. What properties does a mineral require to be regarded as a gemstone? Demarcate the boundary between a precious and a semi-precious stone. Optical properties of gemstone play a special part in ascertaining its value. Explain.
2. Gemstones represent varieties of origin. Give a short note on their genesis. Pearl, coral or agate, though considered as a gemstone, is different from other gemstones. Why?
3. Explain the term 'gem enhancement'? Why is it necessary? Give a short note of various techniques required for gem enhancement.
4. Why gem-cutting is an important aspect in the science of gemstones. Style of cutting varies from one gemstone to another. Why and how? Give a note on some common cuts of gemstones.Why cuts have given special emphasis in case of diamonds?
5. Synthetic preparation of gemstones is widely in use. What are the common methods for such preparation? How can one distinguish between synthetic and natural gemstone?

FURTHER READING

Anderson, B.W. Gem Testing. Butterworths, London. 1980.

Hurlbut, C.S. and R.C. Kammerling. Gemology. John Wiley & Sons, Inc., NY. 1991.

Karanth, R.V. Gems and Gem Industry in India. The Geological Society of India, 1st ed. 2000.

Liddicoat, Richard T., Jr. Handbook of Gem Identification.12th ed. Gemological Institute of America, Santa Monica. 1993.

Schumann, W. Gemstones of the world. Sterling Publishing Co, Inc., NY. 1997.

Sinkakas, John. Gemstone and Mineral Data Book: A Compilation of Data, Recipes, Formulas and Instructions for the Mineralogist, Gemologist, Lapidary, Jeweler, Craftsman and Collector. Geosciences Press, Prescott, AZ. 1988.

Webster, Robert. Gems: Their Sources, Descriptions and Identification. Fourth ed. Rev. B.W. Anderson. Butterworths, London. 1983.

Internet Data Retrieved from:

www. gemenhancementgemstone. com

www. asterism. com

USGS. Gemstones - Synthetic and Simulant. com

www. optical properties of gemstones. com

Allan Taylor. Gemstones and their industrial uses. Reproduced from the Field Geology Club of SA Bulletin, Vol. 34, No. 7, pp 7-9, August 2005

Origins of Minerals - The Gemology Project: A Gemologist & Gemstone Database. htm

CHAPTER

15

MINERALOGY IN EXPLORATION OF MINERAL DEPOSITS USING MAGNETIC, ELECTRICAL AND GRAVITATIONAL PROPERTIES

Exploration geology is the process and science of locating valuable mineral or petroleum deposits, i.e., those which have commercial value. The term "prospecting" is almost synonymous with the term "exploration". The exploitation of the vast mineral reserves to meet the growing requirements for a variety of applications has been a major economic activity, contributing significantly to a country's industrial development and export trade. If bedrock is exposed anywhere at or around a prospect, then surface bedrock mapping is an essential beginning step for an exploration programme. This would include mapping and sampling (field geologic methods). This work focusses on identifying and mapping outcrops, describing mineralization and alteration, measuring structural features (geometry), and making geologic cross sections. Geochemical methods involve the collection and geochemical analysis of geological materials, including rocks, soils and stream sediments. The results mapping and sampling may suggest patterns indicating the direction where an ore deposit could be present underground or at the surface. Geophysical methods focus on measuring physical characteristics (such as magnetism, density or conductivity) of rocks at or near the earth's surface. The measured values are then used to compare with the values and models of known ore deposits. A closer look on all these methods will be given in this chapter.

15.1 INTRODUCTION

Mineral exploration may be defined as systematic investigation of a mineral deposit, encompassing all the geological, geochemical and geophysical data to ascertain the target zone for potential mineralization. Prospecting, on the other hand, involves the detailed examination of surficial materials and underlying bedrock in search of an ore deposit, involving all the geological, geochemical and geophysical tools.

15.2 METHODS OF PROSPECTING—OVERVIEW

The following methods of prospecting are broadly in use:

Geological

Geological maps provide exploration agencies or companies with regional geological and geophysical information so that target areas considered to have a better prospect in terms of mineral deposits may be identified. The cost of undertaking geological surveys, many of which will not prove to be prospective, is high. Geological surveys provide exploration and mining companies with pre-competitive geo-scientific data that is designed to encourage the company to undertake further exploration.

Geological methods rely on the identification of rocks and minerals and an understanding of the environment in which they are formed. These surveys aim to find what rock types occur at or close to the surface and how these rock types are related to each other i. e. their boundaries, ages and structure. Based on known "environments for mineralization" or models for mineralization, regional geological surveys can be used to define smaller areas in which more detailed studies can be undertaken. A geological survey can be undertaken using a number of methods depending on the size of a region and the amount of information that is required.

Geochemical

Geochemical exploration, also known as geochemical prospecting and exploration geochemistry, is the search for economic deposits of minerals or petroleum by detection of abnormal concentrations of chemical elements or hydrocarbons in surficial materials such as soils, waters and plants. For convenience, geochemical exploration is usually divided into two areas of specialization:

1. the search for metallic minerals deposits, and
2. the search for accumulations of crude oil and natural gas.

The object of the search is same in each case—the discovery of some dispersion of chemical elements or hydrocarbon compounds at levels sufficiently above normal to be called a geochemical anomaly. It is hoped that the anomaly might indicate the presence of mineralization or hydrocarbon accumulations at depth.

Geophysical

Geophysical techniques are routinely used in an exploration programme to help the project geologist delineate areas favourable for the type of target being pursued. Geophysical techniques can look beneath alluvial cover. They can be used to directly detect some minerals, indirectly detect others, and to map geological and structural features in exploration programmes. Direct detection includes using induced polarization (IP) to find disseminated sulfides, magnetics to delineate magnetite hosting rocks, and gravity and electrical techniques for massive sulfides.

Examples of indirect detection of targets include using IP to detect pyrite in association with sphalerite and gold (both non-responders to IP geophysical techniques), and copper and molybdenum in porphyry systems. Magnetics are routinely used to search for hydrothermal alteration in association with porphyry systems, and can be used to map buried stream channels (magnetite sands) that might host placer gold.

Remote-sensing Survey

Remote sensing is the science of acquiring, processing and interpreting images and related data, acquired from aircraft and satellites, that record the interaction between matter and electromagnetic energy. Remote sensing images are used for mineral exploration in two applications: (1) map geology and the faults and fractures that localize ore deposits; and (2) recognize hydro-thermally altered rocks by their spectral signatures. Landsat thematic mapper (TM) satellite images are widely used to interpret both structure and hydrothermal alteration. Digitally processed TM ratio images can identify two assemblages of hydrothermal alteration minerals: iron minerals, and clays plus alunite. Hyperspectral imaging systems can identify individual species of iron and clay minerals, which can provide details of hydrothermal zoning. Silicification, which is an important indicator of hydrothermal alteration, is not recognizable on TM and hyperspectral images. Quartz has no diagnostic spectral features in the visible and reflected IR wavelengths recorded by these systems. Variations in silica content are recognizable in multispectral thermal IR images, which is a promising topic for research.

Metallic Minerals

Important metallic minerals include gold, silver, copper, lead, zinc, manganese, iron, tin, nickel, chromite, cobalt, vanadium, molybdenum, tungsten and titanium.

Methods generally used during exploration

Soil surveys are one of the most widely used methods of geochemical exploration for the detailed, local assessment of an anomaly. The method works because weathering and leaching of buried deposits can release anomalous concentrations of heavy metals to the soil and ground water. The released heavy metals spread outwards and create a dispersion halo in the soil which is much larger than the deposit itself. In general, it is the A and B soil horizons (especially the B) that are most effective for soil surveys.

Rock surveys (also known as ***lithogeochemical*** or ***bedrock surveys***) entail the sampling of unweathered bedrock, usually by drilling. The object is to outline host rocks that are favourable for mineralization. Rock surveys are most effective for reconnaissance surveys.

Stream-sediment surveys are used most commonly for reconnaissance surveys. Stream sediment surveys are conducted by sampling the sediments of drainage basins because, if sampled correctly, a stream sediment is an ideal composite of the materials in the drainage basin lying upstream from the sample site. Panning for gold is an example of a stream sediment survey.

Water surveys involve the sampling of surface or ground water. Higher concentrations of chemical elements tend to be found in ground water and for this reason groundwater surveys are preferred in most cases to surface-water tests. Groundwater surveys are often carried out in conjunction with soil surveys and are most effective for detailed surveys. Surface waters tend to be very dilute because they lose much of their dissolved load by adsorption on fine-grained clays in stream sediments.

Biogeochemical surveys are used in cases where vegetation can be used as a test medium. Geobotanical studies have been used since ancient times as a prospecting tool. It has long been known that specific plants or plant communities are indicative of a high concentration of a given chemical element in the soil. It is also true that high concentrations of certain trace elements can cause the malformation of leaves and add colourings in plants. Such clues have long been used as possible indications of buried mineralization. Modern biogeochemical surveys entail the sampling and chemical analysis of plants for unusual concentrations of elements. The system works because the roots of some trees can reach as deep as 50 metres. A large volume of soil can thus be sampled by analyzing appropriate parts of a tree, such as twigs, needles, leaves, or bark. Biogeochemical surveys are less certain than soil or groundwater samples because factors such as the age of a plant, the local climate, and the part of the plant sampled can affect the readings obtained.

Gas surveys are useful to aid in the location of buried ore deposits. These can be found through the detection of gases such as sulphur dioxide, hydrogen sulphide, and vapours of mercury, iodine and radon. Some gases must be measured in the soil but others, such as mercury vapour, can be detected up to 100 metres above the ground surface and can therefore be measured using air-borne techniques.

Non-metallic Minerals

Non-metallic minerals mainly include crushed and broken limestone, magnesite and dolomite, construction sand and gravel; crushed and broken granite and phosphate rock. Other products include gypsum, soda ash and other broken stone. Phosphates and potassium salts are used to make fertilizers. Crushed stone, sand and gravel are also referred to as aggregates.

Most quarries are open-pit mines where the surface is blasted to reach stone mineral and stone deposits. Benches are cut into the walls to enable access to deeper deposits. The rock is blasted from the mine face, loaded into trucks, and carried to the primary crusher, which breaks it into smaller pieces. These smaller pieces are carried to the surface by conveyor and sorted by size. The aggregate is then transported to customers, usually by truck. Some mines further process onsite, whereas smaller mines may ship the aggregate to third-party processing facilities. Processing includes further crushing, sorting, washing and leaching.

Transportation costs can exceed the price of products, so aggregates typically supply local markets. While aggregates have traditionally been transported by truck, some companies also use railroad cars, and to a lesser extent, barges. Federally funded projects, such as road, airport and municipal building construction, can account for as much as 40 percent of aggregates sales. Fluctuations in energy costs can also influence aggregate prices significantly.

Mines and quarries can be quite large, but their use depends on local demand. Each year hundreds of mines are idled, closed, or abandoned and hundreds more opened or reactivated. The changing locations of construction and highway projects drive these decisions. Environmental regulations require companies to return abandoned quarries back to their original look and use.

Research and development expenditures are small, generally less than one percent of revenue.

Non-fuel Minerals

Clay group of minerals is one of the major types of non-fuel minerals, which are also non-metallic in nature. Sillimanite, andalusite and kayanite also fall within this group.

Clay Mineral Exploration

Now-a-days clay minerals have become effectively indespensable in various fields of mineralogy. Clay minerals have wide applicability in interpreting and understanding such problems as tectonics, source, age, boundaries, facies, environments, zonation, correlation, and metamorphism; a relation between the relative abundance of expanded clays and the occurrence of hydrocarbons is suggested.

It is believed that expanded clays retain their pore water to greater depths of burial than do other clays, and that it is this water which transports much of the hydrocarbon. It is shown that whereas time has little effect on the contraction of expanded clays, it does affect the expulsion of pore water from shales.

It is suggested that oil migration into the reservoir took place synchronously with illite diagenesis, as amplified below. The clay mineral relationships indicate that, following late diagenesis, the stable clay mineralogy consists of illite. It follows, therefore, that the kaolinite preserved unaltered in the upper part of the reservoir is a disequilibrium assemblage. Diagenetic processes take place in an aqueous medium, and consequently, as rocks become oilsaturated, diagenesis will be arrested. Inhibition of diagenesis by hydrocarbon accumulation has been described from various localities (e.g. Millot, 1970; Webb, 1974; Marie, 1975). In the middle part of the reservoir, illite began to form before diagenesis was arrested. This demonstrates that progressive filling of the reservoir took place downwards from the top. As the hydrocarbon column extended progressively further downwards, water was excluded and diagenesis halted. The 'frozen' clay assemblages demonstrate that the alkaline waters forming illite must have been present, but they were expelled by the hydrocarbon accumulation before equilibrium could be attained. Oil migration was thus synchronous with the illite diagenesis; this has also been suggested by Sommer (1975). Only near the base of the sandstones did aqueous diagenetic reactions continue long enough for the development of abundant, well crystallized illite.

Outcrop: Clay mineral is the result of chemical changes in ancient volcanic ash. Prevailing winds blew the ash into shallow sea or submerged basin. Present day exposures of the ash deposits are mined to an average depth of 25 feet.

Pre-Operational Monitoring: Prior to mining vegetation and hydrology studies are conducted. Stockpond designs are developed in order to transform excavated areas into ponds.

Drilling and Sampling: Before the actual mining of clay minerals begins, the quality is tested by drilling shallow holes into the deposit and sampling the cuttings for important characteristics.

Soil Profiling: A soil profile is developed to aid in reclamation. During reclamation soils are replaced in a similar order to which they were stripped to enhance revegetation efforts.

Top-soil Removal: Top-soil and subsoil are removed from the mining area and stockpiled for future reclamation.

Overburden Removal: Clay minerals are generally mined on the surface using open pit methods. Overburden is removed in order to gain access to the clay deposits.

Stripping with Scrapers: Machines, called scrapers, carefully remove the layers down to the mineral deposit. If the scrapers are not careful, the underlying deposit can be contaminated. The overburden layers are usually loosely consolidated, so there is no need for explosives and blasting.

Exposed mineral: The clay mineral is mined and stockpiled at the mine site. Then, at the mine site, the mineral is "field dried". Field drying removes approximately 50% of the moisture, and reduces hauling and final drying costs.

Loading mineral: Front end loaders then load the bentonite into trucks, which transport the mineral to stockpiles at processing plants. The stockpiles are separated into several different types and qualities of bentonite.

Feeding the Mill: First, the stockpiled clay mineral is run through a crusher to break the material into small chunks.

Mill: Next, the mineral is processed through coal or gas fired rotary dryers, in addition to calciners to reduce the moisture content. Control of the temperature in the dryer is important because many of the clay properties can be destroyed if too much heat is applied. As a result of the drying, there is only 5-20 percent moisture. The final product is then pulverized and manufactured into customer specifications. After the processing, the clay mineral is either bagged or shipped in bulk by rail or truck.

Bentonite clay: Bentonite is used as binding agent in foundry sand. Specific surface area, oil absorption, water absorption, attenuation of heavy metals and bleaching capacity of edible oil after acid activations were evaluated for the first time in 2001. The bentonite deposit is open for investment mining and exploration companies are invited on the basis of detailed exploration, evaluation and exploitation. Based on the physical and chemical properties the bentonite can be used in the following industries:

- Bleaching of edible oil.
- Oil absorption.
- Odour and liquid absorbent (Cat litter).
- Waste water treatment.
- Cast iron industry.
- Pelletization.
- Filtering and clarification.

15.3 VARIOUS EXPLORATION PROCEDURES

Gravity Methods

The gravity field of the Earth can be measured by timing the free fall of an object in a vacuum, by measuring the period of a pendulum, or in various other ways. Today almost all gravity surveying is done with gravimeters. Such an instrument typically consists of a weight attached to a spring that stretches or contracts corresponding to an increase or decrease in gravity. It is designed to measure differences in gravity accelerations rather than absolute magnitudes. Gravimeters used in geophysical surveys have an accuracy of about 0.01 milligal (mgal; 1 mgal = 0.001 centimetre per second per second). That is to say, they are capable of detecting differences in the Earth's gravitational field as small as one part in 100,000,000.

Gravity differences over the earth's surface occur because of local density differences between adjacent rocks. The variations in the density of the crust and cover are presented on a *gravity anomaly map*. A gravity anomaly map looks at the difference between the value of gravity measured at a particular place and the predicted value for that place. Gravity anomalies form a pattern, which may be mapped as an image or by contours. The wavelength and amplitude of the gravity anomalies gives geoscientists an idea of the size and depth of the geological structures causing these anomalies. Deposits of very dense and heavy minerals will also affect gravity at a given point and produce an anomaly above normal background levels.

Anomalies of exploration interest are often about 0.2 mgal. Data have to be corrected for variations due to elevation (one metre is equivalent to about 0. 2 mgal), latitude (100 metres are equivalent to about 0.08 mgal), and other factors. Gravity surveys on land often involve meter readings every kilometre along traverse loops a few kilometres across. It takes only a few minutes to read a gravimeter, but determining location and elevation accurately requires much effort.

Gravity measurements can be obtained either from airborne (remote) or ground surveys. The most sensitive surveys are currently achieved from the ground. Variations of gravity are due to local changes in rock density and therefore depend on the type of rocks beneath the surface. Sedimentary rocks are, for example, less dense than granite, which is in turn less dense than basalt.

High density
Extrusive igneous rocks, e.g. basalt
Metamorphic rocks
Intrusive igneous rocks, e.g. granite
Sedimentary rocks
Low density

In most cases, the density of sedimentary rocks increases with depth because increasing pressure reduces porosity. Uplifts usually bring denser rocks nearer the surface and thereby create positive gravity anomalies. Faults that displace rocks of different densities also can cause gravity anomalies. Salt domes generally produce negative anomalies because salt is less dense than the surrounding rocks. Such faults, folds, and salt domes trap oil, and so the detection of gravity anomalies associated with them are crucial in petroleum exploration. Moreover, gravity measurements are occasionally used to evaluate the amount of high-density mineral present in an ore body. They also provide a means of locating hidden caverns, old mine workings, and other subterranean cavities.

Density contrasts of different materials are also controlled by a number of other factors. The most important are the grain density of the particles forming the material, the porosity of the material, and the interstitial fluids within the material. Generally, specific gravities of soil and shale range from 1.7 to 2.2. Massive limestone averages 2.7. While this range of values may appear to be fairly large, local contrasts will be only a fraction of this range. A common order of magnitude for local density contrasts is 0.25.

Gravity surveys provide an inexpensive method of determining regional structures that may be associated with groundwater aquifers or petroleum traps. Gravity surveys have been one of the principal exploration tools in regional petroleum exploration surveys. Gravity surveys have somewhat limited applications in geotechnical investigations.

Electrical Methods

Electrical methods are used to map variations in electrical properties of the subsurface. The main physical property involved is electrical conductivity, which is a measure of how easily electrical current can pass through a material. Subsurface materials exhibit a very large range of electrical conductivity values. Fresh rock is generally a poor conductor of electricity, but a select group of metallic minerals containing iron, copper or nickel are very good conductors. Layers of graphite are also very good conductors.

The examples of good conductors mentioned above are quite rare. For most rocks, the electrical conductivity is governed to a large degree by the amount of water filling the pore spaces and the amount of salt dissolved in this water. Pure water has a very low electrical conductivity. On the other hand, seawater, which contains high levels of dissolved salts such as NaCl, is a relatively good conductor of electrical current. Groundwater can vary in salt content from fresh through brackish (slightly salty) to saline (similar in salt content to seawater) through to hyper-saline (more salty than seawater).

Electrical conductivity of rocks is not the only attribute which is of value to exploration geologists. A number of different electrical properties of rocks are measured and interpreted in mineral exploration. They depend on:

(a) Natural currents in rocks—Self-potential method
(b) Polarizability of rocks—Induced polarization method
(c) Electrical conductivity or resistivity of rocks—Resistivity method
(d) Induction—Electromagnetic method

Self Potential Method

Some materials tend to become natural batteries that generate natural electric currents whose effects can be measured. The self-potential method relies on the oxidation of the upper surface of metallic sulfide minerals by downward-percolating groundwater to become a natural battery; current flows through the ore body and back through the surrounding groundwater, which acts as the electrolyte. Measuring the natural voltage differences, usually 50-400 millivolts (mV), permits the detection of metallic sulfide bodies that lie above the water table. Other mineral deposits that can generate self-potentials are graphite, magnetite, anthracite and pyritized rocks.

Induced Polarization

The passage of an electric current across an interface where conduction changes from ionic to electronic results in a charge buildup at the interface. This charge builds up shortly after current flow begins, and it takes a short time to decay after the current circuit is broken. Such an effect is measured in induced-polarization methods and is used to detect sulfide ore bodies.

Resistivity Method

Resistivity methods involve passing a current from a generator or other electric power source between a pair of current electrodes and measuring potential differences with another pair of electrodes. Various electrode configurations are used to determine the apparent resistivity from the voltage/current ratio. The resistivity of most rocks varies with porosity, the salinity of the interstitial fluid, and certain other factors. Rocks containing appreciable clay usually have low resistivity. The resistivity of rocks containing conducting minerals such as sulfide ores and graphitized or pyritized rocks depends on the connectivity of the minerals present. Resistivity methods also are used in engineering and groundwater surveys, because resistivity often changes markedly at soil/bedrock interfaces, at the water table, and at a fresh/saline water boundary.

Electromagnetic Methods

The passage of current in the general frequency range of 500-5,000 hertz (Hz) induces in the Earth electromagnetic waves of long wavelength, which have considerable penetration into the Earth's interior. The effective penetration can be changed by altering the frequency. Eddy currents are induced where conductors are present, and these currents generate an alternating magnetic field, which induces in a receiving coil a secondary voltage that is out of phase with the primary voltage. Electromagnetic methods involve measuring this out-of-phase component or other effects, which makes it possible to locate low-resistivity ore bodies wherein the eddy currents are generated.

A number of electrical methods described above are used in boreholes. The self-potential (SP) log indicates mainly clay (shale) content, because an electrochemical cell is established at the shale boundary when the salinity of the borehole (drilling) fluid differs from that of the water in the rock. Resistivity measurements are made by using several electrode configurations and also by induction. Borehole methods are used to identify the rocks penetrated by a borehole and to determine their properties, especially their porosity and the nature of their interstitial fluids.

Magnetic Methods

One of the most important tools in modern mineral exploration methods is magnetic survey. Magnetic surveys are fast, provide a great deal of information for the cost and can provide information about the distribution of rocks occurring under thin layers of sedimentary rocks—useful when trying to locate ore-bodies.

When the Earth's magnetic field interacts with a magnetic mineral contained in a rock, the rock becomes magnetic. This is called induced magnetism. However, a rock may itself be magnetic if at least one of the minerals it is composed of is magnetic. The strength of a rock's magnetism is related not only to the amount of magnetic minerals it contains but also to the physical properties, such as grain size, of those minerals. The main magnetic mineral is magnetite (Fe_3O_4)—a common mineral found disseminated through most rocks in differing concentrations.

Measurements of the Earth's total magnetic field or of any of its various components can be made. The oldest magnetic prospecting instrument is the magnetic compass, which measures the field direction. Other instruments, which are appreciably more accurate, include magnetic balances, fluxgate magnetometers, proton-precession and optical-pumping magnetometers.

Magnetic effects result primarily from the magnetization induced in susceptible rocks by the Earth's magnetic field. Most sedimentary rocks have very low susceptibility and thus are nearly transparent to magnetism. Accordingly, in petroleum exploration, magnetic surveys are used negatively—magnetic anomalies indicate the absence of explorable sedimentary rocks. Magnetic surveys are used for mapping features in igneous and metamorphic rocks, possibly faults, dikes, or other features that are associated with mineral concentrations. Data are usually displayed in the form of a contour map of the magnetic field, but interpretation is often made on profiles.

It must be remembered that rocks cannot retain magnetism when the temperature is above the Curie point (≈500°C for most magnetic materials), and this restricts magnetic rocks to the upper 40 kilometres of the Earth's interior.

When exploring for petroleum, magnetic surveys are usually made with magnetometers borne by aircraft flying in parallel lines spaced two to four kilometres apart at an elevation of about 500 metres. When searching for mineral deposits, the flight lines are spaced 0.5 to 1.0 kilometre apart at an elevation of roughly 200 metres above the ground. Ground surveys are conducted to follow up magnetic anomalies identified through aerial surveys. Such surveys may involve stations spaced only 50 metres apart. A ground monitor is usually used to measure the natural fluctuations of the Earth's field over time so that corrections can be made. Surveying is generally suspended during periods of large magnetic fluctuation (magnetic storms).

Seismic Methods

Seismic methods are based on measurements of the time interval between initiation of a seismic (elastic) wave and its arrival at detectors. The seismic wave may be generated by an explosion, a dropped weight, a mechanical vibrator, a bubble of high-pressure air injected into water, or other sources. The seismic wave is detected by a Geophone on land or by a hydrophone in water. An electromagnetic Geophone generates a voltage when a seismic wave produces relative motion of a wire coil in the field of a magnet, whereas a ceramic hydrophone generates a voltage when deformed by passage of a seismic wave. Data are usually recorded on magnetic tape for subsequent processing and display. Seismic methods are of two kinds: Refraction methods and Reflection methods.

Seismic Refraction Methods

Seismic energy travels from source to detector by many paths. When near the source, the initial seismic energy generally travels by the shortest path, but as source to geophone distances become greater, seismic waves travelling by longer paths through rocks of higher seismic velocity may arrive earlier. Such waves are called head waves, and the refraction method involves their interpretation. From a plot of travel time as a function of source to geophone distance, the number, thicknesses, and velocities of rock layers present can be determined for simple situations. The assumptions usually made are:

(a) Each layer is homogeneous and isotropic (i.e., has the same velocity in all directions);
(b) The boundaries (interfaces) between layers are nearly planar; and
(c) Each successive layer has higher velocity than the one above.

The velocity values determined from time-distance plots depend also on the dip (slope) of interfaces, apparent velocities increasing when the geophones are updip from the source and decreasing when downdip. By measuring in both directions the dip and rock velocity, each can be determined. With sufficient measurements, relief on the interfaces separating the layers also can be ascertained.

High-velocity bodies of local extent can be located by fan shooting. Travel times are measured along different azimuths from a source, and an abnormally early arrival time indicates that a high-velocity body was encountered at that azimuth. This method has been used to detect salt domes, reefs and intrusive bodies that are characterized by higher seismic velocity than the surrounding rock. Seismic waves may be used for various other purposes. They are employed, for example, to detect faults that may disrupt a coal seam or fractures that may allow water penetration into a tunnel.

Seismic Reflection Methods

Most seismic work utilizes reflection techniques. Sources and geophones are essentially the same as those used in refraction methods. The concept is similar to echo sounding; seismic waves are reflected at interfaces where rock properties change. The round-trip travel time, together with velocity information, gives the distance to the interface. The relief on the interface can be determined by mapping the reflection at many locations. For simple situations the velocity can be determined from the change in arrival time as source to geophone distance changes.

In practice, the seismic reflection method is much more complicated. Reflections from most of the many interfaces within the Earth are very weak and so do not stand out against background noise. The reflections from closely spaced interfaces interfere with each other. Reflections from interfaces with different dips, seismic waves that bounce repeatedly between interfaces ("multiples"), converted waves, and waves travelling by other modes interfere with desired reflections. Also, velocity irregularities bend seismic rays in ways that are sometimes complicated.

The objective of most seismic work is to map geologic structure by determining the arrival time of reflectors. Changes in the amplitude and waveshape, however, contain information about stratigraphic changes and occasionally hydrocarbon accumulations. In some cases, seismic patterns can be identified with depositional systems, unconformities, channels and other features.

The seismic reflection method usually gives better resolution (i.e., makes it possible to see smaller features) than other methods, with the exception of measurements made in close proximity, as with borehole logs. In most exploration programmes appreciably more money is spent on seismic reflection work than on all other geophysical methods combined.

Concluding Remarks

Mineral deposits may only be a few hundred square metres in area. However, effective and ongoing mineral exploration requires access to large areas of land in order to find these relatively small mineral deposits; this is comparable to finding a "needle in a haystack". Land with high mineral potential must remain accessible for exploration so that new deposits of metallic minerals can be discovered and developed; however, the same land is commonly required for a variety of other competing land uses as well.

Think for a while

1. Explain the term mineral exploration and relate it with mineral prospecting. Different minerals require different methods for exploration. Justify.
2. Delineate the differences among the geological, geochemical and geophysical methods. Ascertain the role of remote sensing in mineral exploration.
3. How do the exploration methods for metallic and non-metallic minerals vary? Non-fuel minerals follow a different technique for exploration. Explain.
4. Describe the exploration of minerals using seismic methods. Between seismic refraction and seismic reflection, which method is more beneficial and why?
5. Describe the exploration of minerals using gravity methods and also explain the exploration of minerals using electrical methods.

FURTHER READING

Evans, A.M. An introduction to Ore Geology. Elsevier, New York. 1980.

Jensen, M.L. and A.M. Bateman. Economic Mineral Deposits, 3rd ed. John Wiley and Sons, New York. 1981.

Dodd, C.G., F.R. Conley and P.M. Barnes. Clay minerals in petroleum reservoir sands and water sensitivity effects. Continental Oil Company.

Padmanabhan, B.S. Managing the minerals. *India's National Magazine*, **20(09)**, April 26-May 09, 2003.

Zonge, Ken. Geophysical prospecting methods. Zonge Engineering & Research Organization, Tucson, AZ.

CHAPTER

16

SYNTHESIS OF SELECTED MINERALS (CRYSTALS) IN LABORATORY AND INDUSTRY

Experimental work involves simulating igneous or metamorphic conditions with specialized machines. Samples may include synthetic minerals or mixes of minerals, fluids such as H_2O and/or CO_2 or actual rock compositions. In the lab, samples of a known composition can be held at known pressure and temperatures, and results can be analyzed. Certain processes (melting, crystallization, diffusion) can also be simulated. All these processes leading to synthetic mineral preparation will be discussed in this chapter.

16.1 INTRODUCTION

The need for manufacturing synthetic minerals initiated from the growing demands of minerals in daily life. Synthetic minerals are utilized both for academic purpose, i.e. for laboratory experiments and scientific research and for industrial utilization. The required pressure-temperature and environmental conditions for mineral synthesis, in accordance to the natural geological environment, are produced in the laboratory. The pressure-temperature range required for synthetic mineral preparation has been fixed by the combined study of thermobarometric and thermodynamic principles. For academic purpose, the proportion of synthetic minerals required is quite less. But for industrial purpose bulk amount of synthetic minerals are needed, so synthetic mineral preparation is gaining wide importance now-a-days.

16.2 LABORATORY METHODS OF MINERAL SYNTHESIS

Preparation of Starting Materials

A technique to improve the homogeneity and reactivity of starting materials for mineral synthesis has been developed. Aqueous solutions of any major and many minor elements of geologic importance are mixed in the correct proportions and frozen by spraying as fine droplets into stirred liquid nitrogen. The resulting material is dried under vacuum and subsequently fired into an intimately-mixed oxide powder.

In many types of experimental mineralogic or petrologic investigations it is suggested to start with materials of the highest possible homogeneity. Uniformity in the composition is especially important in order-disorder studies, in which the phenomena measured can be due to compositional variations on a scale of several unit cells. Homogeneity is particularly difficult to achieve in preparation of trace-element doped mixtures.

Edgar (1973) reviewed techniques commonly used in silicate mix preparation. The techniques have disadvantages like: (i) in achieving effective dispersal of minor components; (ii) in possible loss of volatile components such as Na during high-temperature decomposition, or (iii) in possible minor-element contamination during grinding.

Schnettler et al. (1967) applied a technique based on rapid freezing of aqueous solutions, followed by removal of water by sublimation under vacuum to silica-free solutions of salts. A similar approach is applicable to silicate systems. The silica component is mixed as an aqueous colloidal suspension that has been stabilized against the precipitation or gellation that usually results upon addition of electrolytes.

The freeze-dry technique described here has several important advantages over other methods:

- Mixing of all elements in aqueous solution ensures initial homogenization. Since the freezing of the solution is instantaneous and only solid-state diffusion is operational in the mix during subsequent freeze-drying, the initial homogeneity is retained in the freeze-dried solid.
- With the use of organic anions, the mixes may be fired at temperatures sufficiently low to prevent loss of volatile elements such as Na, and to minimize recrystallization.
- No mechanical attrition of the mix is required at any stage, eliminating contamination from the apparatus used for grinding.

Synthetic forsterite was crystallized from freeze-dried mixes at 800°C under hydrothermal conditions in 10 percent of the time required when a mix prepared by the gel technique is used. Homogeneous ferroan periclases (magnesiowiistites) were formed from freeze-dried mixtures at 1000°C in 25 percent of the time required when using mechanically-mixed ground-oxide starting materials. Low structural states in albite were achieved by hydrothermal crystallization as quickly from freeze-dried mixtures as from gel mixtures.

The compatibility of water-soluble compounds of Na, K, Ca, Mg, Al, Si, Fe, Mn, Co, Ni, Zn, Ge, Sr and Ba has been investigated, and the freeze-dry method has been successfully applied to feldspar, olivine, pyroxene, and oxide mineral syntheses.

The freeze-dry method results in homogeneous mixtures for silicate and oxide compositions. These mixtures appear to crystallize to stable crystalline phases more rapidly than those made by conventional techniques, although one must always be aware of possible formation of metastable phases from high-energy starting mixes. Contamination from grinding may be entirely avoided by using the freeze-dry method.

The techniques listed above have been developed for each new composition as the need in the laboratory arose. A systematic search for compatible solution chemistry for all cations of geologic interest, for use on a routine basis, has been initiated. Preliminary results indicate that solution chemistry based on mixed salts of formic and citric acids may be satisfactory.

Preparation of Glass

Techniques of Schairer (1950 and 1959), Schairer and Bowen (1955 and 1956) are followed for glass preparation.

- Appropriate chemical oxide(s) are properly weighed and then dried at 60° C for at least 24 hours in a watch glass.
- The weighed and dried material is transferred into an agate mortar.
- The watch glass is cleaned with pure methanol and the material adhering to it is obtained.
- In the next step all the chemicals are mixed carefully for 30-40 mins in order to make the mixture as homogeneous as possible.
- The mixture is transferred to a platinum crucible which is first kept in an oven at 40°-50° C for one hour to vapourize the methanol and then it is kept in a silicon carbide furnace (high-temperature furnace up to 1400° C) to heat the material at t° C for about two hours, where t° = expected liquidus temperature, which varies from mineral to mineral + 200° C
- The melted material in the platinum crucible is quenched in the water and again heated for an hour to ensure homogeneity.
- Prolonged heating is avoided to restrict loss of alkalis to a minimum.
- Final molten glass is again quenched in water and kept in the dessicator.
- Glasses thus obtained are checked for their homogeneity by determining their refractive indices. As the indices vary systematically with composition,

analogous composition homogeneity can be determined by plotting of *isofract* curves.

Glass preparation is considered to be a difficult job, since the components that comprise the glass are usually not stable together in solution. The common problems that may arise are: (i) most typical acidic digestions that solubilize silicon will cause calcium fluoride to precipitate; (ii) if no fluoride is present, the silicon will not be fully decomposed, and (iii) if the solution is heated, boron can be lost, etc.

In the laboratory, mostly pure chemicals are used. Care must be taken that the raw materials have not reacted with moisture or other chemicals in the environment (such as alkali oxides and hydroxides, alkaline earth oxides and hydroxides, or boron oxide), or that the impurities are quantified (loss on ignition). Evaporation losses during glass melting should be considered during the selection of the raw materials, e.g., sodium selenite may be preferred over easily evaporating SeO_2. Also, more readily reacting raw materials may be preferred over relatively inert ones, such as $Al(OH)_3$ over Al_2O_3. Usually, the melts are carried out in platinum crucibles to reduce contamination from the crucible material. Glass homogeneity is achieved by homogenizing the raw materials mixture (glass batch), by stirring the melt, and by crushing and re-melting the first melt. The obtained glass is usually annealed to prevent breakage during processing.

In order to make glass from materials with poor glass forming tendencies, novel techniques are used to increase cooling rate, or reduce crystal nucleation triggers. Examples of these techniques include aerodynamic levitation (the melt is cooled whilst floating in a gas stream), splat quenching (the melt is pressed between two metal anvils) and roller quenching (the melt is poured through rollers).

Photochromatic Lens

The glass version of these lenses achieve their photochromatic properties through the embedding of microcrystalline silver halides (usually silver chloride), or molecules in a glass substrate.

16.3 COMMON PROCESSES OF SYNTHETIC MINERAL PREPARATION

Synthetic mineral preparation depends on the origin of the minerals—growth from solution or growth from melt. The common processes, with descriptions, are listed in Table 16.1.

Table 16.1: Common processes of synthetic mineral preparation

Origin	*Processes*	*Description*
Growth from solution	Hydrothermal method	For hydrothermal crystal growth of calcite, organic salt solutions were used as a mineralizer for the first time and a solution of ammonium acetate was found to be an excellent mineralizer to obtain high growth rate under mild hydrothermal conditions. Single crystals of hydroxyapatite were prepared from chlorapatite single crystals by ion exchange in alkaline solutions under hydrothermal conditions. The transformation mechanism was explained by dissolution and precipitation of clusters.
	Melt-diffusion method	In 2001, stones treated by a new process entered the market. The colour of these stones resulted from diffusing beryllium into them from the outside, just like dyeing cloth. This process allowed one to manufacture yellow, padparadscha, and orange sapphire from low-value starting material. Later, it was shown that beryllium diffusion could also lighten overly dark blue sapphire.
	Flux-transport method	The constituents dissolve and recrystallize, usually at a much lower temperature compared to the melting point of the gem material. Once the crystals have grown to the size targetted, the flux is poured out while still molten by ingenious methods, including puncturing the Pt crucible to drain the flux. Subsequent boiling in nitric acid will release the crystals.
Growth from melt	The Flame Fusion or Verneuil Process	Originally developed (1902) by a French chemist, Auguste Verneuil, the process produces a boule (a mass of alumina with the same physical and chemical characteristics as corundum) from finely ground alumina (Al_2O_3) by means of an inverted oxyhydrogen torch that opens into a ceramic muffle.
	Pulling or Czochralski's technique	The Czochralski's Crystal Pulling method involves having a small seed on a rotating rod dipped into a pool of molten ruby. The rod is pulled up as the crystal grows. The end crystals result in very high quality material. The product from this method is more expensive, as the process is tedious and requires an expensive iridium crucible.

(*Contd.*)

(Contd.)

	Bridgman-Stockbarger technique	The Bridgman method (named after the American scientist Percy Williams Bridgman) is also widely used for growing large single crystals. The molten material is put into a crucible, often of silica, which has a cylindrical shape with a conical lower end. Heaters maintain the molten state.
	High temperature/pressure method	After over a hundred years of attempts, in the early 1950s, General Electrical (GE) created the first officially recognized synthesis of a diamond. The process used to create this first man-made diamond was HPHT, high pressure, high temperature. This process is an attempt to reproduce the conditions that create natural diamonds. Large presses that can weigh a couple of hundred tons are used to produce a pressure of 5 GPa (giga pascals) at 1500 degrees Celsius. There are two main press designs used to supply the required pressure and temperature: the belt press and the cubic press.
	Skull-melting process	The 'skull' melting technique has been used only in the case of cubic zirconia, because of the very high melting temperature of the ZrO_2.

Multi Anvil Methods

High-pressure experiments have been carried out on a number of beamlines using a diamond anvil cell or a multi-anvil press. The multi-anvil press is a relatively rare research tool. This tool has been used for a variety of material property studies, including diffusion and deformation of ceramics and metals, deep-focus earthquakes, and the high-pressure stability of mineral phases. The experiments using the multi-anvil press generated a high pressure of 11 GPa, or 110,000 times the atmospheric pressure at sea level. This corresponds to 380 kilometres deep into the Earth, or pressures at the centre of moons or asteroids 2500 kilometres in radius (about the size of Mercury). The sample was also heated to a temperature of 1500° C. Under those conditions, the metal melts and the olivine remains solid.

The geometry of the press is key to creating these enormous pressures. For the 11-GPa experiment, a ceramic octahedron had a 10.3-millimetre-long hole with a tiny rhenium furnace, a thermocouple to measure temperatures, and a graphite capsule containing the olivine and iron-nickel-sulfur-oxygen sample inserted in it. The octahedron rested in the centre of eight 32-millimetre tungsten carbide cubes whose inside corners were truncated to accommodate the sample. (Tungsten carbide is used for the cubes, or anvils, because of its hardness, which is close to that of diamonds but at a much lower cost.) Tiny ceramic gaskets were placed at the edges of the carbide cubes to contain the pressure. This assembly of an inner octahedron and eight carbide anvil cubes was put in the press's split-cone, steel buckets as shown in Fig. 16.1. In several stages, the steel buckets pushed on the carbide cubic anvils, which pushed on the octahedral volume inside.

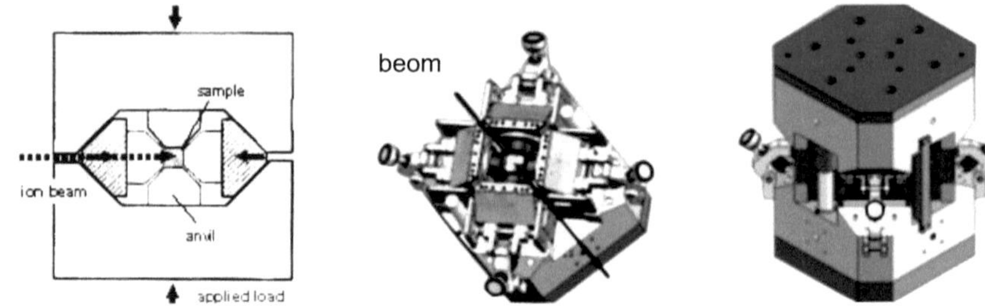

Fig. 16.1: Multi-anvil apparatus.

The process will primarily focus on:

- *Heavy ion-induced modifications of solids that are exposed to extremely high pressures:* For this purpose the sample to be investigated must be enclosed between anvils causing extreme pressure conditions. In order to irradiate pressurized samples with ions, the beam energy has to be sufficiently high to penetrate through one of the anvils of thickness in the range of mm to cm. When entering the solid, the projectiles deposit an enormous amount of energy within a very short time and in a very small volume (corresponding to very high power densities) and trigger many different processes including phase transitions, thermal spikes and pressure waves. The response of solids under extreme pressure conditions is completely unknown, but may have direct implications in the field of geosciences with respect to geological formation and radioactive decay processes in the crust and upper mantle of the Earth.
- *Analysis of material modifications induced by relativistic heavy ions:* It concerns both short-time processes stimulated by the projectiles and final modifications of structure and other characteristics of the material. The signature of short-time processes comprises the emission of various particles such as electrons, ions, atoms and molecules, and of electromagnetic radiation (such as X-rays and Cerenkov light). Their properties as for example intensity, energy, development in time, and spatial distribution, provide valuable insight in track formation processes. Furthermore, short and intense ion pulses are expected to stimulate new processes unreachable under standard irradiation conditions.
- *Radiation hardness of materials:* These studies aim to investigate the stability and specific modifications of different materials exposed to particle beams of high energy and intensity. This will for example allow us to test insulating materials exposed to high-dose environments or to select materials with the most favourable radiation shielding properties.

Hydrothermal Technique

Hydrothermal techniques have been widely used in the syntheses of advanced minerals, the treatment of wastes, and the preparation and extraction of special chemicals. They have also been studied for setting up of geothermal processes. In recent days hydrothermal studies and science lay an important role both in academics and industry.

The hydrothermal unit contains an externally heated cold seal pressure vessel at 1 kbar and various temperatures.

The crystallization vessels used are autoclaves. These are usually thick-walled steel cylinders with a hermetic seal which must withstand high temperatures and pressures for prolonged periods of time. Furthermore, the autoclave material must be inert with respect to the solvent. The closure is the most important element of the autoclave. Many designs have been developed for seals, the most famous being the Bridgman seal. In most cases steel-corroding solutions are used in hydrothermal experiments. To prevent corrosion of the internal cavity of the autoclave, protective inserts are generally used. These may have the same shape of the autoclave and fit in the internal cavity (contact-type insert) or be a "floating" type insert which occupies only part of the autoclave interior. Inserts may be made of carbon-free iron, copper, silver, gold, platinum, titanium, glass (or quartz), or teflon, depending on the temperature and solution used.

Furnace Set-up: Each furnace has the capacity of about 1000 watts. There are two guiding steel pillars along which vertical movement of the furnace is possible and can be fixed at optimum point with the help of the clamps.

Micro-reactor Assembly: They are also known as cold-seal pressure vessels and can be used upto 5 kbar and 1000° C, but the functionability decreases at higher temperatures and pressures.

Concluding Remarks

Experimental studies of minerals provide a useful tool to the understanding of the mechanism of thermodynamic reactions and thereby mineral paragenesis. Laboratory experiments allow scientists to unravel the successive stages of a complex reaction. To mimic nature is not a simple task, as Earth crust processes are complex and many different variables are involved, as temperature, pressure, time, activity of water, gases and dissolved components. Although the experimental approach is a simplification of the natural reaction, this method permits to focus the interest on a particular aspect and to perform a detailed monitoring of the system reactivity and transformation. In this chapter, various methods are described to illustrate the contribution of experimental studies in the fields of mineralogy and geochemistry.

Think for a while

1. Why preparation of synthetic minerals have become prevalent these days? Justify your answer stating all the significance.
2. Define starting materials for synthetic mineral preparation and explain their variability with change of target minerals.
3. With the help of a flow chart, describe the preparation method of glass.
4. How are photochromatic lens being made?
5. What is the significance of multi-anvil methods? Explain with diagram.

CHAPTER

17

INDUSTRIAL MINERALOGY: MINERAL PROCESSING, BENEFICIATIONS AND OTHER RELATED MINERAL USAGE

The growing need of minerals in industries and development lead to the need of understanding the subject more intensely. Various industrial applications and broad beneficiation techniques of different minerals are discussed in this chapter which will help to comprehend the potentiality of minerals in the realm of industrialization.

17.1 INTRODUCTION

The science of examining and evaluating the mineralogical properties for industrial applications is the subject of *industrial mineralogy*. Based upon its distinct physical properties and chemical compositions, the commercially mined geological materials have direct industrial applications, which are not fuel (fuel minerals or mineral fuels) or ores (sources of metals) and are known as *industrial minerals*.

Every mineral has its applicability in industry decided by economic and technical factors mainly. These minerals may be used in their natural state or after *beneficiation* as a raw material in a wide range of applications.

Examples of the *typical mineral properties* for industrial uses are: for its hardness diamonds are used as abrasives, graphite is used as lubricant for its softness, and high temperature melting point of kyanite is useful for refractory material etc. Depending on the type of use, the same mineral may be called as industrial one or an ore, e.g. rutile is an ore for the titanium mineral and an industrial one when for its high refractive index and colour, used as a white pigment.

Diversified industries need minerals as raw materials with a wide price range, e.g. the high market rate of industrial diamond is in per gm or kg whereas low priced materials, e.g. silica, sands and clays are marketed in per tonnes. Most of the other minerals have their price range in between.

Industrial minerals can be divided into two categories: (i) industrial minerals used for commodities, e.g. phosphates, salt etc. and (ii) industrial minerals with performance specifications suitable for end user in mind, e.g. micronised mica (electronics industry), kaolinite (ceramic industry), glass grade feldspar etc. Depending on their use, industrial minerals can be grouped like abrasives, cements, ceramic, drilling fluid, electronics, fertilizers and soil conditioners, filters and absorbants, glass, insulation, lubricants, pigments, refractories etc. Industrial minerals are mostly used as raw materials in metallurgical and chemical (e.g. bromine and sulfur productions) but some times used as such after a little benefication (like gravels etc.).

17.2 SPECIFIC MINERAL INDUSTRIES AND CORRESPONDING INDIVIDUAL MINERALS

An industrial mineral can have a wide range of applications. Depending on the type of application a mineral can be corroborated within a number of different industrial mineral groups listed in Table 17.1. In this section emphasis has been laid upon the specific mineral groups and in the later part of this chapter, the beneficiation procedures and industrial applications of individual industrial mineral will be discussed in details.

Table 17.1: Specific mineral industries and corresponding individual minerals

Industrial use	*Minerals or rocks*
Aggregates	Dolomite, limestone, pumice, vermiculite, diatomite etc.
Abrasives	Industrial diamond, garnet, bauxite etc.
Ceramics	Barite, bauxite, bentonite, silica, kaolinite, feldspar, pyrophillite, talc, zircon, wollastonite, minerals rich in manganese, beryllium, lithium and rare earths, silica etc.
Cement	Dolomite, limestone, gypsum, asbestos
Fertilizers	Nitrates, phosphates, potassium salts, dolomite
Fillers and absorbants	Kaolinite, mica, sapiolite, zeolites, talc, silica, attapulgite, diatomite
Foundry sands	Graphite, silica, bentonite, chromite, olivin, zircon
Fluxes	Limestone, fluorites
Drilling fluids	Bentonite, attapulgite, sepiolite, barite
Electronics	Graphite, manganese minerals, silics, rare earth minerals, berellium minerals
Insulation	Mica, asbestos, pumice, perlite
Lubricants	Graphite, molybdenite
Glass	Borax and borates, celestite, feldspar, fluorite, limestone, lithium minerals, silica, soda ash, zircon
Pigments	Titanium minerals, iron oxide pigments, barite
Refractories	Bauxite, chromite, graphite, pyrophylite, silliminite, andalusite, kyanite, zircon, magnesite
Plasters	Gypsum

The mostly utilized industrial minerals are mainly: asbestos, barite, bauxite, bentonite, borax and borates, beryllium minerals, celestite, chromite, diatomite, dolomite, feldsfar, fluoresper, garnet, graphite, gypsum, industrial diamond, iron oxide pigments, kaolinite and kaolinite bearing clays, lime stone, lithium minerals, magnesite, manganese minerals, mica, molybdenite, nitrates, olivin, perilite and pumice, phosphates, potassium salts, pyrophylite, rare earth minerals, salt, sapiolite and attapulgite, silica, silliminite andalucite and kyanite, soda ash, talc, titanium minerals, vermiculite, wollastonite, zeolites, zircon.

17.3 MINERAL PROCESSING AND BENEFICIATION PROCESSES

It is the first process that is done to separate useful minerals from the waste rock or gangue. This produces a more concentrated material for further processing. Concentrating the needed minerals in the mined material is known as the beneficiation process. Sometime, the beneficiation at the primary level can be done during the mining.

Beneficiation needs the characterization of the materials of interest, e.g.

(a) Petrographic microscopes, chemical analyses, image analysis, QEMSCAN and electron microprobe systems; and

(b) Characterization of the deposits type.

In modern mineral processing plant, sampling, analysis and dewatering operations are included along with communion and concentration.

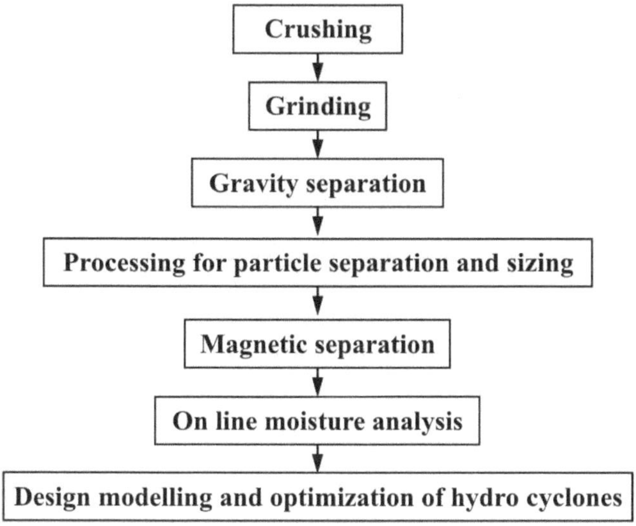

Fig. 17.1: Flow chart showing the beneficiation process.

Sampling and analysis: To estimate the economic potential of the materials of interest, routine sampling and analysis is required for raw material and concentrates. Modern plants usually have fully automatic control system that conducts in stream analysis of the materials and during processing. Necessary adjustments at any stage of the analysis is done to get the richest possible concentrate at minimum operating cost.

Sampling is separating a representative part of the whole material to a convenient size for analysis. It may be done manually or mechanically. The manual process is slow and inaccurate and hence expensive but used only when machine sampling is not possible. The machines used are shovel, pipe sampler and automatic machine samples. A representative sampling depends on the choice of the quantity, total number and kinds of the samples. In order to estimate the appropriate criteria of the sampling, a number of mathematical sampling models have been devised.

Analysis: Representative samples collected from the material stream (conveyor belt etc.), are analyzed using various methods, e.g. chemical, mineralogical, particle size etc. Conventional methods of chemical analysis are used to estimate and detect quantities of elements/minerals in the ores but the process is not very accurate at low concentration and not suitable for process control. Consequently, the use of sophisticated analytical instruments is increasing. Mostly used such techniques are X-ray fluorescence (detects and estimates the elemental concentration upto ppm level), X-ray diffraction for mineral phase identification and quantification, thermal analysis to know the decomposition temperature and other thermal behaviour, gravity separation of different mineral phases by flotation method, particle size analysis by sieving through the internationally standard sieves (50 μm), and for smaller sizes, optical measurements (with laser beams of different frequencies) are used.

Communion is the process of splitting physically different minerals present in the ore from their interlocking state. It is done by crushing the ore below a certain size and then grinding to a powder. The fineness of the powder depends upon the dispersion of the mineral of interest. Crushing is done mostly using mechanical crushers. Crushed materials produced are usually below 10-14 mm (0.4 to 0.6 inch dia) so as to directly feed into the grinding milling machine. To get this grain size, crushing is done in stages. First few crushing is done using jaw crushers of upto 2 metre opening and the grain size is reduced to less than 150mm. Next, the ore is crushed using cone crushers to less than 10-15 mm. Grinding operation starts with materials of these grain-sizes for further disintegration where it can be directly feed from the crushers to the milling machines. These milling machines are cylindrical containers of varying length to diameter ratio, axis mounted almost horizontally and may be partially filled (10% by volume) with grinding bodies like iron/steel balls, flint stones etc. (semi autogeneous). Autogeneous grinding mills do not use the grinding bodies.

In roll crushers, the crushing and grinding are combined together where two cylinders mounted as horizontal shafts drive in opposite directions. When the cylinders are pressed under high pressure, disintegration operation of the materials between the cylinders continues. Separation of useful minerals from others, received from the grinding mill, is the concentration process. In large scale operation of concentrating the desired mineral, different properties of the mineral are used, e.g. colour (optical sorting with colour contrast), density (gravity separation), physiochemical (flotation method), magnetic, electrooptic (using UV and IR detectors) and also radiometric (gamma ray response) separation.

Artificial Synthesis of Industrial Minerals

To produce valuable commercial products, minerals can be synthesized under controlled conditions, with the information of how they are naturally formed. Under thermodynamic equilibrium, growth may occur. But usually the first products are metastable and slowly go to stable materials. Care must be taken in the application of natural minerals, even when the equilibrium conditions of mineral growth are known. Kinetic factors are extremely important for the synthetic mineral preparation processes.

Process and Products of Mineral Synthesis

If appropriate conditions are used, almost all the minerals can be expected to be synthesized. For mineral with complex atomic ordering, synthesis is difficult, but yet can be done. It may be due to extreme low atomic movements at low temperature, e.g. microcline and low-temperature varieties of plagioclase. The high temperature disordered structures of the same minerals e.g. sanidine and plagioclase can be easily synthesized. Minerals are synthesized both for industrial purpose and gem-use and special emphasis need to be given to the synthesis of gem minerals.

Technical development of high pressure research made it possible to synthesize diamond in 1950. Diamonds are formed at temperatures from 1200° C to 2400° C and under pressures from 55,000 to 120,000 atm. The raw materials used are graphite, lampblack or charcoal obtained from sugar, and mixed with catalysts like Cr, Mn, Fe, Co, Ni, Pt, Os, Zn, Rh, Pd and Ta.

From economic viewpoint, synthetic production of zeolites, molecular sieves revolutionized industrial catalytic process. For many different processes, utilizing petroleum, the synthetic zeolites, e.g. faujasite is an extremely rare, natural product of great importance. Some minerals like ruby, used in lasers or in electronic equipments, synthetic alumina used in clock movement, synthetic quartz used in optical equipment and piezoelectric oscillators, are of great importance for industrial purposes.

Table 17.2: Methods for the artificial synthesis of minerals

Methods	*Description*	*Minerals synthesized*
Crystals growth from solution	Synthesized by precipitation from supersaturated solutions under normal pressure.	Ethylene diamine tartarate (EDT), ammonium dihydrogen phosphate, lithium sulphate, sodium chlorate, lead chromate, potassium sodium tartarate, alums, lead chloride, lead nitrate, cadmium iodide, beryllium sulphate, guanidine aluminium sulphate hexahydrate, iodic acid, copper sulphate, sodium chloride, potassium chloride, saccharose, sulphur, zinc sulphate, potassium ferrocyanide, ammonium bromide, ammonium chloride, ammonium orthoarsenate, ammonium orthophosphate
Hydrothermal synthesis	Under high temperature and pressure, in closed system, crystals precipitate from solution on cooling.	Quartz, aluminium orthophosphate (berlinite), alumium arsenate, emarald, calcite, corundum, zinc oxide, zinc silicate, chrysotile, orthoclase, topaz, cryolite, tourmaline, tremolite asbestos.
Crystal growth from the melt	At temperatures from 273° C to 3,700° C, the crystals separate from melt on cooling. A variation of this method is 'flame fusion process'.	A, Ag, Al, Au, Be, Bi, Cd, Co, Cu, Fe, Ge, Hg, K, Mg, Na, Ni, Pb, Se, Sb, Sn, Te, Zn, Si; alloys (Ag-Zn, Au-Ag, Au-Sn, Cu-Al, Cu-Mg, Cu-Si, Cu-Sn, Cu-Zn, Ni-Te); inorganic compounds LiF, NaF, NaCl, NaBr, NaI, KCl, KBr, KI, CsF, CsI, CsBr, CsCl, CaF_2, $NaNO_2$, AsI_2, MgF_3, Fe_3O_4, $BaTiO_2$, micas (muscovite, phlogopite, biotite), ice, PbS, PbSe, PbTe, CdS, CdTe, SnSe, ZnF_2, ZnS, ZnSe, Sb_2T_3, Bi_2T_3, GaSb, AlAs, GaAs, ZnAs, AlSb, corundum, spinel, mullite, chrysoberyll, nepheline, perovskite, scheelite, pyrites, hausmannite, rutile; organic compounds (anthracene, acenapthene, acridine, benzoic acid, quaterphenyl, phenazine, fluorine, naphthalene, pyrene, stilbene, terphenyl)
Crystal growth from vapour phase	Crystals prepared by condensation of the vapour or by reactions in vapour phase.	Ag, BaO, Cd, CdS, Ta, Ti, S, Se, benzophenone, hexamethylenetetramine, Mg, Mo, Si, Zn, ZnS, ZnO, Ge, HgSe, HgTe
Crystal growth by recrystallization	From polycrystalline aggregates, crystals can be synthesized at a temperature near their melting points.	Al, BaO, Cu, Fe, Mg, Pb, Se, Sn, TiO_2, Al_2O_3, α-U, Zn, ZnO, MgO
Crystal growth by electrolysis	Crystalline phases precipitate from the electrolytes at the cathode.	Ag, Cu, Cd, Zn, Ag, Co, Ni

17.4 SOME CHARACTERISTIC DETAILS ABOUT THE INDUSTRIAL ASPECTS OF THE MINERALS/ROCKS

Asbestos

Asbestos is a group of fibrous silicate minerals. The serpentine asbestos, known as, chrysotile has the fibrous form and about 95% of world asbestos production is crystotile. The amphibole asbestos are crodiolite chamosite, anthophylite, trimolite and actinolite. Trimolite and actinolite are commercially unimportant.

Asbestos fibres are incombustible and heat resistant. The strong tensile strength of the asbestos material is used to spin into yarn and then made into textiles. Fibre length is used to define the grades of asbestos.

Crysotile fibres decompose rapidly in strong acids but the amphibole fibres show different degrees of resistance with acids as a structural difference between the two asbestos groups. Concentrated acetic acid decomposes crysotile and all the asbestos are decomposed hydrofluoric acid. All asbestos can resist prolong attack by alkalis, even stronger ones.

Amphiboles thermally decompose in three stages, e.g. loss of physically combined water, loss of chemically combined water and breakdown into pyroxene, iron oxides, and quartz over the temperature range 400 to 1050° C. Asbestos fibres do not have any melting point, only the decomposition products melt. On heating, chrysotile and anthophyllite produce magnesium pyroxene which melts above 1450° C, iron pyroxenes, having melting points around 1000° C produced from crocidolite and amosite. The decomposition products form a poorly conductive layer on the hot surface providing protection to the successive layers of insulation.

The surface area of asbestos is very important in its industrial applications. Raw amphibole asbestos has an absolute surface area (6 to 20) ¼10^3 cm^2/g, whereas for chtysotile, it is upto 30¼10^3 cm^2/g. In the industrial process, the typical surface area of (30 to 90) ¼10^3 cm^2/g is used for asbestos minerals. The required degree of fiberization is achieved by milling raw asbestos. Each process of industrial use requires a critical degree of fiberization.

About 3000 various types of industrial use exist. Largest use is in the fibre composite materials for reinforcement cement (inorganic) and pvc, rubber and plastic (organic bonding agents). Increase in tensile strength with temperature upto 400° C is significantly used in many industries. Widely used asbestos cement for building and industry is produced by adding chrysotile to hydrated Portland cement. This mixture with 10 to 20 wt% asbestos is compressed into sheets or molded into boards, pipes and other shapes.

The asbestos are specified in several groups by Quebec mining association as Gr1 (crude), Gr2 (crude), Gr3 (textile or shipping fibres), Gr4 (asbestos cement fibres), Gr5 (paper stock grades), Gr6 (paper or shingle fibres) and Gr7 (shorts and floats). To evaluate the fibre market, various tests have also been developed by the users.

For flexible asbestos cement sheets production, Gr6 is used and for corrugated sheets some Gr5 fibre is used to increase adhesion of the wet sheets during formation process. Asbestos reinforced polyvinylchloride (PVC) is used for floor coverings whereas asbestos-reinforced rubber for sealant applications. But these are now getting replaced by glass and carbon fibres.

Various blends of Gr4, 5 and 6 are used for production of millboard, a most versatile production. Typical uses of millboard are rollers for transport of hot material used for gas transport, flange gaskets for joints in ducting and trunking, stoppers and plugs for molten metal containers. For the high frictional coefficient of the asbestos a mixture of Gr6 and 7 is used in the break and cluch linings and pad production.

The length and flexibility of longer grade crysotile fibres are used to spinning into yarn and cloth weaving is possible along with organic fibres and has a wide application in heat and fire protective clothing, blankets, curtains, and aprons for electrical insulation. Also as a filtration material, e.g. clarifying filtration of beverages, removal of viruses and pyrogenes from pharmacological solutions.

In particular, for the special properties, crysotile is used enormously in various fields of applications.

Abrasives

Abrasive materials are used for sharpening, polishing, cutting, grinding and cleansing of soft or less resistant materials, natural (minerals) or synthetic (e.g. industrial diamond etc.). Essential physical characteristics of abrasives are grain shape, grain size, cleavage and structure properties, purity or homogeneity, suitable hardness, capacity to form a strong bond with the surface of the support and in some cases, heat resistant.

Various forms of abrasives are used in grinding wheels, sand papers, cut off wheels, pulp stones, ball mills, honing stones etc. If the hardness or grain size etc. of the abrasive material is not chosen properly, it may develop scratches in the finished product, producing damages due to trapped dirt, water etc. Increased surface area permits more chemical reactivity, while increased friction erode coatings (chemical, water-resistant paints).

Table 17.3: Important natural abrasives, classified on the basis of hardness

High (>7)	*Medium (5.5-7)*	*Low (<5.5)*
1. Diamond	1. Quartz	1. Apatite
2. Corundum	2. Chlcedony	2. Magnetite
3. Emery	3. Flint	3. Calcite
4. Garnet	4. Sandstone	4. Hematite
5. Staurolite	5. Quartzite	5. Diatomaceous earth
	6. Basalt	6. Pumice
	7. Feldspar	7. Clay
	8. Granite	8. Talc
		9. Dolomite
		10. Gypsum

The high hardness abrasives can be formed artificially; the processes for certain minerals are as follows:

Diamond: Diamonds have been artificially synthesized from carbon by using chromium, manganese and other metals as catalysts at 55,000-120,000 atm and 1200-2400° C.

Corundum: Corundum is produced artificially by fusing bauxite in an electric arc or by calcinations of alumina.

Emery: Emery is a kind of metamorphic rock which constitutes 60-75% corundum and 10-35% magnetite.

Garnet: Almandine and spessartine garnets are mainly used as abrasives. They are mainly used as abrasive papers and clothes.

Synthetic abrasives are manufactured using minerals as raw materials. Some important synthetic abrasives are electrocorundum (Al_2O_3), produced from bauxite, sintered corundum (Al_2O_3), silicon carbide (SiC), boron carbide (B_4C_3), boron nitride (BN), tungsten carbide, synthetic diamond, chromic oxide, synthetic graphite etc.

Table 17.4: Types of abrasive materials

Materials	*Properties*	*Uses*
Flint	Flint paper is a medium-hard grit material.	It is used for coarse work such as removing paint, etc.; also used on soft woods and is not very useful for finished work.
Aluminium oxide	It is used for sanding hard woods and metals.	It is used for sanding hard woods and metals.
Garnet	It is not as hard as aluminium oxide and is brittle in nature, so it is self-sharpening and fairly durable.	It is used in woodworks.
Silicon carbide	Silicon carbide paper is an expensive, quite hard paper.	It is used for sanding all types of metals and for smoothing finishes on woodworking projects.
Emery cloth	It is an inexpensive material.	It is used for smoothing metals.
Crocus	Crocus powder is medium-expensive, very soft and finely grounded powder.	It is used primarily for metal polishing.

Table 17.5: Types of abrasives

Abrasive types	*Properties*	*Uses and Examples*
Bonded	Consist of abrasive grits that have been mixed with clay, resin, glass or rubber.	Common shapes of this kind are wheels, cylinders, blocks and cones.
Coated	They are made up of abrasive grits bonded to flexible substrates such as paper, cloth, fibre or film.	Common examples are belts, sheets and flap disks.

(*Contd.*)

Table 17.5 (*Contd.*)

Abrasive types	*Properties*	*Uses and Examples*
Non-woven nylon	They appear as random three-dimensional networks of open weave nylon materials coated with abrasive.	Common forms are pads, wheels and brushes.
Metal bonded	Some abrasives, as in diamond, are held together in a metal matrix in the form of a precision grinding tool.	Common shapes are cutting wheels or hole cutting tools for machining of tiles, glass and other ceramic materials.
Burs	They are similar to metal bonded and usually have teeth on the cutting faces; they are generally smaller in size.	They are used for fine machining and precision operations. Commonly used abrasives are tungsten carbide and diamond. Used mainly for enlarging holes.
Lapping compound	It is actually a very finely grit material. It can be supplied in dry powder or paste form.	Used in fine polishing and finishing operations.

Ceramics and Refractories

Ceramics may be defined as *man-made articles which have been first shaped or moulded from a wide range of natural earths, minerals and rocks and clays, and then permanently hardened by heat.* So ceramic is an inorganic and non-metallic substance, which may be crystalline, semi-crystalline or even amorphous.

The simplest process of ceramic-manufacture can be put forward in the following way. The natural clays are regarded as almost coherent rocks which, when wet, are soft and plastic and when dry, they are hard and brittle. Their general composition is one or more clay minerals and quartz, mica and other minerals. This clay when sufficiently heated at elevated temperature, chemical reactions between the clay minerals and other minerals in the clay are initiated. These chemical reactions are followed by melting of the less-resistant materials, recrystallization of some and thereupon bonding of all constituents. Then on cooling the final product is a hard and durable ceramic body. For different ceramic products, different types of raw materials, both plastic and non-plastic are required.

Ceramic Products

There are three main group of ceramic products:

- Pottery and whiteware—bone china, porcelain and earthenware
- Structural clay—building bricks, pipes, floor and roofing tiles
- Refractories and technical ceramics—heat-resistant furnace bricks and components for atomic reactors.

Pottery and Whiteware

Pottery and whiteware can be best studied under three main classified groups:

- Chinaware
- Stoneware and porcelain
- Earthenware

For the manufacture of these types of ceramic products, both plastic and non-plastic raw materials are used. While the plastic raw materials are specific for each of the types, the non-plastic raw materials are common for all. They will be elaborated in the following discussion.

Table 17.6: Types of ceramics

Types	*Properties*	*Uses*	*Plastic raw materials*
Chinaware	It is a special type of white, ceramic, made from china clay and china stone with high amount of calcined animal bone, which is calcium phosphate. Due to this it is highly translucent, dense and impermeable in nature.	It is used to make bone china products like figurines, dinner services, tea and coffee sets. It is also widely used in the art of pottery and ceramics.	China clay: it is mainly consisted of mineral kaolinite, which is hydrated aluminium silicate.
Stoneware	They distinguish themselves from other varieties in the absence of translucency, as they contain no bone. Stoneware is formed from plastic clays which contain natural fluxes that promote vitrification of clay at 1200° C and 1250° C; for porcelain the finishing temperature is between 1200° C and 1400° C.	Stonewares are used to make domestic ovenware and tableware. Porcelain is used for table and laboratory purposes.	Ball and stoneware clays: Ball clay and stoneware clay, which is more siliceous in nature originate from decomposition of feldspathic igneous rocks and are secondary in origin. They are mainly formed of kaolinite, alongwith quartz and hydrous mica.
Earthenware	It covers a wide range of composition and appearance and its colour varies from brownish red to cream and white. It is porous in nature. Its finishing temperature is between 1000° C and 1200° C.	Certain types are used for the manufacture of glazed fireplace tiles and kitchen sinks. Ordinary red clays are used for glazed and decorated ware. A special type of impervious earthen ware is used for wash basins, closet bowls etc.	Red clay: it generally contains kaolinite, quartz and hydrous mica alongwith oxides of iron. Fire Clay: low-grade fireclays are formed as seat earths under coal seams. It is hard, blocky and compact in nature, being composed of kaolinite, quartz and hydrous mica.

Non-plastic Raw Materials

These materials cannot be moulded into permanent shapes without the addition of some binding materials. The most important non-plastic raw materials are:

- Stone: a granitic rock, containing fluxing materials like feldspar, white mica and quartz; helps to increase density, impermeability and strength of the ware.
- Feldspar: predominantly aluminium silicate which is an important fluxing agent in the preparation of vitreous bodies and glazes; the varieties used in ceramic manufacture are orthoclase, microcline and soda-rich plagioclase.

Table 17.7: Structural clay

Types	*Properties*	*Use*	*Raw materials*
Structural or heavy clay product	Supply of suitable clay is necessary; clays are plastic enough to mould easily; can be retained in both wet and dry condition; fire without shrinkage at temperature between 900 and 1200° C.	Used extensively in manufacture of building bricks, roofing and floor tiles, chimney and certain types of glazed and unglazed pipes; also employed in manufacture of some acid-resistant bricks, tiles, tanks and pipes, used in chemical and associated industries.	Almost any clay, also shale, which sinters at a low temperature, is used. Most suitable variety is clay with 25-50% fine-grained non-clayey material and upto 30% kaolinite.

Refractories

Refractories may be defined as inorganic substances which are able to retain their physical and chemical properties when subjected to high temperatures. The primary requisite of a refractory material is resistance to temperature above 1500° C. The other properties are ability to bear mechanical wear and rapid changes of temperature, have low coefficients expansion and to bear considerable structural loads.

Refractories can be grouped under three broad categories. They along with their sub-categories have been summarized in Table 17.8.

Table 17.8: Types of refractories

Refractories	*Sub-categories*	*Properties*
Acid refractories	Fireclay	It is a clay of sedimentary origin; can be heated to temperature above 1700° C without melting. Composed of fine-grained kaolinite, hydrous mica and free quartz.
	Silica rock	Principal source of silica are sandstones, quartzites, grits and sands with more than 95% silica.
Basic	Dolomite	Calcium magnesium carbonate, commonly found in sedimentary rocks.
	Magnesite	Magnesium carbonate
	Serpentine	
Neutral	Chrome	
	Alumina	
	Carbon	

Cements

Cement is a hydraulic binder and is defined as a finely ground inorganic material which, when mixed with water, forms a paste which sets and hardens by means of hydration reactions and processes, which after hardening retains its strength and stability even under water. The main raw materials used in the cement manufacturing process are limestone, sand, shale, clay and iron ore. The main material, limestone, is usually mined on site while the other minor materials may be mined either on site or in nearby quarries. Another source of raw materials is industrial by-products. The use of by-product materials to replace natural raw materials is a key element in achieving sustainable development.

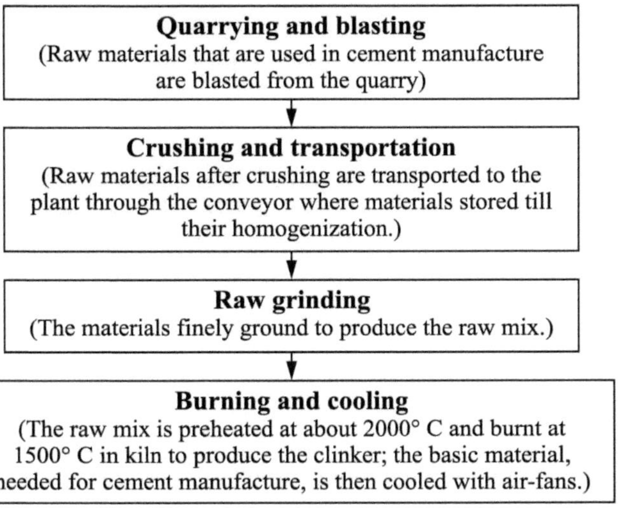

Fig. 17.2: Flow-chart showing cement manufacturing process.

Table 17.9: Types of cement

Simple: Cements that revert to original composition of the raw materials after setting.	**Lime cement:** Manufactured from limestone and dolomite; used as mortar for binding bricks and for plastering the walls.	
	Gypsum cement: Mainly Plaster of Paris, gypsum plaster, manufactured from gypsum.	
Complex: Cements that do not revert to original composition of the raw materials after setting.	**Silicate type**	**Hydraulic lime:** Manufactured by burning impure limestone with 5-30% impurities like calcium silicate, alumina, iron oxide and silica; it sets better under water, than in under air. **Grappier's cement:** Manufactured by grinding the lumps of unburnt $CaCO_3$ and partly overburnt lime silicates. **Natural cement:** Hydraulic lime without any free lime is natural cement. It contains $CaO/(Al_2O_3+SiO_2)$ in the proportion of 2:1 or 3:1. It is better grade cement than the former two. **Portland cement:** Consists a mixture of silicate, aluminate, ferroaluminate of lime, obtained by finely grinding a mixture of limestone and clay or shale; the specification being 75% $CaCO_3$, 20% (Al_2O_3+ SiO_2+ Fe_2O_3) and 5% magnesia and alkalis. **High alumina cement:** Manufactured from crushing and firing limestone/ chalk and bauxite, using coal as a fuel.
	Oxychloride type	**Sorel cement:** Manufactured by using 2.5-3.5 parts of MgO and one part of $MgCl_2$.

Plasters

Plaster is not a singularly defined material, but a group of different materials that harden by different mechanisms. The common plasters are:

(a) **Plaster of Paris:** $CaSO_4$,H_2O, a partially dehydrated form of gypsum ($CaSO_4$, $2H_2O$). When Plaster of Paris is mixed with water, it redehydrates and turns back to gypsum, which then hardens to a material, similar to mortar or concrete, but much softer.

(b) **Lime plaster:** When limestone is heated, quicklime (CaO) is produced. Addition of water in it, produces slake lime [$Ca(OH)_2$]. When slaked lime is mixed with a filler (e.g. sand or other inert materials, a paste is obtained, which on reaction with CO_2 of air, reforms the limestone and that makes the lime plaster harder.

(c) **Cement plaster:** It is a mixture of Portland cement plaster, sand or other filler, water and other Plaster of Paris or lime plaster. As the composite minerals of Portland plaster is of great variety and also cement plaster, the hardening is a complex process and cannot be described as the action of any particular reaction.

Fillers and Absorbents

The powdered minerals which are used for filling any gaps to improve the properties and to modify the composition of fabricated substances during the manufacturing processes of finished products are called *fillers*. An *absorbent* is formed of fillers and a binder. It is produced by mixing at least one binder with atleast one filler and used for absorbing organic and/or inorganic substances from a liquid or gaseous phase. The fillers are used in bituminous compositions, rubber technology, flooring materials, fertilizers, plastics and textile fibre processeing, in paper and paint industries and also for insecticides. Mineral fillers are primarily used to enhance the optical properties of the finished products, such as colour index, opacity, brightness and surface texture and physical properties like viscosity, specific gravity, tensile strength, resistance to abrasion, toughness, shrinkage, resistance to heat, firing and electricity. The chemical properties, mixability, lubrication and thereby the overall stability of the products are improved by the application of the fillers. The minerals commonly used in this industry are kaolinite, mica, sapiolite, zeolites, talc, silica, attapulgite and diatomite.

Table 17.10: Common fillers and absorbents

Minerals	*Properties*	*Uses*
Kaolinite	White or colourless clay, plastic; composed of fine-grained plate-like particles; inert and long-lasting	Used as filler for paper, paint, rubber and plastics.
Bentonite	Fine-grained, firm, earthy, plastic when wet; consists primarily of hydrated silicates of aluminium	Used for making drilling muds and pet waste; also for foundry sand, iron ore pelletizing, sealing and water proofing and filtering liquids.
Mica	Transparent, flexible and elastic; can be ground to very fine particles	Used as ceramic fillers.

(Contd.)

Table 17.10 (*Contd.*)

Minerals	*Properties*	*Uses*
	with high aspect ratio; chemically inert, thus can resist action of heat, light, water, oil, solvents, alkalis and various chemicals; moisture-proof, anti-friction property; low density, sp. gravity and hardness.	Used as an excellent plastic filler; as a mineral filler for polymers; as a filler in building materials, paints, welding electrodes, foundry works, cosmetics and oil drilling.
Sepiolite	A type of clay mineral with complex magnesium silicate structure; can be present in fibrous, fine-particulate and solid forms; non-swelling, light, porous, high surface area; non-flammable, chemically inert, suitable mechanical strength; excellent sorption capacity and colloidal properties.	Used as an industrial absorbent by absorbing liquid spills, keeping work and transit areas dry and safe; used as an adsorbent in waste treatment, moisture control; used as a filler in polymer and elastomer processing.
Zeolites	An inorganic, porous substance	Used as an absorbent for oil and spills; used for water purification.
Talc	Very light, resistant to heat, electricity and acid	Used as an absorbent and helps cut down on friction on the skin. Talc used in powder, helps to absorb excessive sweating on skin.
Silica	Silica present in polymorphic form. Quartz is most common. It is colourless and transparent.	Used as adsorbents for removing moisture and for controlling humidity of rooms; for adsorption of water vapour in natural gas and other industrial gases.
Attapulgite	A crystalloid hydrous magnesium-aluminium silicate with a laminated chain structure; light with high absorption capacity; plastic and adhesive when wet.	Used as an absorbent in oil-well drilling muds; as a pharmaceutical intestinal absorbent; as floor absorbents; as chromatographic adsorbent.
Diatomite	Consists of fossilized remains of diatoms, a type of algae; soft, powdery, siliceous; has an abrasive feel and highly porous.	Used as an absorbent for spill clean-up; used in facial masks to absorb excess oils.

Fertilizers

Fertilizers are organic or inorganic substances that can be applied directly onto the soil in crude state or can be chemically synthesized before use. The main elements essential for plant growth are nitrogen, potassium and phosphorous. So any rock-body or mineral, containing either of these elements, can be estimated as fertilizers; hence on that basis fertilizers can be classified into four groups: nitrogen-bearing, phosphorous-bearing, potassium-bearing and fertilizers of secondary importance, i.e. containing minerals like dolomite, limestone, gypsum and sulphur.

Table 17.11: Types of fertilizers

Types	*Properties*	*Uses*
Nitrogen fertilizer: Ammonium sulphate, sodium nitrate and calcium nitrate	High nitrogen content; high preservation capacity.	Can be used safely in regions of heavy rainfall since it does not get leached out of the soil easily.
Phosphorous fertilizer: Raw materials-rock phosphate, phosphatic nodules, apatite (natural); Basic slag from blast furnace (artificial)	Basic slag with about 20% of P_2O_5 largely available from steel plants.	Used as natural or manufactured super-phosphate fertilizer in phosphorous-deficient soil to increase the crop-yielding capacity of the soil.
Potassium salt fertilizer: Potassium nitrate or salt petre	Found wherever organic nitrogenous substances decay in contact with potassium salts.	Applied in soil, which is poor in potassium salts.
Fertilizers of secondary importance: Dolomite, limestone, gypsum and sulphur	Limestone: fine-grained, pulverized.	Used directly to soil but not as plant food. Dolomite: used in magnesium-poor soil, also neutralizes acidity of the soil, caused by other fertilizers. Limestone: supplies required amount of calcium and magnesium to the soil. Gypsum: used in soil of saline and alkaline types with superphosphate. Sulphur: used in sulphur-poor soil; neutralize alkalinity; used to make superphosphate from phosphate rock in the form of H_2SO_4.

Fluxes

A flux is a substance, which finds its application in accelerating the fusion of minerals and in refining metals by combining with impurities for forming an easily removable molten material. The minerals which are widely used in this respect are limestone and fluorite.

Table 17.12: Types of fluxes

Types of fluxes	*Properties*	*Uses*
Limestone	Sedimentary rock with chemical composition of $CaCO_3$. Some common minerals present in limestone are silica (chert, chalcedony and quartz), clays (illite, smectite, chlorite and kaolinite), iron oxide (hematite, magnetite and limonite), iron carbonates (siderite) and iron sulphides (pyrite).	Used as flux in steel industry, where it helps in lowering of melting temperature and production of a good fluid slag and also for scavenging undesirable elements such as phosphorus, silicon, aluminium and sulphur.
Fluorite	Metallurgical grade fluorite is used as flux and it consists of 75-85% CaF_2 and less than 0. 5% sulphur or sulphide. Fluorite is natural crystalline calcium fluoride (CaF_2); transparent to translucent, glassy in appearance; pure fluorite is crystal clear, but can show many different colours depending on tiny amounts of other elements replacing calcium in the molecular structure; often forms beautiful cube-shaped crystals.	Used as flux in basic open hearth and electric furnaces for lowering the melting point of raw materials in steel production where it helps to accelerate fluidity of the slag and thus assists to remove the impurities like phosphorous and sulphur from the steel to slag; used in iron foundries where it serves the same function as above, and also applied in the production of aluminium.

Foundry Sands

Foundry sands are comprised of clean, uniform, high-grade silica sand that is bonded to form moulds for ferrous (iron and steel) and non-ferrous (copper, aluminum, brass) metal castings. The minerals commonly used in foundry sands are graphite, silica, bentonite, chromite, olivine and zircon. These minerals in varying combinations and varying proportions are used to form different types of foundry sands, which serve different functions.

Table 17.13: Types of foundry sands

Types	*Properties*	*Uses*
Green sand	Consists of high-quality silica sand, about 10 percent bentonite clay (as the binder), 2 to 5 percent water and about 5 percent sea coal (a carbonaceous mold additive to improve casting finish).	Used for ferrous castings
Spent foundry sand	Consists primarily of silica sand, coated with a thin film of burnt carbon, residual binder (bentonite, sea coal, resins) and dust.	Used for moulding and casting
Quartz sand	Generally contains undesirable impurities of feldspar (15-20%), mica, hornblende, magnetite and partially calcite.	Used in sintering of cast iron
Chemically bonded sand	Consists of a mixture of different clays, like bentonite (predominantly montmorillonite), plastic clays (kaolinite).	Used as bonding materials in moulding system, which is due to the adherence of clay particles to quartz grains and to each other

Drilling Fluids

Drilling fluids are the fluids that are applied in the bore-holes within the earth crust where its main purpose is to provide hydrostatic pressure to prevent formation fluids from entering into the well bore. The other functions, served by the drilling fluids, are to keep the drill bit cool and clean during drilling, to carry out drill cuttings and to suspend the drill cuttings while drilling is paused and the drilling assembly is brought in and out of the hole. The drilling fluids are also used to avoid formation damage and to limit corrosion. The main minerals used in drilling fluids are bentonite, attapulgite, sepiolite, barite etc.

Table 17.14: Types of drilling fluids

Types	*Properties*	*Uses*
Air	Compressed in nature	Pumped either down the bore hole's annular space or down the drill string itself.
Air/water	Same as above	To increase viscosity, flush the hole, provide more cooling, and/or to control dust.
Air/polymer	A specially formulated chemical, added to the water and air mixture; example: a foaming agent.	To create specific conditions.
Water		Serve specific functions under specific conditions.
Water-based mud (WBM)	A homogeneous blend of water, clays (like bentonite) and different chemicals (potassium formate).	Clays like bentonite forms a gel which helps in the case that it is free-flowing while pumping and gets into a gel-like mass after pumping. Chemicals are used to obtain properties like viscosity control, shale stability, enhance drilling rate of penetration, cooling and lubricating of equipment.
Oil-based mud (OBM)	Here the base fluid is a petroleum product such as diesel fuel; the clays used are Na-bentonites with thixotropic property; can withstand greater heat without breaking down.	Used for many reasons, some being increased lubricity, enhanced shale inhibition, and greater cleaning abilities with less viscosity, to prevent contamination of the deposits, to prevent the cuttings settling to the bottom of the bore-holes.
Synthetic-based fluid (SBM)	Here the base fluid is a synthetic oil.	Generally used in the off-shore rigs; cool, lubricate and support the bit and drilling assembly.

Insulation

The raw materials needed for insulation purpose are mainly mica, asbestos, pumice, perlite; their properties and uses are given in Table 17.15.

Table 17.15: Types of insulating materials

Insulating Materials	*Properties*	*Uses*
Mica	Mainly sheet mica is used; muscovite and nicanite are widely used.	Used for many insulating purposes like gyro parts, sonar gear, computers, radar equipments and nuclear devices.
Asbestos	Capable of hardening when used with phenol and melamine resins.	Used to make moulding materials, which are used in insulation.
Pumice and perlite	Formed by the frothing of the gas-rich magma shortly before or at the time of ejection from the vent.	Used in low-temperature insulation.

Electronics

Now-a-days a number of minerals have been used in electronic industries, among which graphite, manganese minerals, silica, rare earth minerals and beryllium are important. Minerals are now applied in computer, clocks etc. Piezoelectric minerals are used on the principle of converting mechanical energy to electrical energy. Minerals are also used in acoustoelectronics for generating, transforming and transmitting radio signals. Other fields of electronics, where minerals find their applications are electro-optics, laser technology etc. A brief scrutiny in each of the minerals, used in electronic industry is done in Table 17.16.

Table 17.16: Types of electronic materials

Materials	*Properties*	*Uses*
Graphite	Heat-resistant, light weight, elasticity and radiation resistance.	Used for making electrodes for fine applications; used for making graphite moulds for production of diamond tools; used for making jigs, fixtures etc.
Silica	Hard, chemically inert and has high melting point; piezoelectric (quartz).	Used for pressure gauges, oscillators, resonators and wave stabilizers; used for making computer chips.
Rare earths	Silver, silvery-white or grey metals; high electrical conductivity.	Used in lasers, some types of lamps, magnets, motion picture projectors, X-ray screens etc.
Beryllium	Solid with high melting point of 1560K; high modulus of elasticity; resists attack by concentrated nitric acid and permeable to X-rays; thermally conductive.	Industrial and nuclear defence application, used in light; very strong alloys for aircraft industry; salts are used in X-ray tubes; used in the making of gyroscopes, various computer equipment, watch springs and instruments that require light weight, rigidity and dimensional stability.
Gallium	A rare, soft, silver metal, which is liquid at above room temperature.	Used in integrated circuits with electro-optical devices; to dope semi-conductors and produce solid-state devices like transistors.
Rubidium	An active element; combines violently with halogens; has high boiling point and low melting point; low density and electronegativity.	Used in the manufacture of photo-cells; used to make atomic clocks.
Tantalum	Hard, inert, grey metal; high melting point; highly heat resistant.	Used in construction of electronic equipments, like high-power transistors, mobile phones, onboard electronic systems for automobiles and computers; tantalum pentoxide used in capacitors and in camera lenses to increase refracting power.
Uranium	Silvery white radioactive metal; pyrophoric, when finely divided; malleable, ductile and slightly paramagnetic.	Used in inertial guidance devices, in gyro compasses, as counterweights for aircraft control surfaces, as ballast for missile re-entry vehicles.

Piezoelectric Minerals

These minerals are used in instrumentation for metrology, radio and television and communication. Also used for determining temperature, humidity, stresses and strains. Used in devices for frequency filtration.

Properties
- Characterized by the tensors of dielectric permittivity, elasticity and piezoelectric modulus.
- Electro-mechanical coupling coefficient (k2) should be in the range of 0.01% to 99%.
- Temperature frequency coefficient (TFC) should be zero.
- Indices of frequency stability should lie in the range of 10-5 to 10-12, from some milliseconds to several years.

The minerals exhibiting this property are quartz, tourmaline, $LiNbO_3$, $LiTaO_3$, berlinite ($AlPO_4$) and lithium tetraborate ($Li_2B_4O_7$).

Acoustoelectronic Minerals

These minerals are used to generate, transform and transmit radio signals. These are also used in communication and information systems, instrumentation engineering and in various other systems for generating and filtering signals and controlling their frequencies.

Controlling factors
- Bulk acoustic waves (BAW)
- Surface acoustic waves (SAW)
- Magnetostriction waves (MSW)
- Subsurface acoustic waves (SSAW)
- For piezoelectric materials' signal generation, the properties are elastic wave velocity, intrinsic loss, insertion loss, thermal variations in velocity and time delay.

Properties of acoustoelectronic minerals
- Flat
- Clean
- Acoustic velocity should be very slow for very low frequencies and very fast for high frequencies.

The minerals exhibiting this property are bismuth germinate and silicate, thallium sulvanite, quartz glass, lithium tetraborate and berlinite.

Laser Minerals

These are used in the generation and modulation of electromagnetic radiation by induced emission of monochromatic radiation.

Properties
- The material must be highly transparent (> 80%)
- Optically homogeneous

- Mechanically strong
- Thermally conducting to remove heat

The minerals used for this purpose are ruby, garnet group, scheelite, spinel, alexandrite, andalusite, fluorite and complex borates.

Lubricants

A lubricant may be gas, liquid or solid, which are used to reduce the friction and wear between moving parts of machines. Circulating lubricants are also used for cooling the system, carrying away the internally generated debris or externally introduced contaminations during operational procedures, and also sometimes as a sealing material. The minerals used in particular are graphite (C), molybdenite (MoS_2), tungstite (WS_2), talc, pyrophyllite etc.

Types of lubricants are gas, liquids including emulsions and suspensions, greases, adhesives and solids.

Table 17.17: Types of lubricants

Types	*Minerals*	*Uses*
Gases	Air, technical gases, steam or liquid-metal vapours.	Used in fluid bearings.
Liquids	Lanolin (wool grease, natural water repellant), water, mineral oils (paraffinic, naphenic and aromatic), vegetable (natural oil), synthetic oils, other liquids.	Lanolin is used as a natural water repellant; water is used in engineering processes, such as milling and lathe turning; mineral oils are used in internal combustion engines; vegetable oils used to form specialist synthetic esters.
Solids	Graphite (C), molybdenite (MoS_2), tungstite (WS_2), talc, pyrophyllite etc.	Mostly used as additives to oils and greases, as well as dry powder or as coating materials.
Greases		Greases are used where thick layer of lubrication is necessary and, so oil is unsuitable. These type of lubricants are used in various machinery bearings, e.g. in roller bearings in railways and car wheels, steam turbines, and spindles jet engine bearings.

Glasses

Glass is an undercooled liquid of very high viscosity, generally lustrous and transparent and, on exposure to natural elements, exhibit high durability. Glass is formed by cooling molten liquid in such a way, so as to prevent the ordering of atoms into a crystalline formation. The common minerals, required to form glass, are borax and borates, celestite, feldspar, fluorite, limestone, lithium-bearing minerals, silica, soda ash and zircon. A broad classification of glasses has been enumerated in Table 17.18.

Table 17.18: Types of glass

Types	*Sub-types*	*Properties*	*Uses*
Oxide glasses	Silica-based	Silica glass: Produced from quartz or beach sands, hence chemically SiO_2.	Used where requirements are of high service temperature, very high thermal shock resistance, high chemical durability, very low electrical conductivity and good ultraviolet transparency; used to make products like containers, windows and lightbulbs.
		Soda-lime glass: Main constituent is soda (Na_2O) and lime (CaO).	By adding sodium fluoride or calcium fluoride, a translucent opal glass is produced.
		Borosilicate glass	Used to form chemical glasswares and automobile headlamps.
		Lead glass: Chemically lead oxide (PbO).	Used in glass solders or in sealing glasses with low firing temperatures.
		Aluminosilicate glass: Intermediate between vitreous silica and soda-lime glass in thermal properties.	Used in fibre-reinforced plastics.
	Non-silica	Generally phosphates and borates.	Used in bioresorbable products such as surgical mesh and time-release capsules.
Non-oxide glasses	Heavy metal fluoride glasses	Contain generally fluorides of zirconium, barium, lanthanum, aluminium and sodium; have poor chemical durability.	Used in telecommunication fibres.
	Glassy metals	Essentially a compound of iron, nickel, phosphorous and boron.	Used in flexible magnetic shielding and power transformers.

(*Contd.*)

(*Contd.*)

	Semi-conducting solids	Chalcogenides: Formed by melting the chalcogen elements sulphur, selenium or tellurium with antimony or arsenic and germanium; have semiconducting properties.	Used in threshold and memory switching devices and in xerography.
		Amorphous silicon and amorphous germanium: Liquid like atomic order, but not considered as true glass.	Used in most photovoltaic applications, such as the solar cells in pocket calculators.
Glass ceramics	-	Exhibits certain degree of crystallization in normally random atomic structure.	Used to make cooking vessels and dental implants.
Glass composites	-	Formed by mixing ceramic, metal and polymer powder.	Used to make glass-fibre reinforced plastics, thick film conductor, resistor and dielectric pastes.
Natural glasses	Obsidian (volcanic glass)	High chemical durability	Are of significant commercial interest, like microtektites used for hazardous waste immobilization or conversion.
	Fulgarites (formed by lightning strikes) Tektites and microtektites (found from land and ocean-bottom) Moldavites (from central Europe) Libyan desert glass (from western Egypt)		

Minerals for Colour Industry (Pigments)

Earliest known pigments were natural minerals. Pigment colour is observed as a consequence of selective absorption and following the reflection of light. Depending on the type of light source, natural/artificial, the same pigment may show a colour variation. Also, binders and fillers added to pure pigment chemicals will influence the final colour outcome due to their own absorption and reflection pattern. This is a distinctly different physical process from fluorescence, phosphorescence etc. where the material itself emits light. For any industrial use, the essential properties are performance and stability over time. Fugitive pigments fade over time and long exposure of light. Pigments must have high tinting strength relative to the material it colours and should be in solid form over time and a temperature range including ambient temperature. Pigments are used for colouring plastics, ceramics, cements, bricks, wood strains, paper and other materials.

The basic properties to be considered for any use are:

1, Heat stability
2. Tinting strength
3, Opacity/transperancy
4. Dispersion
5. Lightfastness
6. Staining capacity
7. Acid and alkali resistance
8. Interaction between the pigments
9. Colour stability over time and weather.

Depending on the type of use, physical properties and cost is the deciding factor for pigment choice e.g. for colour glass production, the pigment must be suitably heat resistant for the manufacturing process. Pigments may be insoluble, resulting in a suspension and also a soluble or liquid one, called dye. When a pigment is manufactured from dye by precipitating, it is called a lake pigment.

With the development and refinement of the techniques for extracting mineral pigments for industrial purposes, standardized methods for pigment chemistry became available by mid 20th century. Colour standardization is matured with international industrial colour standards.

Table 17.19: Minerals suitable for pigment production

Materials	*Properties*	*Uses*
Zinc pigments	Sphalerite (ZnS), the ore mineral of Zn, widely used white pigment (ZnO) and is known as zinc white.	Zinc white is widely used in the paints. Zinc-sulphate is a chemical in dyes and pigments and zinc-sulphide is used in luminescent pigments.
Cadmium pigments	The pigments are mainly yellow, orange and red colours with the compositions from Cd-sulfides to sulfoselenides. Cd-green is obtained by mixing Cd yellow (CdS) with viridian, a blue green pigment of hydrated chromium oxide.	These pigments have cadmium as one of the chemical components and used mainly for coating and painting. By colloidal dispersion, Cd pigments are also used to colour glass, ceramic glazes and red stop lights.
Carbon pigments	Carbon black is a form of amorphous carbon that has a high surface area to volume ratio. Ivory black is produced by grinding charred ivory in oil.	As a pigment carbon black is used as reinforcement in rubber and plastic products. Ivory black is used by artists as pigment.
Chromite pigments	Chrome yellow is a yellow pigment of lead chromate, formed naturally. Since the pigment gets oxidized and darkens on exposure to air and contains toxic lead, it is largely replaced by cadmium yellow. Chromium oxide is the principal component of chrome green.	It is used in paints, inks and glasses. It is the colourant in chrome green.
Cobalt pigments	Cobalt blue is a cool, slightly desaturated blue colour substance. Chemically it is a cobalt oxide-aluminium oxide, or cobaltaluminate, $CoAl_2O_4$. Cerulean blue may be applied to a range of colours from deep blue, sky-blue, bright blue or azure colour through greenish blue colours. Cobalt yellow or aureolin is chemically potassium cobaltnitrate.	Cobalt blue was the primary blue pigment used in Chinese blue and white porcelain. It is used as a pigment in blue concrete due to its chemical tolerance in presence of alkali. Cerulean was used to describe blue pigments, particularly mixtures of copper and cobaltous oxides. Aureolin is used as a pigment in oil and watercolour painting.

(*Contd.*)

Table 17.19 (*Contd.*)

Materials	*Properties*	*Uses*
Copper pigments	Han purple ($BaCuSi_2O_6$) is synthetic barium copper silicate pigment, which is dark blue in colour. Egyptian blue is chemically calcium copper silicate ($CaCuSi_4O_{10}$). It is actually a synthetic blue pigment of calcium, copper, silica, lime and alkali. Paris green is chemically copper acetoarsenite. It is an extremely toxic, blue green chemical. Verdigris is formed when copper, brass or bronze is weathered and exposed to air or seawater over a period of time. It is usually a basic copper carbonate.	Han purple, as a pigment is used in beads, painted figurines, ceramic vessels, metal objects and wall paintings. Egyptian blue was earlier used by the Egyptians. Earlier Paris green was used as a pigment, called Emerald Green in artistic paints. The modern, significantly less vivid, substitution is a mixture of phthalocyanine green (blue shade), an organic lemon yellow and white. Verdigris is lightfast in oil paint. It has the property in oil painting that it is initially bluish-green, but transforms into a rich foliage green over the course of about a month.
Iron oxide pigments	Caput mortuum is a purple variety of haematite iron oxide pigment. Ochre when used to create a reddish tint is called red ochre. It is Fe_2O_3, the anhydrate of yellow ochre, which turns red when heated, as this drives off the water molecules. Prussian blue is a dark blue pigment and it is chemically $Fe_7(CN)_{18}(H_2O)_x$ where $14 \leq ¼ \leq 17$.	Caput mortuum is used in oil paints and paper dyes. It was a very popular colour for painting. Red ochre can be used in oil paints that dries quickly and covers surface thoroughly. Prussian blue is used as a pigment on cyanotypes. It is also used in oil paints.
Clay earth pigments	Yellow ochre is term for a form of earth pigment which produces the colour. It is chemically hydrated iron oxide. Raw sienna is a form of limonite clay and is chemically ferric oxide and yellow-brown in colour. Burnt sienna is a warm mid-brown, iron oxide pigment. Raw umber is a natural brown clay pigment of iron and manganese oxides. Burnt umber is a dark colour pigment, formed by heating raw umber.	Yellow ochre can be used in oil paints. Sienna is mostly used in the production of oil paint pigments. Umber is used as a pigment, mostly in association with ochre and sienna since prehistoric times. Burnt umber is mostly used in oil and water colour paints.

(*Contd.*)

(*Contd.*)

Lead pigments	Lead paint contains heavy metal lead which is used as a pigment, with lead chromate and lead carbonate. Naples yellow is chemically lead antimonite with chemical formula $Pb(SbO_3)_2/Pb_3(Sb_3O_4)_2$. Red lead is bright red or orange amorphous or crystalline pigment and composition-wise it is lead tetroxide, Pb_3O_4 or $2PbO{\cdot}PbO_2$.	Lead paint is mostly used in artistic field, where it is applied in oil painting on linen and cotton canvas. Naples yellow is originally highly toxic, so its use is becoming increasingly rare. Red lead, as a pigment, is generally used in undercoat paints for iron objects.
Mercury pigments	Vermillion is a naturally occurring, toxic, opaque, orangish-red pigment, which has been formed from powdered mineral cinnabar. Chemically it is mercuric sulphide (HgS). Nowadays, vermilion is artificially produced by reacting mercury with molten sulfur.	Vermillion is used by Hindu women as a sign of marriage and is also used in many religious Hindu rituals.
Titanium pigments	Titanium yellow is a complex, inorganic, yellow pigment with chemical composition of $NiO.Sb_2O_5.20TiO_2$. Titanium white is chemically titanium dioxide (TiO_2). It is the most widely used white pigment due to its brightness and high refractive index.	Titanium yellow is used mainly as a pigment for plastics and ceramic glasses, and in art painting. Titanium white is used as reflective optical coating for dielectric mirrors and some gemstones. As a pigment it also provides whiteness and opacity to products such as paints, coatings, plastics, papers, inks, foods, medicines (i.e. pills and tablets) as well as most toothpastes.
Ultramarine pigments	Ultramarine is a highly complex, blue-coloured pigment with chemical composition of $Na_{8-10}Al_6Si_6O_{24}S_{2-4}$.	Ultramarine was highly in use in earlier times in cave paintings and so on.

17.5 MINERAL BENEFICIATION AND BRIEF INDUSTRIAL APPLICATION OF SOME SELECTED MINERALS

In this section, for each industrial mineral, its beneficiation process and industrial applications are elaborated. Depending on the different properties for different minerals, their beneficiation procedures also vary considerably, so they are needed to be studied separately. In case of some minerals, certain problems are encountered during beneficiation. For example in graphite, especially in low-grade graphite due to its interlocking property, moisture gets adhered to the grains, leading to the loss in quality. This phenomenon is known as clogging. For such type of problems, special treatments are required during beneficiation.

Usual Methods of Beneficiation

1. Crushing and grinding using (i) ball mills, stirred mills and a tower mill (batch and continuous), (ii) jaw crushers, (iii) high pressure grinding rolls: 0.5 to 60 tonnes per hour and (iv) Hopkinson bar facility: three different diameters available up to 100 mm (the world's largest).
2. Processing for particle separation and sizing: Dry and wet separators based on particle size, density and shape, spirals, and sizing, laboratory-scale cyclones and transportable hydrocyclone pilot plants, laboratory cyclosizers (various size fractions).
3. Gravity separation (including flotation technique): Flotation cells (laboratory, and transportable hydrocyclone pilot plants and fixed pilot plants) also using spirals, eccentric jig, banana screen and dewatering screens and vacuum belt filter.
4. Magnetic separation: Using low intensity magnetic separators (LIMS) and wet high intensity magnetic separators (WHIMS).
5. On line moisture analysis.
6. Design modelling and optimization of hydro cyclones.
7. Designing the flow sheets utilizing the crushers, grinding mills, hydro cyclones, spirals, jigs, magnetic separation and flotation and specifying the machineries of each type.

Asbestos

Asbestos is the general name for a wide variety of silicate minerals, mostly silicates of calcium, magnesium and iron.

Beneficiation

- Primary crushing
- Large jaw crushers and gyratory crushers
- Rotary dryers, vertical tower dryers and fluid-bed dryers (drying process)
- Successive stages of crushing and comminution impact (separation of fiber from gangue)
- Finer fraction screening prior to air separation
- Cleaning operations
- Air-filtration for maximum minimization of dust content

Industrial Application

- Building and construction materials (asbestos cement pipe, insulating cement, insulating block, pipe covering, acoustical panels/plaster, fire brick, vinyl-asbestos and asphalt-asbestos floor tile, linoleum backing, ceiling tile, duct insulation for heating, ventilation and air conditioning (HVAC) systems, roofing felt, transite furnace flue, transite shingles and sidings, insulated electrical wire and panels, fireproofing spray, fire door interiors, refractory and boiler insulation material),
- Textiles (e.g. asbestos cloth in fireproof aprons, glassblower mitts),
- Caulking compounds and panes (mastics, adhesives, coatings, joint compound, putty, acoustical textures),
- Friction products (brake linings, clutch assemblies and gaskets),
- Fire protection and heat or sound insulation,
- Fabrication of papers and felts for flooring and roofing products,
- Pipeline wrapping, electrical insulation, thermal and electrical insulation,
- Friction products in brake or clutch pads,
- Asbestos-cement products,
- Reinforcement of plastics,
- Fabrication of packings and gaskets,
- Friction materials for brake linings and pads,
- Reinforcing agents,
- Vinyl or asphalt tiles, and
- Asphalt road surfacing.

Of these, asbestos-cement products, roof coatings, brake pads and shoes, and clutches are the major markets for asbestos.

Barite

Barite is an orthorhombic mineral, formed by sharing oxygen between S and Ba polyhedra. It is a sulfate of barium, $BaSO_4$, found in nature as tabular crystals or in granular or massive form and has a high specific gravity.

Beneficiation

- Feed
- Feed pan
- Rotating breaker
- Oversize (+5") → waste
- Undersize → selected
- Log washer
  - Overflow → tailing pond → overflow
  - Product
- Trammel screen
- Undersize
- Dewaterer

- Product
- Jigs
  - Tails → dewaterer → road and dam building
  - Product → dewaterer → product

Industrial Applications

- Barite is used as a commercial source of barium and many of its compounds.
- Ground barite is used as filler in the manufacture of linoleum, oilcloth, paper and textile manufacturing, rubber and plastics.
- Finely ground barite is used to make a thixotropic mud for sealing oil wells during drilling.
- *Drilling:* Weighing material with a specific gravity of 4.37 is used to increase the apparent density of a liquid drilling fluid system. Barite [$BaSO_4$] is the most common weighing agent used today. It is a mined material ground to an API specification such that particle sizes are predominantly in the 3 to 74 micron range.
- *Drilling fluids:* A dense mineral comprising barium sulfate [$BaSO_4$] that commonly occurs with a range of accessory minerals, such as quartz, chert, dolomite, calcite, siderite and metal sulfides. Used as a weighting agent for all types of drilling fluids, barites are mined in many areas worldwide and shipped as ore to grinding plants in strategic locations.
- *Barite industrial applications:* It is mainly used in oil-well drilling and it is estimated that 90% of the barites produced in the world is used only for drilling purposes. It is also found useful in the manufacture of white pigment or extender in pigments and barium chemicals, in the glass industry and as filler in rubber, leather, textile and paper goods.
- *Painting industry:* In dope and painting, it can be used as filling and can take the place of some expensive material such as basofor, crypton, titanium dioxide, activity, monox etc. Control the viscosity of the paint compatibly to make the products with bright colour and good stability.
- *Plastic industry:* In plastic industry, it is used as the filling of plastic ABS to make the products with bright colour. It also can improve the intensity, stiffness and abrasive resistance.
- *Rubber industry:* In rubber industry, the products less than 500 mesh can be used as the filling of rubber products to save cost. It can improve the intensity, and acid, alkali, and water proof, and also has benefit for the natural rubber and synthetic rubber.
- *Pharmaceutical industry:* In medicine, it can be used as barium meal material for the intestines and stomach reflection, and also the filling for plaster and dope to extend the time limit.
- *Paper-making industry:* In the papermaking industry, high-refined barites powder can be used as the filling of white paperboard and coat paper, improve the whiteness and percentage of coverage.
- *Cosmetics industry:* In cosmetics, it can be a substitute of titanium dioxide, for its white and gentle, and harmless to the skin.

Bauxite

Bauxite is a sedimentary rock that is an aluminum ore. It is formed in weathered volcanic rocks. Bauxite is composed of three main minerals; gibbsite (Al_2O_3. $3H_2O$), boehmite (Al_2O_3. H_2O), and diaspore.

Beneficiation

- Bauxite feed
- Wet grinding disaggregation
- Classification (mud removing)
  - Wet magnetic filter separation
- At-rich magnetic product (concentrate)
- Silicate and carbonate rich non-magnetic product (reject)
  - Selective agglomeration and floatation of carbonate or silicate
- At-rich non floating product (concentrate)
- Carbonate rich foam product (reject)
- Silicate rich foam product (reject)

Industrial Applications

- Bauxite is the main ore mineral responsible for the production of alumina, and ultimately aluminium. The production of alumina consumes over 90% of global bauxite mined.
- The remainder is used in the refractory, abrasive and chemical industries.
- Bauxite is also used in the production of high-alumina cement,
- as a catalyst in the oil industry,
- in welding rod coatings, and as a flux in making steel and ferroalloys.
- Alumina is used to produce aluminium, which has a wide range of uses in the consumer and capital goods markets. Some uses of aluminium include electrical equipment; car, ship, aircraft construction; metallurgical processes; industrial and domestic construction and packaging materials.

Clay Group of Minerals

Bentonite

Bentonite is an absorbent aluminium phyllosilicate, generally impure clay consisting mostly of montmorillonite. There are different types of bentonites and their names depend on the dominant elements, such as potassium (K), sodium (Na), calcium (Ca), and aluminum (Al).

Beneficiation

- Oxidation of structural ferrous iron.
- For rapid beneficiation grinding.
- Subjected to moist air.
- Clay is agitated, preferably by tumbling within a rotating drum, to provide rapid exchange of moisture and oxygen in the air.

- Increase the speed of oxidation of ferrous iron in the bentonite clay.
- Drying of thus treated bentonite clay may be then achieved by agitating the clay, again preferably by tumbling within a rotating drum, in the presence of dry air.

Industrial Applications

- Bentonite is used in pharmaceuticals, medical and cosmetics markets.
- Major use for bentonite is in well drilling. Bentonite is utilized in drilling mud to extract drill cuttings from the bit.
- Bentonite is used as a filler in pharmaceutical drugs.
- Due to its absorption-adsorption capabilities, it allows paste formation.
- Bentonite is used in industrial protective creams, wet compresses and anti-irritant lotions for eczema.
- In medicine, bentonite is used as an antidote in heavy metal poisoning.
- Personal care products such as mud packs, baby powder, sunburn paint and face creams may contain bentonite.
- Bentonite is used in an assortment of foods, such as candy bars and yogurt.
- Bentonite is very popular in the metal casting industry, where it is used as a bonding material for molding processes. When bentonite is mixed with certain types of molding sands, a bond is formed that can withstand very high temperatures.
- Bentonite is used in the ceramic and building industry where it is mixed with clay and sand to form an adhesive.
- In cosmetics, bentonite provides an active thickening substance for shampoos, calamine lotions, facial creams and lipsticks, where it keeps the preparations in homogenous form, maintains the moisture levels and applies non-greasy stickiness.
- Bentonite's unique bonding abilities have been used for cleansers, sprays, detergents, rubber tyre production and paper in addition to numerous other products.
- Bentonite was used by local Indians as a bleaching clay and as a washing agent.

Kaolonite

Kaolinite is a clay mineral, with the chemical composition $Al_2Si_2O_5(OH)_4$, Kaolin is a clay mineral more correctly known as kaolinite. It is also called China clay.

Beneficiation

- Chemical and mechanical dispersion of raw ore.
- Screening or sieving to remove sand and mica.
- Reverse flotation to remove discolouring impurities.
- Selective flocculation or reverse selective flocculation to remove discolouring impurities.

- High gradient magnetic separation to remove discolouring impurities.
- Centrifugation using solid bowl and/or disc nozzle centrifuges to control particle size distribution.
- Wet attrition milling to delaminate platy kaolins thereby producing high aspect ratio products.
- Chemical flocculation and leaching processes to solubilize iron and prepare kaolin slurries for subsequent dewatering.
- Dewatering processes including the use of hydroseparators, rotary vacuum filters, membrane filters and pressure filters.
- Use of high temperature processing equipment including spray dryers, rotary dryers, rotary and vertical calciners.
- Use of bulk loading and packaging equipment including screw conveyors, pneumatic conveyors, rotary valves, large and small bagging systems.
- Dry size reduction equipment including cage mills, roller mills and high speed pulverizers.
- High speed slurry dispersion equipment including dispersers and rotor stator mixers of various types.

Industrial Applications

- The rheological properties of deflocculated suspensions of kaolin in water are important in modern paper coating applications.
- The main use of kaolin is as a paper filler and a paper coating pigment.
- Kaolin is a major component in most ceramic bodies.
- Kaolin is used as an extender in aqueous based paints and as filler in rubber and synthetic polymers.

Sepiolite and Attapulgite

Sepiolite is a clay mineral, a complex magnesium silicate, a typical formula for which is $Mg_4Si_6O_{15}(OH)_2 \cdot 6H_2O$. It can be present in fibrous, fine-particulate, and solid forms. Attapulgite is a magnesium aluminium phyllosilicate with formula $(Mg, Al)_2Si_4O_{10}(OH) \cdot 4(H_2O)$ which occurs in a type of clay soil. It is one of the types of fuller's earth. Sepiolite and attapulgite have crystalline structure incorporating channels like zeolite. They are also thixotropic (a clay that liquefies when kneaded and solidifies again when left standing) and calcinable (solidifies when heated) like kaolinite. That means, sepiolite and attapulgite have three properties at the same time: they can control moisture content as well as being thixotropic and calcinable.

Beneficiation

- Air drying (moisture reduction)
- Size reduction to 2.5 cm in roller crushers
- Rotary drying for moisture reduction to about 6 to 10%
- Product

- Regular volatile material (RVM)
- Light volatile material (LVM)

Industrial Applications

- Building material: Reinforcement of rockwool building insulator, interlaminar bonding of calcium silicate boards and viscosity conditioning of the exterior walls
- Flame retardant paper: Reinforcement
- Automobile: Damping material and viscosity conditioning
- Fertilizer: Binder
- Paint/coating and adhesive: Thickener

Vermiculite

Vermiculite is a member of the phyllosilicate group of minerals. Vermiculite is formed by weathering or hydrothermal alteration of biotite or phlogopite. It is a naturally occurring mineral composed of shiny flakes, resembling mica. Vermiculite is a natural mineral that expands with the application of heat. The expansion process is called exfoliation and it is routinely accomplished in purpose-designed commercial furnaces. When heated to a high temperature, flakes of vermiculite expand as much as 8-30 times their original size. The expanded vermiculite is a light-weight, fire-resistant, and odourless material and has been used in numerous products, including insulation for attics and walls. Sizes of vermiculite products range from very fine particles to large (coarse) pieces nearly an inch long. Vermiculite is always used in exfoliated form. When exfoliated it possesses nearly 10 to 11 times less bulk density than the original volume. In commerce, vermiculite which expands more than 10 times the original volume is regarded of good quality. With an expansion below 10 times the original volume, vermiculite is considered of low grade. The low bulk density, comparative high refractoriness, low thermal conductivity and chemical inertness make vermiculite satisfactory for many types of thermal and acoustic insulations.

Beneficiation

- Separated from gangue minerals
  - Ore broken down into <20 mm or <17 mm granules.
  - Fine material removed by washing or screening.
- Delaminated into flakes
  - Flakes separated by multistage screening in horizontal streams of air, using suspension procedures.
- Classification of flakes according to industrial size fractions.
  - Flakes expanded five- or eight-fold at 250 to 1,500 °C
  - Cooling at 400 °C.
  - Separator for removal of impurities.

Industrial Applications

- One of its large commercial uses is as an aggregate in light weight concrete and hard wall-plaster because of its acoustic and thermal insulating and fire-resisting qualities.
- The density of raw vermiculite is 50 to 90 lbs. per cu. ft. While that of the exfoliated one is 5-10 lbs. per cu. ft. It is therefore extensively used in concrete work to save weight.
- Vermiculite concrete weighs 20-25 per cu. ft. as against concrete which weighs about 100 lbs. per cu. ft. Vermiculite concrete has the same advantages as concrete made with pumice and perlite.
- Refractory insulations both in the form of loose vermiculite fill and vermiculite bricks are used in furnaces and kilns up to 1100°C.
- About 60% of the present world consumption is in the form of loose fill when the expanded material is merely poured like dry sand into wall spaces or applied over ceiling constructions or attics of residential buildings with a view to insulating homes against cold in winter and heat in summer.
- One inch of Unifil, a trade name of a particular expanded vermiculite, holds back as much of 2 ft. brick wall or wall of concrete 3 ft. thick.
- As a light-weight aggregate it is extensively used in prefabricated houses.
- Vermiculite concrete in the form of monolithic cast is used in sound-absorbing panels in aeroplane engine testing sheds.
- Vermiculite, being a granular expanded aggregate with numerous air voids, when mixed with a suitable binder develops sound insulating properties.
- Vermiculite plaster is widely used for better acoustics and reduction of noise in auditoriums, wireless studios, theatres, hospitals etc.
- Vermiculite mixed with three parts of gypsum is used as plaster for sound-absorbing purposes.
- A new building material called Pyrok, consisting of vermiculite bonded with lime and cement is marketed in England.
- More than hundred major and minor uses of vermiculite have been developed in the fields of agriculture, pesticides, lubricants, disinfectants, plastics and light-weight insulating bricks.
- Unexfoliated vermiculate has a few minor uses, such as for circulation in drilling mud and in the annealing of steel.
- When unexfoliated vermiculate is reacted with concentrated H_2SO_4, it produces a pure form of silica in flake form. This product is known as '*samisilite*'. It is used as a dehydrating medium in air conditioning plants since it can absorb about 20 per cent its weight of water. The potency of this product may be revived by heating.

Beryllium Minerals

Out of the many known beryllium minerals, only two minerals, namely, beryl and bertandite have significant industrial applications in terms of production of beryllium and beryllia. The mineral beryl is a beryllium aluminium cyclosilicate with the chemical formula $Be_3Al_2(SiO_3)_6$. Bertrandite is a beryllium sorosilicate hydroxide mineral with composition: $Be_4Si_2O_7(OH)_2$. Bertrandite is a colourless to pale yellow orthorhombic mineral with a hardness of 6-7. It is commonly found in beryllium rich pegmatites and is in part an alteration of beryl. Bertrandite often occurs as a pseudomorphic replacement of beryl.

Beneficiation

- Floatation
- Grinding in a ball mill
- Fines feeding to a conditioner where floatation reagents added
- Muscovite floated off
- Pulp conditioned at pH 5 with tallow amine acetate
- Beryl and feldspar bulk floated with an alcohol frother
- Minerals separated by conditioning with calcium hypochloride
- Floatation of beryl with petroleum sulfonate

Beryl

Sulfate Process

- Beryl heated at 1625° C.
- Quenched in cold water to produce glass.
- Glass dried and ground into –200 mesh powder.
- Leached with sulfuric acid to obtain sulphates of beryllium and aluminium.
- Sulfates extracted with water.
- Ammonia used for separation of aluminium sulfate.
- Solution made alkaline with NaOH to obtain beryllium hydroxide.
- Calcination of beryllium hydroxide produces BeO.

Fluoride Process

- Beryl sintered at 750° C with sodium hexafluosilicate.
- Beryllium oxide converted to water soluble salt.
- Product leached with water at room temperature.
- The last two steps are same as before.

Bertrandite

- Ore is wet ground and screened to form –20 mesh particles.
- Leached with 10% sulfuric acid solution at about 100° C.
- Sulfate solution extracted with 0.3 m solution of ammonium di-2-ethylexyl phosphate/kerosene in a five-stage mixer at 50° C.
- Be, Al and Fe enter organic phase.
- Metal ions removed from organic compounds by ammonium carbonate solution.

- Formation of basic beryllium carbonate solution containing beryllium hydroxide.
- Pure beryllium hydroxide is obtained by filtration and decomposition at 140-170° C.

Industrial Applications of Beryllia (Formed from Beryl or Bertrandite)

- Sintered beryllium oxide, which is very stable, has ceramic characteristics.
- Beryllium oxide is used for rocket engines, catalysts, semiconductors, moderators of atomic reactors, and neutron reflectors.
- Beryllium oxide is used in many high-performance semiconductor parts for applications such as radio equipment because it has good thermal conductivity while also being a good electrical insulator.
- It is used as filler in some thermal interface materials such as thermal grease.
- Some power semiconductor devices have used beryllium oxide ceramic between the silicon chip and the metal mounting base of the package in order to achieve a lower value of thermal resistance than for a similar construction made with aluminium oxide.
- It is also used as a structural ceramic for high-performance microwave devices, vacuum tubes, magnetrons and gas lasers.

Borax and Borates

Borax, also known as sodium borate, sodium tetraborate, or disodium tetraborate, is an important boron compound, a mineral, and a salt of boric acid. It is usually a white powder consisting of soft colourless crystals that dissolve easily in water.

Celestite

Celestine or celestite ($SrSO_4$) is a mineral consisting of strontium sulfate. The mineral is named for its occasional delicate blue colour. Celestine occurs as crystals, and also in compact massive and fibrous forms. It is mostly found in sedimentary rocks, often associated with the minerals gypsum, anhydrite and halite. Celestite has the same structure as the barite ($BaSO_4$), and forms very similar crystals. The two may seem identical by ordinary methods, but a flame test can distinguish them. By scraping the dust of the crystals into a gas flame, the colour of the flame will confirm the identity of the crystal. If the flame is a pale green, it is barite, but if the flame is red, it is celestite. The flame test works because the elements barium (Ba) and strontium (Sr) react with the flame and produce those colours.

Beneficiation

- Crushing in a jaw crusher
- Attrition tumbler

- 15 mm screen
- Stripa heavy medium separation unit
- Impact crusher
- Vibrating screen

  For –3 mm
  - Hydrosizer
  - Shaking tables
  - Concentrates
  - Drying table
  - Total concentrate

  For –10 + 3 mm
  - Jig
  - Dewatering classifier
  - Hammer crusher
  - Trammel screen
  - Hydrosizer
  - Shaking tables
  - Tailing

Industrial Applications

- Principal use of celestite is the manufacture of strontium carbonate. Strontium carbonate in turn is used in the manufacture of glass faceplates or colour television tubes.
- It is present in glass at approximately 12-14% SrO and functions as an X-ray absorber.
- Strontium carbonate when added to special glasses, glass frits and ceramic glazes, increases the firing range and lowers acid solubility.
- It is also used for production of other strontium compounds.

Chromite

Chromite is an iron chromium oxide: $FeCr_2O_4$. It is an oxide mineral belonging to the spinel group.

Beneficiation

- The chromite ore pulverized and preferably mixed with sodium sulfate.
- Ammonium carbonate added for subsequent vapourization to increase porosity.
- Sulphur oxide-oxygen roasting of the chromite ore conducted at elevated temperatures in the approximate range between 300° and 1,000° C or higher at approximately atmospheric pressures.
- After roasting, the ore is preferably leached using an aqueous ferric sulfate solution to remove soluble undesired constituents from the roasted chromite ore.

- The beneficiated chromite ore shows improved chromium concentrations and chromium to iron ratios when compared to the original ore.

Industrial Applications

- Used to harden steel, to manufacture stainless steel, and to form alloys.
- Used in plating to produce a hard, beautiful surface and to prevent corrosion.
- Used to give glass an emerald green colour. It is responsible for the green colour of emeralds and the red colour of rubies.
- Wide use as a catalyst.
- Dichromates such as $K_2Cr_2O_7$ are oxidising agents and are used in quantitative analysis and also in tanning leather.
- Lead chromate as chrome yellow is a pigment.
- Compounds are used in the textile industry as mordants.
- Used by the aircraft and other industries for anodising aluminium.
- The refractory industry uses chromite for forming bricks and shapes, as it has a high melting point, moderate thermal expansion, and stable crystalline structure.
- Tanning leather.

Diatomite

Diatoms are single-celled (unicellular) organisms that live as individuals or in groups called colonies. They exist in all the waters of the Earth, both salt and fresh. They form shells made out of silica (the mineral name of this silica is opal) which they extract from the water. When diatoms die, their silica shells accumulate on the floor of the body of water in which they lived. Thick layers of these diatom shells have been preserved in the rock record. Such layers, or beds, of diatoms are called diatomaceous earth, or diatomite.

Beneficiation

- Diatomaceous earth purified of sand and clay.
- Reduced to a smaller particle size by wet attritioning raw diatomaceous earth to reduce the diatomite particle size. Wet attritioning can be done in the presence of a clay dispersant.
- Hydraulically classifying the wet attritioned material to remove sand and quartz.
- Dispersing clay.
- Centrifugally classifying to remove solid clay. Centrifugal classification can be done before or after the quartz and sand are removed.

Industrial Applications

- Because of its porosity, diatomite has been used extensively to filter impurities out of everything from beer and wine to oils and greases.
- Similarly, diatomite is used to filter impurities from water to produce drinkable (potable) water in public water systems where the diatomite removes bacteria and protozoa.

- The oldest use of diatomite is as a very mild abrasive and for this purpose has been used in toothpaste and metal polishes.
- It is also used in paper, paint, brick, tile, ceramics, plastics, soaps, detergents and other products as filler.
- Diatomite has also found value as an insulating material in high-temperature mechanisms like furnaces and boilers. It has also proved effective as a sound insulator.

Dolomite

Dolomite is a sedimentary carbonate rock and a mineral, both composed of calcium magnesium carbonate $CaMg(CO_3)_2$ found in crystals.

Beneficiation

- Primary crusher
- Secondary crusher
- Grinding
- Heat treatment

Industrial Applications

- Chiefly used as refractory, ramming and fettling material in steel melting shop.
- As fluxing material in blast furnace operation in secondary steel and ferromanganese manufacture.
- To a lesser extent it is used in the glass industry especially in sheet-glass manufacture.
- It also finds use in the manufacture of mineral wool.
- In England, dolomite has become a useful source for the production of magnesite by reacting calcined dolomite with sea-water.
- Dolomite is also a good source of magnesium metal. The magnesium metal is extracted from dolomite by the well-known ferro-silicon process.
- Dolomite decomposes completely above 900° C. The product resulting from this relatively low-temperature calcination is highly porous and reactive and is known as 'calcinated dolomite'. Dolomite is sometimes used both in the raw and calcined form as refractory material for hearth maintenance and for banking door in open hearth furnaces.
- For most refractory uses, it is desirable to subject the dolomite to a heat treatment at a high temperature of the order of 1700° C, to shrink the material thoroughly and render it less reactive.
- Dead burnt (D.B.) dolomite is the term generally used for the refractory made by firing dolomite, with or without additives, at high temperature to produce dense, well-shrunk particles.
- In basic converters, dolomite bricks are kept in the outer lining because it has lower thermal conductivity than magnesite.

- A major source of magnesium, particularly for agricultural and pharmaceutical applications.
- Dolomite is used for manufacturing certain types of refractory bricks used in steel making. The dolomite is heated to a high temperature to drive off the carbonate as carbon dioxide and the remaining material, a mixture of calcium and magnesium oxides, is blended with carbon and other materials and pressed into blocks for the furnaces. The magnesium and calcium oxides have very high melting points and are an excellent, inexpensive refractory material.
- Dolomite is also used as a source of magnesium oxide for making magnesium metal and for chemical uses, such as the common laxative milk-of-magnesia.
- Dolomitic limestone's and dolomites are mined along with limestone and used for crushed stone and aggregates for manufacture of pavement, concrete for construction and as fill material.
- Dolomite is also used in some cement, as a source of magnesium. Of course dolomite is also used as mineral specimens.
- Dolomite rock is used as an ornamental and structural stone, and for extracting certain metals from their ores. It is useful in the chemical industry in the preparation of magnesium salts.

Feldspars

Feldspars ($KAlSi_3O_8$, $NaAlSi_3O_8$ and $CaAl_2Si_2O_8$) are a group of rock-forming tectosilicate minerals which make up as much as 60% of the Earth's crust.

Beneficiation

- Feldspar generally needs beneficiation by grinding, separation and floatation.
- In grinding, in case of pure and uniform crude feldspar, storing and grinding are done separately.
- It is followed by subsequent blending for different grades.
- Henceforth proper milling and beneficiation are followed.
- Froth floatation and high intensity magnetic separation methods are applied.

Industrial Applications

- In making the body composition of several types of procelain, china and earthenware and also in the preparation of glazes and enamel.
- As an important ingredient in the glass sand batch.
- As a bonding agent in the manufacture of bonded abrasives like wheels and discs of garnet, corundum, emery etc.
- In ceramic bodies, the main vitrifying (fluxing) agent is feldspar. The majority of white ware bodies contain good proportions of feldspar. It acts as a flux. In the ceramic industry, the flux is defined as that portion of the body which develops glass phase. This is provided mostly by feldspar. The amount of flux in a ceramic body should be only in such a proportion as to develop the

desired amount of vitrification. If excess of flux is added, the fired body becomes very glassy and consequently, brittle.

- In ceramic bodies, potash feldspar is preferred although soda feldspar works as a good flux and can be satisfactorily used in developing ceramic bodies.
- In the glass industry also, potash feldspar is used in varying proportions. It may contain 10 to 15% of the batch. Feldspar, in this industry, is valued for two purposes, firstly, it acts as a flux and secondly, for alumina content. Alumina provides the resistance in glass to impact, bendings and thermal shock.
- Glass industry prefers to use mainly orthoclase (potash feldspar) because it tends the melt to clear glass while albite (soda feldspar) tends the melt to a translucent glass.
- Feldspar is used as bonding agent along with magnesium oxide, magnesium chloride and other synthetic glue in the manufacture of abrasives, wheels, discs and other shapes.
- In the preparation of glazes the fine powder of feldspar is mixed with silica powder and thin slurry is made. The fired up goods are dipped into the slurry and fired again. This imparts glaze to the surface. Some other ingredients are also added to the slurry to impart glaze and lustre.

Fluorspar

Fluorspar (also called fluorite) is a halide mineral composed of calcium fluoride, CaF_2. It is an isometric mineral with a cubic habit, though octahedral and more complex isometric forms are not uncommon. The term fluorspar, when used as a commodity name, also refers to calcium fluoride formed as a by-product of industrial processes.

Beneficiation

For industrial use

- Hand sorting for high-grade lump or gravel ore
- Crushing
- Screening to finer fractions

Metallurgical grade fluorite

- Crushing
- Screening
- Size reduction
- Fine fractions sent to flotation
- Coarse particles sent to heavy media separation
- Separator to reduce the phosphate content
- Magnetic separation to remove iron oxide and arsenic contaminants

Ceramic and acid-grade fluorite

- Flotation
- pH controlled by soda ash or sodium hydroxide

- Oleic acid used as collector agent
- Temperature is between 25-80° C to avoid freezing of oleic acid
- Fatty acid used as collector of fluorite
- Using depressants ore crushed and ground to proper size
- Fluorite removed in quick pass through a rougher flotation circuit
- Sent to cleaner circuit
- Rougher tailing discarded
- Further cleaning to separate fluorite and gangue
- Final product

Industrial Applications

- Fluorspar was mainly used as a flux in iron and steel making. The principal metallurgical use is as a flux in the reduction of steel and to a lesser extent in the refining of aluminum, nickel, magnesium, copper, lead and silver.
- It is also used in the chemical industry to make fluorine to add to drinking water and toothpaste.
- The chemical use of fluorspar, which usually involves its conversion to hydrofluoric acid, is related to the production of aluminum, high octane gasoline, refrigerants, insecticides, preservatives, dyestuffs, propellants, the separation of uranium isotopes and many other processes.
- Fluorspar was also used to make chloro-fluorocarbons, which used to be used as aerosol propellants and in fridges, but are now blamed for destroying the ozone layer.
- Some fluorspar was used for making lenses for telescopes and other optical equipment.
- It was also used by lead miners to make spar boxes.
- Fluorspar plays an important part in the manufacture of glazed tiles; enamel coatings on steel for stoves, refrigerators, table tops, counters and other equipment for such places as kitchens and bathrooms; opalescent glass for light bulbs, lamp shades and soda fountains; abrasives; and special lenses for the optical industries.
- In the field of ceramics, fluorspar is used to produce opalescent, milk glass, and coloured glass.
- It is an ingredient of flint glass, brick mix, and in the manufacture of fibre glass.
- It is finding increasing use in the manufacture of enamels for metal surfaces.

Garnet

Garnets are nesosilicates having the general formula $X_3Y_2(SiO_4)_3$. Garnet is a common mineral of metamorphic rocks such as gneiss and schist of all description from basic to acid, crystalline limestone and pegmatites.

Beneficiation

- Primary crushing and screening

- A sink-float process with ferrosilicon as the medium
- Secondary screening with screens divided longitudinally to serve both float and sink
- Product further treated in rolls
- Classifier
- Induced magnetic roll separator
- Vibratory feeder
- Non-magnetic particles get separated
- Garnet particles attached to the roll

Industrial Applications

- Garnet is ground to a variety of sizes to be used as an abrasive.
- Garnet sandpaper was the original application of this mineral.
- It is also used to make a number of similar products, including sanding belts, discs and strips.
- Almandine garnet is one of the best natural abrasive and is preferred over other natural abrasives like silica sand, mineral sands and flint.
- Garnet's unique grain hardness/toughness guarantees efficiency in blasting and reduce the breakdown percentage on impact. About 65% of garnet can be recovered and recycled twice or thrice.
- Almandine garnet (sp. gr. 4.0-4.2) makes 3 or 4 times higher impact than quartz grains (sp. gr. 2.5-2.6) of the same size.
- Garnet is well suited for abrasive blasting/surface cleaning in fields like shipyards, petrochemical industry, building industry, non-ferrous surfaces etc.
- Water (abrasive) jet cutting is more prevalent in North America and Europe where fine grade garnet is injected in a stream of high velocity of water to cut marbles, granites, high strength steel, aero space and automotive glass.

Graphite

Graphite is a form of elemental carbon. It is one of the three allotropic forms which the element carbon exists in nature, the other two being coal and diamond. The utility of graphite is dependent largely upon its type viz., flake, lumpy or amorphous. The flake type graphite is found to possess extremely low resistivity to electrical conductance. The electrical resistivity decreases with the increase of flaky particles. Also the bulk density decreases progressively as the particles become more and more flaky.

These days, artificially prepared graphite has replaced natural graphite to a great extent. Artificial graphite is prepared by heating a mixture of anthracite, high grade coal or petroleum coke, quartz and saw-dust at a temperature of 3000° C, out of contact with air. Graphite carbon is deposited as residue.

Beneficiation

- Flotation

- Pulp with pH 8, made alkaline with soda ash.
- Quartz, mica, feldspar and marble which are common associations of graphite are separated from graphite by addition of kerosene and floated with pine oil as frother.
- Cleaning done, if the product is insufficiently clean.

Industrial Applications

- Flake graphite finds a large use in the manufacture of carbon electrodes, plates and brushes required in the electrical industry and dry cell batteries.
- In the manufacture of plates and brushes, however, flake graphite has been substituted to some extent by synthetic, amorphous, crystalline graphite and acetylene black.
- Graphite electrodes serve to give conductivity to the mass of manganese dioxide used in dry batteries.
- The manufacture of crucibles is served best by flake graphite, although crystalline graphite is also used. Coarse-grained flake graphite from Malagasy is regarded as standard for crucible manufacture.
- Flake graphite containing 80 to 85% carbon is used for crucible manufacture; 93% carbon and above is preferred for the manufacture of lubricants, and graphite with 40 to 70% carbon is utilized for foundry facings.
- Natural graphite, refined or otherwise pure, having a carbon content not less than 95% is used in the manufacture of carbon rods for dry baterry cells.
- All grades of graphite, especially high grade amorphous and crystalline graphite having collodial property i.e. remaining in suspension in oil, are used as lubricants.
- Graphite has an extraordinarily low coefficient of friction under practically all working conditions. This property is invaluable in lubricants. It diminishes friction and tends to keep the moving surface cool.
- Dry graphite as well as graphite mixed with grease and oil is utilized as a lubricant for heavy and light bearings.
- Graphite grease is used as a heavy-duty lubricant where high temperatures may tend to remove the grease.
- Manufactured graphite is used for making furnace electrodes and for modes in the manufacture of chlorine and caustic soda.
- A considerable quantity of graphite is used in foundry-facing to prevent the moulding sands from adhering to cast articles. Here too, flake graphite is preferred. Dust or powder of flake, crystalline-graphite are also used.
- Graphite bricks of high purity are used as moderators in an atomic reactor.
- In the nuclear field, graphite is a good and convenient material as a moderator but this is only true if the graphite is low in certain neutron absorbing elements notably boron and the rare earths and is of consistent quality particularly with regard to density and orientation.

- The latest invention in the use of graphite is in the blast furnace operation, experimented by Oesterreichisch-Alpine Mountangesellschaft, Austria. In view of the lack of cooking coal in Australia, it is likely that the use of graphite in blast furnace will be developed on a commercial scale.
- Other uses of graphite are in the manufacture of paints and pencils. Finely powdered lump graphite of 70% purity is generally employed in paint manufacture.
- Graphite is a great water repellant and thus makes an ideal protective coating for wood.
- Amorphous graphite is generally used in the manufacture of lead for pencils.

Gypsum

Gypsum is hydrated calcium sulphate ($CaSO_4.2H_2O$).

Beneficiation

- Primary crushing with gyratory, jaw or impact crusher, depending on the size of the mine-run ore.
- Secondary crushing in either hammer mills or cone-type crusher, usually conducted with vibrating screens to reduce the fines.
- Drying in case of high moisture by rotary dryer.
- Fine grinding by air-swept roller mills with integral air separator or high energy impact mills with air classifier.

Industrial Applications

- Its main application is as a building material, mostly produced as so-called plaster of Paris for plastering walls and making decorative features in buildings.
- Can be used in making writing chalk, soil conditioning for agriculture, making moulds for pottery, as an additive in the manufacture of Ordinary Portland Cement (OPC), in surgical splints, in increasing the hardness of water, as an additive in certain foods such as tofu and in some types of medicines and pharmaceuticals.
- Its construction uses can also include external applications, such as in making walling blocks, where the climate is generally dry and not prone to the dampness that damages common types of gypsum.
- Gypsum is used in repairing the soil. Its main purpose is to precipitate the clay particles in heavy or the layer of hard subsoil type soils and loosen the soil structure. Then this creates air and moisture slots that will loosen and break-up the soil structure.

Limestone

Limestone is a very common sedimentary rock of biochemical origin. It is composed mostly of the mineral calcite.

Beneficiation

- Cleaning of deposit before quarrying
- Stripping, by mechanical or hydraulic means
- Selective recovery
- Chemical changes due to dolomitization, leaching, mineralization, etc. Recovery may be guided by three-dimensional studies of the quarry and may be accomplished by sorting of the product in the quarry.
- "In plant" processing of stone, including:
  Crushing and screening—with material flow arranged to reject certain selected size grades of materials for quality reasons.
  Washing—with discard of selected size grades isolated usually by screening or classifiers.
  Sorting—varying from hand picking to elaborate recently developed photocell-circuit sorting, based on light-reflectivity properties.
  Grinding and classification—discard of selected size grades after sorting by classifiers is common.
  Flotation and concentration—following grinding and size classification, usually with a fairly complicated flow.
  Heavy-media (sink-float) separation—also relatively complicated and of particular interest when wide variations in specific gravity of component minerals is involved.
  Blending is the most highly developed and widely used method of "beneficiation".

Industrial Applications

- Limestone is very popular and renowned in architecture, and has many samples around the world, is made primarily of this material.
- Limestone is further used as a frontage on some of the skyscrapers, in way of thin plates for covering rather than solid blocks.
- Producer of quicklime and slaked lime.
- Used for cement and mortar.
- Pulverized limestone is used as a soil conditioner to offset acid soil conditions.
- Crushed for use as collective—the solid base for many roads.
- Geological arrangements of limestone are amid the best petroleum reservoir.
- For glass making.
- Limestone is quarried for roadbeds; building and other landscape building; and cement manufacture.

Lithium Minerals

Among a number of lithium-bearing minerals, spodumene [$LiAl(SiO_3)_2$], lepidolite [$K_2Li_4Al_2F_4Si_8O_{22}$], petalite [$LiAl(Si_2O_5)_2$], amblygonite [$LiAl(PO_4F)$] and eucryptite [$LiAlSiO_4$] are important.

Beneficiation

- Heavy medium separation with a mixture of ferrosilicon and magnetite to remove K-feldspar.
- Flotation at pH of 9.2 using starvation quantities of tall oil fatty acid and petroleum sulphonate to separate amblygonite from spodumene.
- Upgradation of the ore to 7.25% Li_2O by flotation to remove quartz and Na-feldspar.
- High intensity magnetic separations to remove iron contents.

Industrial Applications

- *Ceramics and Glass:* Lithium minerals are used in glass and ceramic industries for their lithia content. Lepidolite, spodumene and amblygonite are used along with glass sand-batch for the manufacture of lithium glass. It has the lowest melting point and lowest annealing temperature of all alkali glasses. Lithium reduces its coefficient of expansion. It is reported that lithium minerals have also been used in the manufacture of ceramic bodies.
- *Chemicals:* Lithium minerals are mainly utilized in the manufacture of lithium carbonate which is the starting point for the manufacture of various chemicals which are used in lubricating greases, ceramics, air-conditioning and refrigeration, alkaline storage battery, bearing and welding aluminium, bleaching, chlorination of water in swimming pools and several other purposes.
- Lithium chloride is largely used in refrigeration and air-conditioning plants as it is one of the most hygroscopic of all inorganic salts. Lithium hydroxide is used for absorbing carbon dioxide in submarines and as a constituent of respirators. Chloride and fluoride compounds are used in welding and brazing.
- Dry lithium hypochlorite is used as bleach in laundries and for chlorination of swimming pools.
- *Metallic:* It finds small use in alloying with metals of higher melting points like Cr, Al, Mg, Cu, Pb and Zn. It imparts toughness and tensile strength to the alloy. Lithium is also used as a scavenging material for metallic minerals and gases in the metallurgy of several metals.
- *Medicine:* Lithium in the form of lithium carbonate is used in the treatment of gout.

Magnesite

Magnesite is magnesium carbonate, $MgCO_3$. Iron (as Fe^{2+}) substitutes for magnesium (Mg) with a complete solution series with siderite, $FeCO_3$. Calcium, manganese, cobalt and nickel may also occur in small amounts.

Beneficiation

- Crushing

- Scrubbing
- Screening
- Coarse particles sent to heavy medium separation cone.
- Finer fractions stockpiled for future treatment by flotation.
- Finely divided ferrosilicon medium used as the dense medium and separation done in a 6 m diameter cone.
- Impurities discarded.
- Magnesite "sink" crushed, screened and stored in bins.
- Sodium silicate used as a depressor for quartz and serpentine.
- Oleates or oleic acid used as a collector for magnesite.
- Magnesite concentrate removed as a foam, then thickened and brought to residual water content of 8-12% in centrifuges or fillers.
- Product then briquetted and burned.

Industrial Applications

- *Dead-burned magnesia:* Basic refractories.
- *Fused or electrofused magnesia:* Thermal insulation (refractories), Electrical insulation (electric ovens and appliances).
- *Caustic-calcined magnesia:* Extraction of magnesium, production of fused magnesia, cement, insulation, paper, fertilizer and animal feed, stabilizer during vulcanization of rubber, production of uranium, chemical and environmental applications (notably in water treatment and gas desulfurization).
- *Magnesite or magnesium carbonate (precipitated):* Insulation, rubber, paint pigments and ink, glass and ceramics.
- *Brucite or magnesium hydroxide:* Sugar refining, flame and smoke retardants, chemical and environmental applications (notably in water treatment), pharmaceuticals.
- *Magnesium chloride:* Cement, ceramics and refractories, textiles, paper, chemical applications, production of magnesium.

Manganese Minerals

Among a number of manganese minerals, pyrolusite (MnO_2) is the most important one.

Beneficiation

- Flotation using oil emulsions or conventional soap solutions.
- For beneficiation of low-grade manganese ore, ammonium carbonate process is used.

Industrial Applications

- Steel becomes harder when it is alloyed with manganese. Hardened steel is important in the manufacture of construction materials like I-beams (24%

of manganese consumption), machinery (14% of manganese consumption), and transportation (17% of manganese consumption).

- Manganese dioxide is used to: manufacture ferroalloys; manufacture dry cell batteries (it's a depolarizer); to "decolourize" glass; to prepare some chemicals, like oxygen and chlorine; and to dry black paints.
- Manganese sulfate ($MnSO_4$) is used as a chemical intermediate and as a micronutrient in animal feeds and plant fertilizers.
- Manganese metal is used as a brick and ceramic colourant, in copper and aluminum alloys, and as a chemical oxidizer and catalyst.
- Potassium permanganate ($KMnO_4$) is used as a bactericide and algicide in water and wastewater treatment, and as an oxidant in organic chemical synthesis.

Mica

Two forms of mica, namely the sheet mica and ground mica are significant in terms of industrial utilization.

Beneficiation

- Mine-run ore fed into a rod mill with or without washing, hydraulically from a bin.
- Dispersant added to the mill charge of about 40% solids.
- Production of slurry.
- Addition of Humphrey's spiral to the processing plant for the recovery of finer size mica.

Industrial Applications

- Sheet mica is used as an insulating material in equipment like condensers, transformers, sheostats, radio and electronic tubes and radar circuits.
- It is used in the form of washers, discs, tubes and plates.
- Used in equipment that encounters very high temperatures like rockets, missiles and jet engine ignition system.
- Phlogopite is used in spark plugs.
- Great progress has been achieved in making built-up mica called micanite.
- Mica films are placed with alternate layers of binding materials like shellac, alkyl, or silicon resin and then pressed and baked.
- In the electronic-field, natural mica is mainly used in the manufacture of capacitors such as bridge-spacers in electronic valves and as panel-board where heat-resistance and low-loss properties at high frequencies are required.
- Mica powder is used as filler in rubber goods, plant, lubricant and to some extent in plastic industries.
- It is also used in the manufacture of roofing material, welding rod, wall-paper, lamp chimneys, shades etc.
- Waste or scrap mica is used invariably in the form of ground mica.

Molybdenite

Molybdenite (MoS_2) is the principal source for molybdenum and molybdic oxide, which have several chemical and metallurgical applications.

Beneficiation

- Grinding when molybdenite is a major mineral
- Flotation
- Middling with quartz
- Collector used, is a hydrocarbon like kerosene, stove oil or some kind of light oil.
- Emulsifier added to prevent such hydrocarbons from dispersion and dissolution.
- Commonly used frothers are pine oil or an alcohol.
- Bulk flotation, followed by selective flotation in case of molybdenite as a by-product.

Industrial Applications

- Molybdenum is alloyed with steel making it stronger and more highly resistant to heat because molybdenum has such a high melting temperature. The alloys are used to make such things as rifle barrels and filaments for light bulbs.
- The iron and steel industries account for more than 75% of molybdenum consumption.
- Molybdenum also is an important material for the chemicals and lubricant industries.
- Molybdenum has uses as catalysts, paint pigments, corrosion inhibitors, smoke and flame retardants, dry lubricant (molybdenum disulfide) on space vehicles and resistant to high loads and temperatures.
- As a pure metal, molybdenum is used because of its high melting temperatures as filament supports in light bulbs, metal-working dyes and furnace parts.
- Molybdenum cathodes are used in special electrical applications. It can also be used as a catalyst in some chemical applications.

Nitrates

Nitrates are now regarded as an important group of industrial minerals. Synthetic nitrogen compounds are gaining wide acclaimation in this field.

Beneficiation

- $NaNO_3$ is leached from caliche through a complex process.
- Multiple stage crushing and screening are required.

- Leaching by Shanks or Guggenheim process.
- Shanks process leaches by means of a boiling solution (175°-140° C) and cooled to 10° C
- Guggenheim process uses warm solution (40° C) and cools it to 5° C

Industrial Applications

- Potassium nitrate used in explosives, fireworks, matches and fertilizers, and as a preservative in foods.
- It is sometimes used in medicine as a diuretic.
- Sodium nitrate used in making potassium nitrate, fertilizers and explosives.
- Silver nitrate used in the preparation of silver salts for photography, in chemical analysis, in silver plating, in inks and hair dyes, and to silver mirrors.
- Ammonium nitrate used in fertilizers and explosives.
- Calcium nitrate is used in fertilizers.
- Barium and strontium nitrates are used to colour fireworks and signal flares.
- Bismuth nitrate is used in making pharmaceuticals.
- Most nitrates are soluble in water, and a major use of nitric acid is to produce soluble metal nitrates.
- The presence of nitrates in the soil is of great importance, since it is from these compounds that plants obtain the nitrogen necessary for their growth.

Perlite and Pumic Stones

Perlite is an amorphous volcanic glass that has a relatively high water content, typically formed by the hydration of obsidian. Pumice has similar composition to perlite. The term pumice encompasses three types, namely, pumicite, scoria and cinder.

Beneficiation

Perlite

- Primary crushing to reduce the crude ore to approximately 1.6 cm.
- If required, secondary crusher is applied.
- Grinding in a closed circuit
- Vibratory screens
- Air classifiers
- Hammer mills
- Rod mills
- Separation of milled perlites into basic grades
- Storage
- Blending to meet the required specifications.

Expansion process for perlite

- Softening of glass
- Volatilization of the combined water
- Production of a light-weight cellular aggregate.

Pumice

- Sorting of crude ore
- Cleaning by screening or water-floating.
- Grading by size and quality

Industrial Application

Perlite

- Construction and manufacturing fields.
- Lightweight plasters and mortars, insulation, ceiling tiles and filter aids.
- In horticulture, it makes composts more open to air, while having good water-retention properties.
- It makes a good medium for hydroponics.
- It is also used in foundries and cryogenic insulations.
- Other horticultural applications for perlite are as a carrier for fertilizer, herbicides and pesticides and for pelletizing seed.
- Horticultural perlite is used with equal success in greenhouse growing, landscaping applications and in the home in house plants.
- Used in high performance fillers for plastics to cements for petroleum, water and geothermal wells.
- Other applications include its use as a filter media for pharmaceuticals, food products, chemicals and water for municipal systems and swimming pools.
- Used as abrasives in soap, cleaners and polishes and a number of foundry applications.
- Used in the manufacture of refractory bricks, mortars and pipe insulation.

Pumice

- Pumice is popular in making lightweight concrete and is used as abrasive, particularly in polishes and cosmetics exfoliates.
- When pumice is used as an additive for cement, its fine-grained version known as pozzolan is mixed with lime to form a light-weight, smooth, plaster-like concrete.
- Pumice's abrasive nature makes it a useful ingredient in soaps, cleansers and dental products.
- Many polishing and scouring products also contain pumice.
- Manufacturers use pumice to polish and grind television glass.
- The construction industry uses pumice in concrete, insulation, acoustic tile, stucco and plaster.

Phosphates

Among the phosphates, apatite group of minerals form important economic minerals.

Beneficiation

- Disintegration of the rock with water

- Slurry produced
- Screening
- Further reduction of some remaining coarse size
- Anionic flotation with three stages of anionic cleaners and two stages of tails scavenging.
- Air classifier
- Storage

Industrial Applications

- Apatite and rock phosphate are rich sources of phosphorous. These minerals are mostly utilized for the manufacture of fertilizers.
- Phosphorous obtained from apatite and rock phosphate is utilized mainly for the manufacture of phosphoric acid (H_3PO_4) which in turn is utilized for the production of pure chemicals like sodium phosphate, monocalcium phosphate, fluorine-free dicalcium phosphate as animal food supplement.
- Phosphate products are also used in pharmaceuticals, ceramics, silk, textiles, insecticides, sugar refining and in the manufacture of explosives.
- Red phosphorous is used in match industry.

Pyrophyllite

Pyrophyllite is a layered silicate with a composition of $Al_2Si_4O_{10}(OH)_2$. It has good thermal shock resistance and high resistance to corrosion.

Beneficiation

- Crushing
- Washing
- Picking
- Sizing

Industrial Applications

- Used as a machinable ceramic material used for electrical resistors, transducer cores, high vacuum gaskets and insulators in electron microscopes
- Ground pyrophyllite is used in the production of ceramics, heat-resistant products called *fractories*, and paint
- Used in ceramic formulations e.g. tile and refractory compositions
- Used in plastics and rubber as a filler
- Used in paint
- Used in insecticides

Rare Earth Minerals

Monazite and xenotime are two rare earth minerals, considered for their economic importance. Monazite is actually three different minerals technically, but because of lack of great differences between them they are referred to as one mineral, monazite.

Beneficiation

- Finely ground monazite mixed with 70% NaOH solution
- Heated in an autoclave at 140° C-150° C for 7 hrs.
- Soluble Na_3PO_4 obtained as byproduct from the insoluble rare earth hydroxide.
- Dissolution of the hydroxide in HCl or HNO_3.
- Selective precipitation of $Th(OH)_4$ with NaOH.

Industrial Applications

- Rare earth elements are used in high performance magnets.
- Used as pigment in ceramics.
- Used in robot motors, X-ray screens, fibre optics, energy efficient lanthanum lamps and colour television tubes.
- Thorium is radioactive and is used in incandescent gas mantles and in fuel for nuclear reactors.

Silica

Silica is an oxide of silicon with a chemical formula of SiO_2.

Beneficiation

- Drying
- Comminution
- 20-mesh screen
- Air classifier
- Attrition mills
- Wet cyclone
- Flotation
- Drier
- Magnetic separator
- Sand

Industrial Applications

- The commonest use of quartz and glass-sand, also referred to as silica-sand, is in the manufacture of glass.
- Great advancement has been made in the manufacture of translucent, transparent, coloured and clear glass in sheets or in glassware.
- The size of the sand grains is important in glass industry. It should be between 40 to 80 mesh (BSS). It should be of high purity containing a minimum of 98% SiO_2.
- Glass-sand free from organic and clayey impurities is used in the manufacture of sand-paper, abrasive cloth etc.; generally sands crushed from sandstone and quartzite is used.
- Quartzite, sandstone, quartz and other siliceous rocks lime mica schists are used in the manufacture of silica bricks.

- Quartz and quartzite are used in making ferro-silicon for the manufacture of silicon steel.
- Rounded pebbles of chalcedony are used in ball mills for finer crushing of minerals like feldspar, calcite and barytes. Chalcedony being the cryptocrystalline variety of silica is very hard and has good abrasive effect on the material to be pulverized.
- The agate pieces after cutting to requisite sizes and shapes are utilized in the manufacture of fulcra of scientific balances and making edges, planes and bearing of precision instruments.
- Glass-sand in large quantities is used as moulding sand in the foundry industry, as body constituent in the ceramic industry in the preparation of glazes and sometimes added to the raw material for cement manufacture to balance SiO_2, Al_2O_3 and Fe_2O_3 percentages.
- Silica flour made by grinding silica sand is used in paints.

Al_2SiO_5 Minerals

Sillimanite, andalusite and kyanite are three polymorphic forms of Al_2SiO_5 group. Their uses depend on their ability to produce mullite. Mullite is the mineralogical name given to the only chemically stable intermediate phase in the SiO_2-Al_2O_3 system.

Beneficiation

- Ore feeder
- Jaw crusher
- Cone crusher
- Vibrating screens
- Cone crusher
- Rodmill
- 35-mesh vibrating screen
- Flotation cells
- Rake classifier
- Rotary dryer and calciner
- Magnetic separators
- Rodmill
- Storage

Industrial Applications

- Appreciable quantity of sillimanite powder mixed with high-grade china-clay is used in the manufacture of spark plugs and other insulating materials required at high voltages.
- Sillimanite refractories cut into various shapes and sizes or made out of bonded particles are used in industries like cement, ceramics, glass-making, metal smelting, refinery and treatment, tar distillation, coal carbonisation, chemicals manufacture and iron foundries.

- It is very resistant to glass slag and hence widely used as lining material in glass melting furnace.
- By far the largest use of mullite based products is in refractories. The glass and steel industries are two main markets. The steel industry is the largest user, where refractoriness, high creep resistance and thermal shock resistance are important.
- The main use of high-mullite based products is in hot blast stove checker bricks.
- Many refractories in use in the steel industry have varying amounts of mullite based aggregates in them. Steel ladles, lances, reheat furnaces and slide gates are examples of mullite aggregate based products with various alumina contents.
- The glass industry uses mullite based refractories in burner blocks, ports and in checker bricks as well as in the upper structure of the tanks where the glass is melted and in the drawing chambers.
- Mullite based products are also resistant to particulate carryover into the glass melt. This is particularly important in flat glass production, where contamination by low levels of Al_2O_3 is undesirable.
- The next largest user of mullite is the ceramic industry mostly in kiln furniture items such as kiln setter slabs and posts for supporting ceramic ware during firing.
- The aluminium and petrochemical industries also use mullite-based aggregates for applications requiring chemical attack resistance, thermal shock resistance and hot-load strength.
- Also used in electronic substrates and protective coatings.

Talc

Talc is one of the most important industrial minerals and is the most common mineral for daily use as a body and face powder.

Beneficiation

- Talc ore
- Jaw crusher
- Screen
- Undersize
- Roller mill with air separator
- Flotation
- Overflow
- Thickener
- Underflow
- Filter
- Cake
- Flash dryer

- Product
- Additional grinding
- Shipping

Industrial Applications

- *Paper:* Talc both in pulverized and brick forms, are used in this industry. Pulverized material of 200-300 BSS sieve is required as filler.
- *Textiles:* Textile industry also prefers talc powder free from grit and colouration, which is used both for loading and bleaching certain types of cotton goods. Inferior grades are used in back filling textiles.
- *Cosmetics:* In the manufacture of body and face powder, finely pulverized talc of very high purity is required.
- The pharmaceutical industry requires talc containing less than 0.1% FeO. Natural material of such purity is generally not found and pharmaceutical talc is preferred by passing the high purity powdered talc over a magnetic separator to reduce the iron content.
- *Rubber:* In rubber industry talc is used for two purposes, one to prevent rubber moulds from sticking and the other as compunding material in the preparation of certain types of hard rubber.
- *Ceramics:* Talc is increasingly being used in the manufacture of artwares, jars, wall and floor tiles. It serves as a non-plastic ceramic material. The addition of talc in suitable proportions in the body of mixtures for porcelain, jars etc. prevents the crazing (cracking) effect on the glazes.
- *Paints:* In this industry, foliated, fibrous or lamellar talc of fine mesh (300 mesh) is preferred. It is used as a paint or an extender in paint industry.
- *Other uses:* A fair proportion of medium to inferior grades of talc is used in the manufacture of pesticides and insecticides.

Titanium

Among the titanium minerals, rutile and ilmenite are of economic importance.

Beneficiation

- Primary crushing
- Jaw crusher
- Cone crusher
- Grinding
- Flotation

For ilmenite

- Ilmenite-hematite ore reduced by low-ash coal with or without lime at 1500-1700° C in graphite electrode furnace
- Lighter titanium slag floats to top
- Iron slag settles down and drawn off

- Titanium slag cooled, crushed and screened
- Shipping

Industrial Applications

- Titanium dioxide white pigment is used in paints and plastics, high quality paper, rubber, ceramics (e.g. porcelain), fabric, soap, cosmetics, toothpaste, false teeth, and even some foods (e.g. tumeric).
- It is also used in sun screening products (e.g. sun cream) due to its ability to reflect ultra violet light.
- Rutile provides superior "wet" opacity, which is useful in wet strength paper grades, waxed papers, and other saturated applications.
- Rutile is used for coating welding electrodes.
- Titanium and titanium compounds have found uses in desalination plants, electrical components, glass products, artificial gemstones, jewellery and even as smoke screens.

Wollastonite ($CaSiO_3$)

In respect to the industrial use, wollastonite represents a minor level importance.

Beneficiation

- Crude ore
- Jaw crusher
- 1″ Tyler screen
- Fines
- Roll crusher
- 17 mesh screen
- Fines
- Derrick screens
- Dry magnetic separators
- Garnet-diopside
- Wollastonite
- Attrition mills – wollastonite fibrous
- Pebble mills – air separators – wollastonite granular

Industrial Applications

- It is mainly used in the ceramic industry where it is used as a principal ingredient in the ceramic bodies required for the manufacture of floor and wall tiles.
- It is also utilized as an additive in ceramic bodies and glazes.
- It is used as filller in paint and paper and many other products.
- In metallurgy, it is used as a welding rod coating, to control the viscosity of the slag and alloying agents.
- A suitable mixture of finely pulverized wollastonite with china-clay improves the properties of the clay to be used for paper-coating.

- It is found to increase the strength of Portland cement when mixed as an additive.
- The mineral has been found to react readily with various acids like sulphuric, phosphoric and hydrochloric acids and with alkalis such as soda ash.
- Certain types of pastes have been produced by dissolving the mineral in some proportions of sulphuric and phosphoric acids. These pastes possess a good hiding and oil absorption power.
- Its latest use is in the manufacture of mineral wool.

Zeolite

Zeolite is a mineral group with porous structure, which allows some molecules to pass through and causes others to be either excluded, or broken down. More than 150 types of zeolite have been synthesized. There are 48 known naturally occurring zeolites. They are basically a hydrated alumino-silicate minerals with an "open" structure which can accommodate a wide variety of positive ions, like Na^+, K^+, Ca^{2+}, Mg^{2+} and others. These ions are held loosely and can be exchanged readily for others in a contact solution. Some more common mineral zeolites are: Analcime, chabazite, heulandite, natrolite, phillipsite, stilbite.

Beneficiation

- Crushing
- Drying
- Roller mills
- Rotary kilns
- Zeolites for adsorption processing must be formed into aggregates by using an inorganic binder, generally kaolin clay.
- Mix extruded into cylinder pellets or formed into beads.
- Calcined

Industrial Applications

Natural zeolites

- It is widely used as ion-exchange beds in both domestic and commercial water purification, softening, and other applications.
- In chemistry, it separates molecules, only of certain sizes and shapes can pass through, as traps for molecules so that they can be analyzed, or as catalysts by confining molecules in small spaces that causes changes in their structure and reactivity.
- Zeolites even carry the potentiality of providing precise and specific separation of gases including the removal of H_2O, CO_2 and SO_2 from the lower grade natural gas streams. Some other separations includes HCHO, Noble gases, N_2, freon and formaldehyde. However, the true potential for improving the handling of such gases in this manner remains unknown.
- Agriculture: Clinoptilolite, a naturally occurring zeolite, is used as a soil treatment in agriculture. It provides a source of slowly released potassium.

- Decreases ammonia level in ponds and tanks; filtrates water for cleaner tanks.
- Medical: Zeolite-based oxygen generation systems are used in producing medical grade oxygen. It is used as a molecular sieve for extracting oxygen from air, in a process which involves adsorption of atmospheric nitrogen. The usage of zeolite is also being explored for a quick clotting of severe bleeding.
- Heating and Refrigeration: Zeolites can also be used as solar thermal collectors and for adsorption refrigeration. Their high heat of adsorption and ability to hydrate and dehydrate is exploited while maintaining the structural stability. This hygroscopic property along with an inherent exothermic reaction, while transitioning from dehydrated to a hydrated form, make the natural zeolites effective in the storage of solar and the waste heat energy.
- Natural mordenite and clinoptilolite are used in coal gasification and natural gas purification.
- Gemstones: Thomsonites is collected as gemstones from a series of lava flows along the Lake Superior (Minnesota) and in a lesser degree in Michigan (USA).

Synthetic zeolites

- Synthetic zeolite is widely used as the global laundry detergent.
- Primary use is in the cracking, hydrocracking and isomerization of hydrocarbons. In both fluid catalytic cracking and hydrocracking, zeolite catalysts provide vastly superior combinations of strong acid catalytic sites, uniformity of pore structure and stability, all of which provide improved selectivity, yield, durability, and cost over nonzeolite alternatives.
- Zeolites have provided much higher yield of gasoline and other high-quality fuels per barrel of crude oil.
- Synthetic zeolites Y, ferrierite, ZSM-5 and beta in their H^+ and NH^{4+} forms are used for removal of arsenate from water over a wide range of pH.
- Synthetic zeolites are sold as health food products because of their amino acid, mineral and enzyme content. They are often used as an antioxidant to detox the body and sold in crystal or powder form because of their crystalline properties.

Zircon ($ZrSiO_4$)

Zircon is considered as an important industrial mineral.

Beneficiation

- Dredged beach sand treated in a high-capacity gravity concentrator
- Wet concentration process
- Rougher concentrate
- Upgradation

- Bulk heavy mineral concentrate
- Low-density wet drum magnetic separator for removal of magnetite
- Nonmagnetics treated with high-gradient magnetic separator to remove feebly magnetic material
- Resultant concentrate dried
- Dry mill with high-tension roll separators and induced magnetic roll separators
- Production of zircon

Industrial Applications

- Ceramics: Most zircon gets ground by processors into sands of varying grain size. Ceramics manufacturers make the most overall use of zircon, employing the finest zircon sands to give their products high strength and high resistance to heat. Zircon ceramics often serve as protective coatings on metals. Additionally, powdered zircon in ceramic glazes gives them an opaqueness that allows colours to stay fixed in varying light.
- Refractories: Many of the linings of smelters, chemical vats and heat containment units in power plants employ zircon sands as an essential component.
- Industrial abrasives: Coarser zircon sands is used as polishing agents at metal foundries and glass works. Zircon sands also serve as the ammunition for some sandblasting machines used to strip paint and mold from buildings.
- Glass screens: As with ceramic glazes, adding zircon to glass makes it semi-opaque. Glass mixed with zircon sands often gets used for television screens.
- Welding: Some welding rods carry a zircon coating. The mineral protects applied solder from being displaced by subsequent heating.
- Radiographic dating: The trace amounts of uranium and thorium trapped in the crystal lattice of zircon make the mineral useful to geologists who want to determine the age of soils and rocks. By calculating the amount of atomic decay of the radiological elements in a given sample of zircon from a known location or formation, scientists can approximate the age and the location or formation to within a few thousand years. Zircon-reference dating has been used to determine the age of the moon's surface and the time at which water formed on Earth.

Concluding Remarks

The growing demands of industrial minerals in various fields of technology—both small scale and large scale—emphasize to have a detailed knowledge of the properties and processing mechanism of the individual minerals. Depending on the type of mineral and also on the type of the final product beneficiation process varies to obtain the required quality product.

Think for a while

1. What properties make clays suitable for ceramic preparation? What are the other industrial uses of clays? What basic steps are needed for beneficiation of each use?
2. What type of minerals are used for drilling fluids? Why? How it differs from lubricants and filler absorbants?
3. What are the needed properties for minerals used in insulation, electronics and acoustroelectronic industries? Name the mineral for each type of use.
4. With a flow chart explain the mineral beneficiation process. Depending on specific minerals, the beneficiation is to be altered as needed e.g. graphite and diamond. Explain.
5. For cement industry, what basic properties of the minerals are essential? For different types of cement, what are the special types of raw materials used?

CHAPTER

18

ENVIRONMENTAL MINERALOGY

Minerals have a profound effect on environment. Study of minerals in details, i.e. mineralogy help to ascertain different problems and their causes which environment is facing on regular basis. This chapter deals with these major causes and hazards of environment, caused by minerals.

18.1 INTRODUCTION

Environmental mineralogy deals with the minerals taking part in constructing basic environmental systems. Minerals sometimes play crucial role in environmental degradation. Rock layers excavated for mining, mine and industrial waste dumps, abandoned mines etc. are sources of minerals that may lead to environmental pollution which is a matter of serious concern nowadays. Environmental mineralogy, in an interdisciplinary approach, deals with the interaction of the minerals with the biogeochemical environment. The objective of the study of environmental mineralogy is to provide detailed information regarding mineral-related environmental problems like effects of minerals on human health, formation of surfacial acidic environments, microbe-mineral interactions etc. The scope of environmental mineralogical study is to carry out mineralogical research that ultimately contributes to the development of effective solutions to these problems. The primary parameters, needed to determine these minerals, are identification of minerals, determination of quantity of the phases, together with compositional and textural information, and also microstructure or mineral surface chemistry. Apart from natural minerals, synthetic minerals are also taken into consideration to study their effects on environmental mineralogy. To understand and clarify the basic principles of environmental mineralogy, some aspects are needed to be discussed in details, like, various types of pollutions, health hazards from natural minerals, infrastructures of mineral industry and environmental factors and mining and mineral industry.

18.2 MINERALOGICAL EFFECTS CAUSING POLLUTION AND RELATED HAZARDS

Minerals often take a crucial role in environmental aspects. Various sources of minerals and their varied nature lead to environmental pollution which results in propagation of hazards of different kinds. This area of study, environmental mineralogy, has become a matter of serious concern nowadays. From mineralogical point of view, minerals can be categorized as acid-generating or acid-consuming, which are localized either in the deposit, host rock or in wall-rock alteration.

Acid-generating Minerals

Sulfide minerals react with water and atmospheric oxygen to produce acid. Acid-generating minerals which generate acid through chemical and also sometime by biogeochemical reactions are listed in Table 18.1. Generation of acid by these mineral species depend on a number of factors like:

- Presence of iron in sulfides
- Whether oxidized or reduced metal species produced by oxidation
- Whether oxygen or aqueous ferric iron is the oxidant
- Precipitation of hydrous metal oxides or other minerals as a result of oxidation process.

Table 18.1: Minerals acting as acid-generating minerals

Sulfides with oxygen as the oxidant	
Pyrite, marcasite	FeS_2
Pyrrhotite	$Fe_{(1-x)}S$ (x = 0-0.17)
Bornite	Cu_5FeS_4
Arsenopyrite	FeAsS
Enargite	Cu_3AsS_4
Tetrahedrite	$(Cu, Fe)_{12}Sb_4S_{13}$
Realgar	AsS
Orpiment	As_2S_3
Stibnite	Sb_2S_3
Sulfides as ferric iron as the oxidant	
All of the previous group, and	
Sphalerite	ZnS
Galena	PbS
Chalcopyrite	$CuFeS_2$
Covellite	CuS
Cinnabar	HgS
Millerite	NiS
Pentlandite	$(Fe, Ni)_9S_8$
Greenockite	CdS
Minerals generating acids if hydrous oxides are formed	
Siderite	$FeCO_3$
Rhodochrosite	$MnCO_3$
Alunite	$KAl_3(SO_4)_2(OH)_6$

As an example of oxidation reaction of sulphide minerals, consider the most common one, oxidation of pyrite, a very common ore present in mine and industrial pollutants. This complex process goes through many steps and finally represented as

$$FeS_2 + (15/4)O_2 + (7/2)H_2O \longrightarrow Fe(OH)_3 + H_2SO_4$$

Average concentration on pyrite is about 0.2% to 0.5% and sometimes may be more. The above mentioned reaction can release a considerable amount of arsenic to the aqueous phase. Biogeological reactions with different bacteria (e.g. theobacillous ferroxides etc.) acting as a catalyst increase the reaction rate considerably.

Arsenopyrite oxidation by Fe(III) is about 10 times faster than reaction with pyrite and is more rapid in presence of T. ferrooxides. The overall reaction may be written as

$$FeAsS + (7/2)O_2 + 4H_2O \Longrightarrow Fe(OH)_3 + H_3AsO_4 + H_2SO_4$$

Arsenopyrite oxidation can lead to scorodite ($FeAsO_4{\cdot}2H_2O$) precipitation under acidic conditions and then breaks down to iron oxide. Under high pH conditions, iron oxide is the dominant oxidation product which absorbs or co-precipitate arsenic.

Acid-consuming Minerals

Group of carbonate and aluminosilicate minerals can be grouped as acid-consuming minerals. They reside in close proximity with the acid-generating minerals and can consume acid generated during sulphide oxidation. In general it can be expressed as

$$XCO_3 + H_2SO_4 = XSO_4 + CO_2 + H_2O$$

where X is a bivalent cation like Ca^{2+} (calcite, aragonite), Pb^{+2} (cerrusite), Sr^{+2} (strotianite), Mg^{+2} (magnesite) etc.

The variation in acid consumption by these minerals has been depicted in Table 18.2.

Table 18.2: Minerals acting as acid-consuming minerals

Mineral group	*Individual minerals*
Readily dissolving	Cerrusite, calcite, aragonite, strontianite
Less readily dissolving	Rhodochrosite, siderite, dolomite, ankerite, magnesite, brucite
Intermediate weathering	Epidote, zoisite, enstatite, hypersthene, augite, hedenbergite, hornblende, glaucophane, tremolite, actinolite, anthophyllite, serpentine, chrysotile, talc, chlorite, biotite
Slow weathering	Albite, oligoclase, labradorite, vermiculite, montmorillonite, gibbsite, kaolinite, magnetite
Very slow weathering	K-feldspar, muscovite
Inert	Quartz, rutile, zircon

Here it is needed to mention that environmental pollution is the cause with its effects shaping up in a number of serious hazards. The elemental pollution has been discussed in the next section though it should be remembered that it is the minerals which contain these elements and thus, are the ultimate source of pollution. For example though the term lead pollution is more common, but the sources of this lead are galena (PbS) or cerrusite ($PbCO_3$). The minerals causing various environmental effects and serious health hazards have been elaborated in section 18.4.

Table 18.3: Source minerals for various harmful chemical substances

Chemical substances	*Source minerals*
Aluminum	Andalusite, kyanite, sillimanite, zoisite, epidote, staurolite, grosular, pyrope, almandine, spessartine, alunite, corundum, spinel, beryl, cordierite, dravite, gedrite, hornblende, glaucophane, muscovite, lepidolite, glauconite, illite, chlorite, microcline, orthoclase, sanidine, albite, anorthite, nepheline, kalsilite, sodalite, leucite, natrolite, stilbite
Antimony	Native element, antimonite, allemonite
Arsenic	Arsenopyrite, realgar, energite, orpiment, allemonite (SbAs), cobaltite, (CoAsS), nickeline (NiAs), skutterudite (Co, Ni) As 2-3, scorodite ($FeAsO_4.2H_2O$). Sulfide minerals can contain 1% or more arsenic as an impurity (e.g., pyrite, galena, sphalerite, marcasite and chalcopyrite; Onishi, 1978)
Asbestos	Itself a mineral, the fibrous form of the mineral serpentine, chemical composition $Mg_3Si_2O_5(OH)_4$
Barium	Withrite, barite
Beryllium	Beryl
Boron	Borax, kernite, colemanite, dravite, schorl
Cadmium	Occurs in small quantities as impurities in a few minerals and ores, especially zinc ores, e.g. smithsonite, e.g. barquillite, cadmoselite, cadmoindite, greenockite
Calcium	Very common in rock forming minerals, like anorthite, calcite, dolomite, wollastonite etc.
Chlorine	Halite, sylvite, apatite, pyromorphite, leucite
Chromium	Uvarovite, chromite
Copper	Native element, chalcocite, chalcopyrite, enargite, covelite, bornite, cuprite, antlerite, malachite, azurite, chalcanthite
Fluoride	Fluorite, apatite
Iron	Native element, telluride, nickel telluride, kamacite, fayalite, andradite, staurolite, almandine, triphylite, ankerite, siderite, hematite, ilmenite, chromite, magnetite, arsenopyrite, marcasite, pyrite, bornite, chalcopyrite, pyrrhotite, troilite, ferrosilite, hedenbergite, orthopyroxenes, aegirine, augite, hornblende, ferroactinolite, biotite, illite, chlorite

(Contd.)

Table 18.3 (*Contd.*)

Chemical substances	*Source minerals*
Lead	Galena, pyromorphite, cerussite, anglesite
Manganese	Pyrolusite, columbite, tantalite, rhodochrosite, spessartine, rhodonite
Mercury	Cinnabar, also as free metal
Nickel	Telluride, nickel telluride, kamacite, nickeline
Nitrate	Soda-nitre (Chile salt petre)
Selenium	Commonly occurs with sulfides as selenides in ores of such metals as iron, lead, silver and copper, e.g. cadmoselite
Silver	Native element, argentite
Sodium	Halite, kernite, dravite, schorl, glaucophane, montmorillonite, albite, nepheline, leucite, natrolite
Sulfate	Barite, celestite, anglesite, anhydrite, alunite, antlerite, gypsum, chalcanthite, epsomite
Zinc	Hemimorphite, smithsonite, sphalerite, wurtzite

18.3 VARIOUS TYPES OF POLLUTION

Environmental pollutants from many different sources contaminate water, air and land, putting humans and ecosystems at risk, and often pitting people against industry. By viewing competing interests and their implications within a broad ecosystem analysis, ecohealth approaches strive to protect health while balancing the needs of various stakeholders and contributing to safeguarding the system.

Air Pollution

Air pollution can be defined as the presence in the atmosphere of substances that are toxic, irritant, or otherwise harmful to man or damaging to vegetation, animals or property.

Table 18.4: Classification of air pollutants

Major class	*Subclass*	*Examples*
Inorganic gases	Oxides of nitrogen	N_2O, NO, NO_2
	Oxides of sulphur	SO_2, SO_3
	Oxides of carbon	CO, CO_2
	Other inorganics	O_3, H_2S, HF, NH_3, Cl_2, Rn
Organic gases	Hydrocarbons	Methane, butane, octane, benzene, acetylene, ethylene
	Aldehydes and ketones	Formaldehyde, acetone
	Other organics	Chlorofluorocarbons, PAHs, alcohols, organic acids
Particulates	Solids	Fumes, dust, smoke, ash, carbon soot, lead, asbestos and other industrial minerals during mining
	Liquids	Mist, spray, oil, grease, acids

Greenhouse Gases

The major greenhouse gases are CO_2, CH_4, N_2O and the CFCs. Increase in the amounts of these gases lead to an increase in the absorption of outgoing radiation and a potential increase in global temperature. The causes for the temperature increase are that the increased vibrational or rotational energy of the molecules is transferred by collisions to other molecules in the atmosphere and reradiation of absorbed energy, leading to heat flux to the Earth from the atmosphere.

Carbon Dioxide

Carbon dioxide (CO_2) is one of the major pollutants in the atmosphere. Major sources of CO_2 are fossil fuels burning and deforestation. CO_2 is a good transmitter of sunlight, but partially restricts infrared radiation going back from the Earth into space. This produces the so-called greenhouse effect that prevents a drastic cooling of the Earth during the night. Increasing the amount of CO_2 in the atmosphere reinforces this effect and is expected to result in a warming of the Earth's surface. Currently carbon dioxide is responsible for 57% of the global warming trend.

NO_X—Nitrogen Oxide (NO) and Nitrogen Dioxide (NO_2)

- Natural component of the Earth's atmosphere.
- Important in the formation of both acid precipitation and photochemical smog (ozone), and causes nitrogen loading.
- Comes from the burning of biomass and fossil fuels.
- Average residence time in the atmosphere is 9 days.
- Has a role in reducing stratospheric ozone.

N_2O (Nitrous Oxide)

- Natural component of the Earth's atmosphere.
- Important in the greenhouse effect and causes nitrogen loading.
- Human inputs six million tons per year, and 19 million tons per year by nature.
- Residence time in the atmosphere about 180 years.
- Comes from nitrogen based fertilizers, deforestation, and biomass burning.

Sulfur and Chlorofluorocarbons (CFCs)

- Sulfur dioxide is produced by combustion of sulfur-containing fuels, such as coal and fuel oils.
- Also, in the process of producing sulfuric acid and in metallurgical process involving ores that contain sulfur.
- Sulfur oxides can injure man, plants and materials. At sufficiently high concentrations, sulfur dioxide irritates the upper respiratory tract of human beings because potential effect of sulfur dioxide is to make breathing more difficult by causing the finer air tubes of the lung to constrict.

- Power plants and factories emit 90% to 95% of the sulfur dioxide and 57% of the nitrogen oxides in the United States. Almost 60% of the SO_2 emissions are released by tall smoke stakes, enabling the emissions to travel long distances (Miller 494).
- As emissions of sulfur dioxide and nitric oxide from stationary sources are transported long distances by winds, they form secondary pollutants such as nitrogen dioxide, nitric acid vapour, and droplets containing solutions of sulfuric acid, sulfate, and nitrate salts. These chemicals descend to the earth's surface in wet form as rain or snow and in dry form as gases fog, dew, or solid particles. This is known as acid deposition or acid rain.

Chlorofluorocarbons (CFCs)

- CFCs are lowering the average concentration of ozone in the stratosphere.
- Aerosols are still the largest use, accounting for 25% of global CFC use (Miller 448).
- Spray cans, discarded or leaking refrigeration and air conditioning equipment, and the burning plastic foam products release the CFCs into the atmosphere.
- Depending on the type, CFCs stay in the atmosphere from 22 to 111 years. Chlorofluorocarbons move up to the stratosphere gradually over several decades. Under high energy ultraviolet (UV) radiation, they break down and release chlorine atoms, which speed up the breakdown of ozone (O_3) into oxygen gas (O_2).
- Chlorofluorocarbons, also known as freons, are greenhouse gases that contribute to global warming.
- Photochemical air pollution is commonly referred to as "smog".
- Smog, a contraction of the words smoke and fog, has been caused throughout recorded history by water condensing on smoke particles, usually from burning coal. With the introduction of petroleum to replace coal economies in countries, photochemical smog has become predominant in many cities, which are located in sunny, warm and dry climates with many motor vehicles. The worst episodes of photochemical smog tend to occur in summer.

Aerosols

Aerosols are solid particles or liquid droplets ranging in size from clusters of a few molecules to 20 ìm in radius. Sulphuric acid is an important aerosol, causing atmospheric pollution. It results in increasing the albedo of the Earth, due to their efficiency in reflecting shortwave radiation. The common sources of aerosols are:

- Soil and rock dust
- Sea salt
- Biogenic
- Biomass burning
- Volcanic
- Direct emissions—fuel, incinerators, industry

- Gaseous emissions—biogenic, volcanic SO_2, fossil fuel, greenhouse gases, biogenic and anthropogenic hydrocarbons

Smogs

Smog is characterized by hazy and irritating air that forms in response to the emission of pollutants and an inversion which prevents the dispersal of the pollutants. The major impact of industrial smog is the formation of acid aerosols, which corrode building materials. The combination of the acid aerosols and particulates has a major impact on the human respiratory system. The smog results in the formation of ozone, NO_x and other compounds that form significant respiratory illness and the build up of a near-surface haze.

London Smog

Thousands of tons of black soot, tar particles and sulfur dioxide had accumulated in the air from the heavy coal combustion. Estimates of PM_{10} concentrations during December, 1952, range between 3,000 and 14,000 μg/m³ with the high range being approximately 50 times higher than normal levels at the time. PM_{10} is particulate matter less than 10 micrometres in diameter. Estimates also suggest that sulfur dioxide levels during December of 1952 were seven times greater than normal at 700 parts per billion (ppb).

Industrial sources is one of the major points of concern in modern days of industrialization. The problem with industrial air pollution is that the release of toxins from plants and industrial institutions is very high for the per capita allowance. More air pollution problems are caused by industrial institutions than anything else, and many of these institutions are in chronic violation of air pollution regulations. The nature of emissions and effluents from industries are varied and industry specific. Emissions are not just colour based. Gases that are released from these buildings fly up into the atmosphere and react with the natural gases that are found in the atmosphere. Some gases continually react, causing serious damage to the ozone layer, the breathable air, and then contribute to other forms of pollution when it rains.

Sources

- Effluents containing heavy metals generated from electroplating (Ni, Zn, Cu), tanneries (Cr), textiles and chemical industries.
- Organic pollutants from distilleries, sugar, paper pulp, food processings, pharmaceuticals, dye and dye intermediates and tannery industry.
- Particulates and oxides of sulphur and nitrogen from burning of fossil fuels.
- Specific pollutants such as hydrogen sulphide, volatile organic compounds, hydrocarbons, lead etc. generated from industries like pulp and paper, refineries and lead acid battery units.
- Spent catalysts, produced from oil refineries, fertilizers and pharma industries.
- Process dust from iron and steel industries.

Air Pollution Mitigation

1. Healthy working atmosphere keeps sound health of workers and machines leading to their quality output.
2. Avoid the far-reaching effects of pollution problem on people and the other living things, vegetation, buildings and structures surrounding the industries.
3. Assists in keeping a balance in the ecological system.

Water Pollution

When toxic substances enter lakes, streams, rivers, oceans and other bodies, they get dissolved or lie suspended in water or get deposited on the bed. This results in the pollution of water whereby the quality of the water deteriorates, affecting aquatic ecosystems. Pollutants can also seep down and affect the groundwater deposits.

Major Sources and Types of Water Pollution

- *Domestic sewage* refers to waste water that is discarded from households. Also referred to as sanitary sewage, such water contains a wide variety of dissolved and suspended impurities such as organic materials and plant nutrients that tend to rot. The main organic materials are food and vegetable waste; plant nutrients come from chemical soaps, washing powders, etc. Domestic sewage is also very likely to contain disease-causing microbes.
- Too much garbage is dumped into streams, lakes, rivers and seas, thus making water bodies the final resting place of cans, bottles, plastics and other household products, which contain harmful chemicals.
- Sediment pollution resulting from industrial wastes deposited in drainage system. Chemicals such as pesticides, phosphorous and ammonium are transported with sediment in an adsorbed form, which crystallize as minerals and settle from water that contains lot of dissolved particles.
- Moreover today's cleaning products are synthetic detergents and come from the petrochemical industry. Most detergents and washing powders contain phosphates, which are used to soften the water among other things. These and other chemicals contained in washing powders affect the health of all forms of life in the water.
- *Agricultural run-off:* The use of land for agriculture and the practices followed in cultivation greatly affect the quality of groundwater. Intensive cultivation of crops causes chemicals from fertilizers (e.g. nitrate) and pesticides to seep into the groundwater, a process commonly known as leaching.
- Routine applications of fertilizers and pesticides for agriculture and indiscriminate disposal of industrial and domestic wastes are increasingly being recognized as significant sources of water pollution.
- *Industrial causes:* One of the main sources of water pollution is the waste material discharged by industrial units, known as industrial water pollution. Waste materials like acids, alkalis, toxic metals, oil, grease, dyes, pesticides

and even radioactive materials are poured into the water bodies by many industrial units. Some other important pollutants include polychlorinated biphenyl (PCB) compounds, lubricants and hot water discharged by power plants. The pollutants unloaded into the water bodies usually dissolve or remain suspended in water. Sometimes, they also accumulate on the bottom of the water bodies.

- Another important pollutant, affecting marine life is the oil spilled by the oil tanks.
- The industrial effluents contain pollutants like asbestos, phosphates, mercury, lead, nitrates, sulfur, sulfuric acid, oil and many other poisonous materials. In many countries, industrial water is not treated adequately before discharging it into rivers or lakes. This is particularly true in the case of small scale industries that do not have sufficient infrastructure for pollution control equipment.
- Some major industrial pollutants:
  - Asbestos—This pollutant is a serious health hazard and carcinogenic. Asbestos fibres can be inhaled and cause illnesses such as asbestosis, mesothelioma, lung cancer, intestinal cancer and liver cancer.
  - Lead—This is a metallic element and can cause health and environmental problems. It is a non-biodegradable substance so is hard to clean up once the environment is contaminated. Lead is harmful to the health of many animals, including humans, as it can inhibit the action of bodily enzymes.
  - Mercury—This is a metallic element and can cause health and environmental problems. It is a non-biodegradable substance so is hard to clean up once the environment is contaminated. Mercury is also harmful to animal health as it can cause illness through mercury poisoning.
  - Nitrates—The increased use of fertilisers means that nitrates are more often being washed from the soil and into rivers and lakes. This can cause eutrophication, which can be very problematic to marine environments.
  - Phosphates—The increased use of fertilisers means that phosphates are more often being washed from the soil and into rivers and lakes. This can cause eutrophication, which can be very problematic to marine environments.
  - Sulphur—This is a non-metallic substance that is harmful for marine life.
  - Oils—Oil does not dissolve in water, instead it forms a thick layer on the water surface. This can stop marine plants receiving enough light for photosynthesis. It is also harmful for fish and marine birds.
  - Petrochemicals—This is formed from gas or petrol and can be toxic to marine life.

Soil Pollution

Soil pollution is defined as the build-up in soils of persistent toxic compounds, chemicals, salts, radioactive materials, or disease causing agents, which have adverse effects on plant growth and animal health.

Soils are mainly affected by human activities which influence the pattern, amount, and intensity of surface-run off, erosion and sedimentation. Among these, the most significant is the conversion of natural areas to various land uses.

Major Causes of Soil Pollution

- Main cause of soil pollution is the adsorption property of clays.
- *Use of trace elements in fertilizers:* Pesticides such as DDT [dichlorodiphenyl trichloroethane], a colourless chemical pesticide which is a potent nerve poison in insects, was first widely used to combat diseases such as yellow fever and malaria. It was later used to control and/or eradicate disease carrying and crop eating insects. DDT was later on discovered to cause endangerment of species in the same food chain as the controlled insects, particularly birds. DDT prevents the shelling of bird eggs and in humans causes health threats.
- *Dioxin,* a chemical impurity resulting from the production of the auxin. Dioxin is a toxic chemical and was used as a defoliant by the American army. Dioxin was a major constituent of argent orange which was applied on trees which would then fall off reaviling enemy camps. After the war it was found that the chemical cause congenital deformalities and mental effects to the children born to the American soldiers and in the area over which it was applied. In minute amount dioxin has the ability to cause cancer, chloracne, miscarriage, and fetal abnormalities.
- Both organic (those that contain carbon) and inorganic (those that don't) contaminants are important in soil. The most prominent chemical groups of organic contaminants are fuel hydrocarbons, polynuclear aromatic hydrocarbons (*PAHs*), polychlorinated biphenyls (*PCBs*), chlorinated aromatic compounds, detergents, and pesticides. Inorganic species include nitrates, phosphates and heavy metals such as cadmium, chromium and lead; inorganic acids; and *radionuclides* (radioactive substances). Among the sources of these contaminants are agricultural runoffs, acidic precipitates, industrial waste materials and radioactive fallout.
- *Industrial causes:* The glass industries use arsenic to eliminate a green colour caused by impurities of iron compounds, because arsenic is a violent poison, yet it is widely used and therefore is a frequent contaminant. Arsenic is sometimes added to lead to harden it and is also used in the manufacture of such poison gases as lewisite and adamsite, often used for military purpose.
- Pollution of land by heavy metals is a result of the mining of ores to extract metals such as tin, silver, nickel, lead, iron, chromium and copper. Most of these metals occur naturally as ions in the soils. Though some metals, such as copper, iron and zinc, are necessary for plant growth. It is the high concentration of these ions that renders the land unsuitable for plant growth.
- Another major source of pollutants is the sediments. In many areas it chokes the streams, fills in lakes, reservoirs, ponds, canals, ditches and harbours; buries vegetation.

- Eroded soil has various hazardous effects—it depletes soil at its site of origin, reduces the quality of the water it enters and may deposit sterile materials on productive croplands or other useful land.
- Land-use causes: Conversion of forested land for agricultural purposes usually increases run-off and erosion, resulting in muddiness of the streams and their inability to transport sediments. The channels will thus aggrade, thereby increasing the magnitude and frequency of flooding.

Heavy Metal Pollution

Heavy metals are those metals, whose atomic number is 20 and greater, which are mostly obtained from sulphide minerals. The most important of them are arsenic, selenium, cadmium, mercury and lead. These metals, in a strongly reducing environment and in presence of sulphur, form insoluble sulfides. Due to change of Eh and pH conditions, presence/absence of organic compounds, these elements can be released.

For convenience the heavy metals can be divided into three groups: (i) transitional metals (Zn, Cd and Pb); (ii) As and Se and (iii) Hg.

(i) Transitional Metals (Zn, Cd and Pb)

Under a varied Eh and pH conditions these metals occur in solution as divalent or trivalent cationic species. Adsorption is an important process for these metals. With increasing pH, by the process of effective adsorption, the metals are removed from the solution to particles and ultimately to sediments.

Cadmium: Cadmium derives its toxicological properties from its chemical similarity to zinc—an essential micronutrient for plants, animals and humans. Cadmium is biopersistent and, once absorbed by an organism, remains resident for many years (over decades for humans) although it is eventually excreted. In the general non-smoking population the major exposure pathway is through food, via the addition of cadmium to agricultural soil from various sources (atmospheric deposition and fertiliser application) and uptake by food and fodder crops. Additional exposure to humans arises through cadmium in ambient air and drinking water.

Lead: Lead in the environment arises from both natural and anthropogenic sources.

- Greatest source of lead is emission from vehicles, burning leaded petrol, though use of unleaded fuel or reduction in lead content has nowadays mitigated this problem to some extent.
- Smelting and refining of lead give rise to a considerable generation of fumes.
- Exposure can occur through drinking water, food, air, soil and dust from old paint containing lead.
- In the general non-smoking, adult population the major exposure pathway is from food and water. Food, air, water and dust/soil are the major potential exposure pathways for infants and young children. For infants up to 4 or 5 months of age, air, milk formulae and water are the significant sources.

- Lead is among the most recycled non-ferrous metals and its secondary production has therefore grown steadily in spite of declining lead prices. Its physical and chemical properties are applied in the manufacturing, construction and chemical industries.
- It is easily shaped and is malleable and ductile. There are eight broad categories of use: batteries, petrol additives (no longer allowed in the EU), rolled and extruded products, alloys, pigments and compounds, cable sheathing, shot and ammunition.

(ii) Arsenic and Selenium

These metals occur in the solution as neutral and negatively charged species. Arsenic exists in the environment as As^{+5} and As^{+3} and selenium in +6 and +4 under oxidizing condition and in –2 state under reducing condition. In the case of these elements, adsorption property increases with decreasing pH and they also form sulfides in presence of sulphur under reducing conditions.

Arsenic: Under oxidizing conditions arsenate species are formed, which lead to adsorption by oxyhydroxides and removal from the aqueous environment. If the condition becomes reducing, the bound arsenate will be converted to arsenite and can be released from the particles to the solution.

Arsenic contamination: An understanding of factors controlling the distribution of arsenic in ground water requires knowledge of arsenic sources and of processes controlling arsenic mobility. Geochemistry of arsenic is controlled by many factors including:

1. Redox potential
2. Adsorption/desorption
3. Precipitation/dissolution
4. Source
5. pH
6. Arsenic speciation
7. Presence and concentration of competing ions
8. Biological transformation

In general, most naturally occurring arsenic compounds are insoluble in water and adsorption of arsenic compounds to other soluble minerals or organic compounds plays major role in arsenic concentration and transport in ground water. Speciation of arsenic again controls the adsorption behaviour.

Intake of drinking water having arsenic concentration beyond the permissible limit of 0.05 mg/lit has deleterious effects on human health viz., cardiovascular (heart failure) problem, gastrointestinal (burning lips, painful swallowing, thirst, nausea and severe abdominal colic), hematological effects (anemia and leucopoenia), hepatic effects, renal effects, neurological effects (headache, lethargy, mental confusion, hallucination, seizures and coma), dermal effects (skin disorder, hyperkeratosis), carcinogenic effects (lung cancer) etc.

Selenium: Selenium remains in solution under oxidizing conditions but is removed under reducing condition. Selenium exists in several allotropic forms. The most stable, crystalline hexagonal selenium is metallic grey. Crystalline monoclinic selenium is a deep red colour. Amorphous selenium is red in powder form and is black in vitreous form. Gray crystalline 'metallic' selenium conducts electricity better in the light than in the dark (photoconductive) and it can convert light directly into electricity (photovoltaic).

Selenium combines with metals and oxygen to form selenides (such as zinc selenide, ZnSe), selenates (such as calcium selenate, $CaSeO_4$), and selenites (such as silver selenite, Ag_2SeO_3). Selenium is considered to be toxic if taken in excess. It is carcinogenic and teratogenic. Selenates and selenites are very toxic, and hydrogen selenide (SeH_2) is an extremely toxic, corrosive gas.

(iii) Mercury

Mercury can combine with a methyl group to become methyl mercury. This form of mercury is found in a variety of environmental pollution situations and can produce a range of toxicities. Elemental mercury is less labile but produces a similar set of toxic manifestations.

- Mercury is a global pollutant with complex and unusual chemical and physical properties. The major natural source of mercury is the degassing of the Earth's crust, emissions from volcanoes and evaporation from natural bodies of water.
- World-wide mining of the metal leads to indirect discharges into the atmosphere. The usage of mercury is widespread in industrial processes and in various products (e.g. batteries, lamps and thermometers).
- It is also widely used in dentistry as an amalgam for fillings and by the pharmaceutical industry.
- Concern over mercury in the environment arises from the extremely toxic forms in which mercury can occur.
- Mercury is mostly present in the atmosphere in a relatively unreactive form as a gaseous element. The long atmospheric lifetime (of the order of one year) of its gaseous form, which means the emission, transport and deposition of mercury is a global issue.
- Natural biological processes can cause methylated forms of mercury to form which bioaccumulate over a million-fold and concentrate in living organisms, especially fish. These forms of mercury, monomethylmercury and dimethylmercury, are highly toxic, causing neurotoxicological disorders. The main pathway for mercury to humans is through the food chain and not by inhalation.
- *Minemata incident:* In 1952, the first incidents of mercury poisoning appear in the population of Minimata Bay in Japan, caused by consumption of fish polluted with mercury, bringing over 500 fatalities. Since then, Japan has had the strictest environmental laws in the industrialised world.

Table 18.5: Common environmental impacts

Impact	*Damages*
Pollution	
Air	
Respiratory illness	Lost work days, medical expense
Vegetation effects	Lower crop yields
Soiling of materials	Cleaning costs, more frequent painting
Aesthetic degradation	Lowered visibility, odours leading to devalued property
Water	
Pathogenic organisms or toxic materials in drinking water	Lost work days, medical expense, expenses of alternative supply
Fisheries effect	Lowered catch
Affects water-contact recreation	Loss of tourism revenues
Noise	Lowered property value
Ecosystem degradation	
Forest lands	
Harvesting effects	Sedimentation—shortened life of hydro-electric reservoirs
Monoculture plantation	Lowered water quality
	Loss of services from intact diverse forest
	Loss of biodiversity
Wetlands	
Filling and dredging	Greater flood damage, loss of unique habitat
Coral reefs	
Toxic or sediment effects	Reduced fishery production
	Loss of recreation values
	Loss of biodiversity
Ground water	
Contamination	Cost of alternative supply
Lowered water table	Subsidence of land
	Structural damage

18.4 HEALTH HAZARDS FROM NATURAL MINERALS

Mercury

Mercury and all its compounds like *methyl mercury* (organic) and *mercuric chloride* (inorganic) are toxic in nature. Mercury is a by-product of many factories, and is present in the particles spewed from the tops of smokestacks. As the mercury gets into the Earth's atmospheric system, it mingles with rain, which then falls into water and is absorbed by fish. As larger fish (like tuna) prey on the smaller, they receive even higher amounts.

Table 18.6: Effects and health hazards of mercury ores

Minerals	*Effects*	*Health hazards*
Methyl mercury (organic)	Methyl mercury (CH_3Hg^+) is a neurotoxin, and the form of mercury that is most easily bioaccumulated in organisms. It consists of a methyl group bonded to a single mercury atom, and is formed in the environment primarily by a process called biomethylation. It is quickly taken up into higher organisms through the food chain and is retained in their bodies. It reaches the highest levels in large predatory fish and in birds and mammals which consume fish. Levels of methyl mercury in fish are typically 100, 000 times those in the water in which they swim. Methyl mercury is formed from the addition of a methyl group and appears to be more toxic to higher animals than inorganic mercury (Dietz et al., 1990).	Causes damage to the brain and the central nervous system, while foetal and post-natal exposure have given rise to abortion, congenital malformation and development changes in young children. Damages the stomach and large intestine, can cause lung damage, increased blood pressure and heart rate. May cause cancer.
Mercuric chloride (inorganic)	Mercuric chloride is a very poisonous form of mercury. It is a type of mercury salt. It can be found in some antiseptics and dry cell batteries.	Damages the gastrointestinal tract and causes kidney failure.
Elemental mercury	Elemental mercury is a shiny, silver-gray metal that is a liquid at room temperature. It can be found in thermometers, barometers, sphygmomanometers, dental amalgams, batteries, fluorescent lights and electrical switches.	Causes tremors, gingivitis, and excitability. Elemental (metallic) mercury primarily causes health effects when it is breathed as a vapour where it can be absorbed through the lungs. These exposures can occur when elemental mercury is spilled or products that contain elemental mercury break and expose mercury to the air, particularly in warm or poorly-ventilated indoor spaces.

Calcium

Inadequate intake can adversely affect bone health, but evidence suggests it might have other effects also. Low-calcium diets can accelerate weight gain and fat accumulation; high-calcium diets can prevent fat accumulation.

Table 18.7: Effects and health hazards of calcium ores

Minerals	*Effects*	*Health hazards*
Calcium nitrite	This product is a clear, light yellow to colourless solution with a slight sweet odour.	Causes methemoglobinemia, characterized by cyanosis, headache, dizziness, fatigue, nausea, vomiting, drowsiness, stupor, coma and rarely death and also causes irritation to eyes, respiratory system and skin.
Calcium carbonate	This form of calcium is not soluble in water and is not bio-available. When calcium is taken in the bad form like calcium carbonate, the body tends to put it anywhere it wants to.	Causes scale and plaque. May cause gallstones, kidney stones, buildup in the joints as arthritis. It may even be responsible for diabetes and hypoglycemia.
Calcium chloride	It is a salt. It behaves as a typical halide, and is solid at room temperature. It has several common applications such as brine for refrigeration plants, ice and dust control on roads, and in concrete. The anhydrous salt is also widely used as a desiccant. It can be produced directly from limestone.	Causes respiratory problems.
Calcium bicarbonate	Found in meats, fruits and vegetables, this calcium builds strong bones when it is properly bound to enzymes and converted by the body. It is often used as a drug all over the world.	Secreted by the stomach, it is necessary for digestion. When ingested, for example, with mineral water, it helps buffer lactic acid generated during exercise and also reduces the acidity of dietary components. Finally, it has a prevention effect on dental cavities. An ingestion of 300 mg/kg of body weight of bicarbonate before exercising will help to reduce muscular fatigue and so increase the performance of short-term physical exercise. Application of calcium bicarbonate has following side effects:Gastric varices, acute respiratory failure, blood creatinine increase.

Epidemiologic evidence and at least one intervention study have shown that higher calcium intake lowers the risk of developing colon cancer. Calcium deficiency also results in arthritis, high blood pressure and osteoporosis. Osteoporosis is a disease in which the bones become extremely porous, are subject to fracture, and heal slowly, occurring especially in women following menopause and often leading to curvature of the spine from vertebral collapse.

Copper

Copper and its various compounds are quite toxic to living world above a permissible limit. Long-term exposure to copper can cause irritation of the nose, mouth and eyes and it causes headaches, stomachaches, dizziness, vomiting and diarrhoea. Intentionally high uptakes of copper may cause liver and kidney damage and even death. Whether copper is carcinogenic has not been determined yet.

Industrial exposure to copper fumes, dusts, or mists may result in metal fume fever with atrophic changes in nasal mucous membranes. Chronic copper poisoning results in Wilson's disease, characterized by a hepatic cirrhosis, brain damage, demyelination, renal disease and copper deposition in the cornea. Deficiency of copper causes anemia, hair loss, arthritis and aneurysms.

Table 18.8: Effects and health hazards of copper ores

Minerals	*Effects*	*Health hazards*
Copper sulfate	It is a fungicide used to control bacterial and fungal diseases of fruit, vegetable, nut and field crops. It is also used as an algaecide, an herbicide in irrigation and municipal water treatment systems, and as a molluscicide, a material used to repel and kill slugs and snails. Copper sulfate is a naturally-occurring inorganic salt and copper is an essential trace element in plant and animal nutrition. It is available in the following formulations: dusts, wettable powders, and fluid concentrates.	It is only moderately toxic upon acute oral exposure. Some of the signs of poisoning which occurred after 1-12 grams of copper sulfate was swallowed include a metallic taste in the mouth, burning pain in the chest and abdomen, intense nausea, vomiting, diarrhea, headache, sweating, shock, discontinued urination leading to yellowing of the skin. Injury to the brain, liver, kidneys and stomach and intestinal linings may also occur in copper sulfate poisoning. Long-term effects include Wilson's disease and anaemia.

(*Contd.*)

Table 18.8 (*Contd.*)

Minerals	*Effects*	*Health hazards*
Cupric chloride dehydrate	It is a rare inorganic compound with significant health hazards.	Hazardous in case of skin contact (permeator). Corrosive to eyes and skin. Eye contact can result in corneal damage or blindness. Skin contact can produce inflammation and blistering. Inhalation of dust will produce irritation to gastro-intestinal or respiratory tract, characterized by burning, sneezing and coughing. Severe over-exposure can produce lung damage, choking, unconsciousness or death. Inflammation of the eye is characterized by redness, watering and itching. Skin inflammation is characterized by itching, scaling, reddening, or, occasionally, blistering.
Copper (II) nitrate	Copper(II) nitrate is the chemical compound with the formula $Cu(NO_3)_2$. Commonly referred to simply as copper nitrate, the anhydrous form is a blue, crystalline solid. Hydrated forms of copper nitrate, also blue, are commonly used in school laboratories to demonstrate chemical voltaic cell reactions	Hazardous in case of skin contact (irritant), of eye contact (irritant), of ingestion, of inhalation. Prolonged exposure may result in skin burns and ulcerations. Over-exposure by inhalation may cause respiratory irritation.
Copper carbonate (malachite)	Malachite green, also called basic green 4 or Victoria green B, is a toxic chemical primarily used as a dye. When diluted, it can be used as a topical antiseptic or to treat parasites, fungal infections, and bacterial infections in fish and fish eggs. It is also used as a bacteriological stain	Causes carcinogenesis, mutagene-sis, chromosomal fractures, teratogenecity and respiratory toxicity. Histopathological effects include multi-organ tissue injury. Toxicity occurs in some mammals, including organ damage, mutagenic, carcinogenic and developmental abnormalities.
Cupric oxide (cuprite)	It is a secondary mineral which forms in the oxidized zone of copper sulfide deposits. It frequently occurs in association with native copper, azurite, chrysocolla, malachite, tenorite and a variety of iron oxide minerals. It is known as *ruby copper* due to its distinctive red colour.	It is an irritant. It also can cause damage to the endocrine and central nervous system. Contact to the eyes can cause irritation and contact to the skin can cause irritation. Ingesting cupric oxide powder can result in a metallic taste, nausea, vomiting and stomach pain. Inhalation of fumes during smelting of cupric oxide powder can lead to a disease called metal fume fever, which can result in flu like symptoms. It can cause a toxic build-up of copper in a small subset of the population with Wilson's disease.

Selenium

Selenium is needed by humans and other animals in small amounts, but in larger amounts can cause damage to the nervous system, fatigue, and irritability. Selenium accumulates in living tissue, causing high selenium content in fish and other organisms, and causing greater health problems in human over a lifetime of overexposure. These health problems include hair and fingernail loss, damage to kidney and liver tissue, damage to circulatory tissue, and more severe damage to the nervous system. Deficiency causes cancer, cataracts, age spots, multiple sclerosis.

Table 18.9: Effects and health hazards of selenium ores

Minerals	*Effects*	*Health hazards*
Native selenium	Selenium were used in rectifiers, before the development of semiconductor-based diodes. It is also an essential nutrient with a minimum recommended level of 50 mcg/day.	Shows symptoms including hair and nail loss, discolouration and decay of the teeth, and CNS disturbances including pain and anesthesia of the extremities. Animals in the region were also affected, with hoof damage and horn sloughing reported in water buffalo, cattle and pigs. There is possibility of liver dysfunction as indicated by a delay in prothrombin time among persons with intake of 750-850 μg Se/day.
Berzelianite (Copper selenide)	It is bark blue to black coloured solid and is odourless. It is a semi-conducting material whose structural and optical properties are suitable for photovoltaic application.	Causes pre-existing respiratory, gastric disorders and an increased risk for individuals with Wilson's disease. May affect the respiratory systems, skin, liver, central nervous systems and kidneys. May cause ulceration and perforation of the nasal septum and pharyngeal congestion.

Zinc

Zinc is a trace element that is essential for human health. When people absorb too little zinc they can experience a loss of appetite, decreased sense of taste and smell, slow wound healing and skin sores. Zinc-shortages can even cause birth defects.

Although humans can handle proportionally large concentrations of zinc, too much zinc can still cause eminent health problems, such as stomach cramps, skin irritations, vomiting, nausea and anaemia. Very high levels of zinc can damage the pancreas and disturb the protein metabolism, and cause arteriosclerosis. In the work place environment zinc contagion can lead to a flu-like condition known as metal fever. This condition will pass after two days and is caused by over sensitivity.

Table 18.10: Effects and health hazards of zinc ores

Minerals	*Effects*	*Health hazards*
Sphalerite	It is a common mineral, and occurs in many different colours. Iron impurities are almost always present in this mineral, and for this reason it is rarely transparent. Sphalerite is the main ore of zinc. In some localities, it occurs with significant amounts of cadmium, gallium and iridium, and it is also the main ore of those metals. Sphalerite is an important mineral to collectors, and some transparent varieties are occasionally faceted for collectors.	Only poses pollution and contamination of soil during exploration.
Smithsonite	A white crystalline solid or powder that is insoluble in water. Used in pharmaceuticals, to make other zinc compounds, as a feed additive.	Inhalation of dust or fumes: Dry throat, cough and chest discomfort. Fever and sweating. Skin: Astringent. Ingestion: May cause nausea and vomiting.
Franklinite	It represents spinel group and is considered as an important ore of zinc and manganese. It occurs in crystals, dominantly octahedral, isolated grains, and in coarse to fine or compact, massive granular form.	Induces inflammatory response in lungs.
Willemite	Willemite is an orthosilicate of zinc, containing more or less manganese, iron and magnesium in place of part of the zinc. It is generally fluorescent, and some specimens are strongly phosphorescent with a green colour, but the purest white willemite, like troostite, gives little or no reaction. The so-called "black willemite" does not fluoresce.	May induce tumours, but does not have any potential carcinogenic or mutagenic effects.
Zincite	Zinc oxide, may contain manganese and some iron (Zn, Mn)O. It is red to yellow, brittle, hexagonal mineral. It is considered as an ore of zinc. These crystals can be grown even artificially.	Exposures to high concentrations can result in respiratory system effects. Volunteers inhaling 600 mg/m^3 zinc oxide dust for 10 minutes exhibited decreased vital capacity, coughing, upper respiratory tract irritation and substernal pain. Inhalation of zinc oxide fume can result in metal fume fever. Repeated exposures to zinc oxide by skin contact have resulted in papular-pustular skin eruptions in the axilla, inner thigh, inner arm, scrotum and pubic areas.

Zinc can be a danger to unborn and newborn children. When their mothers have absorbed large concentrations of zinc the children may be exposed to it through blood or milk of their mothers.

Lead

Lead is a soft bluish-gray metal that is highly poisonous. It is added to some industrial paints because it helps resist corrosion and add certain colour characteristics. It is also used in certain solders, some autobody fillers, lead fishing weights, batteries, bullets and some ceramic glazes. The toxic effects of lead are well documented in both children and adults.

Workers' exposure to lead can damage the central nervous system, cardiovascular system, reproductive system, hematological system, and the kidney. Workers' lead exposure can also harm development of their children when they bring lead dust home on their clothing. Lead has been shown to be an animal carcinogen, and authors of recent studies suggest that occupational lead exposure increases the risk of cancer. Lead poisoning often goes undetected since many of the symptoms, such as stomach pain, headaches, anxiety, irritability and poor appetite, are nonspecific and may not be recognized as symptoms of lead poisoning.

The notable effects of lead poisoning are:

- Disruption of the biosynthesis of haemoglobin and anaemia
- A rise in blood pressure and kidney damage
- Miscarriages and subtle abortions
- Disruption of nervous systems
- Brain damage
- Declined fertility of men through sperm damage
- Diminished learning abilities of children
- Behavioural disruptions of children, such as aggression, impulsive behaviour and hyperactivity

Table 18.11: Effects and health hazards of Lead ores

Minerals	*Effects*	*Health hazards*
Anglesite	It is a lead sulfate mineral. It occurs as an oxidation product of primary lead sulfide ore, galena. Anglesite occurs as prismatic orthorhombic crystals and earthy masses, and is isomorphous with barite and celestine.	No information on health risk.
Cerussite	Cerussite is easily identifiable by its weight, brilliant lustre, and interesting twinning habits. It also performs interesting reactions during blowpipe testing. Cerussite forms as a secondary mineral from oxidized lead deposits, particularly in arid regions. Specimens are very fragile and should be handled with extreme care.	Exposure may cause decreased physical stamina, fatigue, sleep disturbance, headaches, aching bones and muscles, constipation, abdominal pains and decreased appetite. Inhalation of large amounts may lead to seizures, coma or possibly death.

(Contd.)

Table 18.11 (*Contd.*)

Minerals	*Effects*	*Health hazards*
Galena	It is the most important ore of lead. Galena forms many interesting crystal shapes. Galena specimens must be taken care of very well. They are easily damaged, and well formed crystals may shatter into small crystal fragments if put under slight pressure. Galena is a primary mineral. Impurities in the structure of galena, such as silver and bismuth, may change galena's cleavage properties. Galena containing bismuth may exhibit octahedral cleavage, and silver in galena may cause a specimen to exhibit flaky, slightly bent cleavage fragments.	Induces chronic kidney disease, heart disease, skin cancer, anaemia and hypertension.
Massicot	Massicot, which has been known by the chemically descriptive term of "*lead monoxide*", is an oxidation product of other lead minerals. It forms under oxidizing conditions of lead ore bodies and its presence is very indicative of the degree of oxidation. Litharge is made by a modern process of direct oxidation of lead, wielding a finely divided yellow orange powder of high purity. Litharge is monoxide of lead and usually contains about 93 percent of lead and seven percent of oxygen. It is not used as a pigment in oil painting, but was used as a pigment in manuscript illumination (egg tempera and watercolour) and infrequently in fresco painting.	It is toxic if inhaled as a dust or if ingested. Induces "painter's colic" or "plumbism".
Minium	Minium, which has been known as "*red lead*", is an oxidation product of other lead minerals. It forms in extreme oxidizing conditions of lead ore bodies. Minium is a member of the Spinel group of oxide minerals.	Increasing amounts can build up in the body and may reach a point where symptoms and disability can occur. These may include anaemia, pale skin, a blue line at the gum margin, decreased hand-grip strength, abdominal pain, severe constipation, nausea, vomiting, and paralysis of the wrist joint. Prolonged exposure may result in kidney damage.

Chromium

People can be exposed to chromium through breathing, eating or drinking and through skin contact with chromium or chromium compounds. The level of chromium in air and water is generally low. In drinking water the level of chromium is usually low as well, but contaminated well water may contain the dangerous chromium (IV), hexavalent chromium.

The health hazards associated with exposure to chromium are dependent on its oxidation state. The metal form (chromium as it exists in this product) is of low toxicity. The hexavalent form is toxic. Adverse effects of the hexavalent form on the skin may include ulcerations, dermatitis, and allergic skin reactions. Inhalation of hexavalent chromium compounds can result in ulceration and

Table 18.12: Effects and health hazards of chromium ores

Minerals	*Effects*	*Health hazards*
Chromite	Chromite, as is indicated by its early crystallization, is resistant to the altering affects of high temperatures and pressures. These characteristics explain the use of chromite as a heat conductive foundry moulding sand, which allows localized or general chilling of castings to improve their integrity	Hazard caused due to excessive inhalation of chromite dust, which may result in transitory upper respiratory irritation.
Chromium III	It is an essential nutrient for humans. For most people eating food that contains chromium (III) is the main route of chromium uptake, as chromium (III) occurs naturally in many vegetables, fruits, meats, yeasts and grains.	Shortages may cause heart conditions, disruptions of metabolisms and diabetes. But the uptake of too much chromium (III) can cause health effects as well, for instance skin rashes.
Chromium (VI) (chromates, bichromates, chromic acid)	It is a danger to human health, mainly for people who work in the steel and textile industry. People who smoke tobacco also have a higher chance of exposure to chromium.	Short-term effects: Irritation and inflammation of the nose and upper respiratory tract, burns to the skin possibly leading to ulcers, eye damage from splashes. Long-term effects: Damage to the nose, including ulcers and holes in the flap of tissue separating the nostrils, irritation of the lungs, kidney damage, allergic reactions in the skin and respiratory tract, risk of cancer of the lung and nose from certain processes.

perforation of the mucous membranes of the nasal septum, irritation of the pharynx and larynx, asthmatic bronchitis, bronchospasms and edema. Respiratory symptoms may include coughing and wheezing, shortness of breath, and nasal itch. Deficiency of chromium causes arteriosclerosis, glucose intolerance in diabetics and diabetic symptoms.

Asbestos

All types of asbestos cause fibrosis and an excess of bronchial cancers which are, however, also closely cigarette related. Mining and milling of chrysotile, the most used type of asbestos, have caused few mesotheliomas, despite heavy dust exposures in the past. The diseases induced by asbestos are dose-related.

Table 18.13: Effects and health hazards of asbestos ores

Minerals	*Effects*	*Health hazards*
Crysotile (white asbestos) $Mg_3Si_2O_5(OH)_4$	A white curly fibre, chrysotile accounts for 90% of asbestos in products and is a member of the serpentine group. It is a magnesium silicate.	Diffuse pulmonary fibrosis (asbestosis), lung cancer (bronchial carcinoma) and primary malignant tumours of the pleura and peritoneum (mesothelioma).
Amosite $(Mg, Fe)_7Si_8O_{22}(OH)_2$	Brown or gray, straight amosite fibres belong to the amphibole group, and contain iron and magnesium.	Cancer or birth defects.
Crocidolite (Riebeckite) $Na_2Fe_3^{2+}Fe_2^{3+}Si_8O_{22}(OH)_2$	A member of the amphibole group, crocidolite takes the form of blue, straight fibres. It is a sodium iron magnesium silicate.	Cancers of the surface of the lung (mesotheliomas), asbestosis.
Fibrous tremolite $Ca_2Mg_5Si_8O_{22}(OH)_2$	Tremolite is not the common form of asbestos found in most of the asbestos products that were so popular during the twentieth century.	Lung cancer, pleural tumour, pleural effusion, pleural plagues.
Fibrous anthophyllite $(Mg, Fe)_7Si_8O_{22}(OH)_2$	One of the lesser-known types of asbestos, the amphibole mineral anthophyllite is typically a yellow or brown, fibrous mineral. In its naturally occurring environment, anthophyllite develops in the form of massive, shapeless crystals,	When people come into contact with anthophyllite asbestos and inhale or ingest its fibres, the fibres have the potential to become lodged in the mesothelial lining of the lungs, heart, abdomen, or testicles. The fibres can

(Contd.)

(*Contd.*)

	sometimes marked with streaks of gray that exude a pearly lustre. Anthophyllite is formed when the mineral talc breaks down in the presence of carbon dioxide and water, and for this reason it can be found in many of the same mines where talc is located.	eventually lead to the development of lung cancer, *mesothelioma* cancer, and other life-threatening cancers and illnesses.
Fibrous actinolite $Ca_2(Mg, Fe)_5Si_8O_{22}(OH)_2$	Actinolite is composed of the elements calcium, magnesium, iron, silicon, oxygen and hydrogen. The composition is very similar to tremolite and anthophyllite, and like these other minerals, actinolite is hazardous to human health.	If actinolite asbestos fibres are inhaled they can become lodged into the lining of the lungs, stomach or heart. Over time, the fibres may accumulate and lead to the development of fatal asbestos diseases.

Cadmium

Cadmium is found in the form of cadmium oxide, cadmium carbonate and cadmium chloride.

Table 18.14: Effects and health hazards of cadmium ores

Minerals	*Effects*	*Health hazards*
Cadmium oxide(CdO)	Found with zinc ores, this is a common by-product of zinc refining; produced by pyrolysis of other cadmium compounds.	It is a known carcinogen and inhaling fumes or vapours can be fatal.
Greenockite (CdS)	Main ore of cadmium; crystalline; occurs with other sulphide minerals like galena and sphalerite.	In humans, long-term exposure causes renal disfunction. High exposure can lead to obstructive lung disease and has been linked to lung cancer, although data concerning the latter are difficult to interpret due to compounding factors. Cadmium may also produce bone defects (osteomalacia, osteoporosis) in humans and animals. In addition, the metal can be linked to increased blood pressure and effects on the myocardium in animals, although most human data do not support these findings.
Cadmoseite (CdSe)	Found in sandstones as interstitial grains, formed under reducing alkaline diagenetic conditions; pale grey to opaque crystals and grains.	
Niedermayrite $Cu_4Cd(SO_4)_2(OH)_6 \cdot 4H_2O$	Occurs as encrustrations and well formed vitreous blue-green prismatic crystals.	
Otavite ($CdCO_3$)	Forms encrustrations and small scale nonhedral crystals with pearly to adamantine lustre; remains associated with azurite, calcite, malachite and smithsonite.	

Clay Minerals

Ball clays (Al_2O_3, $2SiO_2$, $2H_2O$) contain free quartz. Prolonged inhalation even at low levels may cause delayed lung injury (silicosis).

Table 18.15: Effects and health hazards of clay minerals

Minerals	*Effects*	*Health hazards*
Bentonite	Bentonite (sodium bentonite, quaternium-18 bentonite) is a soft, porous clay that expands to many times its own volume as it absorbs water. Often found in cosmetic foundations, bentonite's thick consistency enables bentonite to clog pores and suffocate the skin, preventing normal oxygen transport.	Hazardous in case of eye contact (irritant), of inhalation. may block pores, suffocating the skin, prolonged exposure causes damage to lungs. Ingestion of bentonite has serious health effects which may include death.
Kaolinite	It consists of hexagonal flakes of composition $Si_4Al_4O_{10}(OH)_8$. It is formed by alteration of other clays or of feldspar. From kaolinite is made kaolin, or china clay, which is used for filling and coating paper, filling rubber and paints, and for making pottery and porcelain. Kaolin may also be used in the form of an oral suspension to treat chronic diarrhea, and in dusting powders and poultices.	Hazardous in case of inhalation (lung irritant). Slightly hazardous in case of skin contact (irritant), of eye contact (irritant), Repeated or prolonged exposure to the substance can produce target organs damage.
Illite	This clay-like series is typically found as extremely fine-grained masses of greyish-white to silvery-grey, sometimes greenish-grey, material.	May cause mild irritation to eyes, prolonged or repeated skin contact may cause irritation. It contains breathable dust, prolonged inhalation in high concentration may disrupt lung function.

18.5 INFRASTRUCTURES OF MINERAL INDUSTRY AND ENVIRONMENTAL FACTORS

Mineral Industry—Non-ferrous Metals

Non-ferrous metals generally occur as sulphide compounds. So their extraction involves the following processes: large energy inputs to extract the ore and to separate the desired metal from undesired minerals and impurities; determination of the toxicity level by identifying their chemistry in extraction, purification and use; recycling after use in an eco-friendly manner; and stabilizing the effects of associated impurities.

Extraction: Non-ferrous ores are extracted from both open-pit and underground mines. In view of health and safety assurance of the miners, open-pit mines are more favourable than the underground ones. The waste disposal is generally less intrusive in underground than in open-pit mines which have the problem of hiding the hole at a closure. Successful restoration of a worked out underground mine is usually a simpler task than for an open-pit operation.

Environmental Impact

To reduce the effects on the environment, environmentally disposable routes are taken into consideration. Rising power costs also have to be adjusted. Actually these rises are attributed mainly to the cost of developing low-sulphur basic sources of energy and the cost of neutralizing acidic emissions at power stations burning coal of relatively high sulphur content, to minimize 'acid rain'. The common environmental impacts along with their mitigation procedures are outlined in the following discussion.

Location and access: The location of the mine and its ancillaries is determined by the nature of the deposit. The waste disposal system should not contaminate the local streams by run-off nor hinder the restoration plan. The access routes, the tailing areas should also be considered with care.

Dust-blow: To eliminate the total dust produced due to blasting, transportation, handling and storage, following steps are taken:

- Dampening all areas of dust generation.
- Paving haul roads.
- Providing respiratory protection for all exposed workers and ensuring its use.
- Providing mobile equipment operators supply of adequate filtered air.
- Ensuring that residential, offices, schools and hospital areas are located as far as possible in areas of minimum dust exposure.
- Covering dumps, conveyors etc. permanently.

Mine safety: Physical safety standards are always a prime consideration in the design and construction of both open-pit and underground mines.

Erosion of waste-rock dumps: Mines are constructed where economic mineralization naturally occurs. Since large proportion of extracted low-value materials have to be transported for upgrading, concentration plant associated with the mine also has to be located nearby. Extraction operations break up the terrain and hence increase greatly the surface area of material exposed to rainfall which may fall as intense storm of short duration, giving a high risk of flash flooding.

Mineral Industry—Ferrous Metal: Iron and Steel

Iron and steel industry is a heavy industry. All its raw materials are heavy and colossal. They encompass iron-ore, coking coal and limestone. Location of this industry is thus governed by its proximity to the raw materials. The finished products are also heavy and need efficient transport system for their distribution.

Extraction

Iron ores are rocks and minerals from which metallic iron is extracted. Iron ore concentrates are explored and supplied by the iron ore mining companies. Iron ore mining methods vary depending on the mineralogy and geology of the ore deposits.

Extraction of iron is done in many steps like oxidation, combustion, fusion, reduction, etc. During extraction, cast iron is formed first, which is then converted to wrought iron. Hematite is the ore that is used for extraction of cast iron. The ore is crushed into small pieces and is washed with water to remove the sand particles. Then the washed ore is heated in the presence of air. This process removes the moisture and is called as calcination. In this process, sulphur and phosphorus form oxides. Carbonate forms oxide and ferrous oxide converts into ferric oxide.

Blast furnace is used in the extraction of iron, which is made of steel lined with bricks. The ore, coke and limestone are sent into this furnace from top. The furnace has an inlet for hot air and two outlets for slag and molten iron. The coke that is sent to burns, produces carbon dioxide. This CO_2 reduces to CO in presence of coke.

The limestone subjected into the furnace decomposes and produce CaO. This CaO later combines with silica and forms slag.

Molten slag formed is lighter and is removed out by the outlet produced for it. All the above reactions take place from the bottom to the middle region of the furnace. Only the reduction of oxides of iron to iron is done at the top of the furnace. The iron that is formed melts and moves down to the base of the furnace. This is called as cast iron. This cast iron is again heated with hematite to remove the impurities.

Thus, wrought iron which is the purest form of iron is formed.

Environmental Impact

Environmental pollution

- Air emissions include contributions from every stage of the process. Particulates are generated at virtually every stage, but most heavily by coke-making and blast furnace operations. In addition:
  - Coke-making is a major source of VOCs and carbon monoxide emissions.
  - Blast furnace iron-making generates large quantities of sulfur and nitrogen oxides and carbon monoxide.
  - Downstream processes such as casting and rolling account for significant releases of sulfur oxides as sulfur is liberated from the metal surface.
- Wastewater
  - Wastewater from coke-making has high contaminant levels, requiring extensive removal and treatment before disposal.
  - Wastewater generation from iron-making, steel-making, and downstream processes, much of which results from cooling water and from wet scrubbers, is somewhat mitigated by significant rates of reuse and recycling.

- Solid wastes from these industry present problems due more to the volume of the wastes generated than to the hazard presented. In fact, most of the waste produced in the sector is relatively benign.
- The sector is a very large consumer of energy, and as such is a major contributor to greenhouse gas emissions.

Impacts on land
- Changes in topography and land scenario
- Change of land-use pattern
- Changes in drainage pattern
- Changes in top soil composition by run off from overburden dumps
- Damage to archeological/religious monuments

Impacts on water regime
- Changes in surface and ground water regime
- Removal of surface water bodies
- Damage of all aquifers above the mineral deposits
- Depletion of ground water—Pumping of high pressure aquifers, below the mineral deposits
- Pollution of surface water bodies
  - Run off from mines
  - Leaching from dumps
  - Beneficiation activities
- Mineral deposits rich with pyrites/sulfides leach acidic water to the water bodies
- On closure, open pits act as water bodies and recharge ground water table

Impacts on ecolgy
- Removal of vegetation, affects fauna too
- Pollution of water bodies affects aquatic ecology
- High fugitive emissions and release of other gaseous emissions retard growth of vegetation
- Water scarcity and change in water regime devastate agriculture and stunt growth of vegetation
- Noise pollution due to blasting and subsequent vibration drives away wild animals

Mineral Industry—Coal

Coal can be extracted in an environmentally acceptable manner. The principal methods employed in coal mining are: (a) surface mining and (b) underground mining.

(a) Surface mining: Surface, strip mining, or open cast mining is done where the coal seam is nearly horizontal, covered with relatively thin overburden and where the surface topography is of low relief. Coal is mined by blasting the overburden and using draglines, shovels and dumptrucks. The overburden is removed to avoid stratigraphic water problems.

(b) Underground mining: Underground mining is practiced where the coal is deeply embedded. There are two generic extraction methods as follows:

- Room and pillar
- Long-wall methods.

Environmental Impact

Coal mining process can be broadly categorized into five main steps:

- Exploration and extraction
- Preparation
- Handling and supply
- Conversion (if applicable)
- Utilization (including waste disposal)

Table 18.16: Primary environmental impacts due to the exploration, extraction and preparation phases

Types of mining	*Impacts*
Surface mining	Siting, large-scale land use, overburden removal and disposal, disturbance of hydrology and run-off, acid mine drainage, visual intrusion, noise, blast vibration, flyrock, fugitive dust, transportation, highwall stability, restoration of soil fertility, recreating ecosystem diversity, amenity value.
Underground mining	Siting, spoil disposal, lagoon requirements, subsidence, aquifer disturbance, minewater drainage/disposal, methane emissions, fugitive dust, visual intrusion, noise.
Abandoned mines	Methane migration, flooding, groundwater contamination, structural integrity, land rehabilitation.
Underground (workplace) environment	Ventilation-methane dilution, pollutant dispersal, temperature and humidity control, pollutant sources and monitoring – $CO/NO_x/CO_2/O_2$ deficiency/diesel engine emissions/respirable dust/quartz/radon/occupational chemicals, spontaneous combustion/ fire detection, dust explosibility, noise, vibration, illumination, heat stress management, mine drainage, in-rush protection, rock burst/microseismic activity, falls of ground, ergonomic hazard control.

18.6 MINING AND MINERAL INDUSTRY

Minerals constitute the back-bone of economic growth of any nation. There is much evidence of exploitation of minerals like coal, iron-ore, copper, lead-zinc etc. which has been going on from time immemorial. Mining is the extraction (removal) of minerals and metals from earth.

Large Scale Mining Versus Small Scale Mining

- **Large scale mining** usually involves a company with many employees. The company mines at one or two large sites and usually stays until the mineral or metal is completely excavated. An example of a large scale mine is the Serra Pelada mine in Brazil which yielded 29,000 tons of gold from 1980 to 1986 and employed 50,000 workers (Kricher, 1997).
- **Small scale mining** usually involves a small group of nomadic men. They travel together and look for sites which they think will yield gold or another valuable metal or mineral. Small scale mining occurs in places such as Suriname, Guyana, Central Africa, and many other places around the world. Some researchers believe that small scale mining is more harmful to the environment and causes more social problems than large scale mining.

Types of Small-scale Mining

- **Land dredging** involves miners using a generator to dig a large hole in the ground. They use a high pressure hose to expose the gold-bearing layer of sand and clay. The gold bearing slurry is pumped into a sluice box, which collects gold particles, while mine tailings flow into either an abandoned mining pit or adjacent forest. When the mining pits fill with water from the tailings, they become stagnant water pools. These pools create a breeding ground for mosquitoes and other water-borne insects. Malaria and other water-borne diseases increase significantly whenever open pools of water are nearby.
- **River dredging** involves moving along a river on a platform or boat. The miners use a hydraulic suction hose and carry on suction of gravel and mud as they move along the river. The gravel, mud and rocks go through the tailings (pipes) and any gold fragments if present, are collected on felt mats. The remaining gravel, mud and rocks go back into the river, but in a different location than where it was originally collected. This creates problems for the river. The displaced gravel and mud disrupt the natural flow of the river. Fish and other living organisms often die and fishermen can no longer navigate in the obstructed rivers.

Effect of Mining on Environment and People

Effect on the Environment

- **Acid mine drainage:** Acid mine drainage is acidic water (pH <5. 0), laden with iron, sulfate and other metals, that form under natural conditions when geologic strata containing pyrite are exposed to the atmosphere or oxidizing environments. AMD can form from coal mining, both in surface and in underground mines. Alkaline mine drainage is water that has a pH of 6.0 or above, contains alkalinity, but may still have dissolved metals that can create acid by oxidation and hydrolysis. The drainage quality (acid or alkaline) emanating from underground mines or backfills of surface mines is dependent on the acid (sulfide) and alkaline (carbonate material) minerals contained in the disturbed geologic material. In general, sulfide-rich and carbonate-poor materials are expected to produce acidic drainage. In contrast, alkaline-rich materials, even with significant sulfide concentrations, often produce net alkaline water.

- **Fly ash:** Fly ash is a fine, glass powder recovered from the gases of burning coal during the production of electricity. These micron-sized earth elements consist primarily of silica, alumina and iron. Fly ash particles are almost totally spherical in shape, allowing them to flow and blend freely in mixtures. That capability is one of the properties making fly ash a desirable admixture for concrete.

  Coal fly-ash is an important by-product of coal combustion in thermal power-generating stations. It is composed mainly of silt-sized amorphous ferro-aluminosilicate and exhibits the properties of low permeability, low bulk density and high specific area. Though it is mainly considered as waste, but due to its varied pozzolonic, cementitous and alkaline properties, fly-ash has become a valuable by-product in numerous commercial and environmental applications.
- Gold, copper, diamonds and other precious metals and gemstones are important resources that are found in rainforests around the world. Extracting these natural resources is frequently a destructive activity that damages the rainforest ecosystem and causes problems for people living nearby and downstream from mining operations.
- Other toxic compounds are used and generated in the mining process as well. Mining exposes previously buried metal sulfides to atmospheric oxygen causing their conversion to strong sulfuric acid and metal oxides, which run off into local waterways. Oxides tend to be more soluble in water and contaminate local rivers with heavy metals.
- Cyanide, a highly toxic compound, is also often used to separate gold from sediment and rock. While cyanide is supposed to be carefully monitored to prevent its escape into the surround environment, spills do occur—especially when there's no one around to enforce mining regulations. The effects of poisoning can be widespread, especially when a waste-holding pool overflows or breaks.
- Large-scale mining operations, especially those using open-pit mining techniques, can result in significant deforestation through forest clearing and the construction of roads which open remote forest areas to transient settlers, land speculators, and small-scale miners. These settlers and miners are probably a greater threat to the tropical rainforest environment than industrial mining operations. Wildcat miners enter regions rumoured to have gold deposits and clear forest in search of riches. They hunt wildlife, cut trees for building material and fuelwood, and trigger erosion by clearing hillsides and detonating explosives.

- While deforestation and chemical pollution from mining can impact the rainforest environment, downstream aquatic habitats fare worse. Increased sediment loads and reduced water flows can seriously affect local fish populations.
- Construction of infrastructure—roads, mining installations, housing for miners, power plants and dams—often in remote and unspoilt locations.
- Stripping of large areas of topsoil and of all flora and fauna in addition to manmade constructions.
- Digging up chemically-reactive minerals which are harmless when underground, but when extracted react spontaneously with oxygen and water. The extraction of sulphide ores inevitably produces sulphuric acids and metal ions. This emission was known as 'acid mine drainage', occurs in active mines and in places where waste is stored, and remains a risk even after the mine has closed, sometimes even centuries after the closure of a mine.
- Adding potentially toxic chemicals, such as cyanide, to the ore to extract the metal, which increases the toxicity level of the waste.
- Using large amounts of energy and water, which are often scarce; mining is responsible for about 10 percent of the world's energy consumption.
- Conflicts over resources: Mining usually involves the destruction of resources which people rely on such as agricultural land, drinking or fishing waters, hunting grounds, religious or recreational areas as well as even their homes themselves. If these people are not involved in the decision-making process and/or if they do not benefit from the mine, a conflict will arise.
- Dead waste: Most products made from minerals are dumped after one use. In a lifetime each US citizen uses 1520 tonnes of new minerals. People in industrial countries consume 19 times as much aluminium, 18 times as many chemicals and 13 times as much iron and steel as people in less-developed countries.
- The devastating effect of mining on the environment could be reduced by more recycling. US residents throw away 2.3 million tonnes of aluminium each year. The energy saved by recycling this waste instead of producing new aluminium could meet the annual electricity needs of Chicago.

Effect of Mining on People

- The people who are exposed to the toxic waste from the tailings (waste from the mine containing rocks, metals and poisons) become sick. They develop skin rashes, headaches, vomiting, diarrhoea, etc. In fact, the symptoms of mercury poisoning are very similar to the symptoms of malaria. Many people who cannot afford to go to a doctor, or who live in a village where a doctor is not accessible, are often not treated for their illnesses.
- Many rivers have been pronounced 'biologically dead' due to release of mine tailings into lakes and waterways. Aquatic plant and animal life are choked with toxic sediment. If the water is contaminated, the people cannot use it for bathing, cooking, or washing their clothes.

- Cultural degradation also occurs in mining villages. For example, mining often destroys sacred sites and cemeteries. In Guyana, a special fishing event called *Haiari Fishing* unfortunately cannot take place if the river has been dredged for gold. Remember, the displacement of the gravel and mud obstruct the natural flow of the river. As a result, fish and other organisms die.
- Nations where the most mining takes place are also some of the poorest. There is a correlation between the standard of living (measured by the United Nations Development Program Human Development Index, the highest score being one) and the reliance of government on mining. The little money that poor countries get from mining companies is pocketed by international banks, not local people. Fourteen Majority World countries get at least a third of their export revenues from minerals. Their external debts are 1.4 times greater than their Gross National Product (GNP).
- Mining kills and injures more workers than any other industry. The situation is likely to be worse than statistics suggest as corporations and governments keep poor records.

Concluding Remarks

Mineralogy offers a wide range of applications for the environmental strategies of companies in many industries. The subject deals with the interdependence of living things within their environment and provides an insight into the orderly interplay of factors influencing environmental change. The impact of human demands on renewable and non-renewable resources and the limited availability of these resources in nature have been linked to correlate with patterns of human behaviours necessary to evolve a sustainable environmental paradigm. An attempt to cover these various aspects has been made in this chapter.

Think for a while

1. Explain the impacts of different minerals on environment.
2. Elaborate the sources and types of pollution, resulting from mining industries, both ferrous and non-ferrous.
3. Give a major classification of the air pollutants with their respective role in causing pollution. Aerosols are also considered as a source of air pollution. Why? Mention any hazardous incident, caused as a result of air pollution.
4. What are the role of different minerals in causing water pollution? Arsenic pollution is gaining wider importance nowadays. Explain the phenomenon with suitable illustration. Explain the health hazards caused by the different compounds of following minerals: mercury, lead, calcium, copper, zinc, selenium, chromium, cadmium and clay minerals.

FURTHER READING

Breuer, Georg. Air in Danger: Ecological Perspectives of the Atmosphere. Cambridge University Press, New York. 1980.

Ljung, K., E. Otabbong and O. Selinus. Natural and anthropogenic metal inputs to soils in urban Uppsala, Sweden. *Environmental Geochemistry and Health*, **28,** 353–364. 2006.

Miller, G. Tyler. Living in the Environment: an introduction to environmental science. Wadsworth, Belmont. 1990.

Mielke, H. Lead in inner cities. *American Scientist*, **87,** 62–73. 1999.

Salminen, Reijo, Anne Kousa, Rolf Tore Ottesen, Olle Selinus, Eiliv Steinnes, Timo Stewart and T. Charles. Air Pollution, Human Health and Public Policy. Lexington Books, New York. 1979.

Skinner, C. and A. Berger (eds). Geology and Public Health – Closing the Gap. Oxford Press. 2000.

Tarvainen and Björn Öhlander. Environmental Geology. *Episodes*, **31(1).** 2008.

D.J. Vaugan, R.A.D. Pattrick and R.A. Wogelius. *Mineralogical Magazine*, **66,** 653–676. 2002.

Mukhopadhyay, Lekha and Bhaskar Ghosh. Mining-induced desiccation and consequent impact on traditional economic livelihood – an analytical framework. Paper presented at 4th World Congress of Environmental and Resource Economists, Montreal, Canada, from June 28 to July 2, 2010.

Mukhopadhyay, Lekha and Bhaskar Ghosh. Mining-induced land acquisition and soil degradation – impact on sustainability of traditional economic livelihood: An analytical framework. "Ecological Economics: An Approach towards Socioeconomic and Environmental Sustainability". Proceedings of National Seminar and Prof. R. Misra Birth Centenary Lecture – 2009. Sunil Nautiyal and Bibhu Prasad Nayak (eds). Institute for Social and Economic Change, Bangalore, India.

Internet Data Retrieved from:

http://www.e-journals.net

www. idrc.ca/ecohealth

www. helfordgeoscience.co.uk

Heavy Metal Toxicology - Dr_Theodore B_Hoekman.htm

Heavy Metals - Lenntech.htm

American Mineralogist, Introduction to the Environmental Mineralogy Special Issue. Volume 88, pages 1843, 2003.

High incidence of Arsenic in Ground water in West Bengal. Report published by CGWB, GoI, Faridabad (July 1999).

Material available on the website of Govt. of West Bengal, Deptt. of Public Health and Engineering Department (PHED).

Website of School of Environmental Studies, Jadavpur University.

CHAPTER

19

CONCEPT OF GEOMEDICINE AND MEDICINAL MINERALOGY

Mineralogical factors of an environment have mainly some negative (irritating, toxic) effect on human health. Mineral-induced pathogeneses can be caused by a distant effect (radiation), a tactile one (contacting with easy soluble toxic minerals), through pneumonic (pneumoconiosises—silicosis, anthracosis, asbestosis, etc.) or food (toxic and traumatic exposure of mineral particles in the digestive tract) effects of minerals. The most natural reactions of a living organism on the interacting mineral individuals and aggregates are, commonly, destruction, rejection or conservation. The whole field of medical mineralogy includes preventive measures, mineralogical diagnostics of diseases, the mineralotherapy (psychotherapy and physiotherapy), mineralogical pharmacology, application of minerals in traumatology and as constituents of implantants, the mineral tooling and medicinal topomineralogy, which helps in prospecting deposits and resources of drug minerals.

19.1 INTRODUCTION

Humans live in lands. Most of them live in intimate contact with the immediate geological environment, obtaining their food and water directly from it. The unique geochemistry of these environments has a marked influence on their health, giving rise to diseases that affect millions of people.

A varied and large number of minerals can be treated in the branch of medical mineralogy, as they are either directly or indirectly related to various diseases. The origin of these diseases is geologic in nature as exemplified by dental and skeletal fluorosis, and iodine deficiency disorders. Trace element imbalances and excessive exposures to mineral dusts lead to the diseases like asbestosis, mesothelioma and lung cancers (asbestos); silicosis and lung cancer (silica dusts); and coal-workers pneumoconiosis (coal dust). Moreover exposure to dusts or soils containing pathogens has been assumed to be the cause of regionally common diseases such as valley fever (coccidioidomycosis) and much rarer diseases such as anthrax, to name a few.

Medical Geology, the study of the impacts of geologic materials and processes on animal and human health, is a dynamic emerging discipline bringing together the geoscience, biomedical, and public health communities to solve a wide range of environmental health problems. The branch examines the impacts that geologic materials and processes have on human and ecosystem health.

Medicinal Mineralogy and the Geological Impact

The branch of medicinal mineralogy is gaining wider attention and concern nowadays. The development and modernization of technology leads to the precise assessment and correlating the various problems with their related effects. After understanding the problems faced due to the interaction with minerals, trace elements, gases and observing their immediate effects on environment, people and livestock, the need of emerging a science, dealing with the medicinal aspect, related to mineralogy and their detection and solution came into being.

Some common problems related to minerals can be summarized as follows:

- During smelting, soil pollution may take place with lead, zinc, copper, cadmium, arsenic, uranium and others, which in turn leads to water pollution and plant contamination.
- Over-exposure to manganese can lead to nerve and brain damage in miners.
- Use of pesticides can have adverse effects in humans like suppression of immune system, lowering an organism's resistance to disease.

Medical Geology

1. To identify geochemical anomalies in soils, sediments, and water that may adversely impact human and animal health;
2. To identify the environmental causes of known health problems and, in collaboration with biomedical/public health researchers, seek solutions to prevent or minimize these problems;
3. To evaluate the beneficial health effects of geologic materials and process;
4. To reassure the public when there are unwarranted environmental health concerns associated with geologic materials or processes; and
5. To forge links between developed and developing countries to find solutions for environmental health problems.

Among the environmental health problems that geologists and the medical community need to collaborate on include: exposure to natural dust and to radioactivity; exposure to toxic levels of trace essential and non-essential elements such as arsenic and mercury; nutrient trace element deficiencies; naturally occurring toxic organic and inorganic compounds in drinking water;

identification and effects of volcanic emissions, etc. Geoscientists have also developed an array of tools and databases that can be used by the environmental health community to study vector-borne diseases, to model the dispersion of pollutants in surface and ground water and in the air, and can be applied to occupational health problems resulting from exposure to minerals.

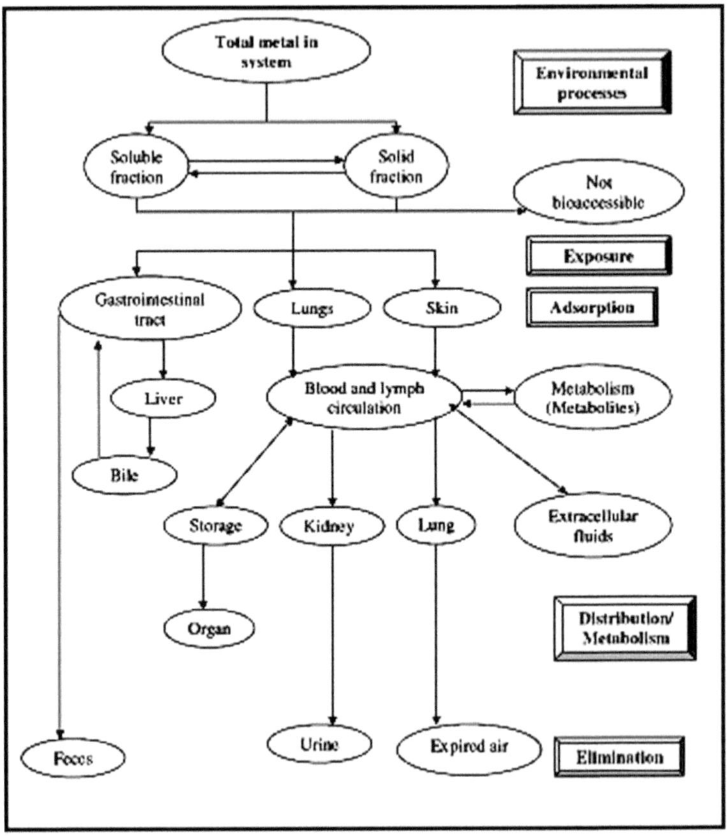

Fig. 19.1: Pathway and effected organs due to metals and pollutants.

19.2 GEOMEDICINE RELATED TO VARIOUS ELEMENTS AND MINERALS

Trace Elements

Trace elements play an essential role in the normal metabolism and physiological functions of animals and humans. Some 22 such elements are known or thought to be "essential" for humans and other animals. "Macronutrients" are required in fairly large amounts (e.g., grams per kilogram of diet), whereas "micronutrients" are required in much smaller amounts (e.g., microgram to milligrams per kilogram of diet). Sixteen elements are established as being essential for good health. Calcium, phosphorus, magnesium and fluoride for example, are required for structural functions in bone and membranes. Sodium, potassium and chloride are required for the maintenance of water and electrolyte balance in cells. Zinc, copper, selenium, manganese and molybdenum are essential constituents of enzymes or serve as carriers (iron) for ligands essential in metabolism. Chemical elements are also important in the functioning of the endocrine system. For instance, iodine is an essential component of the thyroid hormone thyroxine, and chromium is the central atom of the hormone-like glucose tolerance factor. Because these are all critical life functions, the tissue levels of many "nutritionally essential elements" tend to be regulated within certain ranges, and dependent on several physiological processes, especially homeostatic control of enteric absorption and tissue storage. Food is a major source of trace elements in humans and animals. However, other sources such as the deliberate eating of soil (geophagia) and water supplies may also contribute to dietary intake of trace elements.

Arsenic

Overexposure to arsenic is one of the most widespread medical geology problems affecting more than one hundred million people in Bangladesh, India, China, Europe, Africa and North and South America. The arsenic exposure is primarily due to naturally high levels in groundwater but combustion of mineralized coal has also caused arsenic poisoning. Arsenic can produce all three types of toxicity at different dosages: acute, sub-acute and chronic.

Source

- Volcanic emissions and hot springs.
- Found in ground water in dissolved state, resulting from interaction with rock (mobilization from igneous and sedimentary sources, oxidative dissolution of arsenic-bearing sulphide minerals).
- Mining waste, pH-mediated mine effluents and tailing ponds.
- Arsenic containing pesticides (sodium arsenite or lead arsenate).
- Organic arsenic compounds as herbicides, like monosodium methanoarsenate (MSMA), disodium methanoarsenate (DSMA), arsenic acid and dimethylarsenic acid.
- Waste from industrial metal smelting processes.
- Leaching of wood preservatives: chromated copper arsenate (CCA) and ammoniacal copper arsenate.
- Combustion of fossil fuels in electrical power plants.
- Arsenic dusts and gases, released during cement manufacture.
- Animal waste management from feed additives in poultry.
- Arsenic trioxide waste from glass manufacturing process.

Health Effects

General health effects that are associated with arsenic exposure include cardiovascular and peripheral vascular disease, developmental anomalies, neurologic and neurobehavioural disorders, diabetes, hearing loss, portal fibrosis, hematologic disorders (anemia, leukopenia and eosinophilia) and multiple cancers. One sign of acute exposure is edema of the eyelids, and gastrointestinal irritation, and both central and peripheral neuropathies frequently occur. During chronic intoxication "garlic breath", skin sensitivity, dermatitis and keratitus frequently occur. All types of arsenic exposure can cause kidney and liver damage, and in the most severe exposure there is erythrocyte hemolysis. There are significantly higher standardized mortality rates and cumulative mortality rates for cancers of the skin, lung, liver, urinary bladder, kidney and colon in many areas of arsenic pollution.

Precautionary Measures

- Water testing to detect arsenic level in the environment.
- If drinking water contains arsenic, look for an alternative source of water, which has no arsenic contaminants.
- Investing in an arsenic water filter is also a proper safety measure. This can reduce the level of arsenic content from drinking water.
- In a rural area low cost means of using chemical packets for reducing the arsenic content can be used.
- Availability of low arsenic sources of water like rainwater or treated surface water can be a good option for keeping away the health hazards or arsenic contamination in drinking water.
- Arsenic and its compounds should be stored in tightly closed containers in a cool, dry area, under dry atmosphere.

Beryllium

Source

- Beryllium is found in 30 different minerals, the most important of which are bertrandite, beryl, chrysoberyl and phenacite.
- Precious forms of beryl are aquamarine and emerald.
- The most important commercial sources of beryllium and its compounds are beryl and bertrandite.
- Currently, most production of this metal is accomplished by reducing beryllium fluoride with magnesium metal.
- Although the use of beryllium compounds in fluorescent lighting tubes was discontinued in 1949, potential for exposure to beryllium exists in the nuclear and aerospace industries and in the refining of beryllium metal and melting of beryllium-containing alloys, the manufacturing of electronic devices, and the handling of other beryllium-containing material.

Health Effect

- The recommended peak limit for beryllium absorption is 25 $\mu g/m^3$. Absorption above that limit can lead to chest radiographic abnormalities, compatible with interstitial disease, hypoxemia.

- The combination of hypoxemia and radiographic evidence of pulmonary interstitial process is termed as berylliosis.
- Beryllium disease erodes the lungs, making it hard for a victim to even walk across a room without severe pain and exhaustion, and usually results in a slow, painful death by suffocation.
- Some people (1-15%) become sensitive to beryllium. These individuals may develop an inflammatory reaction in the respiratory system. This condition is called chronic beryllium disease (CBD), and can occur many years after exposure to higher than normal levels of beryllium (greater than 0.2 $\mu g/m^3$). This disease can make people feel weak and tired, and can cause difficulty in breathing. It can also result in anorexia, weight loss, and may also lead to right side heart enlargement and heart disease in advanced cases.
- Some people who are sensitized to beryllium may not have any symptoms. The general population is unlikely to develop acute or chronic beryllium disease because ambient air levels of beryllium are normally very low (0.00003-0.0002 $\mu g/m^3$).

Precautionary Measures

- Early researchers tasted beryllium and its various compounds for sweetness in order to verify its presence. Modern diagnostic equipment no longer necessitates this highly risky procedure and no attempt should be made to ingest this substance.
- Beryllium and its compounds should be handled with great care and special precautions must be taken when carrying out any activity which could result in the release of beryllium dust (lung cancer is a possible result of prolonged exposure to beryllium-laden dust).
- This substance can be handled safely *if* certain procedures are followed. No attempt should be made to work with beryllium before familiarization with correct handling procedures.

Fluorine

Fluorine is the most electronegative and reactive of all elements. Fluorine (F) is a constituent of some minerals.

Source

- Fluorine is found in Bengal gram; cereals—especially rice; some leafy vegetables; and most of all in dry tea leaves.
- Fluorine is found in small amounts in normal bones and teeth.
- Since water containing 1 to 2 parts per million (ppm) prevents dental caries and does not do any harm, the fluorine requirements of the body are met by the quantity normally present in drinking water in most regions.

Health Effects

Beneficial

- It is now added to drinking water to minimize the development of dental caries, especially in children.
- Apart from the beneficial effects of maintaining a healthy oral cavity to aid mastication and minimize pain, it is probable that ingestion of fluorine in small amounts (parts per million) over a life-time will stave off osteoporosis, or at least serve to preserve the mineral materials in the skeleton in old age.
- It was the recognition of a connection between high natural fluorine concentrations (100 ppm) in the drinking waters of certain localities in Oklahoma and India and over-abundant calcium phosphate mineral deposition in the skeleton that most clearly illustrates the essential and continuing basic interactions between geology, geochemistry, medicine and biochemistry.
- Fluoride (F^-), the ionic form of fluorine, can stimulate bone formation and it also has been demonstrated to reduce dental caries at doses of at least 0.7 mg/L in drinking water.

Harmful

- Excess fluoride exposure can cause fluorosis of the enamel (mottling of the teeth) and bone (skeletal fluorosis).
- Typical symptoms of fluorosis include mottling of tooth enamel and various forms of skeletal fluorosis including osteosclerosis, limited movement of the joints, and outward manifestations such as knock-knees, bow legs and spinal curvature.
- Fluorosis combined with nutritional deficiencies in children can result in severe bone deformation.
- The etiology of fluorosis is similar to that of arseniasis in that the disease is derived from foods dried over coal-burning stoves.
- Adsorption of fluorine by corn dried over unvented ovens burning high ($\geq$ 200 ppm) fluorine coal is the probable cause of the extensive dental and skeletal fluorosis in southwest China. The problem is compounded by the use of clay as a binder for making briquettes. The clay used is a high-fluorine (mean value of 903 ppm) residue formed by intense leaching of a limestone substrate.

Precautionary Measures

- Fluorine and HF must be handled with great care and any contact with skin and eyes should be strictly avoided.
- Both elemental fluorine and fluoride ions are highly toxic. When it is a free element, fluorine has a characteristic pungent odour that is detectable in concentrations as low as 20 ppb.
- It is recommended that the maximum allowable concentration for a daily 8-hour time-weighted exposure is 1 ppm (lower than e.g. prussic acid).
- Safe handling procedures enable the transport of liquid fluorine by the ton.

Iodine

Low concentrations of iodine (I) characterize the soils and rocks at high elevations and in limestone terrains. This is a natural global phenomenon. Medical acumen and geostatistical and epidemiological investigations have identified iodine as an essential nutrient.

Sources

- Iodine is found in seawater, so any type of seafood is a rich source of this element, particularly seaweed (e.g. kelp).
- Despite coming from the ocean, sea salt is not a good source of iodine. Iodised salt is perhaps the most common source of iodine in the Western diet and can provide enough iodine to avoid low thyroid activity.
- Since an adult requires only around one teaspoonful of iodine over a lifetime, eating fish once a week is enough to fulfil the average iodine requirement.
- Some vegetables also contain iodine, but only if they are grown in iodine-rich soils. Certain regions of Australia, such as Tasmania and the Australian Capital Territory, have low levels of iodine in the soil.

Health Effects

Iodine Deficiency Disorders (IDD) includes *goiter* (enlargement of the thyroid gland), *cretinism* (mental retardation with physical deformities), reduced IQ, miscarriages and birth defects.

Goiter is still a serious disease in many parts of the world. China alone has 425 million people at risk of IDD. In all, more than a billion people, mostly living in the developing countries, are at risk of IDD. In all the places where the risk of IDD is high, the content of iodine in drinking water is very low because of low concentrations of iodine in bedrock.

A recent study has indicated that iodine volatilized from residential coal combustion in Guizhou Province China may be substantially reducing the incidence of IDD in the region. Reduction, but unfortunately not eradication, of this preventable malady is now possible through the use of iodine-enriched table salt and oils.

Precautionary Measures

- Use lead foil (~3 mm thick) to shield mCi quantities of I-125.
- Handle multi-mCi quantities with remote handling tools.
- Use double gloves when working with volatile NaI.
- Handle volatile NaI in fume hood.
- A radioiodine fume hood is available through the Radiation Safety Office.
- Film badges are required for work with I-125, except when using the I-125 kits. Ring badges are required for work with quantities greater than 1 mCi.
- Use GM detector or liquid scintillation counter to monitor for I-125.
- Keep containers of NaI closed when not in use to minimize release of volatile material.

- A thyroid scan is required within 24 hours, if possible, but not later than 72 hours after working with the following limits or greater:
  - processes in open room or bench with possible escape of iodine from process vessels:
    (*a*) 1 mCi, if volatile form and
    (*b*) 10 mCi, if bound to non-volatile agent
  - processes with possible escape of iodine carried out within a fume hood of adequate design, face velocity, and performance reliability:
    (*a*) 10 mCi, if volatile form and
    (*b*) 100 mCi, if bound to non-volatile agent
  - processes carried out within glove-boxes, ordinarily closed, but with possible release of iodine from processes and occasional exposure to contaminated box and box leakage:
    (*a*) 100 mCi, if volatile form and
    (*b*) 1000 mCi, if bound to non-volatile agent
- Seal all waste containers with NaI to prevent release of volatile material.
- Dispose off radioactive waste in accordance with requirements.

Selenium

Selenium is an essential trace element having antioxidant protective functions as well as redox and thyroid hormone regulation properties. However, selenium deficiency (due to soils low in selenium) has been shown to cause severe physiological impairment and organ damage such as a juvenile cardiomyopathy (Keshan disease) and muscular abnormalities in adults (Kaschin-Beck disease). In the 1960s scientists suspected that the diseases were of geological origin and in the 1970s the probable solution was found. The diseases were always located in areas with low selenium soils. The use of selenium in prevention and treatment of the diseases was a great success.

Heavy Metals

Toxic metal pollution can be a much more serious and insidious problem, as these are intrinsic components of the environment. At high concentrations, all the metals are toxic to animals and plants both. Metals are omnipresent in the environment occurring in varying concentrations in parent rock, soil, water, air and all biological matter. Moreover, metals are also released into environment from a wide spectrum of anthropogenic sources such as smelting of metallic ores, industrial fabrication and commercial application of metals, agro-chemicals, pesticides as well as burning of fossil fuels. These metals are redistributed in the biosphere and dispersed in the air, soil, water and consequently in human beings through food chain bio-magnification causing chronic ailments.

Metals are significant to human because some of them are most important trace elements as a co-factor in various metabolic enzymes and constituents of cells. Zinc, copper and iron form important components of cell and co-factor in several metalloenzymes, while organically chelated chromium (Cr^{3+}) compound act as a co-factor in insulin hormone response controlling carbohydrate metabolism in human. Some non-essential metals are present in all tissue and organs of man like under normal conditions more than 90% of lead (Pb) is retained in the skeleton. Human can tolerate 100-200 times its total body content of Cr without harmful effects. However, increased concentration of these metals can affect mineral and enzyme status of human beings. The metals irreversibly bind to active sites of enzymes, thereby destroying normal metabolism producing high-level toxicity. Human can be affected directly by air, water and soil metal pollutants as well as indirectly through contaminated food supplies. In recent past, a large number of silent epidemics have been reported due to metal contamination, which is escalating day by day.

Cadmium

Cadmium is biopersistent and, once absorbed by an organism, remains resident for many years (over decades for humans) although it is eventually excreted. The average daily intake for humans is estimated as 0.15 μg from air and 1 μg from water. Smoking a packet of 20 cigarettes can lead to the inhalation of around 2-4 μg of cadmium, but levels may vary widely.

Source—Natural

- Accumulation in sedimentary rocks, marine phosphates and phosphorites.
- Weathering and erosion of parent rocks and subsequent transport by river water lead to concentration of around 15, 000 mt per annum of cadmium in world's oceans (WHO, 1992; OECD, 1994).
- Volcanic activity, accounting for about 820 mt p.a of cadmium in the atmosphere (WHO, 1992; OECD, 1994; Nriagu, 1980, 1989).
- Forest fires accounting for 1 to 70 mt p. a of Cd to the atmosphere (Nriagu, 1980).

Source—Anthropogenic

- From the manufacture, use and disposal of products, intentionally utilizing cadmium, like Ni-Cd batteries, Cd-pigmented plastics, ceramics, glasses, paints and enamels, Cd-coated ferrous and non-ferrous products, Cd alloys and cadmium electronic compounds.
- From the presence of cadmium as a natural but not functional impurity in non-cadmium containing products, like non-ferrous metals and alloys of Zn, Pb and Cu; iron and steel; fossil fuels, cement and phosphate fertilizers.

Health Effects

In humans, long-term exposure is associated with renal disfunction. High exposure can lead to obstructive lung disease and has been linked to lung cancer, although data concerning the latter are difficult to interpret due to compounding factors. Cadmium may also produce bone defects (osteomalacia, osteoporosis) in humans and animals. In addition, the metal can be linked to increased blood pressure and effects on the myocardium in animals, although most human data do not support these findings.

Chromium

Chromium (VI) compounds (chromates, bichromates, chromic acid) are more dangerous than other compounds and are highly potent human carcinogen.

Source

- Occupational exposure from production of chromate, stainless steel, chromes plating.
- Air emissions and water effluents from
  - ferrochrome production
  - ore refining
  - tanning industries
  - chemical manufacturing industries (like dyes for paints, rubber and plastic products)
  - metal-finishing industries (chrome plating)
  - pharmaceutical manufacturing
  - wood, stone, clay and glass products
  - electrical and aircraft manufacturing
  - steam and air condition supply services
  - cement production
- Air emissions from incineration of refuse and sewage sludge
- Combustion of oil and coal
- Oxidation and leaching from stainless steel into a water-soluble form
- Motor vehicles exhaust and emissions from automobile brake linings and catalytic converters
- Tobacco smoke

Chromium Exposure

A study commissioned by the New Zealand government recommended limits for chromium VI to be 1.1 ng/m^3 as an annual average and 6.7 ng/m^3 for a 1-hour exposure. For chromium metal and chromium II and III, the study recommended a limit of 110 ng/m^3 as an annual average and 670 ng/m^3 for a 1-hour exposure. Overall chromium exposure from domestic coal burning is unlikely to exceed 13 ng/m^3.

Health Effects

Long-term health effects to inhaled chromium are damage to the nose, including ulcers and holes in the tissue separating the nostrils; irritation of the lungs; kidney damage; allergic reactions in the skin and respiratory tract; possible risk of cancer of the lung and nose.

Lead

Lead is one of the few natural substances that have no known use in the human body. At even very low levels, lead has been shown to cause health problems. The difficulty with lead is that once it is mined from the earth, there is no known way to destroy or make it harmless. In most individuals there is a "lead balance", that is one excretes as much as they take in, and the tissue levels are below the concentrations which result in pathological changes. However an increase in the rate of intake will result in accumulation or a "positive lead balance". Since lead is chemically very similar to calcium, it is handled by the body as if it were calcium. Thus the first place to which it is transported is to the plasma and the membrane sites in soft tissues. It is then distributed to the other sites where calcium plays an important role, most notably in the teeth of developing children and in bone at all ages.

Source

Paint: Lead was used in paint to add colour, improve the ability of the paint to hide the surface it covers, and to make it last longer. Lead-based paint becomes dangerous when it chips, turns into dust, or gets into the soil.

Dust: Lead dust is the most common way that people are exposed to lead. Inside the home, most lead dust comes from chipping and flaking paint or when paint is scraped, sanded, or disturbed during home remodeling. Chipping and peeling paint is found mostly on surfaces that rub or bump up against another surface. These surfaces include doors and windows. Young children usually get exposed to lead when they put something with lead dust on it into their mouths. Lead dust may not be visible to the naked eye.

Association with ore: Areas which have deposits of gold, zinc, and other economically useful metals, also have lead as an ore contaminant and the "tailings" of the mining and purification of the ore often have a very high lead content.

Soil: Before 1978, companies used to add lead to gasoline. Lead particles escaped from car exhaust systems and went into the air. This lead fell to the ground and mixed with soil near roads. Today, lead still comes from metal smelting, battery manufacturing, and other factories that use lead. This lead gets into the air and then mixes with the soil near homes, especially if the home is near one of these sources. Flaking lead-based paint on the outside of buildings can also mix with the soil close to buildings.

Drinking water: Lead is unusual among drinking water contaminants in that it seldom occurs naturally in water supplies like rivers and lakes. Lead enters drinking water primarily as a result of the corrosion, or wearing away, of materials containing lead in the water distribution system and household or building plumbing. These materials include lead-based solder used to join copper pipe, brass and chrome plated brass faucets, and in some cases, pipes made of lead that connect houses and buildings to water mains.

Ground water: There are known to be many communities with significant lead, zinc, and manganese concentrations in the ground water.

Imported candies or foods, especially from Mexico, containing chili or tamarind: Lead can be found in candy, wrappers, pottery containers, and in certain ethnic foods, such as chapulines (dried grasshoppers).

The workplace and hobbies: People exposed to lead at work may bring lead home on their clothes, shoes, hair, or skin. Some jobs that expose people to lead include home improvement, painting and refinishing, car or radiator repair, plumbing, construction, welding and cutting, electronics, municipal waste incineration, battery manufacturing, lead compound manufacturing, rubber products and plastics manufacturing, lead smelting and refining, working in brass or bronze foundries, demolition, and working with scrap metal. Some hobbies also use lead. These hobbies include making pottery, stained glass, fish sinkers, and refinishing furniture.

Imported food in cans that are sealed with lead solder: In 1995, the United States banned the use of lead solder on cans. But lead solder can still be found on cans made in other countries. These cans usually have wide seams, and the silver-gray solder along the seams contains the lead. Cans containing lead may be brought to the United States and sold. Over time the lead gets into the food. This happens faster after the can has been opened. Foods that are acidic cause lead to get into the food faster.

Lead-glazed ceramics, china, leaded crystal glassware: Lead may get into foods or liquids that have been stored in ceramics, pottery, china or crystal with lead in it. Lead-glazed dishes usually come from other countries.

Some other common sources of lead: Batteries, radiators for cars and trucks, and some colours of ink also contain lead.

Health Effects

Acute toxicity: Following ingestion of a large amount of lead, there will be direct tissue interaction. This includes tissue dessication, mucosal tissue damage in the GI tract, and convulsion possibly resulting in death. The most sensitive system is the hematopoietic (blood forming) system, with hypochromic microcytic anemia common. The biosynthesis of hemes in general is deranged by the presence of lead. All actively dividing cells are especially susceptible, hence acute intoxication has major potential for GI and renal mucosal damage. In addition there is a high risk of neurological damage. With acute lead poisoning use of intravenous EDTA is the preferred treatment modality, often supplemented with oral penicillamine, and sometimes with intramuscular injections of dimercaprol.

Chronic lead intoxication: With a gradual build-up of a positive lead balance there is no sudden onset of symptoms as seen with acute poisoning. The initial symptoms include clumsiness, ataxia, vertigo, irritability and insomnia. In affected children, they are often considered "slow", the real basis for the difficulty is not recognized. As the lead levels rise, hyper-excitability is seen.

Confusion, delerium and convulsions may occur in some cases, while in others there is progressive lethargy leading to a comatose state.

One of the earliest diagnostic signs present is the appearance of "lead lines" at the gingival border in the mouth. This occurs because the lead following calcium pathways is secreted with the saliva. It then is involved in a reaction with oral bacteria which produce sulfides. The lead reacts with these compounds to form a purplish, or black lead sulfide deposit which precipitates in the region of highest concentration, the "protected area" at the gingival border. Other metals also produce this phenomenon, but with differing colours for the deposit.

Precautionary Measures

- Assess the risk to worker's health to decide whether or not lead exposure is 'significant' (the law explains what this means), and what precautions are needed to protect the health.
- Put in place systems of work and other controls, such as fume and dust extraction, to prevent or control exposure to lead, and keep equipment in efficient working order.
- Provide washing and changing facilities, and places free from lead contamination where workers can eat and drink.
- Concern about the health risks from working with lead and the precautions one should take.
- Train to use any control measures and protective equipment correctly.
- Checking the level of lead in the atmosphere, which should be below the occupational exposure limit.
- Protective clothing should be used.
- Lead level must be regularly measured in the body.

Mercury

Mercury can combine with a methyl group to become methyl mercury. This form of mercury is found in a variety of environmental pollution situations and can produce a range of toxicities. Elemental mercury is less labile but produces a similar set of toxic manifestations.

Source

- Atmospheric deposition is the dominant source of mercury over most of the landscape. Once in the atmosphere, mercury is widely disseminated and can circulate for years, accounting for its wide-spread distribution.
- Natural sources of atmospheric mercury include volcanoes, geologic deposits of mercury, and volatilization from the ocean.
- Although all rocks, sediments, water and soils naturally contain small but varying amounts of mercury, scientists have found some local mineral occurrences and thermal springs that are naturally high in mercury.

- Alkali and metal processing, incineration of coal, medical and other waste, and mining of gold and mercury contribute greatly to mercury concentrations in some areas.

Mercury Exposure

Methylation is a product of complex processes that move and transform mercury. Atmospheric deposition contains the three principal forms of mercury, although inorganic divalent mercury (HgII) is the dominant form. Once in surface water, mercury enters a complex cycle in which one form can be converted to another. Mercury attached to particles can settle onto the sediments where it can diffuse into the water column, be resuspended, be buried by other sediments, or be methylated. Methylmercury can enter the food chain, or it can be released back to the atmosphere by volatilization.

The concentration of dissolved organic carbon (DOC) and pH have a strong effect on the ultimate fate of mercury in an ecosystem. Studies have shown that for the same species of fish taken from the same region, increasing the acidity of the water (decreasing pH) and/or the DOC content generally results in higher mercury levels in fish, an indicator of greater net methylation. Higher acidity and DOC levels enhance the mobility of mercury in the environment, thus making it more likely to enter the food chain.

Mercury and methylmercury exposure to sunlight (specifically ultra-violet light) has an overall detoxifying effect. Sunlight can break down methylmercury to Hg(II) or Hg(0), which can leave the aquatic environment and re-enter the atmosphere as a gas.

People are exposed to methylmercury almost entirely by eating contaminated fish and wildlife that are at the top of aquatic foodchains. The National Research Council, in its 2000 report on the toxicological effects of methylmercury, pointed out that the population at highest risk is the offspring of women who consume large amounts of fish and seafood. The report went on to estimate that more than 60,000 children are born each year at risk for adverse neurodevelopmental effects due to in utero exposure to methylmercury. In its 1997 Mercury Study Report to Congress, the U.S. Environmental Protection Agency concluded that mercury also may pose a risk to some adults and wildlife populations that consume large amounts of fish that is contaminated by mercury.

Health Effects

- Methylmercury [CH_3Hg] is the most toxic form. It affects the immune system, alters genetic and enzyme systems, and damages the nervous system, including coordination and the senses of touch, taste and sight. Methylmercury is particularly damaging to developing embryos, which are five to ten times more sensitive than adults. Exposure to methylmercury is usually by ingestion, and it is absorbed more readily and excreted more slowly than other forms of mercury.

- Elemental mercury, Hg(0), the form released from broken thermometers, causes tremors, gingivitis, and excitability when vapours are inhaled over a long period of time.
- Ingestion of other common forms of mercury, such as the salt $HgCl_2$, damages the gastrointestinal tract and causes kidney failure.
- *Acute toxicity:* Organic or inorganic mercury can both precipitate protein in a local reaction. In the GI tract, acute poisoning produces a sloughing away of the mucosa to an extent where pieces of the intestinal mucosa can be found in the stools. This produces a large loss of fluids and electrolytes. Mercury also breaks down barriers in the capillaries. This results in edema throughout the body. A range of neurological toxicities are also common. These include lethargy (at low doses), excitement, hyper-reflexia and tremor.
- *Chronic toxicity:* Often a psychotic state resulting in hyper-excitability. The expression "Mad as a Hatter" originates from the hat-makers of the 19th century who were chronically exposed to mercury compounds used in making felt hats. In chronic intoxication there is mercury line at the gingival border similar to the "lead line". Mercury is especially poisonous to rapidly growing tissue. A common effect is deterioration of alveolar bone in the jaw, with a subsequent loosening of the teeth. There are also substantial liver and kidney toxicity because of mucosal degeneration.

Precautionary Measures

- Use mercury in uncarpeted and well ventilated areas. Provide troughs on smooth surfaced tables and benches to collect mercury spills. Reserve a room for mercury use only.
- Ask workers to remove all jewellery and watches, especially gold. Mercury readily combines with gold. Workers who handle mercury are to wear mercury vapour respirators and protective clothing: gloves, disposable gowns and shoe coverings.
- Prohibit smoking and eating in or near mercury exposed areas.
- Train employees to understand the dangers and precautions with handling mercury. Also train employees on the properties and hazards of mercury. Proper training on how to dispose off mercury will contribute to the prevention of environmental exposure.
- Clean and calibrate all mercury-containing equipment to the specifications of the manufacturer.
- Properly document and label all containers of mercury in addition to having the proper MSDS (Material Safety Data Sheet) available.
- Have an emergency spill and containment plan in case a spill does occur.
- Ask your safety supply vendor for a mercury vacuum sweeper in addition to an emergency containment and clean up kit.
- Avoid having chemicals such as chlorine dioxide, nitric acid, nitrates, ethylene oxide, chlorine and methylazide in the same area as mercury since they will react violently with mercury.

- Be sure to keep mercury away from biological waste or anything else that will be incinerated since incineration puts mercury vapour into the air.

Manganese

This heavy metal is frequently associated with iron deposits.

Manganese can cause *manganism*, an irreversible neurological disorder similar to Parkinson's disease. Occupational exposures occur mainly in welding, mining as miners are surrounded by manganese dust and air-borne manganese particles, alloy production, processing, ferro-manganese operations especially in which manganese ore or manganese compounds are turned into steel, and work with agrochemicals. The towns and communities surrounding the areas of manganese heavy industry could also become affected by toxic exposure to manganese. It is also hypothesized that long-term exposure to the naturally-occurring manganese in shower water also puts people at risk. Manganese inhibits tyrosine hydroxylation, which is essential for the formation of dopamine. So manganese may cause Parkinson's disease by lowering dopamine levels and is treated with some success using typical anti-Parkinson drugs. While the concentration in the surface water does not exceed official public health limits, it is in the high range of acceptable levels. It would be interesting to see if the epidemiology of Parkinsons syndrome in that locality is different from regions in the province with lower manganese concentrations. Manganese excretion can be facilitated by the use of chelator therapy, but it is not as easily managed as lead or mercury, because it is further down the list of chelator affinities.

Uranium

Source

- It occurs in numerous minerals such as pitchblende, uraninite, carnotite, autunite, uranophane and tobernite.
- It is also found in phosphate rock, lignite, monazite sands, and can be recovered commercially from these sources.
- Uranium can be prepared by reducing uranium halides with alkali or alkaline earth metals or by reducing uranium oxides by calcium, aluminum or carbon at high temperatures.
- The metal can also be produced by electrolysis of KUF5 or UF4, dissolved in a molten mixture of $CaCl_2$ and NaCl.
- High-purity uranium can be prepared by the thermal decomposition of uranium halides on a hot filament.
- Uranium has sixteen isotopes, all of which are radioactive. Naturally occurring uranium nominally contains 99.28305 by weight 238U, 0.7110% 235U, and 0.0054% 234U. Studies show that the percentage weight of 235U in natural uranium varies by as much as 0.1%, depending on the source.

Health Effects

- Uranium toxicity was recognized early. This led to its wide use in experimental pathophysiology, mainly in producing experimental glomerulonephritis.
- Renal damage, both structural and functional, is confined to the terminal third of the proximal convoluted tubule, even after small doses of uranium salts, whereas glycosuria, initially considered a consequence of liver injury, was later thought to be of renal origin. This is also true for hematuria, albuminuria, hyaline and granular casts, azotemia, and tubular necrosis.
- The toxic effects of uranium were shown to be enhanced by the administration of calcium or ephedrine, whereas the administration of adrenaline produced less severe toxic effects.
- Uranium poisoning can be associated with chronic Bright's disease, ranging from nephromegaly with tubular and glomerular alterations to the small granular kidneys, with the functional pattern of polyuria, albuminuria, tubular casts, glycosuria, oliguria and terminal anuria.
- Small doses of uranium caused glomerular injury, with coagulation necrosis glomerular and capsular edema, efferent vessels obstruction, and hyaline degeneration.
- The blood changes in uranium exposure have been well demonstrated. These include: (1) increased nitrogen retention; (2) a decrease of serum albumin and largely unchanged serum proteins; and (3) increased serum ammonia creatinine, blood urea nitrogen and uric acid.
- Sodium and chloride concentrations were lowered, calcium, potassium, phosphate and magnesium were not changed, whereas total lipids and cholesterol, as well as blood glucose, appear to be elevated in uranium poisoning.

Precautionary Measures

- Dust is controlled, so as to minimise inhalation of gamma- or alpha-emitting minerals. In practice dust is the main source of radiation exposure in an open cut uranium mine and in the mill area.
- Radiation exposure of workers in the mine, plant and tailings areas is limited. In practice radiation levels from the ore and tailings are usually very low. At Olympic Dam, direct gamma exposure comprises about half the miners' dose and for those in the mill, a quarter.
- Radon daughter exposure is minimal in an open cast mine because there is sufficient natural ventilation to remove the radon gas. At Ranger the radon level seldom exceeds one percent of the levels allowable for continuous occupational exposure. In an underground mine a good forced-ventilation system is required to achieve the same result, e.g. at Olympic Dam radiation doses in the mine from radon daughters are kept very low, with an average of less than about 1 mSv/yr. Canadian doses (in mines with high-grade ore) average about 3 mSv/yr.

- Strict hygiene standards are imposed on workers handling the uranium oxide concentrate. If it is ingested it has a chemical toxicity similar to that of lead oxide. (Both lead and uranium are toxic and affect the kidney. The body progressively eliminates most Pb or U, via the urine.) In effect, the same precautions are taken as in a lead smelter, with use of respiratory protection in particular areas identified by air monitoring.

Gases

Carbon Disulfide

Humans, exposed to high CS_2 level suffer from abnormality in speed, vigilance, manual dexterity and intelligence. They also show impairment. Latent poisoning was characterized by traits indicative of depressive mood, slight motor disturbances and intellectual impairment, whereas clinically manifested poisoning resulted in lowered vigilance, diminished intellectual activity, diminished rational control, retarded speed and motor disturbances.

Radon

Radon is a naturally occurring colourless, odourless gas that is emitted from rocks containing minerals rich in the transuranic elements. The occupational health effects, in particular lung cancer, suffered by some European coal-miners who work in such rocks were ascribed to radiation, but may equally well have been induced by smoking. Granites that underlie portions of the north-east of the United States (New England) are known to contain minerals that emit radon. Recent epidemiological studies that measured environmental exposure (the average was less than 4 Pico curies for the region) were not able to demonstrate an association between the incidence of lung cancer and sites where radon concentrations (possible doses) were elevated.

In its broadest sense, medical geology studies exposure to or deficiency of trace elements and minerals; inhalation of ambient and anthropogenic mineral dusts and volcanic emissions; transportation, modification and concentration of organic compounds; and exposure to radionuclides, microbes and pathogens.

Minerals

Asbestos

Asbestos is the name given to a group of minerals that occur naturally in the environment as bundles of fibres that can be separated into thin, durable threads. Asbestos minerals are divided into two major groups:

- *Serpentine asbestos:* Serpentine asbestos includes the mineral chrysotile, which has long, curly fibres that can be woven. Chrysotile asbestos is the form that has been used most widely in commercial applications.

- *Amphibole asbestos:* Amphibole asbestos includes the minerals actinolite, tremolite, anthophyllite, crocidolite and amosite. Amphibole asbestos has straight, needle-like fibres that are more brittle than those of serpentine asbestos and are more limited in their ability to be fabricated.

Source—Occupational

- The risks are greatest for workers in industries which produce and use asbestos, such as mining and milling. In the past, workers in these environments were exposed to 100-1000 times more asbestos than today's workers. Today's strict standards limit workers' exposure and the ban of most uses of amphibole asbestos have reduced the risks.
- During renovations and repairs to older buildings, construction workers, tradespeople and other building maintenance workers may be exposed to very high concentrations of asbestos fibres.

Source—Environmental

- Negligible levels of asbestos fibres are found in the soil, water and air, both naturally and from man-made sources. Asbestos concentrations in the air in rural areas are about ten times lower than those in larger cities, which are about 1000 times lower than levels accepted in today's asbestos-related jobs. With such low exposure, environmental risks are negligible.
- Due to natural erosion, high concentrations of chrysotile asbestos fibres may be found in some raw water supplies.

Source—Buildings and Homes

- Because it is a valuable reinforcing, insulating and fire-proofing material, asbestos was used widely in construction materials such as insulation board, asbestos cement, and floor and ceiling tiles. If products containing asbestos are disturbed, tiny asbestos fibres are released into the air. When asbestos fibers are breathed in, they may get trapped in the lungs and remain there for a long time. Over time, these fibres can accumulate and cause scarring and inflammation, which can affect breathing and lead to serious health problems.
- Asbestos fibre concentrations in the air in buildings are usually about the same as in the air outside, and are not a significant risk. However, levels may be higher if friable asbestos materials are disturbed.
- There is also concern about vermiculite insulation which may contain small amounts of amphibole asbestos, principally tremolite or actinolite. These amphibole fibres may cause health risks if disturbed.

Asbestos Exposure

Asbestos poses health risks only when fibres are present in the air that people breathe. Exposure to asbestos depends on:

- the concentration of asbestos fibres in the air;
- how long the exposure lasted;
- how often you were exposed;
- the size of the asbestos fibres inhaled; or
- the amount of time since the initial exposure.

Health Effects

Asbestos has been classified as a known human carcinogen (a substance that causes cancer) by the U. S. Department of Health and Human. When inhaled in significant quantities, asbestos fibres can cause *asbestosis* (a scarring of the lungs which makes breathing difficult), *mesothelioma* (a rare cancer of the lining of the chest or abdominal cavity) and lung cancer. Smoking, combined with inhaled asbestos, greatly increases the risk of lung cancer. In addition to lung cancer and mesothelioma, some studies have suggested an association between asbestos exposure and gastrointestinal and colourectal cancers, as well as an elevated risk for cancers of the throat, kidney, oesophagus, and gallbladder. However, the evidence is inconclusive.

Asbestosis is an inflammatory condition affecting the lungs that can cause shortness of breath, coughing, and permanent lung damage and other nonmalignant lung and pleural disorders, including pleural plaques (changes in the membranes surrounding the lung), pleural thickening, and benign pleural effusions (abnormal collections of fluid between the thin layers of tissue lining the lungs and the wall of the chest cavity). Although pleural plaques are not precursors to lung cancer, evidence suggests that people with pleural disease caused by exposure to asbestos may be at increased risk for lung cancer.

Precautionary Measures

- *Medicinal detections:* Individuals who have been exposed (or suspect they have been exposed) to asbestos fibres should inform their doctor about their exposure history and whether or not they experience any symptoms. It is particularly important to check with a doctor if any of the following symptoms develop:
  - Shortness of breath, wheezing, or hoarseness.
  - A persistent cough that gets worse over time.
  - Blood in the sputum (fluid) coughed up from the lungs.
  - Pain or tightening in the chest.
  - Difficulty swallowing.
  - Swelling of the neck or face.
  - Loss of appetite.
  - Weight loss.
  - Fatigue or anemia.
- A thorough physical examination, including a chest X-ray and lung function tests, may be recommended. Studies have shown that computed tomography (CT) (a series of detailed pictures of areas inside the body taken from different

angles; the pictures are created by a computer linked to an X-ray machine) may be more effective than conventional chest X-rays at detecting asbestos-related lung abnormalities in individuals who have been exposed to asbestos.
- A lung biopsy, which detects microscopic asbestos fibres in pieces of lung tissue removed by surgery, is the most reliable test to confirm the presence of asbestos-related abnormalities. A bronchoscopy is a less invasive test than a biopsy and detects asbestos fibres in material that is rinsed out of the lungs. It is important to note that these tests cannot determine how much asbestos an individual may have been exposed to or whether disease will develop. Asbestos fibres can also be detected in urine, mucus, or faeces, but these tests are not reliable for determining how much asbestos may be in an individual's lungs.
- *Mine safety:* Workers should use all protective equipment provided by their employers and follow recommended workplace practices and safety procedures. For example, National Institute for Occupational Safety and Health (NIOSH)-approved respirators that fit properly should be worn by workers when required.

Clay Minerals

Bentonite, kaolin, and other clays often contain quartz, exposure to which is causally related to silicosis and lung cancer as well as to statistically significant increases in the incidence of or mortality from chronic bronchitis and pulmonary emphysema.

The extensive use of bentonite or kaolin in cosmetics has not been found to produce local or systemic adverse effects. Tests have also revealed that bentonite and kaolin have low toxicity to a wide variety of aquatic species.

The biological effects of clay minerals result not only from their composition but also their particle size. In this respect, the decreasing rank order of quartz, kaolinite, and montmorillonite to damage lung tissue is consistent with their known relative active surface areas and surface chemistries.

Kaolinite

- Long-term exposure to kaolin causes radiologically diagnosed pneumoconiosis but clear-cut deterioration of respiratory function and related symptoms occur only in cases with prominent radiological findings. In this respect, the composition of the non-kaolinite constituents of the clay is important.
- The common health hazard, prevalent among kaolin workers is kaolinosis. It generally results from 12-35 years of kaolin exposure and symptomized by pleural thickening.
- Eye contact may cause irritation, if exposed to large amount of dust.
- Skin contact may cause irritation.
- Inhalation may cause irritation to respiratory tract and lung damage if exposure is repeated or prolonged.

- In case of long-term exposure to extremely high levels of dust, progressive fibrosis may occur with lung function impairment.

Bentonite

Bentonite, unlike other clay minerals has both beneficial and harmful effects, as stated below.

Health Effects—Beneficial

- Bentonite can be applied topically (compress or poultice) to skin and wounds or taken orally to draw toxins and impurities out of the body quickly.
- Bentonite clay removes poisons (bacterial poisoning, salmonella, etc.) and intestinal parasites out of the intestinal tract while alkalizing the body.
- Bentonite clay binds herbicides and other potentially harmful substances by adsorbent action so they do not remain in the body.
- It cleanses the colon and aids constipation and diarrhea, as well as intestinal issues, including ulcers.
- Bentonite clay works well for irritable bowel syndrome.
- Bentonite clay is excellent to rejuvenate and cleanse the skin.
- Topical applications are helpful for acne, eczema, psoriasis, rashes, wound healing, and help draw poison from ant, bee, wasp, and similar bug stings and bites. This helps reduce inflammation, irritation, itching, and swelling and is excellent applied to hot spots.

Harmful

- Long-term occupational exposure to bentonite dust may cause structural and functional damage to the lungs, but currently available data do not conclusively establish a dose-response or even a cause-and-effect relationship, due to limited information on period and intensity of exposure and to confounding factors, such as exposure to silica and tobacco smoke.
- Eye contact may cause irritation.
- Skin contact may cause irritation.
- Exposure to dust levels, exceeding a permissible limit may cause irritation of respiratory tract, resulting in dry cough.
- Over-exposure of free silica containing air-borne bentonite dust may cause silicosis or other respiratory problems with persistent dry cough and troubled breathing.

Graphite

Graphite usually occurs in flakes in metamorphosed rocks rich in carbon, but it can also be found in veins and in pegmatites. Where large deposits are found it is mined and used as an industrial lubricant and for 'lead' in pencils.

Source

The major sources of graphite are in gneisses and schists, where the mineral occurs in foliated masses mixed with quartz, mica and so on. Noteworthy localities include the Adirondack region of New York, Korea and Ceylon. In Sonora, Mexico, graphite occurs as a product of metamorphosed coal beds. Graphite is also observed in meteorites.

Health Effects

- The men and women, working in graphite environment suffer from radiographic abnormalities with respiratory symptoms, the most common being dyspnea. Finger clubbing is also noted.
- It is toxic to upper respiratory tract. The substance may be toxic to cardiovascular system. Repeated or prolonged exposure to the substance can produce target organs damage.
- *Inhalation:* May cause irritation to the respiratory tract. Symptoms may include coughing and shortness of breath. May produce black sputum, decreased pulmonary function and lung fibrosis.
- *Ingestion:* Not expected to be a health hazard via ingestion. May cause irritation to the gastrointestinal tract. Symptoms may include nausea, vomiting and diarrhoea.
- *Skin contact:* May cause mild irritation and redness.
- *Eye contact:* May cause mild irritation, possible reddening.
- *Chronic exposure:* Chronic inhalation exposure to natural graphite is associated with the development of pneumoconiosis. Chronic symptoms include cough, dyspnea, wheezing, increased susceptibility to tuberculosis, decreased chest expansion, and repeated nonspecific chest illnesses.
- Progressive respiratory and cardiopulmonary impairment may be fatal. Chronic inhalation of crystalline silica is a lung cancer hazard.
- *Aggravation of pre-existing conditions:* Persons with pre-existing respiratory or cardiopulmonary problems may be more susceptible to the effects of this substance.

Precautionary Measures

Engineering controls: Use process enclosures, local exhaust ventilation, or other engineering controls to keep air-borne levels below recommended exposure limits. If user operations generate dust, fume or mist, use ventilation to keep exposure to air-borne contaminants below the exposure limit.

Personal protection: Safety glasses. Lab coat. Dust respirator. Be sure to use an approved/certified respirator or equivalent. Gloves.

Personal protection in case of a large spill: Splash goggles. Full suit. Dust respirator. Boots. Gloves. A self contained breathing apparatus should be used to avoid inhalation of the product. Suggested protective clothing might not be sufficient; consult a specialist before handling this product.

Silica

Source

- Silica used in commercial products is obtained mainly from natural sources (IARC, 1997). U.S. production of silica sand was estimated at 27.9 million metric tons (61.4 billion pounds) in 2001 and 28.5 million metric tons (62.7 billion pounds) in 1997.

- In the past, natural quartz crystals were mined in the United States; however, mining no longer occurs today. U.S. production of high purity quartz was 315,000 lb (143 metric tons) in 1979, decreased to 174,000 lb (79 metric tons) in 1981, and rose to 800,000 lb (364 metric tons) in 1983 (IARC, 1987).
- Today, synthetic quartz crystal (hypothermally cultured quartz crystals) are used as the raw material for quartz production. Lascas, the precursor material for synthetic quartz crystals, was mined in the United States for many years, but mining and processing ended in 1997.
- Lascas mining production was estimated at one million pounds (455 metric tons) in 1985 and 600,000 lb (273 metric tons) in 1988. Currently, three U.S. firms continue to produce cultured quartz crystals using imported and stockpiled lascas.
- In 2002, exports of cultured quartz crystals were estimated at 38 metric tons (84 million pounds) and imports at 14 metric tons (31 million pounds) (USGS, 2003).

Health Effects

Quartz can have potentially serious respiratory effects following long-term inhalation (one year or more). In general, quartz dust is not expected to be irritating to the skin. However, foreign-body reactions (granulomas) have been observed after crystalline silica accidentally got under the skin as a result of an injury. Often the effects are delayed for periods ranging from weeks up to more than 50 years.

In general, the dust is not expected to be irritating except as a "foreign object". Some tearing, blinking and mild temporary pain may occur as the solid material is rinsed from the eye by tears. One unconfirmed case is described in which foundry workers with silicosis experienced deterioration in eyesight due to corneal opacities and there was evidence of abnormally high silicon content in the cornea.

Inhalation: Prolonged or repeated exposure to fine air-borne crystalline silica dust may cause severe scarring of the lungs, a disease called silicosis. The risk of developing and the severity of silicosis depends on the air-borne concentration of respirable-size silica dust to which an employee is exposed and duration of exposure. Silicosis usually develops gradually over 20 years or more of exposure. Particles with diameters less than one micrometre and freshly cleaved particles (for example, those produced by sandblasting) are considered most hazardous. Several reliable studies have found silicosis in employees with exposure to considerably less than 1 mg/m^3 respirable quartz.

The early symptoms of silicosis (cough, mucous production and shortness of breath upon exertion) are nonspecific, so the development of silicosis may not be detected until advanced stages of the disease. Silicosis may continue to develop even after exposure to crystalline silica has stopped. Evidence of silicosis can normally be seen on an X-ray.

Silicosis can vary in severity from minimal to severe. In cases of mild silicosis, there is typically no significant respiratory impairment, although there is X-ray evidence of lung injury. In severe cases, significant and increasingly severe respiratory impairment develops. There is no proven effective treatment for the disease. Life expectancy may be reduced, depending on the severity of the case. Death is not usually a direct result of silicosis, but cardiac failure (cor pulmonale) may occur as the heart has increasing difficulty pumping blood through the scar tissue in the lungs. Silicosis may be complicated by the development of bacterial infections, including tuberculosis.

"Accelerated" silicosis results from exposure to high concentrations of crystalline silica over a period of 5 to 10 years. The disease continues to develop even after exposure stops and is often associated with autoimmune diseases, for example, scleroderma (a skin disease involving thickening of the skin).

"Acute" silicosis (also referred to as "silicotic alveolar proteinosis") is rare in humans, but can develop if very high concentrations of crystalline silica dust are inhaled over a relatively short period (1-2 years) and has occurred in occupations such as sandblasting or tunnelling where exposure controls were minimal. Acute silicosis may result in death within a few years, often with tuberculosis as a complication.

Inhalation of quartz has also been associated with a number of other, less well characterized, harmful effects including effects on the kidney (glomerulonephritis), the liver, the spleen and immune system disorders (progressive systemic sclerosis, scleroderma or rheumatoid arthritis).

The International Agency for Research on Cancer (IARC) has concluded that crystalline silica in the form of quartz or cristobalite from occupational sources should be classified as carcinogenic to humans (Group 1), upgraded from its previous classification as probably carcinogenic to humans (Group 2A). This conclusion was drawn on the basis of a relatively large number of human population studies that together provide sufficient evidence in humans for the carcinogenicity of inhaled crystalline silica. In many (although not all) of these studies, lung cancer risks were elevated and could not be explained by other factors. Recent reviews have tended to conclude that if exposures are controlled to prevent silicosis, they will probably also prevent cancer. The risk of developing silicosis depends on the air-borne concentration of crystalline silica, the particle size and the duration of exposure.

The US National Toxicology Program (NTP) identifies crystalline silica (respirable size) as a substance which may reasonably be anticipated to be a carcinogen. The American Conference of Governmental Industrial Hygienists (ACGIH) has not assigned a carcinogenicity designation for crystalline silica.

Talc

Talc is closely related to the potent carcinogen asbestos. Talc particles have been shown to cause tumors in the ovaries and lungs of cancer victims. Numerous studies have shown a strong link between frequent use of talc in the female genital area and ovarian cancer. Talc particles are able to move through the reproductive system and become imbedded in the lining of the ovary.

Talc poses a health risk when exposed to the lungs. Talc miners have shown higher rates of lung cancer and other respiratory illnesses from exposure to industrial grade talc, which contains dangerous silica and asbestos. The common household hazard posed by talc is inhalation of baby powder by infants.

Precautionary Measures

Engineering controls: Use process enclosures, local exhaust ventilation, or other engineering controls to keep air-borne levels below recommended exposure limits. If user operations generate dust, fume or mist, use ventilation to keep exposure to air-borne contaminants below the exposure limit.

Personal protection: Safety glasses. Lab coat. Dust respirator.

Personal protection in case of a large spill: Splash goggles. Full suit. Dust respirator. Boots. Gloves. A self contained breathing apparatus should be used to avoid inhalation of the product. Suggested protective clothing might not be sufficient; consult a specialist before handling this product.

Mineral Dust

Exposure to mineral dusts can cause a wide range of respiratory problems. These exposures can be due to local conditions such as the dusts generated by mining hard rocks or coal, use of fine-grained mineral matter in sand-blasting, and formation of smoke plumes from fires (both natural and man-made). Dust exposure can affect broad regions such as the dust stirred up by earthquakes in the arid regions of the southwestern U.S. and northern Mexico. This dust carries spores of a fungus (coccidioidis immetus) that causes Valley Fever, a serious respiratory problem that can lead to fatigue, cough, fever, rash, including damage to internal organs and tissues such as skin, bones and joints. Dust exposure can even take on global dimensions. Ash ejected from volcanic eruptions can travel many times around the world and recent satellite images have shown wind blown dust picked up from the Sahara and Gobi deserts blown more than halfway around the world. Of greatest concern for effects upon human health are the finer particles of the respirable (inhalable) dusts. In this regard, considerable work is being conducted in identifying dust particles derived from soils, sediments and weathered rock surfaces.

Coal

Coal pollutants damage human health at every stage of the coal life cycle.

Source

- Coal smoke has a varied profile of metal emissions and domestic coal emits metals in quantities potentially harmful to human health. Those of concern are: lead, arsenic, chromium and mercury but there are others which are poisonous and domestic smoke exposure combined with that from other sources may mean an exceedance of safe limits for many.
- Coal fly ashes (CFA) are complex particles of a variable composition, which is mainly dependent on the combustion process, the source of coal and the precipitation technique. Toxic constituents in these particles are considered to be metals, polycyclic aromatic hydrocarbons and silica. For most of the effects coal mine dust was chosen as a reference, since it contains up to 10% of crystalline silica (α-quartz) and is well studied. Epidemiological studies in fly ash exposed working populations have found no evidence for effects commonly seen in coal workers (pneumoconiosis, emphysema) with the exception of airway obstruction at high exposure. In conclusion, the available data suggest that the hazard of coal fly ash is not to be assessed by merely adding the hazards of individual components.
- Coal combustion releases mercury, particulate matter, nitrogen oxides, sulfur dioxide, and dozens of other substances known to be hazardous to human health.

Health Effects

- The acute health effects, resulting from coal mining are psychoses, psychoneuroses and acute bronchitis and chronic respiratory problem and pneumoconiosis.
- *Black lung disease*, also known as Coal workers' pneumoconiosis (CWP), is caused by long exposure to coal dust. It is a common affliction of coal miners and others who work with coal. Inhaled coal dust progressively builds up in the lungs and is unable to be removed by the body; that leads to inflammation, fibrosis, and in the worst case, necrosis. Coal workers' pneumoconiosis, in its most severe state, develops after the initial, milder form of the disease known as *anthracosis* (*anthrac*—coal, carbon). This is often asymptomatic and is found to at least some extent in all urban dwellers due to air pollution. Prolonged exposure to large amounts of coal dust can result in more serious forms of the disease, simple coal workers' pneumoconiosis and complicated coal workers' pneumoconiosis (or Progressive massive fibrosis, or PMF). More commonly, workers exposed to coal dust develop industrial bronchitis, clinically defined as chronic bronchitis (i.e. productive cough for three months per year for at least two years) associated with workplace dust exposure. The incidence of industrial bronchitis varies with age, job, exposure and smoking. In non-smokers (who are less prone to develop bronchitis than smokers), studies of coal miners have shown a 16% to 17% incidence of industrial bronchitis.
- *Respiratory effects:* Air pollutants produced by coal combustion act on the respiratory system, contributing to serious health effects including asthma,

lung disease and lung cancer, and adversely affect normal lung development in children.

- *Cardiovascular effects:* Pollutants produced by coal combustion lead to cardiovascular disease, such as arterial occlusion (artery blockages, leading to heart attacks) and infarct formation (tissue death due to oxygen deprivation, leading to permanent heart damage), as well as cardiac arrhythmias and congestive heart failure. Exposure to chronic air pollution over many years increases cardiovascular mortality.
- *Nervous system effects:* Studies show a correlation between coal-related air pollutants and stroke. Coal pollutants also act on the nervous system to cause loss of intellectual capacity, primarily through mercury. Researchers estimate that between 317,000 and 631,000 children are born in the U.S. each year with blood mercury levels high enough to reduce IQ scores and cause lifelong loss of intelligence.
- *Global warming:* Even people who do not develop illnesses from coal pollutants will find their health and well-being impacted due to coal's contribution to global warming. The discharge of carbon dioxide into the atmosphere associated with burning coal is a major contributor to global warming and its adverse effects on health and well-being worldwide, such as heat stroke, malaria, declining food production, scarce water supplies, social conflict and starvation.

Precautionary Measures

- Methods that are effective in controlling worker exposures to coal dust, depending on the feasibility of implementation, are as follows:
    - Process enclosure
    - Local exhaust ventilation
    - General dilution ventilation
    - Personal protective equipment
- Workers responding to a release or potential release of a hazardous substance must be protected.
- *Workplace monitoring and measurement:* Determination of a worker's exposure to the air-borne respirable fraction of coal dust containing less than five percent silica is made using a tared low ash polyvinyl chloride (LAPVC) filter (five microns), preceded by a 10 mm cyclone. Samples are collected at a maximum flow rate of 1.7 litres/minute until a maximum collection volume of 816 litres is reached.
- *Storage:* In the event coal dust requires storage, it should be stored in a cool, dry, well-ventilated area in tightly sealed containers that are labelled in accordance with OSHA's Hazard Communication Standard [29 CFR 1910. 1200]. Containers of coal dust should be protected from physical damage and ignition sources and should be stored separately from oxidizing agents.
- *Spills and leaks:* In the event of a spill or leak involving coal dust, persons not wearing protective equipment and clothing should be restricted from contaminated areas until cleanup has been completed.

19.3 VARIOUS TYPES OF DISEASES CAUSED BY DIFFERENT ELEMENTS AND MINERALS

Table 19.1: Possible effects of different minerals as health hazards

Health disease	*Description*	*Causitive elements or minerals*	*Precautionary measures*
Arsenicosis	Chronic arsenic poisoning results from drinking water with arsenic over a long period of time. It results in various health effects including skin problems, skin cancer, cancers of the bladder, kidney and lung, and diseases of the blood vessels of the legs and feet, and possibly also diabetes, high blood pressure and reproductive disorders.	Arsenic	Monitoring by health workers, people need to be checked for early signs of the disease, like skin problems; health education.
Berylliosis	The pulmonary and systemic granulomatous lung disease, in which, on prolonged exposure, the lungs become hypersensitive to it causing the development of small inflammatory nodules, called granulomas. Other symptoms include chest pain, joint aches, weight loss and fever.	Beryllium	Beryllium exposure should be kept below permissible limit.
Fluorosis	Ingestion of excess fluoride, most commonly in drinking water, can cause fluorosis which affects the teeth and bones. Moderate amounts lead to dental effects, but long-term ingestion of large amounts can lead to potentially severe skeletal problems. Clinical *dental fluorosis* is characterized by staining and pitting of the teeth. In more severe cases all the enamel may be damaged. Chronic high-level exposure to fluoride can lead to *skeletal fluorosis*. In skeletal fluorosis, fluoride accumulates in the bone progressively over many years. The early symptoms of skeletal fluorosis, include stiffness and pain in the joints. In severe cases, the bone structure may change and ligaments may calcify, with resulting impairment of muscles and pain.	Fluoride	Find a supply of safe drinking water with safe fluoride levels. De-fluoridation, methods include: use of bone charcoal, contact precipitation, use of Nalgonda or activated alumina. Health education regarding appropriate use of fluorides. Mothers in affected areas should be encouraged to breastfeed since breast milk is usually low in fluoride.

(Contd.)

Table 19.1 (*Contd.*)

Health disease	*Description*	*Causitive elements or minerals*	*Precautionary measures*
Iodine deficiency disorder (IDD)	The term IDD refers to all the ill-effects of iodine deficiency: 1. Goiter, the medical term for enlargement of the thyroid gland. 2. Brain damage and irreversible mental retardation. 3. Cretinism resulting from inadequate secretion of thyroid hormones during foetal life or early infancy. The brain and skeleton fail to develop properly, resulting in mental retardation and dwarfism.	Iodine	Cosumption of the required iodine in diet.
Parkinson's disease	It is an age-related deterioration of certain nerve systems, which affects movement, balance, and muscle control.	Manganese	All workers should take proper precautions when working in an environment with heavy manganese exposure
Mesothelioma	Mesothelioma is a rare type of cancer that typically affects the lining of the lungs, heart and abdomen.	Asbestos	Avoid asbestos exposure, protective work clothing.
Asbestosis	Asbestosis is a breathing disorder caused by inhaling asbestos fibres. Prolonged accumulation of these fibres in lungs can cause scarring of lung tissue and shortness of breath.	Asbestos	Asbestos-workers should wear a protective mask or a hood with a clean-air supply and obey recommended procedures to control asbestos dust. Workers in asbestosis-related industries should have regular X- rays to determine whether their lungs are healthy.
Pneumo-coniosis	It is a chronic lung disease caused due to the inhalation of various forms of dust particles, particularly in industrial workplaces, for an extended period of time. Hence it is also said to be an occupational lung disease, which are a particular subdivision of occupational related diseases that are related primarily to being exposed to harmful substances, whether they are gas or dusts, in the work place, and the pulmonary disorders that may result from it.	Coal dust, silica dust, asbestos dust	A protective mask should be worn when working with any of the above-mentioned mate-rial, such as coal, asbestos, silica, and so on. Also, governments should ensure that companies comply with the permitted levels of dust regulations.

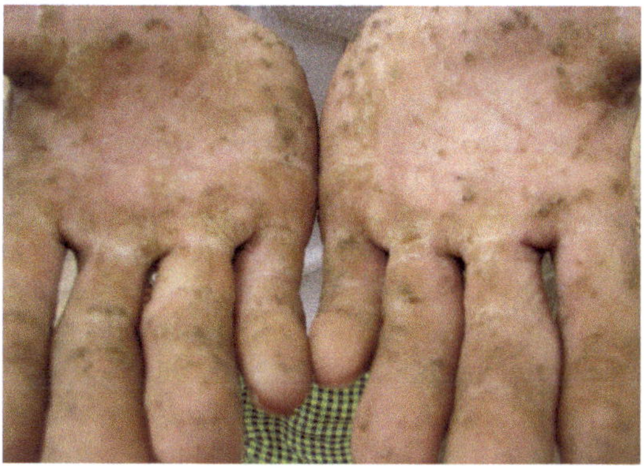

Fig. 19.2: Arsenocosis.

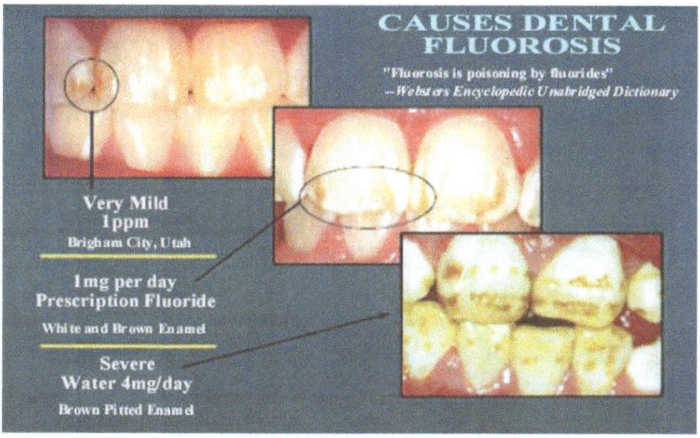

Fig. 19.3: Fluorosis.

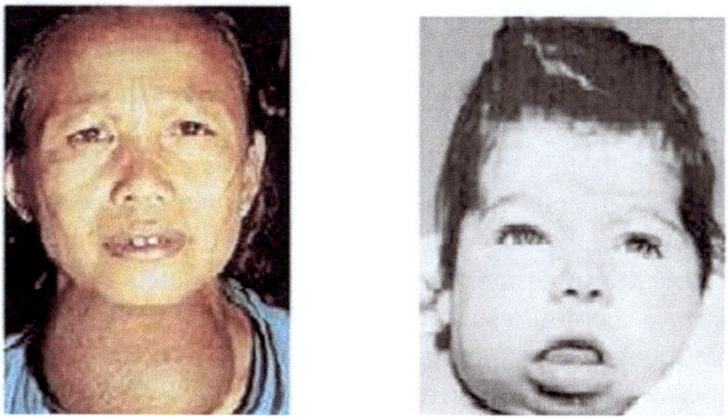

Goiter Cretinism

Fig. 19.4: Iodine deficiency disorder (IDD).

19.4 SOME TERMINOLOGIES RELATED TO GEOMEDICINE

Chelation

Chelation (pronounced key-LAY-shun) is the use of a chemical substance to bind molecules, such as metals or minerals, and hold them tightly so they can be removed from the body. Chelation has been scientifically proven to remove excess or toxic metals before they can cause damage to the body.

The first rational antidote to heavy metal intoxication was developed during World War II in anticipation of a re-initiation of gas warfare by the Germans. It was developed specifically for the British Lewisite gas and was thus known as British Anti-Lewisite, or BAL. The chemical compound is *dimercaprol*. The rationale for its use was based on the fact that arsenic binds quite specifically to sulfur groups in the affected tissues. Dimercaprol has very active and relatively non-toxic sulfhydry groups that interact with the arsenic to inactivate it. This phenomenon is known as *chelation*. Chelators are the class of compounds which are used in the treatment of heavy metal intoxication. A chelator is a flexible molecule with two or more electronegative groups that can form stable co-ordinate covalent bonds with cationic metal atoms. The complexes are then excreted by the body. The efficacy of a chelator is determined in part by the number of ligands available for metal binding; in general the greater number of ligands, the more stable the chelator-metal complex. Depending on the number of metal-ligand bonds, the chelator is designated as mono-, bi- or polydentate. The chelator ligands include groups such as -OH, -SH, or -NH. Dimercaprol is bidentate and forms a single heterocyclic ring with a metal ion. Unfortunately, chelators are relatively non-specific as to the metal ions they isolate, hence they also slurp up things like calcium and zinc which are vital for normal physiological function.

Toxicity from inorganic lead can be treated with chelators, but organic lead compounds such as tetra-ethyl lead produces a similar symptomology, but cannot be treated with these agents because they already have formed strong ligands with their organic constituents. The alkyl lead eventually is converted to inorganic lead, which can be treated with the chelators.

Mercury can be chelated in the peripheral tissues with EDTA and penicillamine. Even though the plasma levels can be reduced very efficiently, mercury forms alkyl ligands in the tissues. This is especially persistent in CNS tissue. Recovery from mercury poisoning can require months or years even with efficient chelator treatment, and is often incomplete.

Dimercaprol is a colourless oily liquid with the odour of rotten eggs. It is used in the treatment of lead and mercury intoxications as well as arsenic. It is far from innocuous, producing frequent side effects of hypertension and tachycardia, as well as headache, nausea, vomiting, lacrimation, salivation, parathesia and pain. It is generally given by intramuscular injection. Currently several other analogs are being investigated which have less toxicity and the possibility for oral administration.

The most widely used chelator is ethylenediamine-tetra-acetic acid or EDTA. It has four ligand sites that focus on the metal atom and thus is a very efficient and stable chelator. It works well on many metals, the most notable of which are calcium, magnesium and lead. The ligand sites consist of two nitrogen and two oxygen groups. EDTA has relatively low toxicity, the major toxic response being impaired renal function. This may in fact be a result of the isolation and concentration of the toxic metal which is then excreted by the kidney in association with the EDTA. Because there is frequently a substantial difference in pH between plasma and kidney tubule, this can change the relation between chelator and metal with a consequent release of some chelated metal which then can react with the renal tissue. This is often controlled by balancing urine pH with iv infusion of bicarbonate to maintain neutral or alkaline conditions which prevent chelator dissociation. The action of EDTA is non-specific chelating many metals, but is especially valuable in treatment of lead intoxication.

Mercury and other important metal toxicant appear to have restricted access because of either "tighter" ligands to tissue sites, or primary intracellular locations which are inaccessible to EDTA.

Penicillamine is the only important chelator which can be administered orally. It is prepared by hydrolytic degradation of penicillin and only the D-isomer is recommended for clinical use. It is an effective chelator of copper, mercury, lead and zinc promoting their excretion in the urine, ligands forming with the sulfhydryl, amine, and possibly the carboxyl group. The major toxicity is related to inhibition of pyridoxal dependent enzymes, and it is usual to supplement patients with pyridoxine to compensate for this effect. Individuals with sensitivity to penicillin will also react to penicillamine. Penicillamine is often used in conjunction with EDTA to treat lead and mercury intoxication, especially because of the oral administration property. The major use of the drug other than for heavy metal poisoning is the treatment of Wilson's disease. This hereditary disorder involves a storage defect in the handling of copper. Normal individuals have plasma protein which is the transport mechanism (and storage depot) for copper. This protein is defective in Wilson's patients and the copper cumulates producing neurotoxicity. Since these individuals must be treated throughout their life to keep copper levels within normal limits, the low toxicity and oral route of administration of d-penicillamine are especially useful. In addition penicillamine is used in the treatment of rheumatoid arthritis. Here the mechanism is obscure but probably involves a reduction of inflammatory response by reduction of tissue calcium and magnesium levels.

Deferoxamine is a very specific naturally occuring chelator. It is synthesized by a streptomyces organism, and has a specific high affinity for iron, with virtually no effect on calcium and magnesium. It is the agent of choice for treatment of iron intoxication, which mainly occurs in small children who get into their parents supplement pills. Full-blown acute iron intoxication can produce GI damage, with convulsions and coma. Since the mortality for untreated severe iron intoxication is about 50%, and the use of deferoxamine in these types of cases produces significant increases in survival, it is an important and useful drug. It is also used in the treatment of relatively rare iron storage disorder analogous to Wilson's disease.

Geophagy

Geophagia is also of concern in medical geology. Geophagy or geophagia can be defined as the deliberate ingestion of soil, a practice that is common among members of the animal kingdom, including certain human populations. Soil may be eaten from the ground but in many situations there is a cultural preference for soil from special sources such as termite mounds. Geophagia is considered by many nutritionists to be either a learned habitual response in which clays and soil minerals are specifically ingested to reduce the toxicity of various dietary components or as an in-built response to nutritional deficiencies resulting from a poor diet. Geophagy is attaining renewed and serious interest within the scientific research community.

Concluding Remarks

Medical geology, a long-recognized but perhaps underutilized discipline, presents the geoscience community with tremendous opportunities for collaborative work with the biomedical and ecological research communities. Among the environmental health problems that geologists are working with the medical community to address are: exposure to toxic levels of trace essential and non-essential elements such as arsenic and mercury; trace element deficiencies; exposure to natural dusts and to radioactivity; naturally occurring organic compounds in drinking water; volcanic emissions, etc. Geoscientists have also developed an array of tools and databases that can be used by the environmental health community to address vector-borne diseases, to model pollution dispersion in surface and ground water, and also can be applied to some aspects of industrial pollution and occupational health problems. Such collaborations have great potential to help understand, mitigate and possibly eradicate environmental health problems that have plagued humans for thousands of years.

Think for a while

1. Classify the elements and minerals into various groups on the basis by which they are related to geomedicine.
2. Trace elements are intermittently related to geomedicine. Illustrate with few examples. Is there any distinction between the health hazards caused by trace elements or heavy metals?

3. Some elements have both beneficial and harmful effects on human health. Give examples of such elements, citing both the effects.
4. Diversity is observed in health effects caused by different members of the silicate class. Justify. State the precautionary measures needed to combat the hazards, caused by mineral dust.
5. Explain the term chelation. State its impact in geomedicine. Name some commonly used chelators, stating their individual application. Define geophagy.

FURTHER READING

Belkin, H.E., B. Zheng, D. Zhou and R.B. Finkelman. Preliminary results on the Geochemistry and Mineralogy of Arsenic in Mineralized Coals from Endemic Arsenosis in Guizhou Province, P.R. China. Proceedings of the Fourteenth Annual International Pittsburgh Coal Conference and Workshop. CD-ROM p. 1-20. 1997.

Belkin, H.E., D. Kroll, D.-X. Zhou, R.B. Finkelman and B. Zheng. Field test kit to identify arsenic-rich coals hazardous to human health. Abstract in Natural Science and Public Health – Prescription for a Better Environment. U. S. Geological Survey Open-file Report 03-097. Unpaginated. 2003.

Bencko, V. and J. Vostal. Air pollution by solid particles and public health: When can we conclude on causality. *Central European Journal of Public Health.* **7(2)**, 63-66. 1999.

Borm, Paul J.A. Toxicity and occupational health hazards of coal fly ash (cfa). A review of data and comparison to coal mine dust. *Ann. Occup. Hyg.*, **41(6)**, 659-676. 1997.

Boyle, R.W. and I.R. Jonasson. The geochemistry of arsenic and its use as an indicator element in geochemical prospecting. *J. Geochem. Explor.*, **2**, 251-296. 1973.

Centeno, J.A., F.G. Mullick, L. Martinez, N.P. Page, H. Gibb, D. Longfellow, D. Thompson and E.R. Ladich. Pathology Related to Chronic Arsenic Exposure. *Environmental Health Perspectives*, **110(5)**, 883-886. 2002a.

Centeno, J.A., F.G. Mullick, L. Martinez, H. Gibb, D. Longfellow and C. Thompson. Chronic Arsenic Toxicity: An Introduction and Overview. *Histo-pathology*, **41(2)**, 324-326. 2002b.

Colbourne, P., B.J. Alloway and I. Thornton. Arsenic and heavy metals in soils associated with regional geochemical anomalies in southwest England. *Sci. Total Environ.*, **4**, 359-363. 1975.

Cronin, S.J. and D.S. Sharp. Environmental impacts on health from continuous volcanic activity at Yasar (Tanna) and Ambrym, Vanuatu. *International Journal of Environmental Health Research*, **12(2)**, 109-123. 2002.

Feder, G.L., Z. Radovanovic and R.B. Finkelman. Relationship between weathered coal deposits and the etiology of Balkan endemic nephropathy. *Kidney International*, **40(Suppl. 34)**, s-9-s-11. 1991.

Forstner, U. and G.T.W. Wittmann. Metal Pollution in the Aquatic Environment. Springer-Verlag, Berlin, Heidelberg, New York. 1979.

Harlow, B.L., D.W. Cramer, D.A. Bell and W.R. Welch. Perineal exposure to talc and ovarian cancer risk. *Obstetrics & Gynaecology*, **80:** 19-26. 1992.

Hodge, H.C. Mechanisms of uranium poisoning. Proceedings of International conference for peaceful uses of atomic energy; 1955; Geneva, Switzerland. New York: United Nations; 1956; 13:229-32, P/73.

Hollinger, M.A. Pulmonary toxicity of inhaled and intravenous talc. *Toxicology Letters*, **52**, 121-127. 1990.

National Toxicology Program. Toxicology and carcinogenesis studies of talc (GAS No 14807-96-6) in F344/N rats and B6C3F, mice (Inhalation studies). Technical Report Series No. 421. September 1993.

Priyadarshi, N. Distribution of arsenic in Permian Coals of North Karanpura coalfield, Jharkhand. *Jour. Geol. Soc. India*, **63**, 533-536. 2004.

Radhakrishna, B.P. Medical Geology. *Jr. of The Geological Society of India*, **66(4)**, 395. 2005.

Internet Data Retrieved from:

Asbestos Exposure and Cancer Risk - National Cancer Institute.htm

Bentonite Clay Health Benefits.htm

Carbon (C) - Chemical properties, Health and Environmental effects.htm

http://madrad 2002.word.press.com

EnviroNews Archives - Health Hazards of Heavy Metals.htm

It's Your Health - Health Risks of Asbestos.htm

U.S. Department of Housing and Urban Development Office of Healthy Homes and Lead Hazard Control - www.hud.gov/offices/lead

Environmental Health Perspectives, The Prevalence of Lead-Based Paint Hazards in U.S. Housing - www.hud.gov/offices/lead/techstudies/lead.PaintHousing Survey.pdf

http://www.usgs.gov

Silicon (Si) - Chemical properties, Health and Environmental effects.htm

http://energy.er.usgs.gov/health_environment/medical_geology/

http://www.mindfully.org/Nucs/DU-Medical-Effects-Mar99.htm

http://www.wateraid.org/documents/plugin_documents/iodine1.pdf.pdf

http://gsa.confex.com/gsa/2009AM/finalprogram/abstract_161522.htm

http://www.agiweb.org/geotimes/nov01/feature_medgeo.html

http://science.jrank.org/pages/47843/medical-geology.html

www.geologica-acta.com. Medical Geology: a globally emerging discipline.

PROBLEMS

Question 1: Calculate the density of (i) Aragonite (orthorhombic) (ii) Calcite (hexagonal) and Strontianite (orthorhombic) from the following data:

Mineral species	*Chemical formula*	*aÅ*	*bÅ*	*cÅ*	*Number of formula units per unit cell (Z)*	*Avogadro number (N)*
Aragonite	$CaCO_3$	4.96	7.97	5.74	4	6.02×10^{23}
Calcite	$CaCO_3$	4.989		17.062	6	6.02×10^{23}
Strontianite	$SrCO_3$	5.107	8.414	6.029	4	6.02×10^{23}

Answer: $\text{Density} = \dfrac{\text{Formula wt} \times Z}{V \times N}$

(i) Formula wt of aragonite (orthorhombic)

$$= CaCO_3 = 40 + 12 + (3 \times 16) = 100$$
$$V = (4.96 \times 7.97 \times 5.74) \times 10^{-24} = 226.91 \times 10^{-24}$$
$$\text{Density} = (100 \times 4)/(226.91 \times 10^{24}) \times (6.02 \times 10^{23}) = 2.92$$

Thus for isostructural minerals like strontianite ($SrCO_3$), cerrurite ($PbCO_3$) etc. the density variation is caused by the unit cell dimensions and the variation of cation weight.

(ii) Formula wt of calcite = $CaCO_3$ = 100

$$V = 4.989 \times 4.989 \times 17.062 \times 10^{24} \times \sin 60^{\circ}$$
$$= 424.675 \times 10^{24} \times 3^{1/2}/2 = 367.76$$
$$\text{Density} = (100 \times 6)/(376.76 \times 10^{-24}) \times (6.02 \times 10^{23}) = 2.710$$

(iii) Formula wt of Strontianite = $SrCO_3 = 87.62 + 12 + (3 \times 16) = 147.62$

$$V = 4.989 \times 4.989 \times 17.062 \times 10^{24} \times \sin 60^{\circ}$$
$$= 424.675 \times 10^{24} \times 3^{1/2}/2 = 367.76$$
$$\text{Density} = (147.62 \times 4)/(259.07 \times 10^{-24}) \times (6.02 \times 10^{23}) = 3.43$$

This indicates that the polymorphs (aragonite and calcite) have different densities due to their different crystal structures, while isostructural minerals (aragonite and strontianite) have different densities due to their different cation weights.

Question 2: Calculate the weight % of (i) Fe in Pyrite (FeS) and (ii) Pb in Cerrusite ($PbCO_3$).

Answer:

(i) Molecular wt. of FeS = 55.85 + 32.07 = 87.92

Wt % of Fe = (55.85/87.92) × 100 = 63.52%.

(ii) Molecular wt. of cerrusite ($PbCO_3$) = 207.21 + 12 + (3 × 16) = 267.21
Wt.% of Pb = (207.21/267.21) × 100 = 77.54%.

Question 3: From the measured chemical data, calculate the formula of the mineral

Element	*Wt.%*	*Atomic wt.*
S	34.82	32.06
Cu	34.30	63.6
Fe	30.59	55.9

Answer: Total weight % = (34.82 + 34.30 + 30.59) = 99.71
Ratios of the three atoms per formula unit =

S:Cu:Fe = (34.82/32.06) : (34.30/63.6) : (030.59/55.9)
= 1.086: 0.539 : 0.547 ≈ 2:1:1

Therefore the formula = $CuFeS_2$, Chalcopyrite.

Question 4: From the data below calculate the formula of replacement for Sphlarerite (ZnS).

Formula	*ZnS*		*(ZnFe)S*			*(ZnMnFe)S*			
	Zn	*S*	*Zn*	*Fe*	*S*	*Zn*	*Mn*	*Fe*	*S*
Wt %	67.13	32.78	60.80	6.20	33.0	51.20	1.61	14.32	32.60
Atomic proportion	1.026	1.024	0.929	0.111	1.029	0.783	0.018	0.256	1.0
Atomic weight	65.39	32.06	65.39	54.94	32.06	65.39	54.94	55.85	32.06
Decimal fraction for cation	1	-	0.893	0.107	-	0.741	0.017	0.242	-
Formula	ZnS		$(Zn_{0.90}Fe_{0.11})S$			$(Zn_{0.74}Mn_{0.2}Fe_{0.24})$			

Answer: Depending on the environmental condition temperature and pressure, the replacement varies. The estimation of this variation in minerals can be used for geothermometric and geobarometric studies.

Question 5: Calculate the Packing Index (P.I.) for (i) Anatase and (ii) Rutile from the data given.

Mineral	*Unit cell parameter*		*Volume of ions*	*Z*	*N*
	a	*c*			
Rutile (Tetragonal)	4.594	2.958	41.22	2	6.02×10^{23}
Anatase	3.793	9.51	53.35	4	6.02×10^{23}

Answer:

$$\text{Packing Index (P.I.)} = \frac{\text{Volume of ion}}{\text{Volume of unit cell}} \times 10$$

(i) P.I. for anatase = (53.35 × 10)/(3.793 × 3.793 × 9.51) = 3.9
(ii) P.I. for rutile = (41.202 × 10)/(4.594 × 4.594 × 2.598) = 6.6

Question 6: Calculate the % of Galena in an ore body containing quartz.

Mineral	*M(% of mineral)*	*Density*	*Volume*
Quartz	X	2.65	$x/2.65$
Galena	$100–X$	7.58	$(100–x)/7.58$
Vein material	100	3.65	100/3.65

Answer:

$(X/2.65) + (100 - X)/7.58 = 100/3.65$

$X = 57.88$ i.e. the % of quartz = 57.88% and the % of galena = 42.12%.

Question 7: Comment on the internal structure of fluorite, using Pauling's 1st and 2nd rules. The ionic radii of Ca^{++} and F^{-} are 0.99 Å and 1.31Å respectively.

Answer outline: To tell the internal structure, you need to find out the geometries of the cation-centred polyhedra, anion-centred polyhedra and nature of bond. (isodesmic/anisodesmic/mesodesmic).

You may attempt to answer it as follows:

(i) To determine the geometry of the cation-centred polyhedron from the 1st Rule of Pauling, determine the radius ratio of Ca^{++} with respect to F^{-}. From the *range* in which that radius ratio belongs, you'll be able to find out the geometry of the cation-centred polyhedron.

(ii) From the 2nd rule of Pauling and the equation of electrostatic valency, find out the strength of each bond reaching the F^{-}. Then you'll know the total number of bonds reaching each F^{-} from the neighbouring Ca^{++}. Thus the number of Ca^{++} surrounding each F^{-} is known, and the geometry of the anion-centred polyhedron is also determined. (Note that the polygons may or may not be regular.)

(iii) The charge of $F^{-} = 1$. Therefore bond is anisodesmic if bondstrength > ½, bond is mesodesmic if bondstrength = ½ and bond is isodesmic if bondstrength < ½,

Question 8: Comment on the internal structure and nature of bond (iso/aniso/mesodesmic) of the following minerals using Pauling's 1st and 2nd rules. The ionic radii of all the anions and cations are given.

$NaCl$, KCl, CuS, CaF_2, ZnS, $CaCO_3$, Fe_2O_3, As_2S_3, $BaSO_4$

Answer outline: To tell the internal structure, you need to find out the geometries of the cation-centred polyhedra, anion-centred polyhedra and nature of bond.

Firstly, to determine the geometry of the cation-centred polyhedron, calculate the radius ratio of cation with respect to the anion. From the range in which that radius ratio belongs, you'll be able to tell the geometry of the cation-centred polyhedron.

Now find out the strength of each bond reaching the cation form. Then you'll know the total number of bonds reaching each anion from the neighbouring cations. Thus the number of cations surrounding each anion is known, and the geometry of the anion-centred polyhedron is also determined. (Note that the polygons may or may not be regular.)

Bond is anisodesmic, mesodesmic or isodesmic if bond strength > ½ , = ½, or < ½ of the charge of anion respectively.

Example: The internal structure and nature of bond of CaF_2:

Radius ratio of Ca^{++} with respect to F^-: 0.99/1.31=0.755. It falls in the range 0.732 – 1.000. Therefore the coordination number is 8, and the cation-centred polyhedron is cubic.

From the electrostatic valency principle, strength of each bond reaching the F^- from the neighbouring Ca^{++} is: 2/8=1/4.

Therefore the total charge of F^- (= 1) is shared by: 1/(1/4) = 4 Ca^{++}. Therefore the anion-centred polyhedron is tetrahedral.

Since the strength of each bond is < ½ of total charge of F^-, therefore the bond is isodesmic.

Question 9:

(a) Complete solid solution exists between enstatite and ferrosilite. But there is no *complete* solid solution of these two mineral species with wollastonite. Explain with reason. (Refer to table for ionic radii.)

(b) Complete solid solution exists between albite and anorthite. But there is no solid solution between anorthite and orthoclase. Explain with reason. (Refer to table for ionic radii.)

Name of the ion	Ca^{++}	Na^+	K^+	Fe^{++}	Mg^{++}	O^{2-}
Radius (in Angstrom)	0.99	0.95	1.33	0.80	0.65	1.40

Answers:

(a) The chemical formulae of the three minerals are as follows:
Enstatite: $Mg_2(SiO_3)_2$, Ferrosilite: $Fe_2(SiO_3)_2$, Wollastonite: $Ca_2(SiO_3)_2$
The difference of ionic radii between Fe^{++} and Mg^{++}: (0.80 – 0.65) × 100/0.80 = 18.75%. Since the difference is small, Fe^{++} and Mg^{++} can easily substitute each other.
The difference of ionic radii between Ca^{++} and Mg^{++}: (0.99 – 0.65) × 100/0.65 = 52.30%. Since the difference is very large, Ca^{++} and Mg^{++} cannot easily substitute each other.

(b) The chemical formulae of the three minerals are as follows:
Albite: $NaAlSi_3O_8$, Anorthite: $CaAl_2Si_2O_8$, Orthoclase: $KAlSi_3O_8$.
The difference of ionic radii between Ca^{++} and Na^+: (0.99 – 0.95) × 100/0.99 = 4.04%. Since the difference is small, Ca^{++} and Na^+ can easily substitute each other.
The difference of ionic radii between Ca^{++} and Mg^{++}: (1.33 – 0.99) × 100/0.99 = 34.34%. Since the difference is very large, Ca^{++} and K^+ cannot easily substitute each other.

Question 10: Project the following faces of an isometric crystal:

$$(100), (010), (001), (\bar{1}00), (0\bar{1}0), (110), (1\bar{1}0), (\bar{1}10), (\overline{11}0), (101), (011), (\bar{1}01),$$
$$(0\bar{1}1), (111), (1\bar{1}1), (\bar{1}11), (\overline{11}1)$$

Answer: (100) and (010) are plotted on the circumference of the primitive circle at its points of intersections with positive A and B axes respectively. (001) is plotted at the centre of the primitive circle. (100) and (010) are plotted on the circumference of the primitive circle at diametrically opposite points of (100) and (010) respectively. (110) is plotted on the circumference of the primitive circle where the bisector of positive A and B axes intersects the circumference. It lies in between (100) and (010). Similarly $(\overline{11}0)$, $(\overline{11}0)$ and (110) may be plotted.

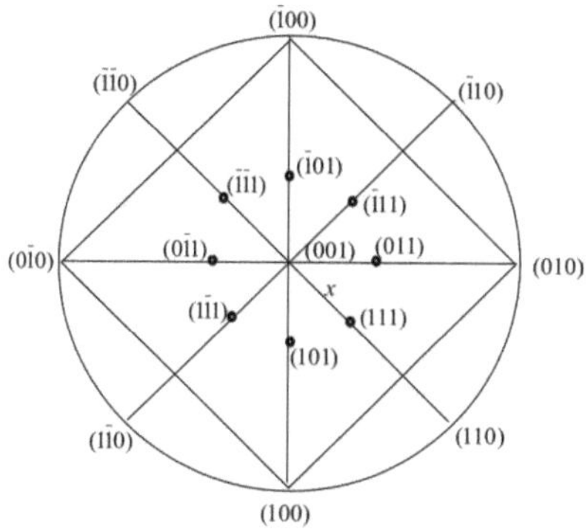

Fig. 1: Primitive Circle

An auxiliary circle is constructed through C axis and (100). The face normal of (101) is constructed, which makes an angle of 45° with A axis. Thus the spherical pole and the stereographic projection of (101) are obtained. The distance of the stereographic projection of (101) from the centre of the auxiliary circle is measured and plotted on the primitive circle. Similarly (011), (101) and (011) may be plotted.

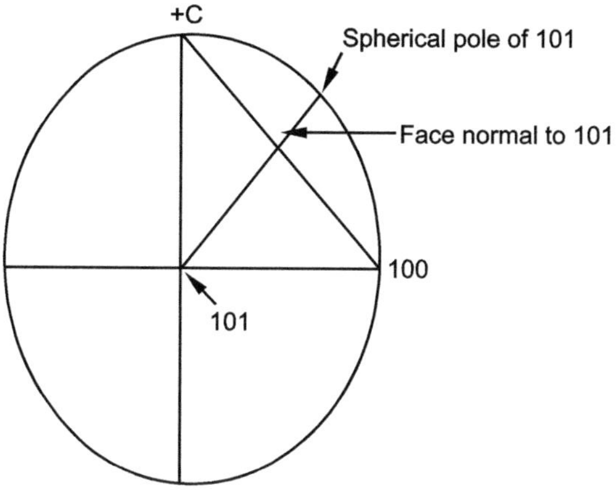

Fig. 2: Auxilliary circle passing through (100) and C.

An auxiliary circle is constructed through C axis and (110). The vertical trace of (111) is constructed on the auxiliary circle by measuring the distance *x* from the primitive circle. From this the face normal of (111) is constructed and the spherical pole and the stereographic projection of (111) are obtained. The distance of the stereographic projection of (111) from the centre of the auxiliary circle is measured and plotted on the primitive circle. Similarly (111), (111) and (111) may be plotted.

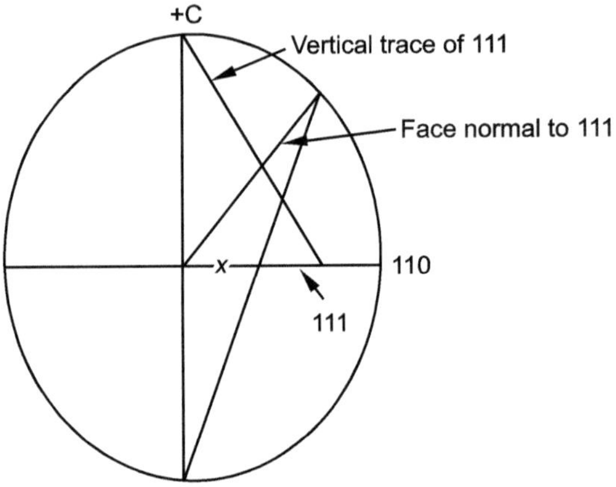

Fig. 3: Auxilliary circle passing through (110) and C.

Question 11: Composition of garnet and biotite from a politic migmatite is given below:

	Garnet	*Biotite*
Fe^{2+}	1.897	0.55
Mg	1.07	2.122
Al	2.0	1.263
Si	3.0	2.881
O	12	10

Find out the temperature of equilibrium using the expression

$$T\,(^0\mathrm{C}) = 2109/(\ln K_D - 0.782)$$

where $$K_D = [(\mathrm{Mg/Fe^{2+}})^{\mathrm{Grt}}/(\mathrm{Mg/Fe^{2+}})^{\mathrm{Bt}}] - 273$$

Answer: Using the compositions of the phases $K_D = 0.146$

Therefore, $\ln K_D = -1.923$

$$T = 506\ ^\circ\mathrm{C}.$$

Question 12: Using thermo analytical technique, quantify both in a mixture of calcite and magnesite. Given data

Minerals	*Decomposition temperature of*	*Total sample weight*	*Wt. loss observed*
Calcite ($CaCO_3$)	870°C	1 gm	0.264 gms at 870°C
Magnesite ($MgCO_3$)	560°C		0.174 gms at 560°C

Answer: For only (100%) calcite

$CaCO_3$ decomposes to CaO + CO_2 (this causes wt. loss)

i.e. (40 + 12 + 3 × 16) (40 + 16) (12 + 32)

or wt. loss is 44 for 100% calcite i.e. 0.44 gms in 1 gm of the sample.

Thus, for 0.264 gms wt. loss = (100/44) × 0.264 = 60%

and magnesite = 40% (this can also be calculated separately)

Question 13: Unit cell parameter of garnet varies with its composition. From the experimental X-ray powder diffraction data of a garnet sample estimate the unit cell parameters of the mineral and also nearest composition of garnet sample.

Given data

Garnets belong to isometric (cubic) crystal structure and the unit cell parameter is given by $d_{\mathrm{hkl}}^{-2} = (1/a^2)(h^2+k^2+l^2)$ (in Å)

Garnet	*Almendine*	*Pyrope*	*Spessertine*
Composition	$Fe_3Al_2(SiO_4)_3$	$Mg_3Al_2(SiO_4)$	$Mn_3Al_2(SiO_4)$
Unit cell dimension	a = 11.53Å	a = 11.455Å	a = 11.63Å

Answer: XRD data

Measured d_{hkl} values	2.594	2.900	1.608
h, k, l	4, 2, 0	4, 0, 0	6, 4, 0
Estimated unit cell a(Å) $d_{\mathrm{hkl}}^{-2}=(1/a^2)(h^2+k^2+l^2)$ (in Å)	11.60	11.60	11.595

Observations: The unknown garnet phase has the unit cell dimension nearer to the spessertine (Mn-rich) variety.

INDEX

MIX
Papier aus verantwortungsvollen Quellen
Paper from responsible sources
FSC® C105338

If you have any concerns about our products,
you can contact us on
ProductSafety@springernature.com

In case Publisher is established outside the EU,
the EU authorized representative is:
Springer Nature Customer Service Center GmbH
Europaplatz 3, 69115 Heidelberg, Germany

Printed by Libri Plureos GmbH
in Hamburg, Germany